Symmetry, Causality, Mind

Symmetry, Causality, Mind

Michael Leyton

A Bradford Book
The MIT Press
Cambridge, Massachusetts
London, England

This book was set in Times Roman by Asco Trade Typesetting Ltd., Hong Kong, and was printed and bound in the United States of America.

Library of Congress Cataloging-in-Publication Data

Leyton, Michael.
Symmetry, causality, mind / Michael Leyton.
p. cm.
"A Bradford book."
Includes bibliographical references and index.
ISBN 0-262-12163-8
1. Time perception. 2. Symmetry—Psychological aspects. 3. Causation. I. Title.
BF468.L487 1992
153.7′53—dc20 91-24605
CIP

Contents

Acknowledgments

I wish to acknowledge, with considerable gratitude, the feedback which the following people gave me, either on a draft of this book, or on topics related to the book: Peter Dodwell, Julie Gerhardt, Jim Higgins, Jurgen Koenemann, Barbara Landau, Ann O'Leary, Nancy Nersessian, Eleanor Rosch, Azriel Rosenfeld, Louis Sass, Chuck Schmidt, Wayne Wickelgren, and Rob Zimmer. Considerable gratitude also goes to NSF for a Presidential Young Investigator Award (IRI-8896110)—which funded me throughout the writing of this book.

Symmetry, Causality, Mind

Introduction

There is perhaps nothing more singular, in our relationship to time, than the fact that we live only in the present. At any moment in time, we cannot have contact with the past because the past no longer exists, and we cannot have contact with the future because the future has not yet come into being. We are, inescapably and completely, prisoners of the present.

When one stands in an empty subway station in the early morning, one sees all around one objects that hold for one the memory of events at which one has not been present. The dents in a bin are unavoidably seen as the result of kicking. The graffiti records the swift movements of a young artist. The scratched surface of the platform is seen as the result of active and impatient feet. A crumpled newspaper evokes the compacting actions of hands. The large splash of coffee, on the floor, points to an unintended spilling. The broken corner of a jutting wall testifies to the impact of a moving vehicle. A squashed beer-can recalls a compressing grip. A torn shirt, on the ground, signifies some hasty scuffle.

Like the subway station, the present is a silent chamber that has a history we cannot experience. It is only from the contents of this chamber, that we might be able to infer prior events. Indeed, in this chamber, we have no contact even with our own past. For, because we exist only in the present, any event that has happened to us is now out of reach. A wall stands between us and our own past. We can examine only what we possess within the present, the relics that surround us—and, only from these relics, can we infer what we have undergone.

But, you might argue, we have *memory*.

Memory, however, does not solve this problem; it defines the problem. To see this, observe first that memory, like anything experienced by the mind, exists only in the present. In fact we shall argue that memory is always some *physical* object, in the present—a physical object that some observer *interprets as holding information about the past*. The object can exist in the external environment of the observer, e.g., it can be a squashed beer-can; or it can exist within the observer, e.g., it can be a piece of neuronal material. In either case, the interpretation of the object as memory belongs to the observer.

But, how can an object be interpreted as holding information about the past? The only possible answer, we argue, is this: The past, about which the object is holding information, is the past *of the object itself*. In fact, an object becomes memory for an observer when the observer examines certain features of the object and *explains* how those features were *caused*.

We shall argue, in this book, that all cognitive activity proceeds via the recovery of the past from objects in the present. Cognitive activity *of any type* is, on close examination, the determination of the past.

Surprisingly, our argument will center on an analysis of the psychological relationship between *shape* and *time*. It will be argued that an important means by which the mind recovers the past is *shape*. As such, shape forms a basis for memory. The mind assigns to any shape a causal history explaining how the shape was formed. It is by doing this that the mind *converts shape into memory*. Furthermore, we will reduce the study of shape to the study of *symmetry*, and thus we will show that symmetry is crucial to everyday cognitive activity: *Symmetry is the means by which shape is converted into memory*.

Our argument will be elaborated as follows: The first three chapters will give an analysis of *perception*. Whereas perception is conventionally understood to be the study of how the mind recovers the *spatial* layout of the environment, we shall argue that perception concerns the recovery of *time* that is locked into the environment. In doing so, we shall show that Computational Vision can be reduced purely to symmetry principles. In Chapter 4, we will examine *cognition* in general, and it is here that we will argue that all cognition is the determination of the past. Chapter 5 will present preparatory material so that we are able to amplify our view further with respect to perception in the lengthy Chapter 6. In Chapter 7, we shall extend our basic rules to *linguistics*. We will argue that any sentence is understood psychologically as a piece of causal history. That is, a sentence is an "archaeological" relic—one that is dis-interred by the listener such that it reveals the past. Chapter 8 then presents a theory of *art*. We analyze a number of paintings in detail, and argue that an art-work is an object from which a person can extract the maximal amount of history. Finally, in Chapter 9, we examine *political subjugation*, because we argue that this reveals further the particular relationship we propose between history and mind.

1 Recovering Process-History

1.1 Seeing Objects as History

One often has the sense that the shape of an object tells one something about the object's history; i.e., the processes that formed it. For example, the shape in Fig 1.1 might be understood as the result of various processes of growth, pushing, pulling, resistance, etc. This type of understanding is, in fact, a remarkable phenomenon. A shape is simply a single state, a frozen moment, a step outside the flow of time; and yet we are able to use it as a window into the past. How is this possible?

Let us return to the scene with which we opened the book: the subway station in the early morning. We noted that, in this scene, one finds many objects that seem to contain the memory of past actions—actions that one did not in fact witness. The objects that hold this memory include dented garbage cans, graffiti, scratched surfaces, crumpled newspapers, splashes, broken walls and bottles, squashed containers, and torn clothes. The actions that are represented in these objects might have occurred long before one entered the scene. Furthermore, the present, as one confronts it in the early morning, is static. It must therefore be concluded that some ingredient within the static present allows one to recapitulate time. What could this ingredient be?

Close consideration reveals that, in each of the objects mentioned above, the source of time is *shape*. But how is shape linked to time? One can search in vain through mathematics texts to find some mention of a connection between shape and time. Historically, the mathematical investigation of shape has been prolific—from the first simple theorems of Greek geometry to the intricate details of modern differential geometry—and yet none of these investigations has mentioned *time*. Can it be that something essential has been missed about shape?

The central proposal of this book is this:

SHAPE IS TIME

We live in a world constructed by history. The claim to be made here is that shape is a powerful record of that history. Shape is, after all, the central tool of the archeologist. In fact, the range of disciplines that extract time from shape is considerable: medicine, geology, astrophysics, meteorology, evolutionary biology, etc. However, we shall argue that the extraction of time from shape is not just the *technical* activity of

Figure 1.1
A complex shape.

countless scientists, but is central to human perception generally. As such, we shall claim that it is a major organizing principle that explains perceptual data ranging from the local and low-level to the holistic and conceptual.

1.2 The Process-Recovery Problem

All the examples given in the previous section have the following form: An individual observes a single state, which we will call the *present moment*. A certain structural ingredient of that moment allows the person to "run time backwards" and recapitulate the processes that lead to that moment. That is, the person is able to *recover history from x*, where x is atemporal and completely contained within the present. We shall say that the person is solving the *history-recovery problem*, which we will also call the *process-recovery problem*. Later, in Chapters 3 and 4, we shall argue that the history-recovery problem defines all perception, and in fact is basic to cognition generally, e.g., to linguistics.

In perception, the ingredient x, from which one can recover the past, has something to do with shape. We shall later define more precisely what this ingredient is. For the moment, however, we can simply regard our recovery problem as this: Recover history from shape.

Now, the recovered history consists of past events that are of course prior to the present; e.g., the kick that was responsible for the dent in the door. Thus, by definition, the recovered past—the actual kick—can no longer be observed. That is, since actual observations can be made only in the present, the past is always inaccessible. This inaccessibility forces a fundamental requirement on the ability to recover the past. We shall now turn to this requirement.

1.3 Process Directionality

Because the past is inaccessible, one can quite validly hypothesize any possible history. This is because there is no way of actually going back into the past to test a historical hypothesis. One therefore needs some consistent way of eliminating the arbitrariness; i.e., of guaranteeing a unique choice of history. We shall say that one requires a means of guaranteeing *unique recoverability.*

The requirement of unique recoverability forces historical processes to be psychologically represented in a particular way. To understand this means of representing processes, let us examine an alternative means of representation that does not allow unique recoverability.

Barr (1984) has developed representations of a number of common deformations; e.g., twisting, bending, tapering, scaling. Consider his representation of *twisting*. Although this representation is expressed mathematically, it can be intuitively described as follows: *Turn one end of the object with respect to the other.*

For the process-recovery problem, i.e., the problem of recovering history, there is a crucial difficulty with Barr's representation. Consider the object shown in Fig 1.2a. Now twist it so that the the right-hand end of the object is rotated by 90° in the direction shown by the arrow, while the left-hand end is held fixed. The result is shown in Fig 1.2b. However, people, presented with Fig 1.2b, are unlikely to regard the latter as a twisted shape. Nevertheless, the progression from Fig 1.2a to 1.2b was obtained by a twist in Barr's sense.

The problem is that deformation is psychologically a *uni-directional* phenomenon; i.e., it goes from straight objects to non-straight objects, not the reverse. Barr's representation of twisting can be used in both directions.

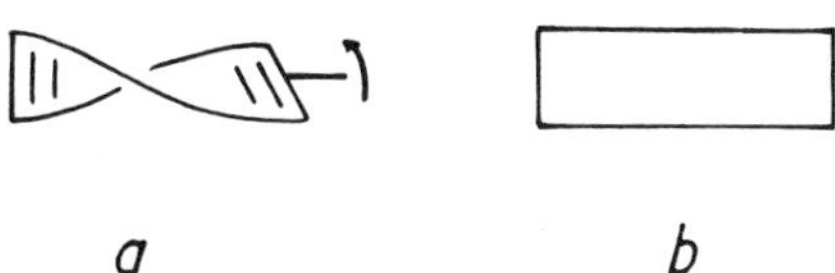

Figure 1.2
The shape in (a) is understood as twisted and that in (b) is understood as not twisted.

The uni-directional nature of deformation can be understood as conforming to a principle of *prototypicality* due to Rosch (1975, 1978), as follows: Rosch found that non-prototypical objects are seen in terms of prototypical ones, but the reverse tends not to occur. For example, off-red is seen in terms of red, diagonal orientations in terms of the vertical (or horizontal), the number 99 in terms of 100. However, the reverse relationships are not conceived; e.g., red is not seen in terms of off-red. Now, returning to Fig 1.2, we find that deformation also has this structure. That is, although one sees Fig 1.2a as a twisted version of Fig 1.2b, one does not see Fig 1.2b as a twisted version of Fig 1.2a.

We shall now see that, in order to carry out the recovery of process-history, the mind requires a uni-directional definition of processes. To illustrate: If processes were to be understood as bi-directional, then both Fig 1.2a and Fig 1.2b could be understood as the starting state for each other. Suppose then that one wanted to recover the history of Fig 1.2a. Then, one could hypothesize Fig 1.2b in its past. But due to bi-directionality, Fig 1.2b could itself contain Fig 1.2a in its past. However, this would mean that Fig 1.2a could have both Fig 1.2b and Fig 1.2a in its past. Thus, one could regard Fig 1.2a as originating from Fig 1.2b or from Fig 1.2a. The choice would be arbitrary. That is, bi-directionality prevents one having *unique recoverability*.

The violation of unique recoverability, created by bi-directionality, is still worse if one has more than two objects. That is, suppose that Fig 1.2 contained several objects that were all bi-directionally related. Then any one of the objects could be a starting state for any of the others in that set. Thus, suppose one were presented with one of the objects from the set, and one tried to recover the history of the object. Because any of the other objects could act as a starting state, any history through that set would be possible. Again bi-directionality would prevent unique recoverability.

To summarize: The goal of the process-recovery problem is to recover events that are in the past and are therefore inaccessible. The inaccessibility forces one to possess some means, other than going into the past, of guaranteeing unique recoverability. Unique recoverability requires processes to be defined uni-directionally. When we look at the way that processes are defined psychologically we do indeed find that they are understood as uni-directional.

1.4 The Fundamental Proposals

Having identified a basic requirement that a process-recovery system must satisfy, we shall now state our two fundamental proposals for the solution of the process-recovery problem. The entire book will be an elaboration of these proposals.

It is first necessary to explicitly note the fact that the recovery of a process is possible only if it leaves a *memory*. While many types of processes leave memory, there are many types of processes that do not. In fact, there are many types of processes whose effect is to actually wipe out memory, e.g., by removing a stain. Observe that, in these latter cases, the processes can not be recovered by an examination of their results. Only processes that leave memory can be recovered. This notion will soon be prolifically illustrated.

On a concrete level, there are many forms of memory that can be left on objects: scars on the surface of the moon, chips on vases, graffiti on subway trains, etc. However, we will claim that a single *abstract* property characterizes *all* perceptual situations of memory:

ASYMMETRY IS THE MEMORY THAT PROCESSES LEAVE ON OBJECTS

Furthermore, we will also claim the following:

SYMMETRY IS THE ABSENCE OF PROCESS-MEMORY

The first statement will be called the *Asymmetry Principle*, and the second will be called the *Symmetry Principle*. They will be stated more precisely at the end of this section.

We shall now illustrate these two proposals in detail. Consider first a situation that is much discussed in physics: A tank of gas. Suppose that a tank of gas stands on a table in a room. Suppose also that the gas has settled to equilibrium; i.e., the gas is uniform throughout the tank, as is shown at TIME 1 in Fig 1.3. Uniformity is an example of symmetry. In the case being considered, every position in the tank is equivalent, i.e., symmetric, to every other position. In particular, the gas configuration is reflectionally symmetric about the middle vertical line. Now use some means, e.g., a magnet, to cause the gas to move to the left side of the tank, as shown in TIME 2 in Fig 1.3. The gas has become *asymmetrically* ar-

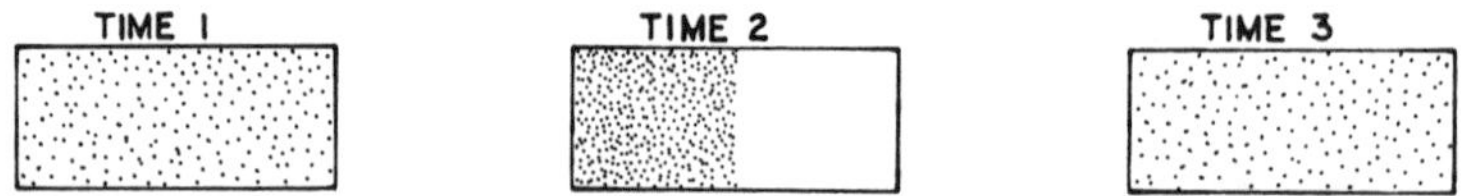

Figure 1.3
A tank of gas in three successive states.

ranged. If a person came into the room at this point and saw the gas compressed into the left side like this, he or she would conclude that the gas had undergone movement to the left side. Thus, even though the person did not see the movement, the asymmetry acts as a *memory* of the movement.

Now let the gas settle again to equilibrium; i.e., uniformity throughout the tank, as shown in TIME 3 in Fig 1.3. Suppose that, at this point, a person, who has not yet been in the room, now walks into the room. He or she would not jump to the hypothesis that anything had happened to the gas. This is because, in returning to the symmetrical state, the gas had wiped out any memory of the causal event that had previously occurred. That is, if one has symmetry in the present, one cannot deduce a past that is any different from it.

Now let us go to another illustration.

The following phenomenon comes from a series of five psychological experiments that are described in Leyton (1986b). In those experiments, we found that, when human subjects are presented with only a parallelogram oriented in the picture plane as shown in Fig 1.4a, they see it in terms of a non-rotated one shown in Fig 1.4b, which they then see in terms of the rectangle shown in Fig 1.4c, which they then see in terms of the square shown in Fig 1.4d. That is, given *only* the first shape, that shown on the far left of Fig 1.4, their minds go through the entire sequence of shapes as shown in Fig 1.4. It must be emphasized that the sequence is generated not by the experimenter, but by the subjects.

Examining this sequence, it seems that what the subjects are doing here is providing a process-explanation for the first shape: That is, the shape in Fig 1.4a is understood as having been obtained by rotating that in Fig 1.4b, off its horizontal base. This latter shape is understood, in turn, as having been obtained by skewing the rectangle in Fig 1.4c. And the rectangle, in turn, is understood as having been obtained by stretching the square in Fig 1.4d.

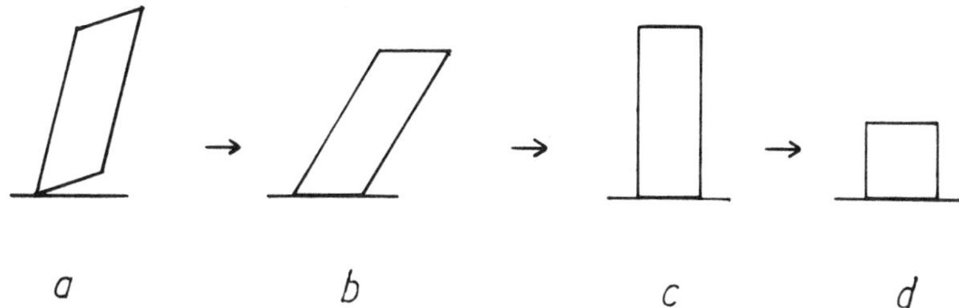

Figure 1.4
Subjects presented with (a), reference it to (b), then to (c), and then to (d).

What is crucial to observe here is that the process-history, backwards in time (from left to right), is given by successive symmetrization. In fact, if one compares any successive pair of shapes, the left-hand member of the pair has a single asymmetry that the right-hand member does not have. For example, the second parallelogram has unequal angles, whereas the rectangle, to the right of it, does not. Again, the rectangle has unequal sides whereas the square, to the right of it, does not. According to our view, it is an *asymmetry* at any stage that indicates that some process has acted; i.e., each asymmetry is a *memory* of a prior event. By removing each asymmetry, one is able to go backwards through the events; i.e., successively recover the past.

These examples enable us to understand how an asymmetry in the present can act as a memory of a process. The asymmetry is a memory of the process that *created* the asymmetry. That is, at some point in the past, the asymmetry did not exist, i.e., there was symmetry, and the process replaced the symmetry with the current asymmetry. Realizing this allows us to state the Asymmetry Principle more precisely:

ASYMMETRY PRINCIPLE. *An asymmetry in the present is understood as having originated from a past symmetry.*

Now observe, in the sequence of shapes in Fig 1.4, that, once a symmetry has been gained, e.g., equal angles as in the rectangle, it is retained throughout the rest of the sequence, backwards in time; e.g., the equal angles, gained in the rectangle, are retained by the square. This means that, where a symmetry exists, it is assumed to have existed at all prior points. We saw that the tank of gas also illustrates this phenomenon. When the gas is uniform, as at TIME 3 in Fig 1.3, a person entering the room for the first time will not hypothesize any past configuration that is different from what he or she sees. That is, a symmetry in the present is not a memory of anything other than itself. One can state this concept as follows:

SYMMETRY PRINCIPLE. *A symmetry in the present is understood as having always existed.*

An investigation of these two principles will be the subject of the entire book. We shall argue that they are fundamental to the recovery of history.

1.5 Conforming to the Uni-Directionality Requirement

Recall now that, because the past states, i.e., the goal of the process-recovery problem, are inaccessible, one must have some means of ensuring their uniqueness. Recall also that this implies that recoverable processes must be represented uni-directionally.

We should therefore check that the Asymmetry and Symmetry Principles do indeed allow unique recoverability and define processes uni-directionally. In fact, it is easy to see that they do, as follows:

To see this, we need first to give a definition of *symmetry*. Basic to the standard mathematical definition of symmetry is this: *Symmetry is indistinguishability*. For example, human faces are usually regarded as symmetric because the left side tends to be indistinguishable from the right side. Notice that, in this case, the indistinguishability could be defined only because one used a reflection transformation to map the left side of the face onto the right side. Thus the indistinguishability often exists by virtue of a transformation. Conversely: *Asymmetry is distinguishability*. For example, some faces are actually asymmetric because, for example, the left eye-lid droops more than the right.

With this basic definition of symmetry and asymmetry, let us return to the successive reference phenomenon shown in Fig 1.4. Consider the first parallelogram (Fig 1.4a). This shape has three crucial distinguishabilities:

(D1) Its orientation is distinguishable from the gravitational orientation.

(D2) Its vertices have two distinguishable sizes.

(D3) Its sides have two distinguishable lengths.

Now observe that, in moving from this parallelogram to the next, the first distinguishability is removed. Next observe that, in moving from the second parallelogram to the rectangle, the distinguishabilities between the vertex-sizes is removed; i.e., all vertex-sizes become indistinguishable. Finally, observe that, in moving from the rectangle to the square, the distinguishabilities between the side-lengths is removed; i.e., all side-lengths become equal.

Thus each successive symmetrization is the removal of a distinguishability. The term we can also use is *homogenization.* The rectangle homogenizes the sizes of the angles, and the square homogenizes the lengths of the sides. That is, the square represents complete homogenization with respect to the distinguishabilities noted in D1, D2, and D3 above.

The above illustrates the Asymmetry Principle, which claims that situations, in which distinguishabilities can be found, point to situations in which the distinguishabilities do not exist. It also illustrates the Symmetry Principle which claims that an indistinguishability can be a memory of only itself; i.e., at any point in the sequence of Fig 1.4, a recovered indistinguishability remains for the rest of the sequence.

Now let us turn to the issue of unique recoverability. Observe that, to each situation with a distinguishability, one can define a unique situation that removes the distinguishability. Thus the Asymmetry Principle—which states that, from a situation with distinguishability, one recovers a situation that removes the distinguishability—allows unique recoverability. So does the Symmetry Principle because, quite simply, it says that, from a situation with indistinguishabilities, one recovers only the same.

Observe also that, according to the two principles, processes are defined uni-directionally: i.e., a process moves *from symmetry to asymmetry but not vice versa.* This can be seen by looking at both principles. According to the Asymmetry Principle, a process moves from a past symmetry to a present asymmetry. If the reverse could happen, then a process could move from a past asymmetry to a present symmetry. However, the Symmetry Principle forbids this. It says that a symmetry in the present corresponds to the same symmetry in the past, i.e., the absence of a process. Thus processes can go only from symmetries to asymmetries. Therefore, using the Asymmetry and Symmetry Principles, together, one sees that processes conform to our uni-directionality requirement.

1.6 Atemporal-Temporal Duality

Our basic proposal, that asymmetry, in the present, is the memory left by processes, refers to asymmetry that is completely contained within *a single moment*, the present moment. That is, the single moment corresponds to a single snapshot; and, by examining this snapshot, alone, one can recover the past. Thus, the asymmetry is completely contained within the frozen snapshot. This asymmetry will therefore be referred to as *atemporal asymmetry*.

Besides this atemporal asymmetry, there is another asymmetry that is crucial. We have seen that, under the Asymmetry and Symmetry Principles, processes are *uni-directional*; i.e., they are understood as going from symmetries to asymmetries, but not the reverse. Uni-directionality is, in fact, an example of asymmetry. However, this asymmetry is *not* the asymmetry mentioned in the previous paragraph; i.e., the asymmetry within the snapshot. The uni-directional asymmetry is in the structure of time. In contrast, the asymmetry, within the snapshot, exists when you remove time.

Our basic proposal is that these two completely different asymmetries are brought together in the solution to the process-recovery problem. That is, there are two asymmetries, one within a single moment, and one within time, and process-recovery connects them in this way:

FIRST DUALITY PRINCIPLE. *Asymmetry within the moment implies asymmetry within time (and vice versa).*

To put this statement simply: If you find asymmetry within a moment, then the moments themselves have to be organized asymmetrically with respect to each other. Again, correspondingly, we have:

SECOND DUALITY PRINCIPLE. *Symmetry within the moment implies symmetry within time (and vice versa).*

The two Duality Principles can be regarded as roughly equivalent to the Asymmetry and Symmetry Principles, respectively (section 1.4), except that they concern both past and future time whereas the Asymmetry and Symmetry Principles concern only past time.

A corollary of these two principles can be stated in terms of the notion of *stability*. One can define stability as temporal symmetry; i.e., the absence of change through time. The corollary is this:

STABILITY PRINCIPLE. *The more symmetric a configuration is, the more stable it is understood to be.*

The advantage of involving the term *stability* is that the notion of stability links us closely to the phenomenon of *dynamics*, which will be useful to us later.

One can understand temporal asymmetry simply as the assignment of direction to time. However, it is important to understand that we have given this asymmetry a more *substantive* characterization. The substantive characterization is that the temporal asymmetry is between two situations that differ *structurally*; one with distinguishabilities and one without. A reader familiar with thermodynamics might, at this point, think that the above principles have been anticipated by the standard claim that the Second Law of Thermodynamics determines the direction of time. But this is not the case. The above principles would be true of a world defined purely by classical mechanics; i.e., in which thermodynamic factors did not exist. This will become evident in Chapter 2.

1.7 The Second System Principle

The Asymmetry Principle states that an asymmetry in the present is explained by a process that introduced that asymmetry from a previous symmetry. By *process*, we mean, for the first few chapters, a *sequence of states* of the system undergoing the transition from past to present.

Given this definition of process, the Asymmetry Principle makes no mention of the notion of *causality*; i.e., what actually made the system undergo this change. We introduce the notion of causality via this next principle:

SECOND SYSTEM PRINCIPLE. *Increased asymmetry over time can occur in a system only if the system has a causal interaction with a second system.*

It is crucial to understand that the Asymmetry Principle and the Second System Principle are completely independent principles. In order to understand this, let us see what the two principles would infer from an asymmetry in the present.

Observe that, on its own, the Second System Principle concludes absolutely nothing from an asymmetry in the present. It can form a conclusion only if it knows that the asymmetry originated from a symmetry. However, if it does not know whether the asymmetry originated in this way, then it can conclude nothing. In particular, it cannot say anything about events prior to the present. Thus, we can say that given the *entire* present, the Second System Principle cannot conclude anything from it.

In contrast, the Asymmetry Principle, on its own, does produce a conclusion from an asymmetry in the present. It says that the asymmetry originated from a symmetry in the past. That is, the Asymmetry Principle makes a fundamental statement about the structure of the preceding history. Thus we can say that given the *entire* present, the Asymmetry Principle can conclude something about every single asymmetry in the present.

To summarize so far: Given only the present, the Second System Principle can conclude nothing about it, and in particular nothing about any of the asymmetries in the present. In contrast, the Asymmetry Principle can conclude something about every single asymmetry in the present.

The Second System Principle can perform its job only after the Asymmetry Principle has been used. That is, once the Asymmetry Principle has established that there was a transition from symmetry to asymmetry over time, the Second System Principle can be brought in to infer something from this new piece of information; i.e., that there must have been a causal interaction with a second system.

Besides the fact that the Second System Principle is subordinate to the Asymmetry Principle, we should observe that there is a clear division of labor between the two principles. The Asymmetry Principle deduces change in the structure of a system over time. However, it does not concern causality. In contrast, the Second System Principle does concern causality. But at an expense: it cannot deduce change in the structure of a system.

1.8 Uni-Directionality and the Second System Principle

We shall now describe a rather different theoretical role that can be played by the Second System Principle. In the previous section, we looked at the relationship between the Second System Principle and the Asymmetry Principle. In this section we shall look instead at the relationship between the Second System Principle and the phenomenon of uni-directionality. To fully clarify this relationship we shall forget that the Asymmetry Principle exists.

We observed, in section 1.5, that, given an asymmetry in the present, one can recover a *unique* symmetry from it; i.e., that corresponds to its neutralization. The reverse is not possible: That is, given a symmetry there are many asymmetries that could be derived from it. This means that the direction of unique recoverability is from asymmetry to symmetry, and not the reverse. Thus, when the present contains an asymmetry, this points to a unique symmetry. However, is this symmetry in the future or in the past? Without the Asymmetry Principle, the symmetry could be just as easily in the future as in the past.

To illustrate, consider once again the successive reference example shown in Fig 1.4. We had previously regarded this sequence, i.e., of successively recovered symmetries, as going backward into the past; i.e., describing how the rotated parallelogram originated. However, one could easily regard the same sequence of recovered symmetries as going forward in time; i.e., showing what *will* happen to the rotated parallelogram. For example, consider the second parallelogram: It is recovered from the first. However, it can be given two interpretations: (1) The second parallelogram was in the past. This means that, at some time in the past, the second parallelogram was *raised* and became the first parallelogram. (2) An alternative interpretation is that the second parallelogram is in the future: It is the result that will be obtained when one lets the first parallelogram *fall*.

The example just given illustrates the fact that a recovered symmetry can be a memory or a prediction. That is, the transition from asymmetry to symmetry can represent backwards or forwards time. In conclusion, we seem to have arrived at a deep paradox. The ability to recovery unique symmetry gives time a direction. But either direction is possible: into the future, or into the past.

It turns out that this paradox is a theoretical benefit to us. By being careful enough to let the paradox emerge, we can see that a further ingredient is needed to distinguish the two uni-directional interpretations of time; i.e., to distinguish between the two possible interpretations of the recovered symmetry—in the past or in the future.

The distinction is in fact supplied by using the Second System Principle, as follows: If the recovered symmetry is in the past, then the temporal transition from the past to the present must involve increased asymmetry. By the Second System Principle, one will therefore invoke an additional system to explain the transition. However, if the recovered symmetry is in the future, then the transition from the present to the future involves decreased asymmetry. No additional system will be hypothesized in this case.

To illustrate: Consider again the two parallelograms in Fig 1.4. If one assumes that the recovered parallelogram, i.e., the second one, was in the past, then the obvious *physical* interpretation of the transition from past to the present—i.e., from the second parallelogram to the first—is that the second parallelogram was lifted by the action of some additional system; e.g., a hand that gripped and raised the object. Conversely, if one assumes that the recovered parallelogram, the second one, is in the future, then the obvious physical interpretation of the transition from the present to the future—i.e., the first parallelogram to the second—is that the hand that held up the first parallelogram was removed; e.g., the hand let go of the object. Thus, in accord with the Second System Principle, given the two interpretations for the recovered symmetry, an additional system is hypothesized only in the interpretation in which asymmetry increases over time.

1.9 Curvature and Process

We will now apply the above principles to understand how one recovers process-history from complex shapes such as those of embryos, tumors, and clouds. The recovery of history from such shape will concern us for most of the rest of the chapter. We shall argue that a remarkably rich process-history is extracted from a type of asymmetry known as *curvature variation.*

What exactly is curvature variation? Observe first that *curvature* is the amount of bend. For example, consider the series of lines in Fig 1.5a. The top-most line has no bend and thus is regarded as having zero curvature. The amount of bend (curvature) increases in the progression of lines from top to bottom. Finally, observe that the bottom-most curve has a special point E where the curvature on that line is greatest. That is, at any other point on that line, the curve is flatter, i.e., has less bend, than at E. Thus E is called a *curvature extremum.*

If the curve bends one way, the extremum is called a *maximum.* If the curve bends the other way, the extremum is called a *minimum.* To illustrate, consider the face in Fig 1.5b. The minima are the "valleys". Each one is marked on the figure with a dot. The maxima are the peaks in

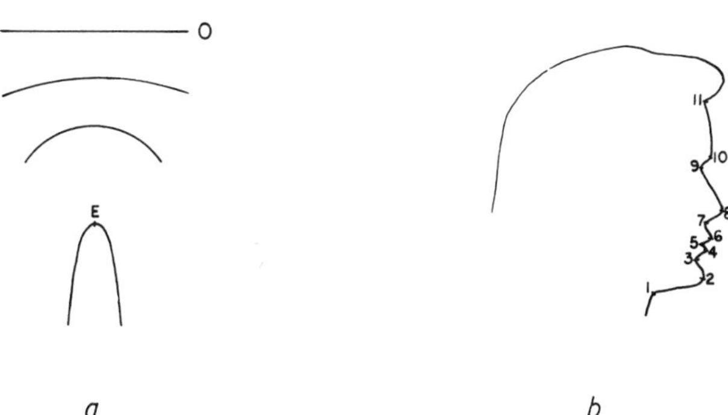

Figure 1.5
(a) The successive lines downward have greater curvature, and point *E* is a curvature extremum. (b) The numbered points are curvature extrema.

between the valleys. By simple mathematical considerations, minima and maxima have to alternate along the curve.

Curvature extrema are known to have two psychologically salient roles: First, as was demonstrated by Attneave (1954), people choose curvature extrema when making a visual summary of a shape. For example, when people are asked to summarize an entire shape using only a few dots, they will choose the dots to be the curvature extrema. This phenomenon is the basis of the childhood activity of "drawing by numbers": The child is given a sheet of paper with some numbers scattered on it and the child has to connect the numbers with lines. The numbers are in fact the curvature extrema of the shape.

The second psychological role of curvature extrema was discovered by Hoffman & Richards (1985; Richards & Hoffman, 1985). These researchers offered compelling evidence that contours (curves) are psychologically segmented at negative minima. For example, on the face in Fig 1.5b, these extrema are the *valleys*. Going down the face, one finds that the consecutive valleys demarcate the ends of the forehead, the ends of the nose, the ends of the upper lip, the ends of the lower lip, and the ends of the chin. In other words, the minima are the ends of the psychological *parts* of the face.

We will now argue that curvature extrema have a third very powerful psychological role: They are used to infer *process-history*. In order to see this, we need to carefully apply the Asymmetry and Symmetry Principles proposed in the previous sections. Throughout sections 1.10 to 1.18, we will be looking at the process-history of closed two-dimensional curves where curvature can be defined at all points. Such curves do not have corners or end points. They are closed smooth curves, and they can represent the outlines of many biological shapes, e.g., amoebas, embryos, biological organs, human beings, etc., as well as non-living entities such as spilt coffee, rain puddles, etc. It is not a difficult matter to generalize our analysis to objects with non-smooth outlines, as we will see later.

1.10 Symmetry in Complex Shape

Since symmetry and asymmetry are central to the application of our principles, we have to understand what is symmetric and asymmetric in complex shape.

How can one define symmetry in a complex shape? Note that a symmetry axis is usually defined to be a straight line along which a mirror will reflect one half of a figure onto the other. However, observe that, in complex natural objects, such as Fig 1.1, a straight axis might not exist. Nevertheless, one might still wish to regard the figure, or part of it, as symmetrical about some *curved* axis. For example, a branch of a tree tends not to have a straight reflectional axis. Nevertheless, one understands the branch to have an axial core that runs along its center.

How can such a *generalized* axis be constructed? As we shall see in Chapter 6, there are good reasons to proceed in the following way (Blum, 1973; Brady, 1983; Leyton, 1988). Consider Fig 1.6. It shows two curves, c_1

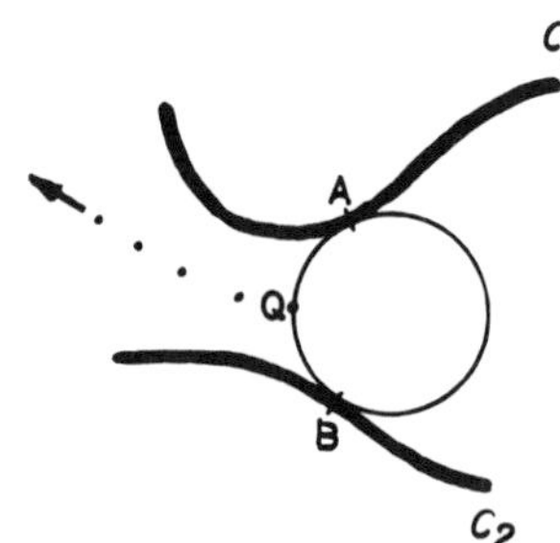

Figure 1.6
Defining a symmetry axis (the dotted curve) between the two bold curves.

and c_2, which can be understood as two sides of an object. Now introduce a circle that is tangential simultaneously to the two curves, as shown. The two tangent points, *A* and *B*, are then defined as symmetrical to each other.

Now move the circle continuously along the two curves, c_1 and c_2, while always making sure that it touches the two curves. The circle might have to expand or contract to maintain the double-touching. One can then define a symmetry axis to be the trajectory of the midpoint *Q* of the arc *AB*, as the circle moves. For example, in Fig 1.6, the trajectory of *Q* is represented by the locus of dots shown. This definition was proposed in Leyton (1988), and we shall argue, in Chapter 6, that it is more suited for the specific task of process-inference than are other symmetry analyses, such as those proposed by Blum (1973) and Brady (1983). The intuitively satisfying results of our alternative analysis will be seen later.

1.11 Symmetry-Curvature Duality

We shall see that *curvature* contains the information by which one unlocks the history of a shape. However, our main principles all take *symmetry* (and asymmetry) to be the key to process-inference. Thus we need to have a means of linking curvature to symmetry.

However, forming such a link presents a problem that underlies the structure of mathematics. Symmetry and curvature are two very different ways of describing shape. Curvature, or bend, is measured within a region; symmetry is constructed across regions. Curvature is a smooth property (bending); symmetry is a discrete one (reflecting). Curvature requires the curve to be continuous; symmetry does not require continuity. Ultimately the difference between curvature and symmetry mirrors a deep schism in the foundations of mathematics. Mathematics is comprised of two branches: topology and algebra.[1] Curvature is a topological notion and symmetry is an algebraic one. In fact, this schism can be seen also in perceptual psychology: Curvature and symmetry have been investigated as two separate phenomena involving two separate areas of research; the school of Attneave, Hoffman, and Richards examining curvature, and the Gestalt tradition examining symmetry.

Given the crucial difference between symmetry and curvature, it comes as rather of a surprise that a recent mathematical result has shown that there is an intimate relationship between these two descriptors. This relationship is a theorem that was proposed and proved in Leyton (1987b), and it will be a crucial step in our argument:

SYMMETRY-CURVATURE DUALITY THEOREM (Leyton, 1987b). *Any section of curve, that has one and only one curvature extremum, has one and only one symmetry axis. This axis is forced to terminate at the extremum itself.*[2]

To illustrate, consider the curve shown in Fig 1.7. There are three extrema on the curve: m_1, M, and m_2. The section of curve *between* extrema m_1 and m_2 has only one extremum, M. Therefore, according to the theorem, this section of curve yields one and only one axis. Furthermore, according to the theorem, this axis has to terminate at the extremum M. The axis is shown as the dotted line in Fig 1.7.

Now let us apply the theorem to the complex shape with which we started this chapter. It is presented again in Fig 1.8. The shape has eight curvature extrema. Thus, by the above theorem, there are eight unique symmetry axes associated with, and terminating at, the extrema. These are the dotted lines shown in Fig 1.8.

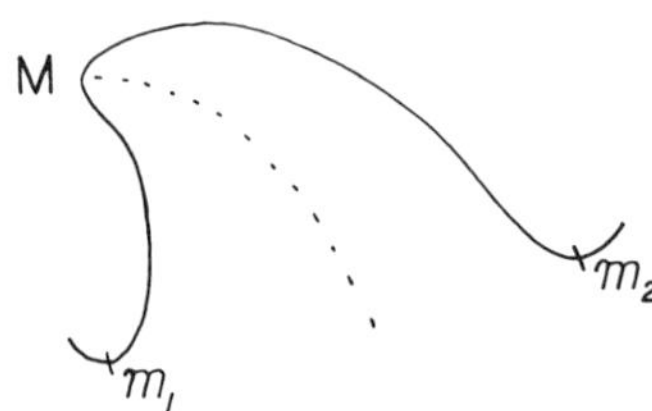

Figure 1.7
An illustration of the Symmetry-Curvature Duality Theorem.

Figure 1.8
The shape in Fig 1.1 with the symmetry axes predicted by the Symmetry-Curvature Duality Theorem.

1.12 Process Direction

The Symmetry-Curvature Duality Theorem is crucial to our analysis because it allows us to lock into place the principles we have developed for the recovery of process-history. Consider first the Symmetry Principle, which states that a symmetry in the present is understood as having always existed. That is, *in running time backwards, one does not destroy a symmetry in the present.*

How can symmetry be preserved over time? In Leyton (1984), it was shown that a symmetry is most likely to be preserved if the hypothesized process is directed along the symmetry axis. For example, consider the rectangle in Fig 1.4c. It has a vertical symmetry axis. However, the rectangle is supposed to have been derived from the square by stretching along the vertical symmetry axis evident in the resulting rectangle. That is, the vertical symmetry axis of the rectangle gives the direction along which the process took place. This example, and many others to be considered later, imply that, if one is given some object, the direction of the process hypothesized as most likely to have occurred is along its symmetry axis. This is because, in running time backwards, that process is the one that preserves the symmetry; i.e., conforms to the Symmetry Principle. That is, we have:

INTERACTION PRINCIPLE (Leyton, 1984). *Symmetry axes are the directions along which processes are hypothesized as most likely to have acted.*

The Interaction Principle will be corroborated later with several types of psychological data, in both shape and motion perception.

1.13 From Curvature Extrema to History

Recall our argument concerning complex shape: The Symmetry-Curvature Duality Theorem allows us to convert curvature considerations into symmetry considerations. It states that, to each curvature extremum, there is a unique axis that terminates at that extremum. The advantage of converting curvature concepts into symmetry concepts is that we can then use the Symmetry Principle which states that, in running time backwards, one cannot destroy a symmetry in the present. This rule becomes re-expressed as the Interaction Principle, which says that processes are most likely to have occurred along symmetry axes. Thus, returning to Fig 1.8, our conclusion is that the processes acted along the symmetry axes.

However, one final component is needed to understand how process-history is recovered from complex shape. Although the Interaction Principle says that the processes must have acted along the symmetry axes, the principle does not determine which of the two directions, along an axis, the process must have acted. We therefore require another rule to disambiguate direction along an axis. The Interaction Principle is based on our Symmetry Principle. However, we have not so far used our Asymmetry Principle which states that, in running time backward, asymmetry is removed.

In the present case, the asymmetry we shall consider is distinguishability in curvature; i.e., curvature variation. (The reason for using this will be examined more fully later.) Thus running time backwards removes curvature distinguishability. But this means that one eventually arrives at a circle, because the circle is the only smooth closed curve without curvature distinguishability; i.e., every point on a circle has the same curvature as every other point. The Asymmetry Principle therefore implies that the past started with a circle and, in going from the past to the present, the processes created the distinguishability in curvature.

Now according to the Interaction Principle, the processes moved along the symmetry axes evident in the present shape. This means that the processes moved along the axes in the direction of creating greater distinguishability in curvature. Thus, for example, each protrusion in Fig 1.8 was the result of *pushing the boundary out* along its associated axis, and each indentation was the result of *pushing the boundary in* along its axis. That is, each axis is the *trace* or *record* of boundary-movement, in the direction of greater curvature distinguishability.

Observe the role of the extrema. By the Symmetry-Curvature Duality Theorem, they are the points at the end of the symmetry axes. The processes are therefore understood as creating the curvature extrema; e.g., they introduce protrusions and indentations into the shape boundary. This means that, if one were to go backwards in time, undoing all the inferred processes, one would eventually remove all the extrema. One would of course arrive at the circle because the circle is the only curve without extrema.

Thus the Asymmetry Principle and the Symmetry Principle have led to the following conclusion:

Each curvature extremum implies a process whose trace is the unique symmetry axis associated with and terminating at that extremum.

1.14 Application

To see that our rules consistently yield highly appropriate process-analyses, let us obtain the analyses that these rules give for a large set of shapes. We shall take the set of all possible shapes that have eight extrema or less. In Fig 1.9 we have presented all these shapes (as given in a paper by Richards, Koenderink & Hoffman, 1987), and we have superimposed, on each shape, the processes that are inferred under the inference rules. The arrows are the inferred processes. As the reader can see, the results accord remarkably well with intuition.

Let us call a shape, together with the set of processes inferred on it under the inference rules, a *process-diagram*. Surveying the process-diagrams in Fig 1.9, another important feature emerges: Purely structural considerations yield strong *semantic* constraints, as we shall now see. First of all, note that the curvature along a contour can be represented by a graph of the type shown in Fig 1.10. That is, position along the shape's outline is represented by the horizontal axis in the graph, and the curvature at each point of the outline is represented by height in the graph. Let us now represent the extrema on the graph in this way: Let M and m denote a local maximum and local minimum respectively, and $+$ and $-$ denote positive and negative curvature respectively.[3] Then there are four types of extrema: M^+, m^-, m^+, M^-, as illustrated in Fig 1.10.

Now turn back to Fig 1.9. All the extrema on all the shapes have been labeled according to this classification.

It is important now to observe the following result: In surveying the shapes in Fig 1.9, it becomes evident that the four types of extrema, given by the above purely structural characterization, correspond to *semantic* terms that people use to classify *processes*: The correspondence is as follows:

SEMANTIC INTERPRETATION RULE

M^+ protrusion

m^- indentation

m^+ squashing

M^- internal resistance

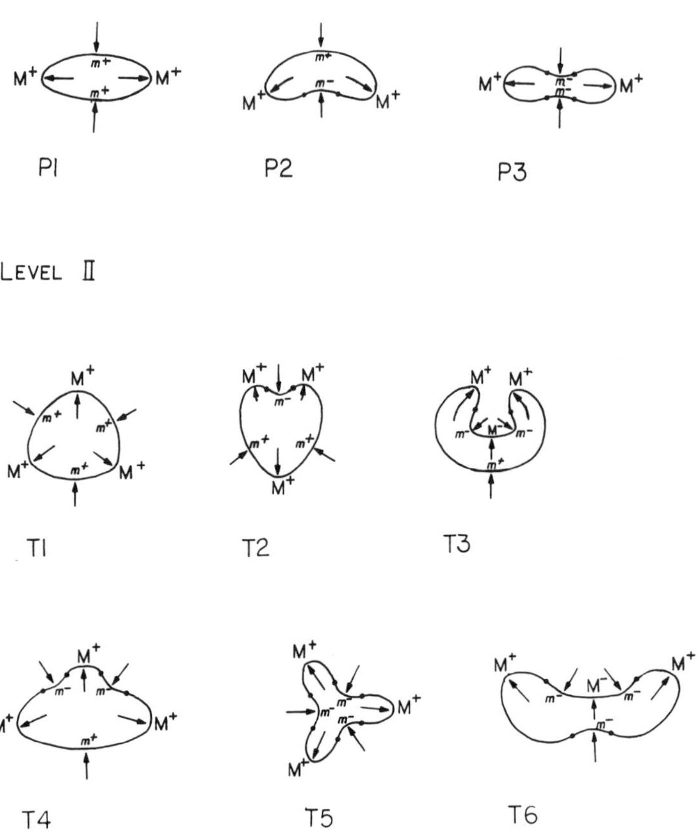

Figure 1.9
The process-histories inferred by our rules when applied to the the shapes of Richards, Koenderink, and Hoffman (1987).

LEVEL III

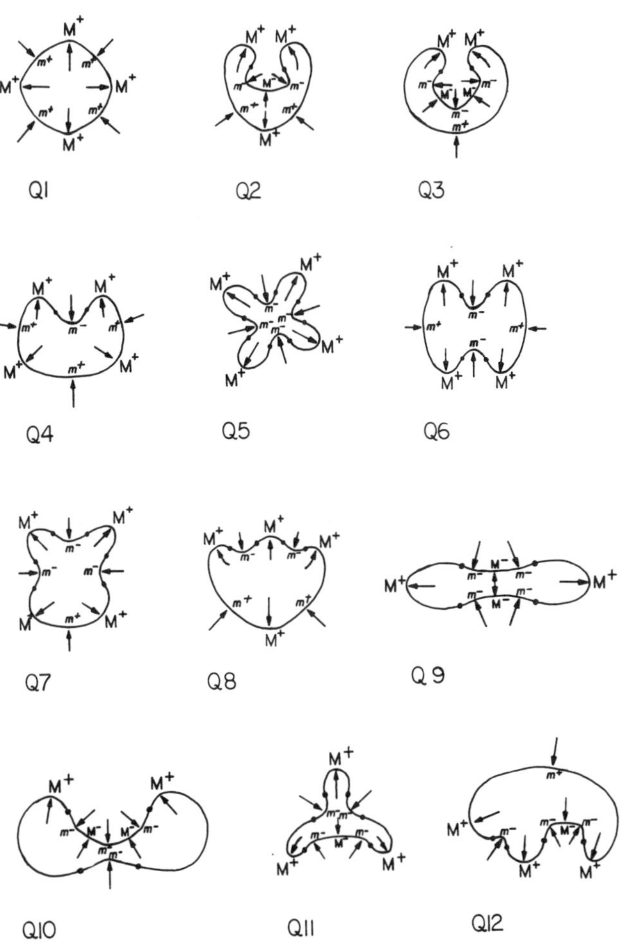

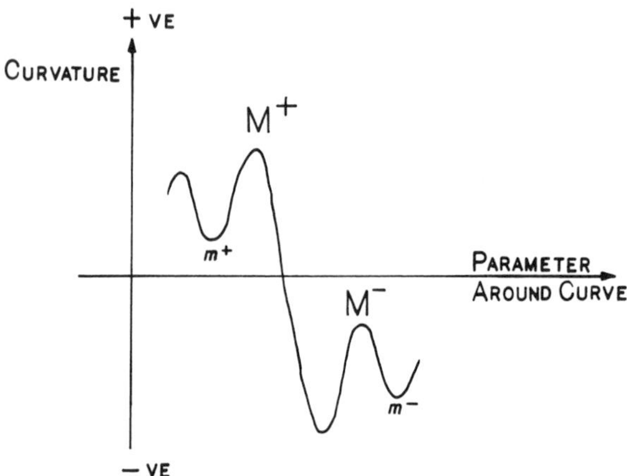

Figure 1.10
The four possible kinds of extrema in a curvature graph.

A final comment should be made: Fig 1.9 is stratified into levels according to number of extrema. The reader might wonder why, in the set of shapes with up to eight extrema, there are only three levels. To answer this, first observe that any shape must have an *even* number of extrema because maxima and minima alternate, and the curves that we are considering are closed. Furthermore, the Four-Vertex Theorem, in differential geometry, states that any smooth curve that is not a circle must have at least 4 curvature extrema (Do Carmo, 1976, p37). Thus, any shape with 8 extrema or less must have either 4, 6, or 8 extrema. These are the three levels in Fig 1.9.

1.15 Partitioning into Asymmetry and Symmetry

Consider any individual shape in Fig 1.9. It is important to understand that our two main principles, the Asymmetry and Symmetry Principles, partition the shape into its asymmetric and symmetric components, as follows:

(1) *Asymmetric Component*: The distinguishability in curvature.

(2) *Symmetric Component*: The symmetry axes leading to the extrema.

Having partitioned the shape thus, the Asymmetry and Symmetry Principles then relate these components to *time*, as follows: The curvature distinguishability represents history; and the symmetry axes represent the absence of history. In particular, the Asymmetry Principle says that the curvature distinguishability is removed backwards in time. In contrast, the Symmetry Principle does the following, which will be of great concern to us in the next section: The principle *constrains* the way in which the Asymmetry Principle can be used to remove the distinguishability backward in time. It says that this temporal change cannot affect the symmetry component; i.e., the processes must move along the axes.

1.16 The Asymmetry in Curvature Variation

Fig 1.9 exhibits the process-history of several smooth shapes. The derivation of the histories of these shapes rests crucially on our observation that the distinguishability in curvature is a form of asymmetry. This section will be devoted to understanding more clearly the asymmetric nature of curvature distinguishability.

To proceed, we now need to recall that, while symmetry is indistinguishability, one might often need to define this indistinguishability *relative to a transformation*. For example, a face is symmetric only relative to a reflection transformation that sends the left side of the face onto the right side.

Given this concept, we shall now see that curvature distinguishability can be regarded as existing relative to two kinds of transformation. That is, curvature distinguishability contains *two* forms of asymmetry, as follows:

Type 1 Asymmetry

In order to understand the first type of asymmetry inherent in curvature distinguishability, consider the shape without any curvature distinguishability; i.e., the circle. Imagine that the circle is a track around which one is driving a car. When one starts driving, one sets the steering wheel at a particular angle. However, because the track is circular, after one has set the initial angle of the wheel, one does not have to alter the wheel as one drives. This is a result of the fact that the curvature of the track is the same

Figure 1.11
A complex shape.

at every point—curvature corresponding to the amount of turn made by the wheel. Thus, one's actions are indistinguishable, i.e., symmetric, as one moves around the circle.

Suppose now that one is driving around a track that has distinguishability in curvature, as in Fig 1.11. In this case, one continually has to readjust the wheel—because the curvature is continually altering. Therefore, in this case, one's actions are distinguishable; i.e., asymmetric, as one moves around the track.

We have said that asymmetry is distinguishability relative to *transformations*. The transformations we have just considered are in fact *rotations*. Any direction (tangent) on the perfect circle is indistinguishable from any other direction, with respect to rotation around the circle center; i.e., the tangents can all be rotated onto each other. In contrast, the tangents on a complex curve cannot all be rotated onto each other using some common rotation center. Thus the first type of asymmetry embodied in curvature variation is rotational asymmetry.

Type 2 Asymmetry

A circle contains another type of symmetry that is violated when distinguishability is introduced into curvature. This symmetry is very obvious, but it has non-obvious consequences in terms of the analysis of processes, as we shall soon see. The symmetry is reflectional symmetry.

One first has to appreciate the enormity of the reflectional symmetry in a circle. Consider Fig 1.12a. Point A is reflectionally symmetric to point B about the dotted axis. But, as shown in Fig 1.12b, point A is reflectionally symmetric also to point C about the dotted axis shown in that figure. Indeed point A is reflectionally symmetric to *every other point on the circle*.

Now consider a shape with distinguishability in curvature; e.g., Fig 1.12c. Consider an arbitrary point; e.g., the point A on the lower right protrusion in Fig 1.12c. By the Symmetry-Curvature Duality Theorem,

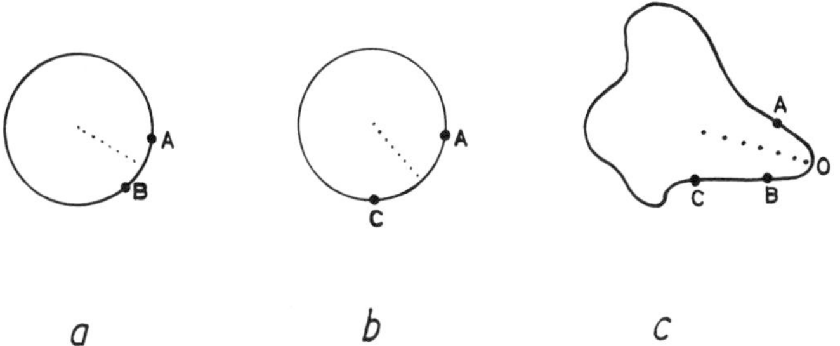

Figure 1.12
(a & b) Both pairs of points $[A, B]$ and $[A, C]$ are reflectionally symmetric on a circle. (c) However, on the protrusion, only the pair $[A, B]$ is symmetrically related.

the protrusion has one and only one axis. Because there is only one axis, *A* can be reflectionally symmetric to only one other point on the protrusion, the point *B* as shown in Fig 1.12c. Thus, consider any other point *C* on the protrusion. Point *A was* reflectionally symmetric to *C* on the original circle from which the shape was derived, because *A was* reflectionally symmetric to any other point on the circle. However, pushing out the boundary into a protrusion has caused *A* to lose its symmetry with respect to *C*. Since there are an infinite number of points on the protrusion—all of which once had symmetry with *A*—the symmetry of *A* has jumped from infinity down to one.

Another aspect of the structure can be seen by considering, in Fig 1.12c, the extremum *O* at which the axis of the protrusion ends. On the original circle, the point *O* would have been reflectionally symmetric to every other point. However, it is now symmetric only to itself. This means that we can now gain a deeper understanding of the *memory* inherent in the symmetry axis of the protrusion in Fig 1.12c. In the original circle, the point *O* had an infinite number of symmetry axes associated with it; i.e., the axes allowed reflectional symmetry with respect to *all* the other points on the circle. One of these axes was the axis about which *O* was self-symmetric. When the protrusion was created, each of the infinite number of axes was lost *except one*: the axis of self-symmetry. This became the axis of the protrusion. That is, the protrusion grew along the axis of self-symmetry of *O*. Furthermore, the self-symmetry axis of *O* became significant for many other points: For all the points such as *A*, that ended up on the same protrusion as *O*, the axis of *O* became the only axis about which they retained any symmetry.

Thus, one of the implications of the Symmetry-Curvature Duality Theorem, under the process analysis, is this: The theorem implies that a process destroys all symmetry axes except one, the self-symmetry axis of the extremum.

1.17 The Generality of Processes

How do people perceptually define a *part* of a shape? Hoffman & Richards (1985) have argued that a part perceptually corresponds to a segment in between two consecutive valley points, m^- extrema, along the shape's outline. To illustrate: In the hand shown in Fig 1.13, the black dots on the outline indicate the m^- extrema. In accord with Hoffman and Richards' theory, any segment, whose endpoints are two consecutive dots, is a finger, i.e., a perceptual part of the shape.

We should now observe that Hoffman and Richards' theory is purely a *segmentation* theory of parts; i.e., parts are defined by their endpoints. We will now argue that our process-analysis gives a more cognitive view of what a part is.

Note first that, in between any two consecutive m^- extrema in Fig 1.13, there is a M^+ extremum: the tip of the finger (e.g., this extremum has been marked on the left-most finger). Recall now that the Symmetry-Curvature

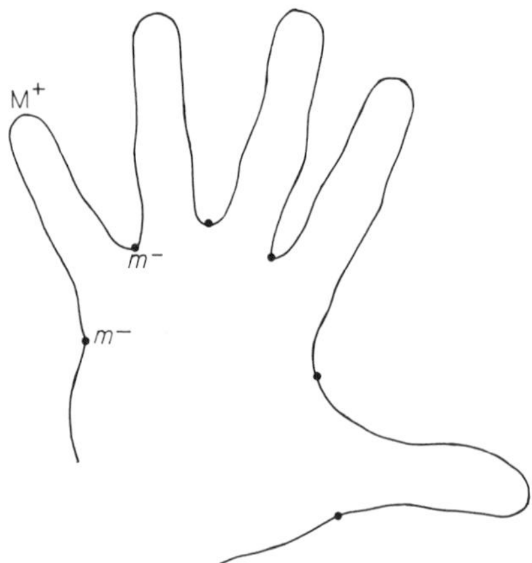

Figure 1.13
A part, the little finger, with its arrangement of curvature extrema.

Duality Theorem states that, because there is only one extremum (M^+) on the segment *in between* two consecutive m^- extrema, there can be only one axis on that segment. Furthermore, the theorem states that the axis must terminate at the M^+ extremum. For example, in the case of a finger, there is only one axis, and this axis terminates at the tip of the finger.

The second rule of our process-analysis, the Interaction Principle, then leads to the crucial conclusion that the axis is the *trace* of a process.

The importance of this conclusion is that it implies that a part is not a rigid segment: A part is a *process*! It is a *temporal* or *causal* concept.

Thus, an important difference exists between the Hoffman & Richards analysis and our process-analysis: The former gives a description of a part simply as a *segment*, whereas the latter views a part as a *causal explanation.* We argue that causal explanation is central to the cognitive representation of shape. That is, one cannot help seeing an embryo, tumor, cloud, geological formation, etc., as a consequence of historical processes.

There is another important difference between the Hoffman & Richards segmentation approach and our process approach: According to the process approach, a part is a process terminating at a M^+ extremum. By our Semantic Interpretation Rule (section 1.14), a part must therefore be a *protrusion.* However, as Fig 1.10 shows, a M^+ extremum is only one of four types of extrema that exist. Furthermore, since, by our analysis, processes are hypothesized to explain extrema, there must be four times as many processes as there are parts. In fact, by our Semantic Interpretation Rule, the full collection of processes associated with extrema are *protrusions, indentations, squashings,* and *resistances.* Parts correspond to only the first of these categories.

We conclude therefore that our process approach to parts is more cognitive and inclusive than a straightforward segmentation approach for two reasons: (1) The process approach considers parts to be causal explanations. Such explanations are, we claim, central to the *cognition* of shape. (2) The process approach accounts not only for *parts*, but, since there are four times as many processes as parts, the process-analysis accounts for many more phenomena than parts.

1.18 Transversality or Symmetry-Curvature Duality?

Hoffman & Richards (1985) justify their definition of a part by what they call the *Transversality Principle.* This principle states that, when two surfaces *interpenetrate*, they tend to do so at valley points (concavities). For example, consider Fig 1.14a. It shows a finger initially as a separate object. The finger is then brought toward the rest of the hand until it penetrates the latter. The result is that there are valley points where the finger penetrates. These valley points are shown as the m^- extrema in Fig 1.14b. Hoffman & Richards argue that their rule, which says that parts are segmented at m^- extrema, is justified environmentally by the Transversality Principle; i.e., that, when *separate* objects are made to interpenetrate, they will result in valley points.

However, is the Transversality Principle a valid justification in the case of biological objects? Biological parts are formed by outward growth not by interpenetration. A finger grows out from the rest of the hand; it is not made separately and then attached to the hand as shown in Fig 1.14. Thus the Transversality Principle seems to be an invalid justification in the case of biological objects.

We shall now argue that the Symmetry-Curvature Duality Theorem is the appropriate justification in the case of biological objects. We have

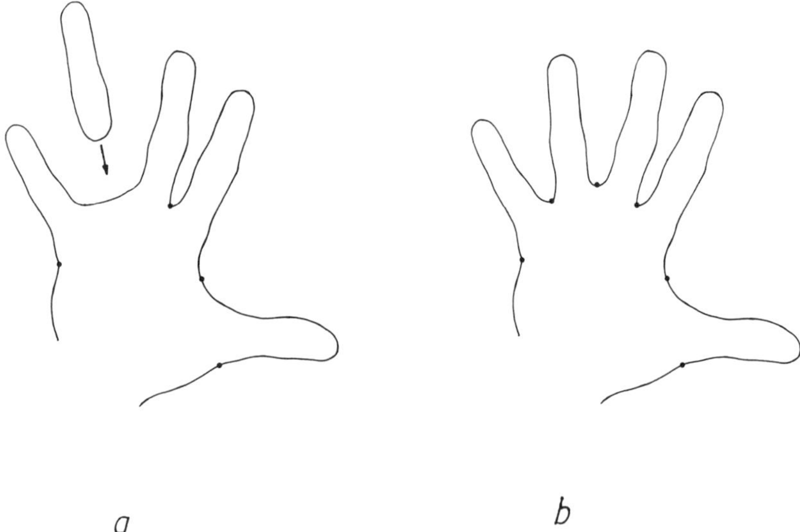

Figure 1.14
The progression from (a) to (b) shows the Hoffman and Richards scenario for the creation of a part.

argued that a part corresponds to a single process terminating at an extremum. What the theorem guarantees is that the single process is bounded by the two adjacent extrema, one on either side of the process extremum. That is, if one extends the curve any further past either of the adjacent extrema, then the additional curve will be associated with another axis, and therefore another process. Thus a single part, i.e., process, is confined within the two bounding extrema. In this way, the Symmetry-Curvature Duality Theorem justifies segmentation at the bounding extrema. This is a very different justification from the Transversality Principle.

The validity of our justification is further emphasized by considering the actual embryology of a part. In the formation of a part, i.e., a limb, the surface of the embryo, in the region of the limb, is initially homogeneous, e.g., chemically. It then becomes circularly organized around a point. In fact, the surface in the region takes on the organization of a polar coordinate system, as shown in Fig 1.15a (French, Bryant & Bryant, 1976; Bryant, French & Bryant, 1981).

Now, the development of the limb consists of "pushing out" this circular plot, as shown in Fig 1.15b. The pole center itself becomes the tip of the limb. However, observe that the tip is the curvature extremum. This means that the limb is defined by a single process terminating at the curvature extremum. But the correspondence of a single extremum and a single process axis is determined by the Symmetry-Curvature Duality Theorem. We therefore conclude that the Symmetry-Curvature Duality Theorem constrains both the biological formation of parts and the perceptual inference of parts.

Figure 1.15
The actual creation of a part.

1.19 History from Non-Smooth Curves

In the preceding sections, we have been looking at curves that are smooth. The reason is that curvature can be defined at each point on the curve. Our system is easily extended to non-smooth curves; i.e., curves that contain corners (first-order discontinuities). Any such curve consists of a number of smooth segments joined at the corners.

Given such a curve, consider one of these smooth segments. The segment might be, for example, a wavy line. Now the Asymmetry Principle states that the history of this curve-segment is one in which the distinguishability is removed backward in time. The removal of distinguishability from such a segment would result in a *straight line*. This is because one removes distinguishability in curvature as before, thus giving the segment constant curvature. However, one can now also remove distinguishability between the two sides of the curve, thus making the curve straight. Therefore, if the entire curve is closed and contains a number of corners, the removal of distinguishability, *within* each smooth segment, results in the entire figure becoming a straight-sided polygon; i.e., each smooth segment becomes a straight side, and each corner is preserved as a vertex in the polygon. This means that one sees a "wavy" polygon in the present as having originated from a straight polygon in the past.

Now the resulting straight polygon might have sides of different lengths; i.e., the polygon might be irregular. However, the Asymmetry Principle states that distinguishability is removed backwards in time. Therefore, the distinguishability *between* the lengths of the sides is removed backward in time. Thus an irregular polygon in the present is seen as having originated from a regular polygon in the past.

By the above discussion, we conclude that one sees an arbitrary closed curve, i.e., with a number of corners, as having originated from an irregular polygon which one sees in turn as having originated from a regular one.[4]

1.20 Goodness

We have seen that the removal of process-history from closed shapes produces regular polygons—essentially what Plato called *good* forms. Thus under our analysis, the Platonic forms are not regarded as they have been for two thousand years, the objects of unmotivated respect or mystical contemplation. Under our system, they have a very real significance: They represent the *absence of history*. From them, one concludes that there has been no change. They are important therefore because they say something crucial about the structure of time.

Thus, under our view, their significance arises from their *inferential* role in perception. That is, as we argue, human cognition is fundamentally concerned with the inference of history from organization. With respect to this inference, the Platonic forms are those entities that produce a particular common verdict: there is *no causal history*.

Of course, realizing that these forms are the objects that lead to this verdict explains how they could be taken to be objects of mystical contemplation. Because perception infers, from their geometric symmetry, that they are *temporally* symmetric, they represent undifferentiated eternity from the past into the future.

Plato used the term *good* to designate symmetrical forms. But perhaps a positive evaluation of symmetry is not appropriate. According to our analysis, symmetry is the destruction of time. It is the absence of process, of development, growth, and action. In contrast, *asymmetry* means change—it means interaction, progressive organization, unfolding. That symmetry could be called "good" reflects a preference for conformity and sameness; and a fear of dissimilarity and exception. In the final chapter, we shall see that life itself requires asymmetry, and that death is, technically, an example of symmetry.

1.21 Summary

In Chapter 1, we introduced the *history-recovery problem*, which we also called the *process-recovery problem*. The problem is this: How can people, given a single moment, recover history from that moment; i.e., run time backwards and recapitulate the processes that lead to that moment? We will argue, in this book, that the history-recovery problem is central to all cognition.

Because the past is inaccessible, and therefore can be arbitrary, given information only about the present, the recovery of a unique past depends on the uni-directional definition of processes. This is a fundamental constraint on any solution to the process-recovery problem.

Our solution to the problem rests on two basic proposals: (1) the Asymmetry Principle, which states that an asymmetry in the present is understood as having arisen from a symmetry in the past; and (2) the Symmetry Principle, which states that a symmetry in the present is understood as having always existed. We observed that the principles together provide unique recoverability of the past and uni-directionality of processes.

Two principles generalize the above two to both the past and the future: (1) the First Duality Principle, which states that asymmetry within the moment implies asymmetry within time (and vice versa); and (2) the Second Duality Principle, which states that symmetry within the moment implies symmetry within time (and vice versa).

The notion of causality was introduced by the Second System Principle, which states that increased asymmetry over time can occur in a system only if the system has a causal interaction with a second system. This principle can be used only after the use of the Asymmetry Principle, because, on its own, it cannot conclude anything from an asymmetry in the present. We observed also that, given that one can recover a symmetry from an asymmetry, the Second System Principle implies that the symmetry is in the past if a second system was involved, and in the future if a second system was not involved.

As an extended case study, we then examined the way in which history can be recovered from the curvature of a shape. The recovery is provided by two simple rules: (1) the Symmetry-Curvature Duality Theorem, which states that any section of curve, that has one and only one curvature extremum, has one and only one symmetry axis; and this axis is forced to terminate at the extremum itself; (2) the Interaction Principle, which states that symmetry axes are the directions along which processes are hypothesized as most likely to have acted.

Combining the two rules leads to the conclusion that each curvature extremum implies a process whose trace is the unique symmetry axis associated with and terminating at the extremum. The Asymmetry Principle then implies that processes acted in the direction of creating the curvature extrema. Application of the rules to a large catalogue of shapes reveals that the rules do indeed give psychologically satisfying results. It also reveals that the four types of curvature extrema consistently correspond to four types of process-terms used in English.

A shape with curvature variation looses two forms of symmetry contained in a circle: (1) the rotational symmetry of the tangents, with respect to a common rotation center, and (2) the reflectional symmetry of each point with respect to all other points. On a section of curve with only one curvature extremum, each point retains only one symmetrical point, and the curvature extremum is the only point that is self-symmetrical.

We found that our process approach to parts is more cognitive and inclusive than a straightforward segmentation approach for two reasons: (1) The process approach considers parts to be causal explanations. Such explanations are, we claim, central to the *cognition* of shape. (2) The process approach accounts not only for *parts*, but, since there are four times as many processes as parts, the process-analysis accounts for many more phenomena than parts. We also saw that the Hoffman & Richards Transversality Rule is not an appropriate environmental basis for biological parts, but the Symmetry-Curvature Duality Theorem is.

Our analysis of history from curvature was quite easily generalized to non-smooth curves; i.e., curves with corners. Using the Asymmetry Principle, we argued that any such curve is seen as having originated from an irregular straight-sided polygon, which one sees as having originated from a regular one.

Finally, we observed that, according to our Symmetry Principle, "good" shapes, such as the Platonic forms, are those shapes in which there is an absence of time.

2 Traces

2.1 Recovery from Several States

In all examples of process-recovery that we have examined so far, the information available in the present moment has been a *single state of a process*. For example, the rotated parallelogram in Fig 2.1a was the *only* state which was presented. The mind inferred the entire history from only that single presented state. Again, each of the complex shapes analyzed in Chapter 1 (i.e., embryos, tumors, clouds, etc.) was a single state in its respective history. For each of these shapes, we showed how the history could be inferred from the single presented state.

In some situations, however, the record that is left in the present moment is that of more than one state. For example, a scratch that has been left on a surface is a record of a sequence of states. Each of the points comprising the scratch is a result that was produced at a different time. For example, Fig 2.2 represents a scratch. The numbers 1 and 2 mark two points on the scratch. These points must have been created at different times. Since the present contains the entire scratch, it contains the whole sequence of moments from the process; i.e., a trace of moments. We shall say therefore that the present moment is assumed to contain *multiple states*. This contrasts with the rotated parallelogram where the present (the rotated parallelogram itself) was assumed to contain only a *single* state of the inferred history.

As another example, where the present is assumed to contain multiple states, consider two strokes that have been made by a pen on a piece of paper, e.g., as shown in Fig 2.3. The two strokes are assumed to have involved two types of processes: (1) the *drawing* movement of the pen in contact with the paper, and (2) the *shifting* movement between the two strokes. Note that we naturally break the entire trace into two phases. Each stroke represents a phase of drawing. The set of phases is itself the trace of the shifting process that moved from one phase to the next.

How does one infer traces and their structure? The answer is that one applies, once again, the rules developed in Chapter 1. To give a foretaste of how this is done, consider the situation of the two pen-strokes. Consider first an individual stroke. Any two points on the stroke are distinguishable by position. Thus, by the Asymmetry Principle—which states that a distinguishability is seen as having arisen from an indistinguishability—the two points must have come from a situation in which they were not distin-

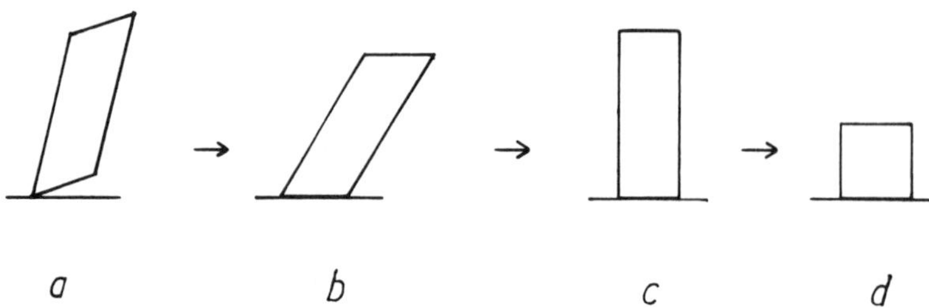

Figure 2.1
The successive reference example discussed in Chapter 1.

Figure 2.2
Two points on a scratch are seen as having been created successively.

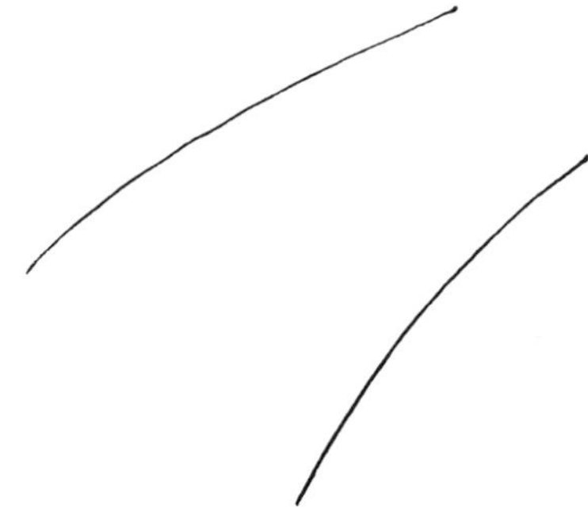

Figure 2.3
Two pen-strokes.

guishable by position; that is, there was only a single position, i.e., one point. Thus, the whole trajectory of distinguishable points must have been traced out from only one point. Now consider the two strokes. They are distinguishable by position. Thus, by the Asymmetry Principle, they must have arisen from a situation in which their position was not distinguishable; i.e., prior to the two strokes, there was one. Therefore, this positional distinguishability corresponds to the shifting movement that created a different position for the second stroke. To recapitulate: The within-stroke distinguishability corresponds to the drawing process. The between-stroke distinguishability corresponds to the shifting process. In accordance with our Asymmetry Principle, each distinguishability corresponds to a process that originated from a moment in time when the distinguishability did not exist.

The example just given is only a sketch of how one uses the Asymmetry Principle to infer history from multiple states. In the following sections, we will begin to go more deeply into this recovery problem, and continue, in still greater detail, in Chapter 6. We complete this section with a crucial characterization of the difference between inference from an assumed single state and inference from assumed multiple states.

(1) The single-state assumption. In this case, the present is assumed to contain the record of only a single state. Asymmetries within this state imply a past state where the asymmetries did not exist. For example, differences between the sizes of the angles of a parallelogram imply a state where these differences were absent; i.e., one had a rectangle. The crucial point is that the past state *is not represented in the present*; e.g., the past rectangle is not part of the parallelogram in the present. Thus, we shall say that the past state, and therefore the process, are *external* to the present. We shall say that the *single-state assumption* corresponds to *external* inference.

(2) The multiple-state assumption. In this case, the present is assumed to contain the records of several states. Distinguishabilities between the states point to a past where the distinguishabilities did not exist; i.e., sufficiently far back, only one state existed. For example, different points along a penstroke are assumed to have started with only one point; again, the shift between two pen-strokes is assumed to have started with one penstroke. Since the hypothesized process will account for the multiple states, these states will be distributed along the process. The crucial point is that the past states *are represented within the present*. We shall say, therefore, that the past states, and thus the process, are *internal* to the present. Thus we shall say that the *multiple-state assumption* corresponds to *internal* inference.

To summarize: With the assumption of a single state, the past states, and thus the process, are *external* to the present. With the assumption of multiple states, the past states, and thus the process, are *internal* to the present.

2.2 Internal Structure

Let us return to the successive reference example of Fig 2.1. The inferred starting state for the history is the square. As observed in section 1.5, the square is the removal of three distinguishabilities: (1) the difference between the orientation of the object and the orientation of the environment; (2) the difference between angle-sizes; (3) the difference between side-lengths.

However, further distinguishabilities still remain in the square. Although the sides have been given equal length, they are nevertheless still distinguishable by *position*; i.e., there is a *top* side, a *bottom* side, a *left* side, and a *right* side. By the Asymmetry Principle this distinguishability must be the memory of a process-history. In fact, according to that principle, there must have been a moment in the history when there was *no positional distinguishability* between the sides. Observe that, since the *size* distinguishability, between the sides, was removed previously, the removal of the *positional* distinguishability makes the sides completely indistinguishable; i.e., they become the same side. This means that a square, as shown in Fig 2.4a, must have originated from a situation in which it was a single side, as shown in Fig 2.4b.

Observe now that this single side (Fig 2.4b) still contains a distinguishability: The *points* on the side are distinguishable by position. In fact, this is the only distinguishability between the points on a side. Therefore, once the positional distinguishability is removed from the points, one obtains a single point. That is, the single side, in Fig 2.4b, originated from a moment in time when it was a single point, Fig 2.4c.

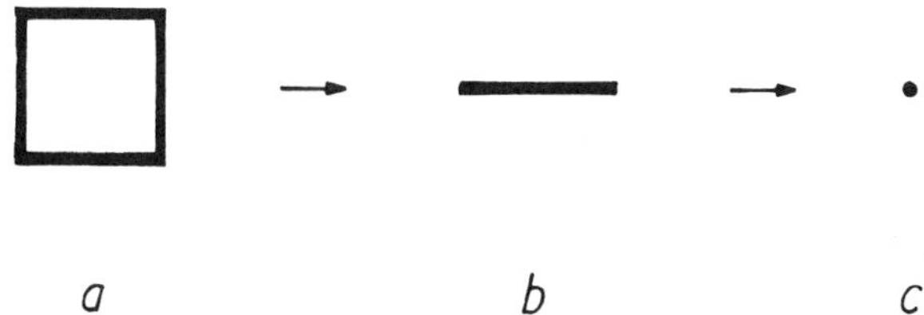

Figure 2.4
(a) A square. (b) The removal of positional distinguishabilities between sides, and (c) between points.

The two inference stages just described are shown as the two stages from Fig 2.4a to 2.4b to 2.4c. These are successive stages backwards in time. Recall that both are obtained by removal of positional distinguishabilities. The transition from Fig 2.4a to 2.4b represents the removal of the positional distinguishabilities between the sides of the square, and the transition from Fig 2.4b to 2.4c represents the removal of positional distinguishabilities between the points on a side.

Since the backwards-time direction, Fig 2.4a to 2.4c, is the removal of positional distinguishabilities, the forwards-time direction, Fig 2.4c to 2.4a, must mean the introduction of positional distinguishabilities. Since, by our principles, a process is postulated to explain the introduction of distinguishability, and since the distinguishability is one of position, the process must have been one of *movement*—because movement is the process that causes change in position.

Thus, consider the forward-time transition from the point, Fig 2.4c, to the line, Fig 2.4b. Because the process of movement is hypothesized to account for the positional distinguishabilities in the line, the line is understood as the *trace* of a point. One is reminded here of Paul Klee's comment that a line is a point being taken for a walk.

Again, consider the forward-time transition from the line, Fig 2.4b, to the square, Fig 2.4a. Because the process of movement is hypothesized to account for the positional distinguishability of the side in the square, the square is understood as the *trace* of movement of a side.

Let us return now to the backwards-time direction; that is, Fig 2.4a to 2.4b to 2.4c. Note that, in each of the two inference stages, the assumption is made that the present contains *multiple* states. That is, the square is assumed to be the multiple states (positions) of a side, and a side is assumed to be the multiple states (positions) of a point.

A multiple-state interpretation is, in fact, forced upon one, in any situation where one attempts to remove all distinguishabilities. This is because, at some stage in the removal of distinguishabilities, one cannot remove any more without removing distinct parts. At this point, backward-time must therefore represent the removal of the parts and forward-time must represent the *process of introducing the parts.* This means that different parts must have been introduced at different times; that is, they represent different moments in the process that introduces them; i.e., different states in that process. Therefore, the collection of parts is given a multiple-state interpretation.

As we have said, the assignment of a multiple-state interpretation becomes the only possible one at some stage in the removal of all distinguishabilities. The backward history from a rotated parallelogram illustrates this. The progression from a rotated parallelogram to a non-rotated one to a rectangle to a square represents the successive removal of three of the distinguishabilities. However, if one wants to remove any more distinguishabilities, i.e., in the square, one has to assign a multiple-state interpretation because removal of these remaining distinguishabilities will remove parts; i.e., one will obtain the sequence shown in Fig 2.4.

Thus, the backward progression from a rotated parallelogram, Fig 2.1a, starts with a *single-state* assumption; i.e., the rotated parallelogram is assumed to be a single state and it is seen as implying a state outside itself, a non-rotated parallelogram. The non-rotated parallelogram, Fig 2.1b, is in turn assumed to be a single state and it is seen as implying a state outside itself, the rectangle, Fig 2.1c. Then, the rectangle, Fig 2.1c, is in turn assumed to be a single state and it is seen as implying a state outside itself, the square, Fig 2.1d. Thus, a single-state interpretation is held through the successive inference stages, in Fig 2.1, until one reaches a square. Then, in order to remove the remaining distinguishabilities, i.e., those in a square, one is forced to switch to a multiple-state interpretation.

The assignment of a multiple-state interpretation necessitates the square being viewed as a trace; i.e., the sides are ordered in time and so are the points. We shall later look at how traces are structured, but it is worth observing, here, that an obvious physical interpretation of the trace is that it was produced by *drawing*. For example, starting with the point in Fig 2.4c, a pen traced out the points along a side, producing Fig 2.4b; and then the pen traced out the successive sides around the square, producing Fig 2.4a.

2.3 Environmental Validity

We have said that a trace interpretation is the inevitable result of attempting to remove all distinguishabilities, and our Asymmetry Principle claims that the perceptual system does indeed remove all distinguishabilities. Therefore, the perceptual system always assigns a trace interpretation to some aspect of the percept. This might at first seem environmentally insupportable; that is, many traces that are hypothesized to account for distinguishabilities might not seem to correspond to traces in the environment. It is, of course, true that there are several types of authentic traces in the environment; e.g., a sequence of footsteps in the snow, a boat trail on a lake, the skid-marks of a car, brushstrokes, handwriting, scratches, etc. But, many situations might seem to be states that should not be broken down into traces because they do not correspond to actual environmental traces.

This problem is epitomized by the interpretation of a two-dimensional geometric shape such as a square. On the one hand, the square can be viewed as drawn by a pen, i.e., points along an edge follow points, sides follow sides. On the other hand, the square can be viewed as a compact object, for example, a sheet of glass in a window, or a sheet of wood in a bookcase, or the sheet of cardboard at the back of a picture frame. Here the sheet represents a single moment in time, and there does not seem any reason to conjecture that the points along the edge come from different times and the sides are also temporally distinguished. The points arrive all at the same time, and so do the sides.

In fact, the opposite is true. A sheet of glass was the result of a glass-cutter cutting sequentially along the perimeter; i.e., the perimeter points were ordered in time and so were the edges. Similarly, a sheet of wood was produced by the sequential action of a saw along its edges. The knife that cut out a sheet of cardboard worked its way sequentially along the perimeter.

In some cases, one might suppose that the perimeter was not *sequentially* cut, but was stamped out by a sharp template. However, even if sequence did not enter at this cutting stage, it entered earlier in the production of the template. For example, the template was itself cut sequentially by a knife; or the template came from another template which was itself cut sequentially. At some stage, we claim, the edge information was created by *sequence*. Therefore, returning to the final produced object, e.g., the square sheet, distinguishability within the edge of the sheet is the memory of the sequential process that occurred at some stage in the past.

We can state this proposal more generally as follows:

TRANSFER PRINCIPLE. *If an asymmetry in an object A did not exist for the first time as a result of a causal interaction of object A with another object, then it was* **transferred** *to object A from another object where it was brought into existence by a causal interaction. There might have been several stages of transfer (i.e., a sequence of intervening objects) from object A back to that initial causal interaction.*

The example we shall often use to illustrate this principle is that of an *ink stamp*: Consider a word appearing in ink on a sheet of paper, and consider two points along the word itself. The points are distinguishable by position. The Transfer Principle implies that, if their positional distinguishability was not created by a pen tracing out the word directly on the paper, then it was transferred from some object where tracing was nevertheless done. For example, this object could have been an ink stamp, and the word was traced out on the stamp by the knife that etched the word onto the stamp. The stamp then transferred the structure of the trace onto the sheet of paper.

2.4 Trace Structure

Let us return to the trace structure of a square. A square is, of course, an extremely simple object. Nevertheless, despite this simplicity, its trace structure has a number of features that are important to identify and understand if one is to gain insight into the process-recovery problem generally. We shall see, later in the book, that we can fully analyze the perceived history in complex shape only after we have carefully defined the trace structure of simple shape. In fact, the *full* history assigned to any complex shape rests on the trace structure of some simple shape. The perceptual system decides which deformational operations created a complex shape by defining a trace structure for a simple shape that is hypothesized to be more deeply in the past.

The first thing to observe about the trace structure of the square is that it has three crucial properties. The square is:

(1) Nested

As we saw, in the previous section, the trace structure has two levels: (1) The drawing action that traces out a single side by moving a point. (2) The shifting action that moves from one side to the next. The first level is *nested* within the second. Thus, an entire side is traced out (Level 1) at each successive shift in Level 2. That is, one goes through the *entire* sequence of movements in Level 1, for each *individual* movement in Level 2. We shall say that the trace is given by a *nested structure of control.*

(2) Repetitive

Each level in the trace structure is purely the repetition of a single action. Level 1, the tracing of a single side, is the repeated *translation* of a point. Level 2, the shift that produces the successive sides, is the repeated clockwise or anti-clockwise movement from one side to the next; i.e., a repeated *rotation.*

(3) Euclidean

As we have just said, the two levels of the trace structure are *translations* (Level 1) and *rotations* (Level 2). These two types of movement are both examples of what are called *Euclidean* motions. The characterizing property of any Euclidean motion is that it preserves size and shape. If one *translates* or *rotates* an object, the size of the object remains the same, and the object remains un-distorted.

We have observed that the trace structure of a square is (1) nested, (2) repetitive, and (3) Euclidean. A further comment should be made about the final property. There are actually three types of motions that are Euclidean:

EUCLIDEAN MOTIONS

(1) Translations

(2) Reflections

(3) Rotations

We have considered the trace structure of the square as comprising two of the motions: translations and rotations. There is actually a different, quite common, way of drawing a square that also involves *reflections*. This alternative scheme has three levels: (1) One first draws a side. (2) One then draws its *opposite* side, i.e., one shifts from one side to its *reflectionally* opposite side. (3) One then rotates 90° and repeats the same sequence, producing the remaining pair of sides. Thus, the three successive levels are (1) translation, (2) reflection, and (3) rotation. It should be observed that this alternative means of drawing a square has the properties listed above: (1) The levels are successively *nested* within each other; (2) each level is the *repetitive* use of a single operation; and (3) the operation in each level is *Euclidean*.

2.5 The Symmetry-to-Trace Conversion Principle

Let us now study how the trace-structures, described in the previous section, arise. In order to do this, we need to go back and consider, once again, the sequence shown in Fig 2.1, from the rotated parallelogram to the square.

In fact, let us consider a rotated parallelogram with one more asymmetry in it than we usually consider: a dent in one of its sides, as shown in Fig 2.5. We have added this dent to illustrate more fully our discussion. Observe that, in accord with our Asymmetry Principle, the dent will be removed backward in time; i.e., the side will be considered to have been straight, in the past. Thus the backward-time direction must remove the dent, as well as the other asymmetries we have considered for a rotated parallelogram. Let us examine the removal of the asymmetries.

Recall first that symmetry is indistinguishability, and that indistinguishability is often defined with respect to some type of transformation. For example, an object is reflectionally symmetric if it cannot be distinguished from a reflected version of itself, or rotationally symmetric if it cannot be distinguished from a rotated version of itself.

Figure 2.5
A dented parallelogram.

Consider now the parallelogram with the dent. Observe that the side with the dent is not translationally symmetric, as follows: If any straight segment of that side is translated along the side to the dent, the segment will be distinguishable in shape from the dent. That is, the points on the straight segment will not lie over the points on the dent. However, after removing the dent, this distinguishability will disappear. That is, when any region within the side is translated to any other region within the side, the former region will be indistinguishable from the latter region because the points of the two regions will lie exactly over each other. Thus, removal of the dent gains translational symmetry for points along the side. Observe that translation is one of the three types of Euclidean motions (translations, rotations, reflections).

Now consider the rotated parallelogram without the dent. We have seen that this shape has at least three asymmetries, and we have psychological evidence that the order of their removal is given by the sequence: rotated parallelogram → parallelogram → rectangle → square. (The order with which the dent is removed in relation to this sequence is not relevant to our present discussion.)

Consider, in particular, the transition from the non-rotated parallelogram to a rectangle; Fig 2.1b to Fig 2.1c. Observe that the parallelogram is not *reflectionally* symmetric. That is, one cannot place a mirror on a parallelogram so that one half of the parallelogram lies over the other, when reflected. In contrast, the rectangle does have reflectional symmetry; e.g., the left half is indistinguishable from the right half after it is reflected. Therefore, in the transition from the parallelogram to the rectangle, reflectional symmetry has been gained. Observe that reflection is one of the three types of Euclidean motions (translation, rotation, reflection).

Now consider the transition from the rectangle to the square; Fig 2.1c to Fig 2.1d. Observe that the left side of a rectangle is not the same size as the top side. Therefore, if one were to rotate the left side by 90° to the top (i.e., rotating around the center of the rectangle), the rotated side would still be distinguishable from the top side. Thus the rectangle is not *rotationally symmetric* under 90° rotations. The opposite is true of a square. If the left side were to be rotated by 90° (around the center of the square) to the top of the square, it would be indistinguishable from the top side of the square. Thus, the transition, from the rectangle (Fig 2.1c) to the square (Fig 2.1d), introduces 90° rotational symmetry. Observe that rotation is one of the three types of Euclidean motions (translation, rotation, reflection).

Let us now list the symmetries we have gained in the transition from the dented parallelogram (Fig 2.5) to the square. (1) In going from the dented parallelogram to the non-dented one, we gained *translational* symmetry of the points along a side. (2) In going from the parallelogram to the rectangle, we gained *reflectional* symmetry between the sides. (3) In going from the rectangle to the square, we gained 90° *rotational* symmetry between the sides. These three kinds of symmetry are defined by the three kinds of Euclidean transformations: translations, reflections, and rotations.

Let us now go back to considering the *trace* structure of a square. Recall that the scenarios for drawing a square all involved levels of Euclidean operations. For example, one scenario consisted of two levels: (1) the *translation* of a point to draw a side; (2) the successive *rotational* shift from one side to the next. An alternative scenario involved an additional, intermediate, level where one used *reflection* to move from one side to its opposite.

What we can see now is that the Euclidean transformations that were used to define these trace-scenarios were made possible by introducing the Euclidean symmetries in the successive transitions from the dented parallelogram to the square. That is, the symmetry that was introduced in going from a dented parallelogram to a non-dented one—i.e., the translational symmetry of points along a side—allows the translational movement that traces out the side from a single point. Again, the symmetry that was introduced in going from a parallelogram to a rectangle—i.e., the reflectional symmetry between opposite sides—allows, in the trace structure, a reflectional movement from one side to its opposite. Again, the symmetry that was introduced in going from a rectangle to a square—i.e., the 90° rotational symmetry between sides—allows the 90° rotational movement that traces one side after the other.

This illustrates a crucial principle, which we introduce at this point.

SYMMETRY-TO-TRACE CONVERSION PRINCIPLE. *Any symmetry can be re-described as a trace. The transformations defining the symmetry generate the trace.*

This principle is rather paradoxical. A symmetry between elements means that the elements are equal, balanced, homogeneous, etc. A trace, on the other hand, is asymmetric; i.e., the elements contain a directionality. The paradox is that the very transformations that define the elements to be symmetric are the transformations that can organize the elements into the asymmetric structure that defines a trace. The way symmetries are converted into traces will be discussed in great detail in Chapter 6.

2.6 The Externalization Principle

Recall our distinction between *external* and *internal* inference (section 2.1). In *external* inference, one infers a past state that is outside the present stimulus. For example, when given a parallelogram in the present, one infers the past to be a rectangle—and this later shape is not contained in the present. Thus the process-history that one hypothesizes, i.e., the process of shearing (slanting) a rectangle, is *external* to the present stimulus. In contrast, under *internal* inference, one infers a past state that is part of the present configuration and this inference yields a trace structure. For example, given some handwriting, one assumes that it was drawn, starting at one end, i.e., one end of it represents a past state and this state has left a record in the present. Thus, the process that one hypothesizes is *internal* to the present stimulus; i.e., the process goes between parts of the present stimulus.

When an external process has been conjectured, we shall say that the history has, to that extent, been *externalized.* We now propose the following principle concerning the purpose of externalization:

EXTERNALIZATION PRINCIPLE. *The external processes that are hypothesized are those that reduce the internal (i.e. trace) structure, as much as possible, to a nested hierarchy of repetitious Euclidean processes.*

The successive reference example of Fig 2.1 illustrates this. The rotated parallelogram is assigned an *external* history leading backward in time to a square. The square has an *internal* (i.e. trace) structure that is nested, repetitive, and Euclidean. That is, the external processes that have been assigned to the rotated parallelogram conform to the Externalization Principle.

The Externalization Principle describes what happens when an external history is chosen. It does not state that an external history is always chosen. In fact, there are situations in which the chosen history is purely internal; i.e., a trace. An example is graffitti. This is seen purely as having been traced out over time; i.e., as a purely internal structure. In order to externalize graffitti, one would have to see it as a *deformed* line, like a curled piece of spaghetti. That is, one would conjecture a history, where one had, originally, a straight line that became bent. There is, in fact, a particular situation where this is the actual history of a hand-written word: a neon sign representing a hand-written word. Such a sign has, in fact, been bent from a straight neon tube. That is, the history contains an external component.

Another example, which illustrates that a choice is possible between a history with an external component and a fully internal one, is the rotated parallelogram in Fig 2.1a. Up till now, we have assigned to it an external history back to the square. However, the figure could have a purely internal history. That is, it could have been drawn directly on a sheet of paper; e.g., the pen started at one corner of the rotated parallelogram and drew out the sides successively.

In conjecturing a history, that proportion which one chooses to make external can therefore vary. However, according to the Externalization Principle, if some of the history is made external, the purpose is always to reduce the internal history, as much as possible, to a nested hierarchy of repetitious Euclidean processes. If sufficient external history is chosen to completely achieve this reduction, we shall say that history has been *fully externalized.* The chosen history in Fig 2.1, from rotated parallelogram to square, is a fully externalized history, i.e., it reaches a figure, the square, which has an internal structure that is nested, repetitive, and Euclidean.

The view that we will take is that perception attempts to fully externalize the history unless additional memory is used (e.g., in the form of knowledge about such situations) indicating that the actual history was not fully external. For example, according to this view, graffitti evokes extra memory that prevents one from seeing it as an external structure, i.e., a deformed line.

2.7 Prototypes

It is often been argued that human beings define an arbitrary shape in terms of a restricted vocabulary of shapes, called *prototypes.* Whereas the arbitrary shape may by highly irregular, the prototypes, with respect to which it is defined, are regular. For example, the Gestalt psychologists advanced much evidence that arbitrary shapes are seen in terms of more symmetrical versions of themselves, called "singular" forms (Goldmeier, 1936/72). In more recent times, researchers have argued that arbitrary natural shapes are perceptually seen in terms of variants of cylinders (Binford, 1971; Hollerbach, 1975; Marr & Nishihara, 1978). In fact, in a famous drawing by Marr & Nishihara (1978), shown here as Fig 2.6, a man is represented in terms of completely regular cylinders. Again, researchers who analyze the planning of motion through a complex environment, often argue that planning becomes much easier if the environment is re-described as a collection of completely regular three-dimensional polygons (e.g., Canny, 1984). One is reminded of Plato's notion that the many forms in the real world should be understood in terms of the few regular polygonal solids. Examples of the shapes that have been suggested as prototypes, by various researchers, are shown in Fig 2.7.

The important thing to observe is that each of the above mentioned prototypes, i.e., cylinders, regular polygons, etc., has a particular property, that we now state:

PROTOTYPE HYPOTHESIS. *Shape prototypes are nested repetitive Euclidean hierarchies.*

In fact, any one of these shapes, is a hierarchy in which, ascending up the hierarchy, the points comprising a single edge, are inter-related either by pure translation (as in a polygon) or pure rotation (as in the cross-section of a cylinder); the edges comprising a single face are inter-related by translations, rotations and/or reflections; the faces comprising the entire shape are inter-related by translations, rotations, and/or reflections.

Consider now the function that prototypes are supposed to serve. As noted above, the usually stated function is that they are the shapes with respect to which *all* general shapes are judged; i.e., all shapes are understood as *versions* of these shapes. We can now understand this function as an example of our concept of *externalization*; as follows: A prototype is the nested repetitive Euclidean structure that underlies any shape. That is, a prototype is a shape that one obtains after one has fully *externalized* a general shape. Therefore, in making the prototypification judgement, that some shape is basically a variant of a prototype, one is performing, on that shape, a full externalization.

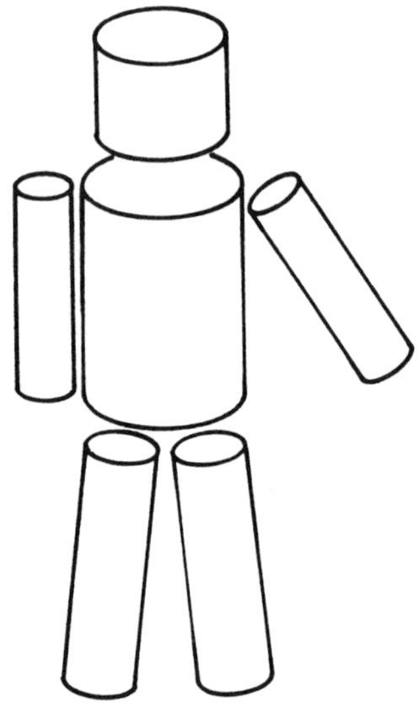

Figure 2.6
The cylinder decomposition of a man.

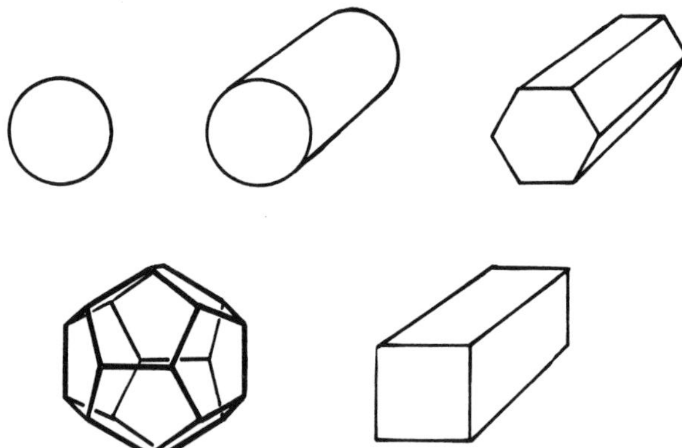

Figure 2.7
Examples of nested, repetitive, Euclidean hierarchies.

2.8 Intervening History

Let us turn to the complex natural shapes—embryos, tumors, clouds, etc.—given in Chapter 1, and shown again here as Fig 2.8. In sections 1.8–1.14, we presented rules by which one can infer the history of such shapes. There were two rules: (1) The first assigns to each curvature extremum a unique axis leading to the extremum. (2) The second says that the boundary had been pushed along each axis and that this movement had created the extremum. These histories are shown as the arrows in Fig 2.8.

We should recall that, under this view, the processes are seen as introducing the curvature variation into a shape. This means that, prior to the action of the processes, the shape had no curvature variation; i.e., the shape was a circle.[2]

Observe now that the points of a circle, the starting shape, can be generated purely by the repetition of rotation (around a single center). This means that the histories produced by the above inference rules accord exactly with the Externalization Principle, as follows. Observe, first, that the histories are *external* ones. That is, the history of each shape implies a shape in the past (the circle) that is not contained in the present complex shape; i.e., the past lies outside the present shape. Now, the Externalization Principle states that, if an *external* history is inferred, then that chosen history is one that attempts to make the internal structure a repetitive Euclidean one. The circle, which is the starting shape for any of the above complex shapes, has an internal structure that is repetitive and Euclidean.[3]

In sections 2.8 to 2.13, we examine a different type of process-recovery problem with respect to complex shape. In Chapter 1, we considered the issue of how one infers processes from a *single* complex shape. However, in many situations, one is faced with another problem: One has two shapes that one knows to be two *stages* in the development of a particular object, and one tries to infer the *intervening* history between those two shapes. For example, a doctor is faced with this problem when he or she has two X-rays of a tumor, taken a month apart from each other, and tries to understand what happened in the intervening time. In this case, rather than assigning an external history, the doctor is forced to devise a purely *internal* history, as follows: The *present* contains two shapes. Nevertheless, the doctor is forced to regard these shapes as representing two different

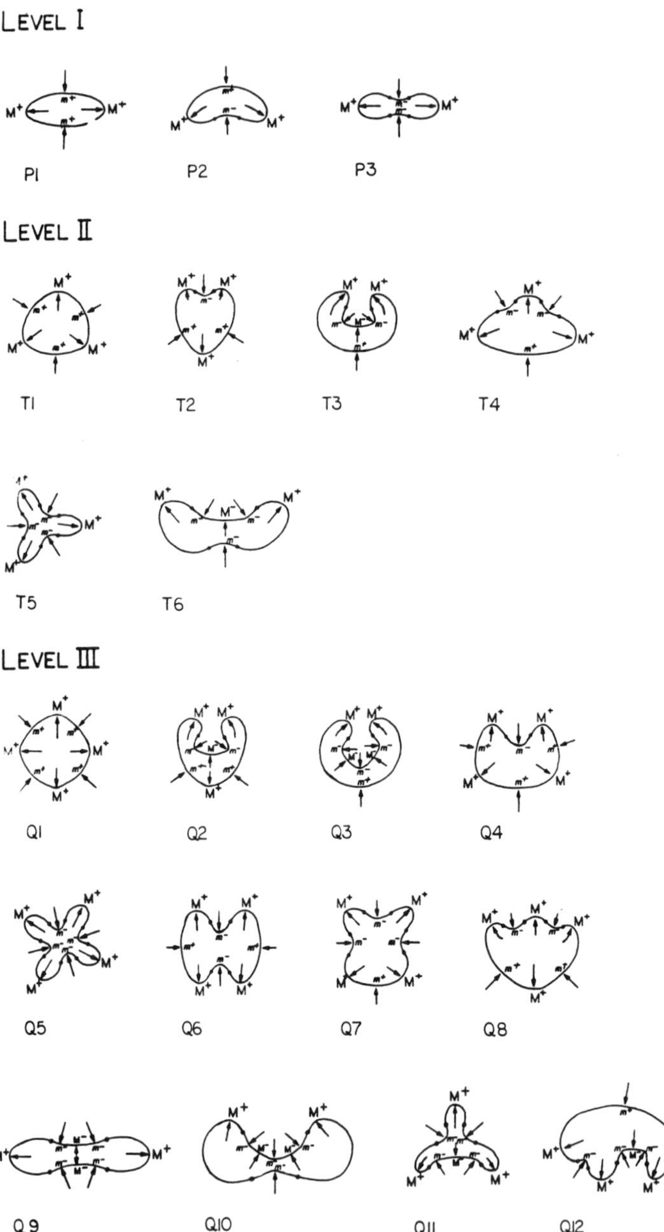

Figure 2.8
The rules of Chapter 1 applied to the outlines of Richards, Koenderink, and Hoffman (1987).

moments in time; i.e., the present contains a sequence or trace. We now examine the problem of constructing this internal history.

2.9 Continuations

The first thing to observe is that, since the later shape is assumed to emerge from the earlier shape, one will wish to explain it, as much as possible, as the outcome of what can be seen in the earlier shape. In other words, one will try to explain the later shape, as much as possible, as the *extrapolation* of what can be seen in the earlier one.

As a simple first cut, let us divide all extrapolations of processes into two types:

(1) Continuations

(2) Bifurcations (i.e. branchings).

What we will do now is elaborate the only forms that these two alternatives can take. We first look at continuations and then at bifurcations.

Consider any one of the extrema labeled M^+ in Fig 2.9. It is the tip of a protrusion, as predicted by our Semantic Interpretation Rule (section 1.14). What is important to observe is that, if one continued the process creating that protrusion, i.e., continued pushing out the boundary in the direction shown, the protrusion would remain a protrusion. That is, the M^+ extremum would remain a M^+ extremum. This means that continuation at a M^+ extremum does not structurally alter the boundary.

Exactly the same argument applies to any of the m^- extrema in Fig 2.9. That is, continuation of an indentation remains an indentation. In other words, a m^- extremum will remain m^-.

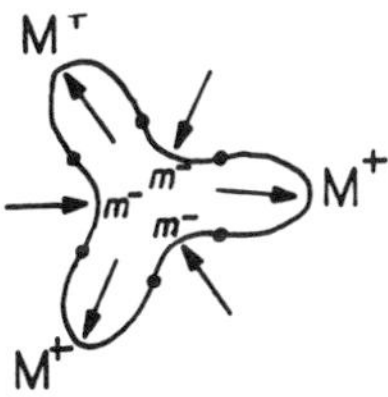

Figure 2.9
Illustrations of two of the curvature extrema.

Now recall that there are four types of extrema, M^+, m^-, m^+, M^-. We have seen that continuations at the first two do not structurally alter the shape. However, we shall now see that continuations at the second two do cause structural alteration. Let us consider these two cases in turn.

Continuation at $\mathbf{m^+}$

A m^+ extremum occurs at the top of the left-hand shape in Fig 2.10. In accord with our Semantic Interpretation Rule (section 1.14), the process terminating at this extremum is a squashing process. Now let us continue this process; i.e., continue pushing the boundary in the direction shown. At some point, an indentation can be created, as shown in the top of the right-hand shape in Fig 2.10.

Observe what happens to the extrema involved. Before continuation, i.e., in the left-hand shape, the relevant extremum is m^+ (at the top). After continuation, i.e., in the right-hand shape, this extremum has changed to m^-. In fact, observe that a dot has been placed on either side of the m^-, on the curve itself. These two dots are points where the curve, locally, is completely flat; i.e., where the curve has 0 curvature. Therefore the top of the right-hand shape is given by the sequence Om^-O.

Thus the transition between the left-hand shape and the right-hand shape can be structurally specified by simply saying that the m^+ extremum, at the top of the first shape, is replaced by the sequence Om^-O, at the top of the second shape. This transition will be labeled Cm^+, meaning *Continuation at* m^+. That is, we have:

$$Cm^+ : m^+ \rightarrow Om^-O$$

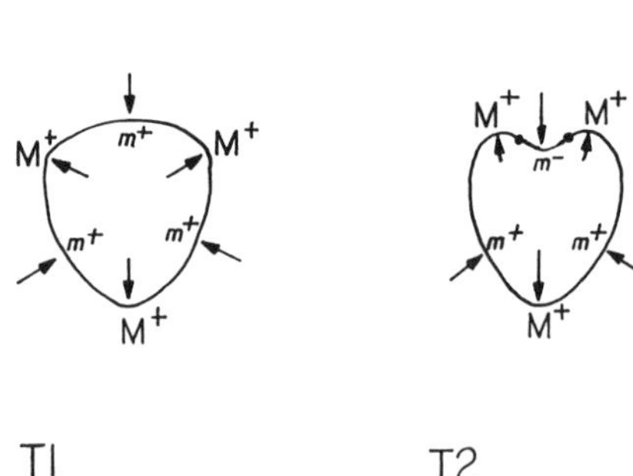

Figure 2.10
Continuation of a squashing process creating an indentation.

This sequence of symbols simply reads: *Continuation at* m^+ *takes* m^+ *and replaces it by* Om^-O.

Finally, observe that, although this operation is, formally, a rewrite rule defined in terms of discrete strings of extrema, the rule actually has a highly intuitive meaning. Using our Semantic Interpretation Rule, we see that it means: *squashing continues till it indents*.

Continuation at M^-

As noted earlier, we need to consider only one other type of continuation, that at a M^- extremum. In order to understand what happens here, consider the left-hand shape in Fig 2.11. The symmetry-analysis defined in section 1.10 describes a process-structure for the indentation that is very subtle, as shown: There is a flattening of the lowest region of the indentation due to the fact that the downward arrows, within the indentation, are countered by an upward arrow that is within the body of the shape and terminates at the M^- extremum.[4] This latter process is an example of what our Semantic Interpretation Rule (section 1.14) calls *internal resistance*. The overall shape could be that of an island where an inflow of water (into the indentation) has been resisted by a ridge of mountains (in the body of the shape). The consequence is the formation of a *bay*.

Now, recall that our interest is to see what happens when one continues a process at a M^- extremum. Thus let us continue the M^- process upward, in the left-hand shape of Fig 2.11. At some point, the process can burst out and create the protrusion shown at the top of the right-hand shape in Fig 2.11. In terms of our island example, there might have been a volcano, in the mountains, that erupted and sent lava down into the sea.

Now let us observe what happens to the extrema involved. Before continuation, i.e., in the left-hand shape, the relevant extremum is M^-, in the

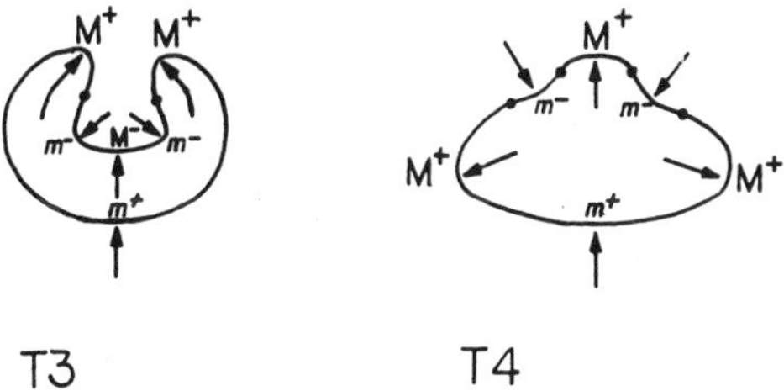

Figure 2.11
Continuation of an internal resistance creating a protrusion.

center of the bay. After continuation, i.e., in the right-hand shape, this extremum has changed to M^+, at the top of the protrusion. In fact, observe that, once again, a dot has been placed on either side of the M^+, on the curve itself. These two dots represent, as before, points where the curve is, locally, completely flat; i.e., where the curve has 0 curvature. Thus the top of the right-hand shape is given by the sequence OM^+O.

Therefore the transition between the left-hand shape and the right-hand shape can be structurally specified by simply saying that the M^- extremum, in the first shape, is replaced by the sequence OM^+O, at the top of the second shape. This transition will be labeled CM^-, meaning *Continuation at* M^-. Thus we have:

$$CM^- : M^- \rightarrow OM^+O$$

This sequence of symbols simply reads: *Continuation at* M^- *takes* M^- *and replaces it by* OM^+O. Finally, observe, once again, that, although this operation is a formal rewrite rule defined in terms of discrete strings of extrema, the rule actually has a highly intuitive meaning. Using our Semantic Interpretation Rule, we see that it means: *internal resistance continues till it protrudes.*

2.10 Bifurcations

We saw above that process-continuation can take only two forms. Our purpose now is to elaborate the only forms that the *bifurcation* (branching) of a process can take. Note that, because there are four extrema, we have to examine bifurcation at each of these four.

Bifurcation at M^+

Consider the M^+ extremum at the top of the left-hand shape in Fig 2.12, and consider the protruding process terminating at this extremum. We wish to examine what would result if this process bifurcated. Under bifurcation, one branch would go to the left and the other to the right. That is, the branching would create the upper lobe in the right-hand shape of Fig 2.12.

Observe what happens to the extrema involved. Before splitting, one has the M^+ extremum at the top of the first shape. In the situation after splitting (i.e., in the second shape), the left-hand branch terminates at a M^+

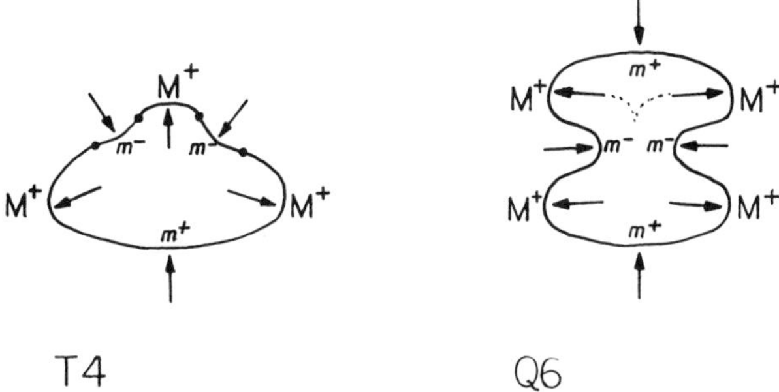

Figure 2.12
Bifurcation of a protrusion creating a lobe.

extremum, and so does the right-hand branch. That is, the M^+ extremum, in the first shape, has split into two copies of itself in the second shape. In fact, for mathematical reasons, a new extremum has to be introduced in between these two M^+ copies. It is the m^+ shown at the top of the lobe of the second shape. Therefore, in terms of extrema, the transition between the first and second shapes can be expressed thus: The M^+ extremum at the top of the first shape is replaced by the sequence $M^+m^+M^+$ along the top of the second shape; that is, we have the transition $M^+ \rightarrow M^+m^+M^+$. This transition will be labeled BM^+, meaning *Bifurcation at* M^+. That is, we have:

$$BM^+ : M^+ \rightarrow M^+m^+M^+$$

Finally, observe that, although this transition has just been expressed as a formal rewrite rule defined in terms of discrete strings of extrema, the transition has, in fact, the following highly intuitive meaning: *a nodule develops into a lobe.*

Bifurcation at m$^-$

Consider now the m^- extremum at the top of the left-hand shape in Fig 2.13, and consider the indenting process terminating at this extremum. We will examine what results if this process bifurcates. Under bifurcation, one branch would go to the left and the other to the right. That is, the branching would create the *bay* in the right-hand shape of Fig 2.13.

Observe, again, what happens to the extrema involved. Before splitting, one has the single extremum m^- in the indentation of the first shape.

Figure 2.13
Bifurcation of an indentation creating a bay.

In the situation after splitting (i.e., the second shape), the left-hand branch terminates at a m^- extremum, and so does the right-hand branch. That is, the top m^- in the first shape has been split into two copies of itself, in the second shape. Again, for mathematical reasons, a new extremum has to be introduced in between these two m^- copies. It is the M^- shown at the center of the bay in the second shape. Therefore, in terms of extrema, the transition between the first and second shapes can be expressed thus: The m^- extremum in the first shape is replaced by the sequence $m^-M^-m^-$ in the second shape; that is, the transition is given by the rule $m^- \rightarrow m^-M^-m^-$. This transition will be labeled Bm^-, meaning *Bifurcation at* m^-. That is, we have:

$$Bm^- : m^- \rightarrow m^-M^-m^-$$

Finally, note that, although the transition has just been expressed as a formal rewrite rule using discrete strings of extrema, the transition has, in fact, a highly intuitive meaning: *an inlet develops into a bay.*

Bifurcation at $\mathbf{m}^+$

The above two types of bifurcation illustrate the general form that bifurcations take: An extremum, E, is split into two copies of itself. Furthermore, for mathematical reasons, an intervening extremum, e, of the opposite type (max or min) has to be introduced in between the two copies.[5] That is, a bifurcation always takes the form $E \rightarrow EeE$.

We now consider bifurcation at m^+. By the discussion in the previous paragraph, one can see, in advance, that a bifurcation at m^+ must change the m^+ into the sequence $m^+M^+m^+$ (two copies of m^+ separated by M^+).

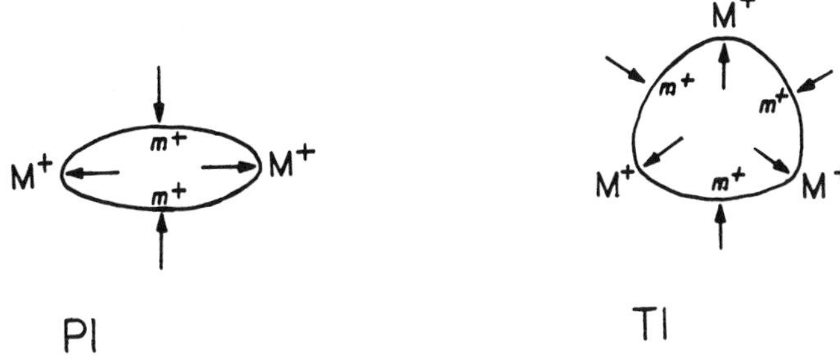

Figure 2.14
Creation of a protrusion.

That is, the bifurcation is given by the rule $m^+ \to m^+M^+m^+$, which we shall label Bm^+, meaning *Bifurcation at* m^+. That is, we have:

$$Bm^+ : m^+ \to m^+M^+m^+$$

This rule is illustrated in Fig 2.14. The m^+ extremum, at the top of the left-hand shape, splits and becomes the two m^+ extrema on either side of the right-hand shape. Simultaneously, a M^+ extremum is necessarily introduced at the top of the right-hand shape. Since a M^+ is always a protrusion, one can characterize this bifurcation simply as: *a protrusion is introduced.*

Bifurcation at M⁻

We now consider bifurcation at M^-. Again, from the above considerations, one can deduce, in advance, that a bifurcation at M^- must change the M^- into the sequence $M^-m^-M^-$ (two copies of M^- separated by m^-). That is, the bifurcation is given by the rule $M^- \to M^-m^-M^-$, which we shall label BM^-, meaning *Bifurcation at* M^-. That is, we have:

$$BM^- : M^- \to M^-m^-M^-$$

This rule is illustrated in Fig 2.15. The M^- extremum, at the center of the bay in the left-hand shape, splits and becomes the two M^- extrema on either side of the lagoon in the right-hand shape. Simultaneously, a m^- extremum is necessarily introduced at the lowest point of the lagoon (in the right-hand shape). Since m^- is always an indentation, one can characterize this bifurcation simply as: *an indentation is introduced.*

Figure 2.15
Creation of an indentation.

2.11 The Complete Grammar

Recall that the problem that we are examining in this part of the chapter is this: Given two views of an object (e.g., a tumor), at two different stages of development, how is it possible to infer the *intervening* process-history? Observe that, since one wishes to explain the later shape as an outcome of the earlier shape, one will try to explain the later shape, as much as possible, as the *extrapolation* of the process-structure inferable from the earlier shape. We have now shown that all possible process-extrapolations are generated by only six operations: two continuations and four bifurcations. Therefore, these six operations form a grammar that will generate the later shape from the earlier one via process-extrapolation. The grammar is as follows:

PROCESS GRAMMAR

$$Cm^+ : m^+ \rightarrow Om^- O$$

$$CM^- : M^- \rightarrow OM^+ O$$

$$BM^+ : M^+ \rightarrow M^+ m^+ M^+$$

$$Bm^- : m^- \rightarrow m^- M^- m^-$$

$$Bm^+ : m^+ \rightarrow m^+ M^+ m^+$$

$$BM^- : M^- \rightarrow M^- m^- M^-$$

Recall however that, although these operations are expressed as formal rewrite rules using discrete strings of extrema, they describe six intuitively compelling situations, as follows:

SEMANTIC INTERPRETATION OF THE GRAMMAR

Cm^+: squashing continues till it indents.

CM^-: internal resistance continues till it protrudes.

BM^+: a protrusion bifurcates; e.g., a nodule becomes a lobe.

Bm^-: an identation bifurcates; e.g., an inlet becomes a bay.

Bm^+: a protrusion is introduced.

BM^-: an indentation is introduced.

These six situations were illustrated in Figs 2.10 to 2.15.

2.12 The Process Stratification of Shape-Space

We shall now see that the grammar gives a highly organized structure to the space of all shapes—a space we will call "shape-space". Recall that Fig 2.8 stratifies shape-space into levels, where each level is defined by the number of extrema. (In fact, Fig 2.8 gives the first three levels of shape-space.) Examining Fig 2.8, one might wish to ask whether shape-space has any further structure than this.

Let us consider what organization is induced by the process-grammar. It might be thought that, if one interlinked the shapes by all possible uses of the grammatical operations, one would obtain a very disorganized network. In fact, the opposite is true. One obtains the highly structured system shown in Fig 2.16. Close examination of this system reveals that it consists of six intersecting strata-systems where each system is a set of parallel planes. One stratification is that which we had before: It is the descending system of horizontal planes, where each plane corresponds to a level in Fig 2.8. However, there are five other strata-systems that the grammar now reveals. Fig 2.17 shows Fig 2.16 six times, each time revealing a different strata-system. For example, Fig 2.17a shows the sequence of planes obtained by the successive iteration of the grammatical operation Cm^+; Fig 2.17b shows a sequence of planes elaborated by a successive iteration of the operations BM^+ and Bm^+; Fig 2.17c shows a sequence of planes elaborated by a successive iteration of the operations BM^- and Bm^-; etc.

Those of us working in shape perception have not previously suspected that shape-space is so highly structured. Nevertheless, the stratifica-

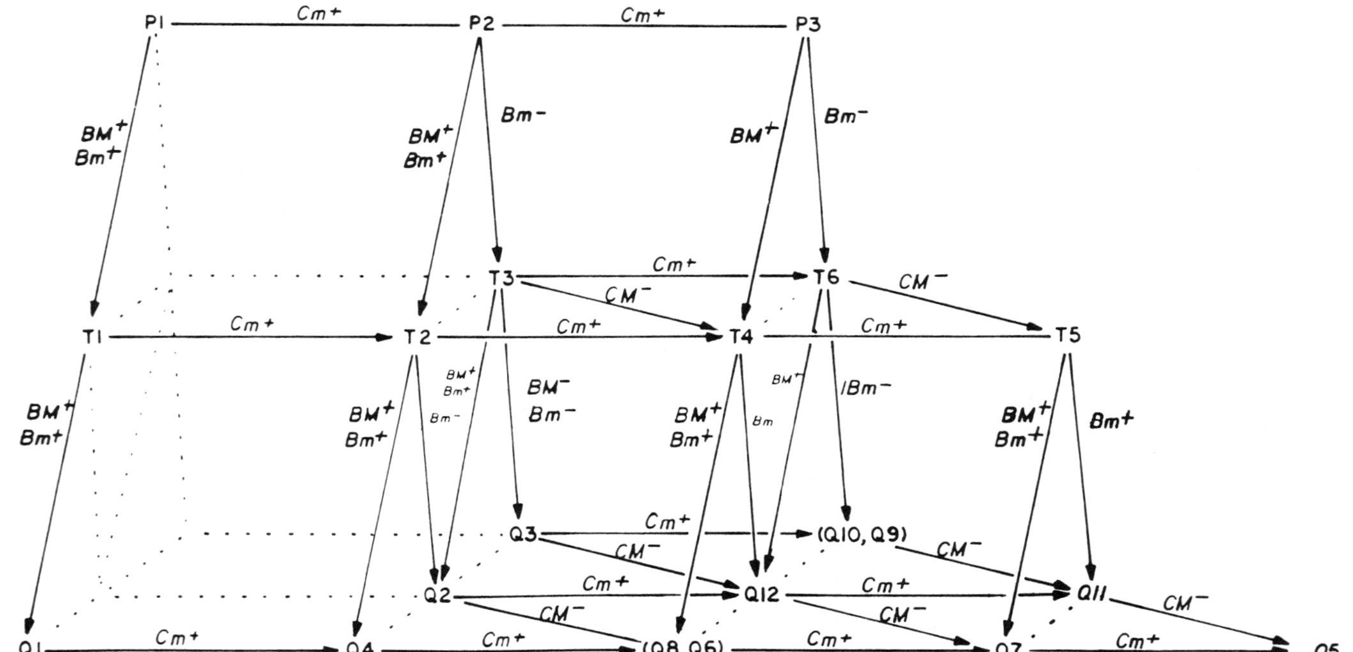

Figure 2.16
The six rules of the grammar used to connect all the shapes in Fig 2.8.

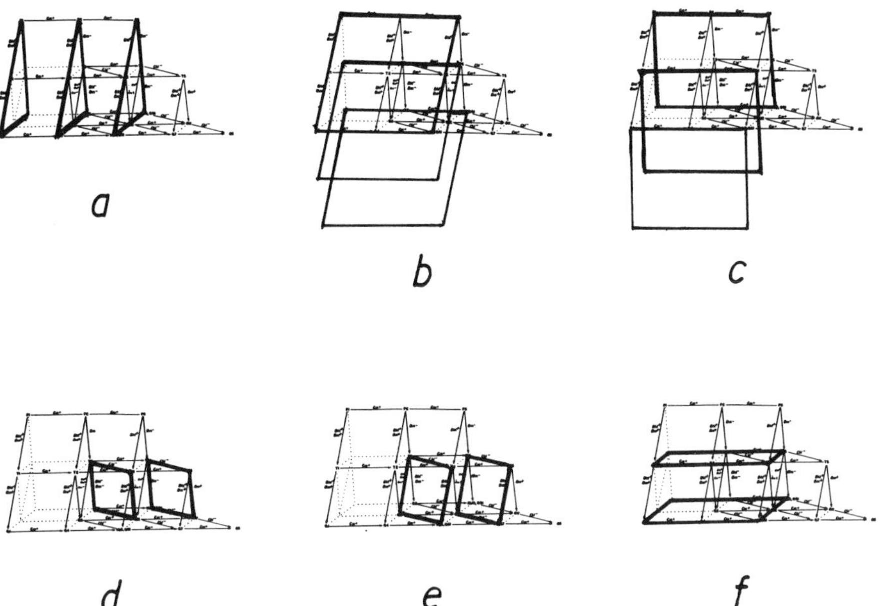

Figure 2.17
The stratification of shape-space by the grammar.

tions embody the psychologically meaningful phenomenon of process-extrapolation.

2.13 Application of the Grammar

Let us now illustrate the intuitive power of the grammar to yield process-explanations. Let us choose two arbitrary shapes, for example, the pair shown in Fig 2.18. The assumption is that the two shapes are two stages in the development of the same object; e.g., a tumor, cloud, island, embryo, etc. We now show that the process-grammar gives a compelling account of the intervening development.

The first step is to locate the two shapes, T6 and Q5, in the state-transition diagram, Fig 2.16. One then identifies a path between them. There are in fact many alternative paths between them. However, in the Appendix to this chapter, we will develop a rule that not only selects a unique path but appears to select the psychologically salient one. Let us suppose that the

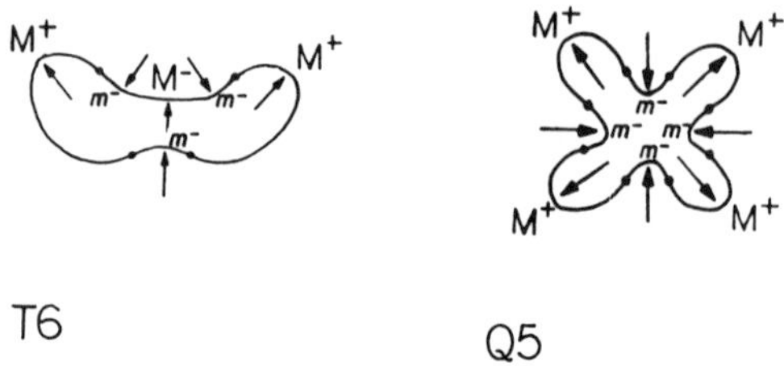

Figure 2.18
Two shapes to be given an intervening history by the grammar.

path chosen by the rule is T6 → T5 → Q7 → Q5. This sequence of shapes is shown in Fig 2.19. The network also reveals that the sequence of grammatical operations is $CM^- * BM^+ * Cm^+$. The sequence is the process-explanation for the intervening development. Using our semantic rules, the explanation can be "translated" into English as follows:

(1) One particular process turns out to be crucial to the entire development. It is the internal resistance represented by the bold upward arrow in the first shape of Fig 2.19 (the arrow terminating at M^-).

(2) This continues upward and creates the protrusion in the second shape of Fig 2.19.

(3) This same process then bifurcates, creating the lobe in the third shape of Fig 2.19, where a downward squashing process has also been introduced from above.

(4) The new squashing process continues, creating the top indentation shown in the fourth shape of Fig 2.19.

2.14 History Minimization

In sections 2.8 to 2.13, we have elaborated a grammar that describes the intervening history between two complex shapes. Let us now return to the issue of internal inference generally.

We have examined, in this chapter, examples from two very different types of internal inference: (1) The first type was the case in which full externalization occurs first, thus giving an internal structure that is a nested, repetitive, Euclidean, hierarchy; i.e., a prototype such as a regular polygon or cylinder. (2) The second type, considered in the last few sections, was that where a person, such as a doctor, is presented with two shapes, known to be of the same object at two stages of development, and the person assumes a purely internal relationship between the two shapes; i.e., an initial phase of externalization is excluded.

Closer examination reveals that both types of situation conform to a rule that we now introduce. Consider first the situation where full externalization takes place. One of our standard examples of this type is the external history from the rotated parallelogram to the square, followed by the description of the square as the result of tracing each individual side and shifting from one side to the next successively. This internal (trace) history might come from drawing with a pen or cutting the square as a sheet from a larger sheet of material.

There are alternative ways in which a square could have been traced. For example, the pen could have randomly skipped between different points, placing a point on some side, then jumping to another random point on another side, etc. However, this is not the type of tracing scenario that one assumes to have happened. One assumes, for example, that the points on each side were traced out by drawing *along* the side, and that

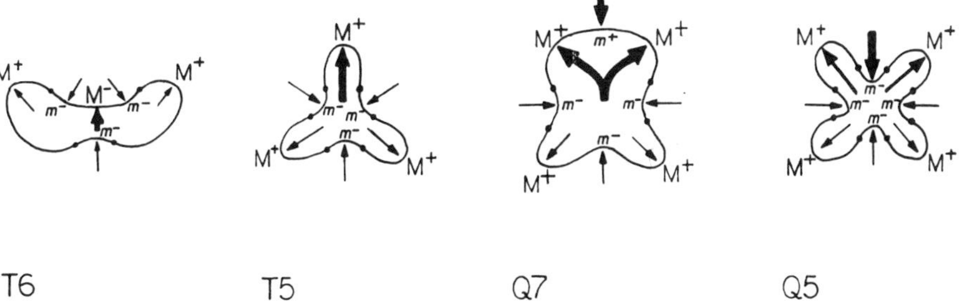

Figure 2.19
The two shapes in Fig 2.18 with the intervening history created by the grammar.

there was some simple order to the sequence with which the successive sides were drawn.

A crucial difference between the randomly-skipping scenario and the orderly scenario is that the former covers much more distance. That is, in the former scenario, one leaves a side and returns to it many times, and on each of these occasions, one has made a skip that can be as long as a side. Thus one's trajectory is many times the length of a side. In contrast, in the orderly scenario, one's trajectory can be the length of only four sides. This latter trajectory is, in fact, the *minimal* trajectory that can be constructed.

Consider now the second type of internal inference we examined—that where a doctor had to conjecture the history linking two shapes. We argued that a grammar of six operations constitutes the set of primitives out of which a history is constructed. Observe now that each operation is, in fact, a minimal event. For example, in the operation BM^+ (Bifurcation at M^+), shown in Fig 2.12, a single extremum defining a nodule splits into two extrema defining a lobe. No events are conjectured to happen other than this minimal one that is required to get from the first configuration to the second. A more complex intervening history could be conjectured, involving the creation and destruction of many extrema, but such a scenario is not conjectured. Similarly, when the operations are put together, one assumes that the actual history that took place did not involve more operations than one minimally requires to construct a history. For example, between any two shapes, a protrusion might have grown and receded a million times. However, if a protrusion is larger on the second shape than on the first, the observer will assume that the protrusion increased in size only once; and similarly, if the protrusion is smaller on the second shape than on the first, the observer will assume that the protrusion decreased only once.

It appears therefore that the history that one conjectures conforms to the following principle:

HISTORY MINIMIZATION PRINCIPLE. *The inferred process-history contains minimal distinguishability.*

It should be observed that the above principle applies not only to internal history but also to external history. For example, in conjecturing the rotation that produced the rotated parallelogram, one conjectured a rotation that was as simple and direct as possible—not a rotation in which the parallelogram moved clockwise for a while, and then anti-clockwise for a while, and then clockwise, etc.

The History Minimization Principle states that temporal distinguishability must be minimal. We shall argue that this principle has two components:

SUBSUMPTION COMPONENT. *The history is composed of a minimal set of distinguishable actions.*

INERTIAL COMPONENT. *The change induced by an action is minimal.*

One can regard the first component as concerning the distinguishability occurring *between* the individual actions, and the second component as concerning the distinguishability occurring *within* an individual action. These two components will be examined in considerable detail later in the book.

The reason why the History Minimization Principle is required is that, without the principle, one would be faced with the problem of *non-uniqueness of recoverability*—a problem we considered in the first chapter. In the present case, the problem would arise as follows: If one were to allow a history that is more complicated than one needs to explain the shapes, then there would be many alternative histories that would be possible and one would have no reason to choose one alternative rather than another. In contrast, the choice of simplest or minimal history greatly narrows the range of histories from which one makes a selection, and, in fact, often narrows that range to a single unique history.

2.15 Two Asymmetries?

At the end of the previous section, it was argued that the basis for the History Minimization Principle is the attempt to derive a unique (or, at least, constrained) history. The unique recoverability argument was used also in Chapter 1 to substantiate the Asymmetry Principle. It was argued that, given an asymmetry in the present, one had to hypothesize its origin to be symmetric because only this would assure a unique past. The argument with respect to the History Minimization Principle is that the conjectured history itself must also have minimal distinguishability, because only that history could be unique.

We have said, in Chapter 1, that there are two types of asymmetry involved in a process-history. One is the asymmetry contained within the present moment. We have called this, *atemporal* asymmetry. The Asymmetry Principle proposes that atemporal asymmetry should be reduced backwards in time. The other asymmetry is that which exists in the conjectured history itself. We have called this, *temporal* asymmetry. The History Minimization Principle proposes that the conjectured history is one in which temporal asymmetry is reduced to its absolute minimum. Thus, the Asymmetry Principle concerns the reduction of atemporal asymmetry, and the History Minimization Principle concerns the reduction of temporal asymmetry.

The question we now wish to raise is whether the two types of asymmetry are actually different. Consider the following. It appears to be the case that what people mean by the term *complexity* is the quantification of asymmetry. Recent research in a number of independent areas has proposed that the complexity of a situation should be regarded as a measure of the amount of *process* needed to produce the situation. Such measures include *algorithmic complexity* (Chaitin, 1969; Solomonoff, 1964; Kolmogorov, 1965); *computational complexity* (e.g., Traub, Wasilkowski & Wozniakowski, 1983); *labor and value added* (Leyton, 1981); *logical depth* (Bennett, 1988); and *thermodynamic depth* (Lloyd & Pagels, 1988).

For example, in Leyton (1981), we argued that the complexity of an object such as a car is equivalent to the amount of labor that produced it from its parts. The labor, of course, usually takes several stages. For example, different parts of the engine are created from different raw materials, and these parts are then assembled into the engine. Then the engine and other parts, such as the tires, doors, etc., are assembled into the whole car. At each stage there is the creation of greater complexity. Furthermore, economists consider that value is added at each stage of the labor. Thus they keep an account of what is called the "value-added" at each stage. Therefore, the value-added can be regarded as corresponding to the increased level of labor or complexity.

If complexity can be considered to be a quantification of asymmetry, then the above approaches to the measurement of complexity would seem to indicate that *atemporal* asymmetry, i.e., the asymmetry within an individual moment, is actually *temporal* asymmetry, the asymmetry within the history that created that moment. This means that one cannot talk about the amount of asymmetry within the present without already having assigned a history to the present.

We are therefore led to a deep paradox: Our Asymmetry Principle states that, given an asymmetry within the present, one assigns to it a history that explains how it arose. However, by the above considerations, the asymmetry can be defined only after the history is defined. More strictly, the history is the asymmetry.

This does not make the Asymmetry Principle invalid. It simply makes an asymmetry and its explanation inextricably bound together. Therefore, rather than invalidating the Asymmetry Principle, it makes the principle more inevitable.

Besides the term *asymmetry*, let us also consider the term *shape*. We have said that, within the present, the ingredient from which one recovers time is *shape*. However, we have also said that, within the present, the ingredient from which one recovers time is *asymmetry*. One might therefore consider that the two terms, *shape* and *asymmetry* have the same meaning. However, our current understanding of these two terms does not allow us to decide fully whether they are equivalent. Nevertheless, we shall for the remainder of the book use these two terms as if they are equivalent.[6]

Now let us return to the hypothesis that atemporal asymmetry is temporal asymmetry; i.e., that atemporal asymmetry is defined as the history that created it. If *shape* and *asymmetry* are the same, then this would mean that shape is inherently a temporal concept. In fact, this would mean that shape is literally the history that created it. The consequence of this would be that our dictum, *shape is time*, would be true in a completely literal sense. We can state this consequence as follows:

SHAPE-IS-TIME HYPOTHESIS. *Shape equals the history that created it.*

That is, rather than regarding shape as some atemporal content of the present, which one can use to extract history, one would have to regard shape as that history.

The identification of shape with time, of course, mirrors and depends on the identification of atemporal and temporal asymmetry—which in turn depends on a thorough understanding of the temporal nature of complexity. Given that the current understanding of complexity is only in its infancy, we shall not base any of the book upon the identification of atemporal and temporal asymmetry, or the complete identification of shape with time. Rather, we will tend to speak of two separate asymmetries, atemporal and temporal asymmetry, and consider the former asymmetry to be the basis for the extraction of the latter.

2.16 Energy and Causal Interactions

In the remaining sections of this chapter, we argue that there is a relationship between perceived *asymmetry* and *energy* in the environment. We begin as follows:

In physics, a *causal interaction*, between two systems A and B, is said to have taken place when energy has been transferred from one of the systems to the other. For example, when a billiard player uses a billiard cue to knock a ball, there is a successive transfer of energy as follows: First, the energy in the moving arm of the player is transferred to the billiard cue, causing the latter to move. Then, when the moving cue interacts with the ball, energy in the cue is transferred to the ball causing the latter to move. The ball can then go on to collide with a second ball, at which point, energy will be transferred from the first ball to the second causing the second to move.

Let us now return to the Second System Principle given in Chapter 1. Recall that the Asymmetry Principle states that, if a system A is asymmetric in the present, it is understood to have been symmetric in the past; and the Second System Principle states that this increase in asymmetry is assumed to have arisen from a causal interaction of system A with a second system B. What we wish to add now is that the *causal interaction* between system A and B is precisely defined as the *transfer of energy from one system to the other*.

2.17 Energy as Memory

We now wish to argue is that there is a close relationship between the notion of *energy* and that of *memory*—the fundamental issue in this book.

To illustrate this relationship, let us consider an experimental situation investigated by Galileo. The situation is illustrated in Fig 2.20, and consists of two inclined planes. A ball is held at position P on one of the planes. The ball is then released, and it rolls down the plane to the bottom

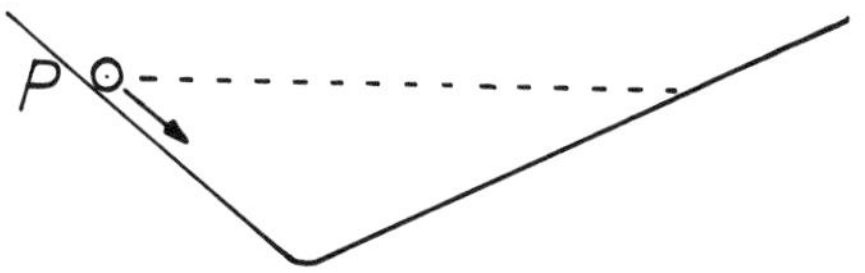

Figure 2.20
A ball leaving from *P* will rise to the same level as *P*.

and then begins to roll up the second plane. Galileo discovered that, at whatever angle the second plane is inclined, the ball ends its movement up the second plane at the same height as the height at which it started on the first plane; i.e., the same level as position *P*.

One can say therefore that the ball has a *memory* of the height at which it started: How is this memory possible and what is the precise form it takes?

The answer is that, at its initial position, *P*, the ball possesses a certain amount of energy and that this energy is conserved. It is the principle that *energy is conserved* that allows, or is equivalent to, the existence of memory. To see this, let us examine the situation in more detail.

In its initial position at *P*, the ball has energy by virtue of its height. Energy is the capacity to do work—that is, to induce action. Because the ball has a certain height, it has the capacity to descend. The energy that an object has by virtue of position or configuration is called *potential energy*. Now, when the ball is released, it begins to roll down the incline, gathering speed on its way, and reaching the maximum speed at the bottom. Since, at the bottom, the ball has no height, it has lost all its potential energy. However, it has energy by virtue of its speed. That is, it has the capacity to induce action by virtue of its motion. The energy that an object has by virtue of its motion is called *kinetic energy*. What has happened is that the potential energy, which the ball had at its starting position, *P*, was converted into kinetic energy while the ball was descending. At the bottom, all the potential energy was converted into kinetic energy.

Now, as the ball begins to ascend, the reverse happens. The kinetic energy, the energy of motion, is converted back into potential energy, the energy of position. This means that the speed decreases corresponding to the decrease in kinetic energy, and the height increases corresponding to the increase in potential energy.

Now the Principle of the Conservation of Energy states that, at each point, the *total* energy, potential plus kinetic, remains the same. That is, even though potential energy can be converted into kinetic energy, and vice versa, there is never any more or less total energy than before.

The crucial consequence for us is that, when the ball reaches the same height as its starting height, the ball will end its upward ascent. It cannot go any higher because, to reach a greater height would mean that it would have a greater energy than it had when it started.[7] Thus it *remembers* the height at which it started because its energy has to equal the energy it had when it started. Therefore the conservation of energy is equivalent to the existence of memory.

What exactly is the ball remembering? It is remembering the totality of its past causal interactions, as follows: Recall that what happens in a causal interaction between two systems is that energy from one system is transferred to the other. In the particular situation we are considering, the ball was placed at the starting position *P*, by the scientist. Prior to the experiment, the ball was resting on the table and had no energy. However, the scientist raised the ball to put it at position *P*. In the causal interaction between the scientist and the ball, energy was transferred from the scientist to the ball. That is, the ball gained height because energy was being transferred to it from the scientist. Then, when the scientist released the ball at *P*, the ball had all the energy it gained from the scientist in the prior interaction. When the ball returned to the same height as *P* on the other plane, it remembered the amount of causal interaction it had with the scientist.

The Principle of the Conservation of Energy makes energy both a prospective and a retrospective concept. Prospectively, energy is the capacity to do work, to create causal interaction. Retrospectively, energy is the memory of how much causal interaction must have taken place. The retrospective view can be stated as follows:

ENERGY-IS-MEMORY PRINCIPLE. *The energy of a system can be regarded as memory of the causal interactions that transferred the energy to the system.*

The view that *energy is memory* allows one to construct a type of argument that is often used in physics. This argument is a *historical* argument, in which one accounts for the energy in an object by going successively backwards in time. For example, a standard argument that is used is the following: The energy that drives an airplane came from potential energy locked in the fuel. This energy, in turn, came from animals and plants which were living millions of years ago producing the fuel-rich deposits. However, the energy in those animals and plants came, in turn, from the sun. Thus, forwards in time, the energy was passed from the sun to the animals and plants, and then became the potential energy of the fuel, and finally became the kinetic energy of the airplane. The motion of the airplane is therefore a *memory* of the energy of the sun.

The ultimate starting point for the energy of our universe was the Big Bang, from which the universe was created. The energy we see around us is traceable back to that event. The entire energy from that event has been conserved, but has been changed into different forms. It is the *conservation* itself that allows us to regard the present forms of energy as *memory* of that primordial event.

2.18 Shape and Energy

We will now argue that there is a psychological relationship between shape and energy.

Recall first that the use of shape as memory is subject to the uni-directional definition of processes as we saw in Chapter 1. For example, the non-straight sheet of paper in Fig 2.21a is assumed to have been twisted from the straight sheet in Fig 2.21b, whereas the straight sheet is not assumed to have arisen from the non-straight one, but from itself. This means that Fig 2.21a is assumed to have undergone causal interaction whereas Fig 2.21b is not.

Now recall that, in physics, a causal interaction can be precisely defined as the transfer of energy. This means that, in twisting an initially straight sheet of paper, one transfers energy from one's hands to the paper. Thus the twisted shape corresponds to an increased state of energy.

Consequently, by inferring the asymmetry in Fig 2.21a to be the result of twisting a straight sheet of paper, one is understanding this asymmetry as corresponding to what a physicist would regard as a higher state of

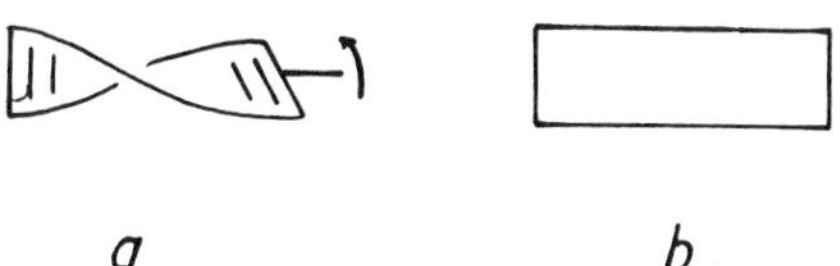

Figure 2.21
The sheet in (a) is seen as twisted, and the sheet in (b) is not.

energy. That is, perceptually, greater asymmetry corresponds to greater energy. In contrast, Fig 2.21b, the straight sheet, is not seen as having a history of deformation, and is thus seen as having gained no energy.

Let us refine this idea further. The energy gained by the paper in Fig 2.21a is *potential* energy; i.e., energy by virtue of configuration. This energy gives the paper a tendency to return to its flattened state. However, suppose one does something more violent to the paper—e.g., crunches it up. Then it will retain this crunched-up shape and not return to its straightened state. What has happened is this: Initially, when it was flat, it had zero potential energy. Then, while it was being crunched, potential energy was being transferred to it. However, the crunching caused damage to the arrangements of molecules in the paper and the paper settled to an equilibrium (balance of forces) that had a deformed shape. As it settled to this deformed equilibrium, a portion of the potential energy gained from the crunching was lost from the paper into the surrounding air and was, for example, heard as sound.

What this situation illustrates is that the asymmetry in the paper is not a correlate of the potential energy that the paper possesses in the present, but of the potential energy that the paper received at some previous time. We can say that it is a memory of the energy that was transferred to the paper, even though a portion of that energy might have dissipated. We can summarize this conclusion as follows:

ENERGY-ASYMMETRY PRINCIPLE. *Asymmetry is taken to be memory of the energy transferred to an object in a causal interaction.*

2.19 The Energy Correlates of External and Internal Inference

Recall now that we have argued that there are two alternative types of inference that can be used in the process-recovery problem: *external* inference, in which one assumes that the present contains a record of only a single state of a process; and *internal* inference, in which one assumes that the present contains records of multiple states of a process.

Now observe the following: In external inference, the asymmetry in the present corresponds to a single *configuration* of the system. This means that the energy represented by the asymmetry is energy by virtue of configuration; i.e., *potential* energy. That is, in external inference, the present asymmetry is memory of potential energy.

Now consider internal inference: Here, the assumption is that the asymmetry is between several states of a process. This means that the energy represented by the asymmetry is energy by virtue of successive states; i.e., *kinetic* energy. That is, in internal inference, the present asymmetry is memory of kinetic energy.

Note that, in internal inference, it is not necessary to see actual movement. For example, the inference of a track in the snow, or a scratch, is an example of internal inference; i.e., different points along the trace are understood as coming from different points in time. Therefore, the trace is memory of movement that once took place. Thus the trace is memory of kinetic energy that once existed but no longer exists.

To restate the two main conclusions of this section: In external inference, the energy remembered by the present asymmetry is potential energy. In internal inference, the energy remembered by the present asymmetry is kinetic energy.

Appendix

2.20 Inferring Temporal Order

In sections 2.8 to 2.12, we developed what we called a "process-grammar", to provide a history between two shapes. However we noted, in passing, that the grammar can provide a number of alternative histories between two shapes. Thus we need to understand how one forms a single *most plausible* intervening history. To do this, we need to examine more closely how one infers *temporal relations* between processes along a history. These relations will determine the selection and order of the grammatical operations such that they will generate the optimal history.

In the real world, the temporal relations between processes can be sequential, overlapping, or simultaneous. For example, in an embryo, the growth of the torso starts *before* that of the arms, but later *overlaps* the growth of the arms. Furthermore, the growth of one arm is *simultaneous* with the growth of the other arm.

Remarkably, we will find that there is a very simple rule that can infer such complex process-relationships. The rule is based on the concept of *blurring*, as follows.

Observe that, as one gradually blurs an outline, the outline becomes smoother and smoother. This means that the curvature variation disappears. If one then takes the blurred outline and applies the reverse procedure, i.e., *de-blurring*, one will gradually introduce greater and greater curvature variation.

Now recall, from sections 1.9–1.13, that we argued that processes are conjectured by human beings to explain how the *curvature variation* is introduced into the boundary. That is, processes are conjectured to have shifted the boundary in the direction of increased curvature variation. What we have found above is that de-blurring incrementally moves the boundary in the direction of increased curvature variation. Thus de-blurring introduces what processes introduce. Conversely, *blurring*, by reducing curvature variation, mimics the backwards direction of time, i.e., the undoing of processes.

Now observe that, when one blurs a shape, smaller details disappear before larger ones. This means that, in the de-blurring direction, i.e., mimicking forwards time, larger details are introduced before smaller ones. Furthermore, as one continues to de-blur, the curvature variation continues to increase, which means that, once introduced, any detail continues to increase while de-blurring is continued. These temporal relationships can be illustrated by considering the formation of an arm in the developing embryo. Since the arm is larger than the fingers, the arm appears earlier, in fact, as a small stump on the embryo. At some point, while it is increasing the stumps for the fingers begin to appear. Furthermore, the arm continues to grow while the fingers grow. De-blurring orders the processes in exactly this way. Thus, one can use de-blurring to obtain the order of processes in a history. We therefore argue that the order is defined by this rule:

SIZE-IS-TIME HEURISTIC. *In the absence of information to the contrary, size corresponds to time. That is, the larger the boundary movement, the more likely it is to have (1) started earlier, and (2) taken longer to develop.*

We have seen that de-blurring mimics the processes that create boundary movement, but to conjecture that de-blurring is a technique that is indeed used by human beings would require us to suppose that human beings actually define a shape at multiple levels of blurring. How valid are we in supposing this? Vision researchers are finding overwhelmingly that human beings represent the environment on such multiple-levels (Marr, 1982). In fact, the importance of blurring hierarchies has emerged in the 1980s as one of the major discoveries in vision (e.g., Koenderink 1984; Mokhtarian & Mackworth, 1986; Pizer, Oliver & Bloomberg, 1986; Oliver, 1989; Rosenfeld, 1984; Terzopoulos, 1984; Witkin, 1983; Yuille & Poggio, 1986; Zucker & Hummel, 1986). Thus we can assume that information about blurring is a very real part of perceptual representations.

Let us now examine the role of de-blurring in relation to the two inference problems that concern us with respect to complex shape: (1) the inference of history from a single shape, i.e., the external-inference problem discussed in Chapter 1; and (2) the inference of intervening history between two given shapes, i.e., the internal-inference problem discussed in Chapter 2. We shall consider these two problems in the next two sections respectively.

2.21 Temporal Order in Inference from a Single Shape

Recall that, in order to solve the problem of inference from a single shape, we proposed two rules: (a) the Symmetry-Curvature Duality Theorem which assigns, to each extremum, a unique symmetry axis terminating at the extremum, and (b) the Interaction Principle which implies that each axis is the trace of a process. As a consequence of applying these two rules, we derived diagrams such as those shown in Fig 2.8. We call such diagrams, *process-diagrams*.

We can now see that process-diagrams do not contain a complete reconstruction of the process-history. They do not encode the temporal *order* of the processes; i.e., which of the processes followed each other, which were overlapping, and which were simultaneous. Thus the process-diagrams must be supplemented with extra information. But how can this information be specified?

Let us consider again the example of the embryological development of the arm. A newly introduced limb of a real body is a bud-like structure with a single curvature extremum at the tip of the limb. This accords with our Symmetry-Curvature Duality Theorem: The bud has a single symmetry axis terminating at a single extremum. Now, as the limb begins to grow, the curvature extremum continues to move along the direction of the symmetry axis, in accord with our Interaction Principle. This is seen in the development of an arm: When the arm begins to appear, it is initially a small bud terminating at a curvature extremum. As the arm begins to grow, the curvature extremum continues to move along the direction of the symmetry axis.

Now, at some point in the growth of the arm, the fingers begin to appear as limbs growing out from the end of the arm. The fingers themselves follow the scenario described above, i.e., each finger is associated with a curvature extremum at its tip, and, as the finger grows, this extremum remains at the tip of the growing axis.

Consider now the very moment at which the curvature extrema of the fingers are introduced. Prior to this point, there was only the arm with its single extremum. This means that the introduction of the fingers is associated with the transition from the single extremum of the arm to the several extrema of the fingers. That is, the emergence of the fingers must involve the bifurcation of extrema.

We see therefore that, in the development of limbs, two things happen to extrema: quite simply, they *continue* to move, i.e., along symmetry axes, and they can also *bifurcate*. Now recall that the order in which these events occur is determined by the Size-is-Time Heuristic; e.g., the fingers, being smaller, emerge later than the arm. That is, the order of events is determined by deblurring.

We can therefore summarize our discussion on deblurring as follows: *When deblurring a shape, two things happen: (1) processes emerge in the order of their prominence, and (2) the processes grow to their full size by continuing and possibly bifurcating.*

The crucial point to observe here is that, in sections 2.8 to 2.12, we developed a means by which the continuation and bifurcation of processes can be specified: the process-grammar. The grammar was in fact developed to solve our second inference problem: inference of *intervening* history between *two* shapes. However, we can now see that the grammar is also relevant to our first inference problem: inference from a *single* shape. It will specify the structural changes that occurred when processes continued and bifurcated in the history leading up to that shape.

Recall now the problem we wish to solve. None of the process-diagrams in Fig 2.8 contains a full reconstruction of the history of that shape. Such a diagram does not encode the temporal order of processes. Thus a process-diagram should be supplemented with extra information that specifies the ordering relationships between the processes. We can now see that the extra information that is needed consists of *operations from the process-grammar*. Furthermore, the order in which operations are given is determined by the Size-is-Time Heuristic, i.e., deblurring.

2.22 Temporal Order in Intervening History

Let us now consider the second inference problem: the inference of *intervening* history between two given shapes; e.g., two X-rays of the same tumor at different stages of development. In using the grammar to reconstruct the intervening history, two problems are involved that did not occur in inference from a single shape:

Ambiguity. There might be alternative ways to correspond processes in the first shape with processes in the second. For example, in Fig 2.22, the protrusion *A* in the left shape might have become either protrusion *B*, *C*, or *D*, in the right shape.

Process-Reversal. Processes might have receded and disappeared in the transition. The evidence for this is that curvature has diminished at certain points. Process-reversal cannot arise in inference from a *single* shape, because a single shape can offer no *evidence* of reversal; i.e., one has to assume that the curvature variation in that shape is the result of processes that grew over time.

Let us consider the ambiguity problem first. There is, in fact, a natural answer to this problem. The reader can easily see, that, in matching the two shapes in Fig 2.22, he or she will automatically match the large features before the small ones. That is, the reader first ignores detail while automatically matching the following: (a) the large-scale protrusion *X* in the first shape with the large-scale protrusion *X* in the second shape; (b) the large-scale indentation *Y* in the first shape with the large-scale indentation *Y* in the second shape; and (c) the large-scale protrusion *Z* in the first shape with the large-scale protrusion *Z* in the second shape.

Figure 2.22
Two shapes to be given an intervening history by the grammar.

Thus, because the reader is initially ignoring detail, and matching overall features, we can conclude that the reader starts by matching two *blurred* versions of the shapes.

This is an essential aspect of a general matching technique that has been elaborated by Witkin, Terzopoulos & Kass (1987). These authors have argued that, when there is possible ambiguity in matching two objects that are deformations of each other, an appropriate technique is to blur the objects until a non-ambiguous match can be made between them, and then to deblur them slowly so that, as the detail gradually appears, it can be matched incrementally across the two shapes.

We can easily see that human perception does something like this with the two shapes in Fig 2.22. The matching of large-scale features provides a basis upon which the matching of smaller-scale features can proceed. The coarse-to-fine matching of features therefore gives a coarse-to-fine order on the grammatical operations. Note that, because some features can disappear, forwards in time, reversals of the grammatical operations can occur in the ordering hierarchy.

Finally, we observe that the coarse-to-fine ordering on the grammatical operations is substantiated again by the Size-is-Time Heuristic, which states that the larger the boundary movement, the more likely it is to have (1) started earlier, and (2) taken longer to develop.

The use of the heuristic is discussed in much greater detail in Leyton (1989).

2.23 Summary

In this chapter, we distinguished between the external and internal inference problems. The former is based on the assumption that the present contains a record of only a single state. Asymmetries within this state imply a past state in which the asymmetries do not exist. Therefore, the past state is external to the present. In contrast, internal inference is based on the assumption that the present is a record of multiple states. The asymmetries are between these states, and the removal of the asymmetries produces past states that have records in the present.

We found that, although the perceiver can choose which proportion of the hypothesized history is internal and external, some internal history must inevitably occur when all distinguishability is removed. We argued that this proposal is environmentally valid: For example, even when one has an object cut by a template, the template itself must have been cut by sequential actions, i.e., it must have internal structure. In fact, generally, we proposed the Transfer Principle, which states that, if an asymmetry in an object A did not exist for the first time as a result of a causal interaction of object A with another object, then it was *transferred* to object A from another object where it was brought into existence by a causal interaction. We argued also that a frequent source of inferred trace-structure is symmetry that has been recovered externally. More generally, we proposed the Symmetry-to-Trace Conversion Principle, which states that any symmetry can be re-described as a trace; the transformations, defining the symmetry, generate the trace. In fact, the symmetries that are recovered externally have the form given by our Externalization Principle which states that the external processes that are hypothesized are those that reduce the internal (i.e., trace) structure, as much as possible, to a nested hierarchy of repetitious Euclidean processes. This proposal allows one to understand what a prototype is: it is a nested, repetitive, Euclidean hierarchy, i.e., a shape that results from full externalization.

We then examined the purely-internal inference problem of bridging two complex shapes by an intervening history. We found that a grammar of only six operations, two continuations and four bifurcations, is sufficient to generate any such history. While the grammar is expressed formally as rewrite rules using discrete strings of extrema, each rule corresponds to a psychologically compelling situation. We also saw that, when imposed on shape-space, the grammar gives the space a strong organization not previously suspected for that space.

We then argued that all process-inference situations we have considered assume the History Minimization Principle in the inference system: i.e., that distinguishability is minimized in the history. Later in this book, we will argue that this principle has two components, the Subsumption Component, which states that a history is composed of a minimal set of actions, and the Inertial Component, which states that the change induced by any action is minimal.

We observed that the Asymmetry Principle concerns the reduction of atemporal asymmetry, and the History Minimization Principle concerns the reduction of temporal asymmetry. The question arose as to whether the two forms of asymmetry are the same. We observed that certain current theories of *complexity* (viewed as the quantification of asymmetry) would indicate that the atemporal asymmetry of a situation is its history, i.e., temporal asymmetry. Since, in this book, we tacitly assume that shape is the same as asymmetry, the identification of atemporal and temporal asymmetry leads to the literal Shape-is-Time Hypothesis which states that shape equals the history that created it.

We then turned to an examination of the relationship between perceived asymmetry, inferred history, and environmental energy. We saw that the Conservation of Energy Principle forces objects to have memory, e.g., an object rolling up and down two facing inclined planes will "remember" the point at which it started. In fact, our Energy-is-Memory Principle states that the energy of a system can be regarded as memory of the causal interactions that transferred the energy to the system.

In addition, our Energy-Asymmetry Principle states that shape, or asymmetry, is taken to be memory of the energy transferred to an object in causal interactions. Furthermore, external inference is the recovery of potential energy; and internal inference is the recovery of kinetic energy.

Finally, in an appendix, we considered the ordering of operations provided by the process-grammar, and argued that the ordering is determined by the Size-is-Time Heuristic, which states that, in the absence of information to the contrary, size corresponds to time. That is, the larger the boundary movement, the more likely it is to have (1) started earlier, and (2) taken longer to develop.

3 Radical Computational Vision

3.1 Introduction

The visual system is able to see the world using remarkably little information. The only information that is available is the two-dimensional array of stimuli on the retina. Yet, given this array, the perceptual system is able to create a three-dimensional representation of the environment. Furthermore, although the stimuli on the retina are independent points of light, the perceptual system is able to unify them into cohesive three-dimensional objects, set in space.

Computational Vision is the main branch of research that attempts to understand how perception achieves this goal of reconstructing, from the two-dimensional array of independent stimuli (on the retina), the spatial environment consisting of three-dimensional cohesive objects.

The purpose of the present chapter is to examine Computational Vision and show that it can be given a new and very different interpretation: It can be understood as studying how perception recovers *history*. That is, we shall argue that Computational Vision concerns the means by which *time* is unlocked from the stimuli on the retina.

In this chapter, we shall go through several of the main principles and results of contemporary Computational Vision, and show that they can more deeply be understood if they are subsumed under our general theory of process-inference. In fact, we shall argue that the main ideas and theoretical results of Computational Vision are simply instantiations of the general principles we have elaborated for the recovery of process-history. Since these general principles are based on the concept of *symmetry*, this will allow us to show that

ALL COMPUTATIONAL VISION CAN BE REDUCED PURELY TO SYMMETRY PRINCIPLES.

We shall call Computational Vision, as it is studied and understood today, *Standard Computational Vision*. What we are investigating in this book is a new problem—that of the recovery of history. The name we will give to this new problem is *Radical Computational Vision*. Our purpose, in the present chapter, is to show that Standard Computational Vision is, on deeper analysis, a branch of Radical Computational Vision.

A consequence of this argument is that one can regard all vision as the recovery of the past. That is, rather than understanding the visual system as recovering the three-dimensional environment, the cohesive objects that constitute the environment, the spatial layout of the environment, etc., we will argue that the visual system serves only the single function of unlocking *time* from the image.

3.2 Standard Computational Vision

As we said above, we shall use the term *Standard Computational Vision* as a label for Computational Vision as it is currently studied and understood. In this section, we describe the overall principles of Standard Computational Vision, for those readers who are not acquainted with the principles. Readers who are acquainted with these principles can omit this section and go directly to section 3.3, where we begin to show, in detail, how Standard Computational Vision can be understood more deeply as part of Radical Computational Vision—the subject of this book.

Standard Computational Vision is based on the following four ideas:

(1) Perception is Recovery

In Standard Computational Vision, perception is understood as the means by which the structure of the three-dimensional environment is recovered from the degenerate two-dimensional image on the retina. That is, perception is understood as a *recovery* process.

This problem of recovery is analogous to the famous *cave story* in Plato's *Republic*: Imagine that one is standing in a darkened cave, staring at a screen. Behind the screen, certain events are happening that one cannot see—people and objects are moving, etc. All that one can see are the changing complex shadows that are cast on the screen. From these shadows, one tries to recover what must be happening behind the screen. This is analogous to the problem faced by the perceptual system. The retina is like the screen across which complex shadows are moving. We have no perceptual access to the environment that lies *behind* the retina. All we have is the retina with its changing patterns of light and dark. We examine the retina and its patterns and try to infer, from the patterns, what the structure of the environment is behind the retina—the environment to which we can never gain perceptual access, except via the clues on the retina.

(2) Image-Formation

A second basic principle of Computational Vision is this: In order to understand how the environmental structure can be recovered from the re-

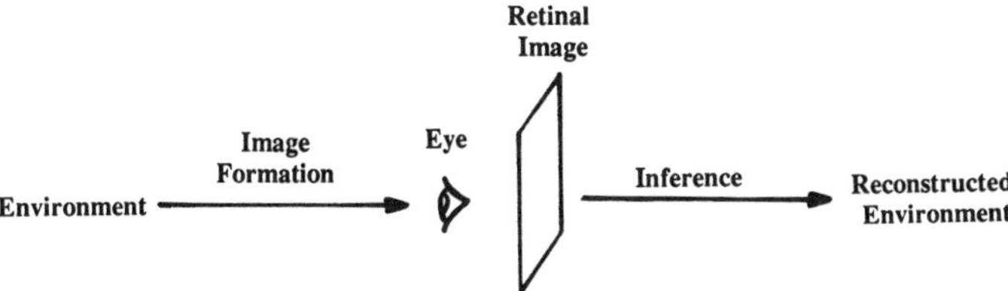

Figure 3.1
The two-stage model of Standard Computational Vision.

tinal image, one should investigate how the image is related to the environment. Consider Fig 3.1. It shows the following sequence of events from left to right. At the far left, one has the environment. This is projected onto the retina in the middle, represented here by the plane. All that the mind has available is the image on this plane. From this image, the mind tries to reconstruct the structure of the environment. That is, the brain produces a reconstructed version of the environment, shown on the far right. Hopefully, the reconstructed version will be similar to the actual environment.

There are two stages involved in going from the far left to the far right of Fig 3.1. The first is the projection of the environment onto the retinal image. This stage is generally called *image-formation.* The second stage is that in which the mind *infers* the structure of the environment from the image. This stage is called the *inference stage.*

A basic principle of Computational Vision is that, in order to understand how the mind *infers* the structure of the environment from the image, i.e., accomplishes the second stage, one should first understand what has happened in the first stage, the image-formation.

(3) Perception as the Recovery of Surfaces

An important idea, due to Gibson (1950), is that perception is mainly devoted to recovering *surfaces* in the environment. This is because, on the human level, environmental structure is primarily a structure of surfaces.

Surface structure is often referred to as *shape*, i.e., the shape of objects and the spaces between them. Thus the primary role of perception is the recovery of environmental shape from the retinal image.

Gibson also argued that shape is recovered from a variety of *cues* in the image. One important cue is the outline or silhouette of an object. That is, the two-dimensional outline of an object on the retina often tells us much about the full three-dimensional object in the environment. However, there are also other cues. Gibson listed at least five main cues, in the retinal image—five cues that act as sources of information about the shape in the environment. They are:

(1)	Shading:	The varying level of brightness across the retinal image.
(2)	Texture:	The small-scale repetitive elements across the retinal image.
(3)	Contour:	The lines (straight or curved) in the retinal image.
(4)	Stereo:	The differences between the retinal image in the left and right eyes.
(5)	Motion:	A moving pattern of light across the retinal image.

That is, these five types of features in the retinal image, enable the mind to recover the three-dimensional shape of the environment.

These five sources remain the five main sources investigated in Computational Vision today. That is, we still believe that they are the five main cues by which environmental shape is recovered from the retinal image. The sources are referred to as the Shape-from-x sources: That is, corresponding to the five cues listed above, one has: (1) Shape-from-Shading, (2) Shape-from-Texture, (3) Shape-from-Contour, (4) Shape-from-Stereo, and (5) Shape-from-Motion.

(4) Perception as Computation

Computational Vision is founded on the above ideas, together with the following additional principle: The recovery of the environmental structure from the retinal image should be regarded as a *computation*, in the same way as one regards the function of a pocket calculator as that of carrying out a computation. For example, in order to divide 23 into 837, the calculator takes the symbols 23 and 837 and performs a sequence of operations on them until the answer is obtained. The sequence of operations can be listed on a sheet of paper in the same manner as a high school student writes, on an exam paper, the sequence of steps in a calculation. The sequence of steps that the students writes out is a sequence of manipulations of symbols.

A crucial idea is that this sequence of steps is independent of the hardware used to carry them out. For example, the steps can be carried out by a human brain, which is made of biological tissue, or by a computer, which is made of silicon chips. Therefore, given a calculation, one can isolate a *level* of action that is independent of any of the alternative hardwares. This level is called the *computational level*. (For a full discussion, see Newell & Simon, 1976; Newell, 1980; Pylyshyn, 1984; Fodor & Pylyshyn, 1988.)

In Computational Vision, the recovery of the environmental structure from the retinal image is understood as starting with the representation on the retina and performing a sequence of operations on this representation until the final three-dimensional representation is produced. This sequence of operations is understood as a calculation. Furthermore, the calculation can be studied on a computational level; i.e., independently of the hardware used to carry it out.

3.3 Radical Computational Vision

According to Standard Computational Vision, that is, Computational Vision as it is currently understood, the purpose of perception is to recover the structure of surfaces that comprise the environment. Radical Computational Vision—the approach developed in this book—takes a very different view:

> *According to Radical Computational Vision, the purpose of perception is not to recover* **surfaces**, *but to recover* **causal structure**.

The difference between the two views is profound. According to Standard Computational Vision, the environment is dead and timeless. According to Radical Computational Vision, the environment is a system of causal processes. The perceptual system can recover the processes via the memory that they leave through time. Thus the role of perception is to unpack time from this memory. Therefore, as we shall see, all the diverse areas of Computational Vision can be reduced to the study purely of *symmetry principles*.

3.4 Layers of Memory

We now argue that the memory, from which perception unpacks time, is created in two stages. These stages are presented as the first and second

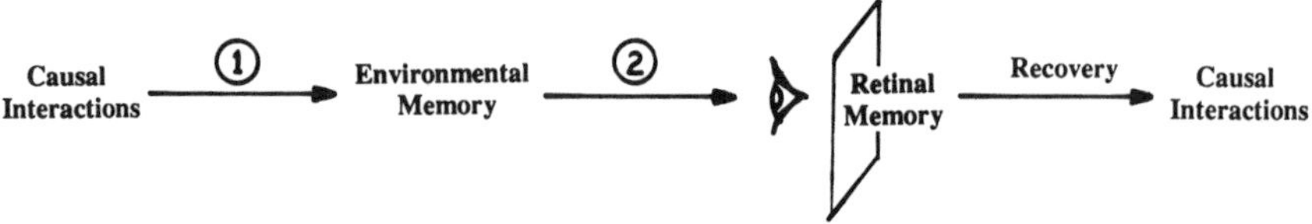

Figure 3.2
The model of perception according to Radical Computational Vision.

arrows from left to right in Fig 3.2. They are, respectively, as follows:

Stage 1: Causal Interactions → Environmental Memory
The first stage is indicated by the number 1 in Fig 3.2. That is, the past consists of causal interactions. These interactions leave environmental memory; e.g., deformed objects, tracks in the snow, graffiti, etc. In fact, we will suppose that *everything* in the environment is the memory of causal interactions.

Stage 2: Environmental Memory → Retinal Memory
The second stage is indicated by the number 2 in Fig 3.2. The environmental memory is transferred onto the retina. This process accords exactly with what we have called *transfer*, in section 2.3. Consider one of our standard examples of transfer, a rubber printing stamp, e.g., of someone's signature. The signature on the stamp is a memory of the trajectory of some knife that cut out the shape of the stamp. The stamp is layered with ink, and this acts as a transfer medium. It will allow the memory of the past cutting trajectory to be transferred from the stamp to a sheet of paper.

In the case of perception, as described in Fig 3.2, the environment corresponds to the stamp. It is the memory of the actions that caused its current shape. The transfer medium consists of the light photons that transfer the environmental structure through space onto the retina. The retina now contains an imprint of the environmental memory, in the same way that the sheet of paper contains the memory inherent in the stamp. That is, the retina now holds memory from the original causal interactions.

According to Radical Computational Vision, the role of perception therefore is to examine the memory contained in the retinal image, and infer the original causal structure, i.e., to carry out the recovery shown as the last stage in Fig 3.2.

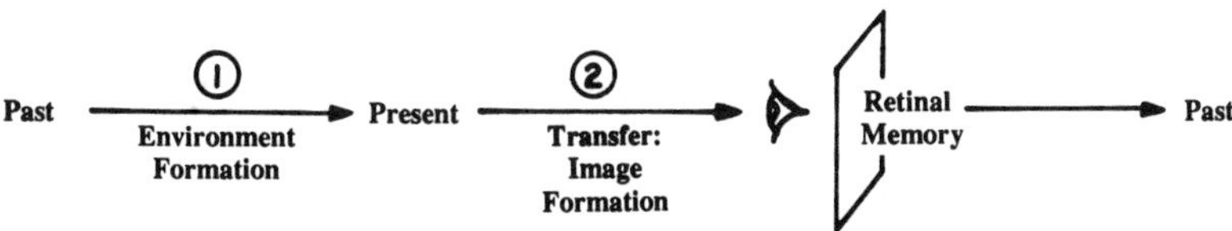

Figure 3.3
A tentative model of the temporal structure of perception.

Let us now contrast Standard Computational Vision with Radical Computational Vision.

According to Radical Computational Vision, the image on the retina is *memory*. The memory is created in two process stages: *Stage 1*, the processes that created the present environment, and *Stage 2*, the processes (i.e., illumination) that transferred the present environment onto the retina. We can call the former type, the *environment-formation processes*, and the latter, the *image-formation processes*.

The situation is redescribed in Fig 3.3, which is virtually the same as Fig 3.2, except that we have emphasized different aspects. On the far left, one begins with the *past*. Progressing right-wards, in Stage 1, the *environment-formation processes* create the *present* environment. Then, in Stage 2, the *image-formation processes* transfer the structure of the present environment onto the *retina*; i.e., creating the image. According to Radical Computational Vision, this image is what is left of the causal past, and is "unpacked" by the brain to reveal the past.

According to Standard Computational Vision, the brain is concerned with only part of the above sequence. It is concerned with only the transfer stage, Stage 2. That is, perception is the recovery of only the *present*.

We can thus re-interpret the issues of Standard Computational Vision in terms of our process-recovery scheme. Standard Computational Vision is concerned with the recovery of the *present* environment, and the present structure of the environment is considered to be its *shape*; i.e., the geometry of objects and the spaces between them. In fact, as we noted in section 3.2, Standard Computational Vision decomposes the recovery of shape into a variety of recovery problems: Shape-from-Shading, Shape-from-Texture, Shape-from-Contour, Shape-from-Stereo, Shape-from-Motion. These recovery problems should be understood as dealing with only Stage 2 in our overall recovery scheme.

Finally, we observe that, because Standard Computational Vision is concerned with the recovery of the *present*, Standard Computational Vision is what philosophers call *positivistic*: it is concerned with the recovery of what is tangible, testable, accessible. Our concern in this book is with the recovery of what is no longer accessible. We are concerned with the remarkable ability of human beings to infer the unavailable past.

3.5 What is the Present Environment?

Researchers in Computational Vision are a pragmatic people. They are concerned with how perception can reconstruct the present environment, the environment that sits immutably and unquestionably in front of the perceiver's eyes. These researchers are not interested in the reconstruction of the past. The past has no practical value to them. The past is no longer a reality and therefore need not be taken into account.

However, is the *present* environment actually in the present? Simple considerations reveal that what is called the present environment is actually in the *past*; as follows:

The structure of the environment is carried by light from the environment to the retina. This light has taken a finite amount of time to travel. Therefore, the image that arrives at the retina is of an environment that existed a finite amount of time previously, i.e., an environment in the past.

The value of understanding this to be the case is as follows. We will argue that, if there is any scientific truth in the theoretical and empirical results of Standard Computational Vision, it is because these results concern the recovery of the past, not the recovery of the present—even though they are currently understood as concerning the recovery of the present. Furthermore, in order that these results become as scientifically fruitful as they can be, they must be understood for what they really are: results concerning the recovery of the past. In fact, we shall argue that these results are simply instantiations of the general principles elaborated in this book.

One might object that the time traveled by light in moving from the environment to the retina is so short that it can effectively be ignored. This is not true. The time involved is long enough to create crucial asymmetries that, in fact, characterize the image.

Consider, for example, a symmetric three-dimensional object, such as a cube, in the environment. The projection of this object onto the retina

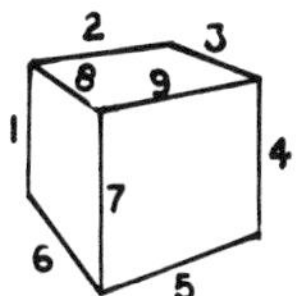

Figure 3.4
Sides of the same length in the environment have different lengths in the image.

makes it highly asymmetric in the image. For example, Fig 3.4 shows the retinal image. There are actually nine edges (lines) visible. In the image (Fig 3.4), these edges are of different lengths. However, on the three-dimensional cube in the environment, these edges were all the same length. Thus, projection onto the retina has created asymmetries that did not exist before.

In accord with our Asymmetry Principle, these asymmetries are the *memory* of the transfer process.

In this chapter, we shall review Standard Computational Vision, and show that the many procedures that have been developed for recovering what has been called the present environment, are in fact instantiations of our general principles for the recovery of the past. In particular, the procedures are instantiations of our Asymmetry Principle; that is, asymmetries in the image are explained as having arisen over time from symmetries in the environment. That is, *we can understand all Computational Vision as the removal of asymmetries backwards in time.*

3.6 Three Crucial Layers of Asymmetry

Recall that we have divided the history leading to the retinal image into two successive stages: Stage 1, the environment formation processes, which create the *shape* of the environment—the dents, snow-tracks, surface-scratches, etc.—and Stage 2, the image formation processes, in which *light* transfers the structure of the environment onto the retina. We will now divide Stage 2, the effects of light, into two stages: Stage 2a and Stage 2b.

Stage 2a. In this stage, light coming from a light-source creates *illumination asymmetries* on the objects in the environment. For example, on each individual object, light introduces a variation in brightness across the ob-

Figure 3.5
The successive creation of asymmetries in the image.

ject, from light to dark. This brightness variation is an extra layer of asymmetry that is added to the surface of the object. That is, the surface was already asymmetric by virtue of its shape, as created in Stage 1. In Stage 2a, the illumination asymmetries are then "painted" onto these shape asymmetries.

Stage 2b. In this stage, the light leaves the environmental objects and travels to the retina, projecting the structure of the objects onto the retina. In this projection process, a new type of asymmetry is added: the *viewpoint asymmetries*. For example, these asymmetries were illustrated in Fig 3.4, where we saw that projection makes the edges of a cube different in size in the retinal image, whereas they were equal in size in the environment.

Thus, as illustrated in Fig 3.5, we have three stages in the overall process-history: Stage 1, Stage 2a, and Stage 2b. Each stage adds its own layer of asymmetry. Stage 1 creates the asymmetries we know as the object's shape. Stage 2a adds the illumination asymmetries; and Stage 2b adds the viewpoint asymmetries.

It is important to observe that, in accord with the Second System Principle (section 1.7), each of these layers of asymmetry is created by an interaction with an additional system. In Stage 1, the object asymmetries, i.e., object shapes, are created by objects interacting, e.g., rock against rock, feet on snow, pen across paper, etc. In Stage 2a, the illumination asymmetries are created by the interaction of the object shapes with a new system, the light rays. In Stage 2b, the viewpoint asymmetries are created when the light rays, reflected from the objects, interact with a new system, the viewer's retina. These successive interactions are diagramed in Fig 3.6. There are three successive pairs of converging arrows, from left to right, corresponding to the three stages. In each pair, the two arrows are converging because they represent the fact that two systems are brought

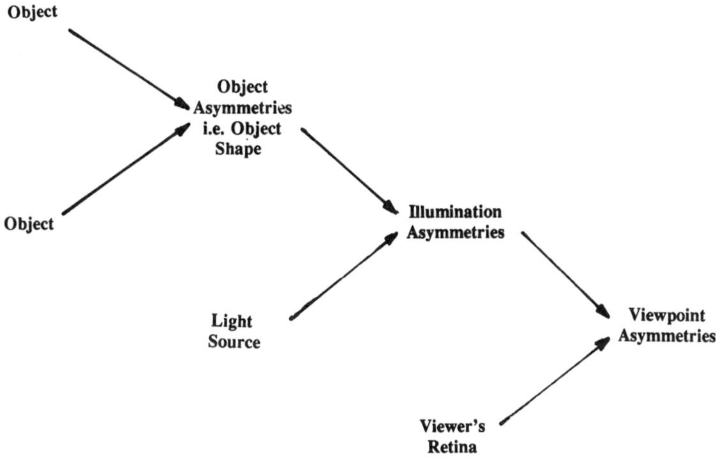

Figure 3.6
Successive causal interactions responsible for image asymmetries.

together in an interaction. The lower arrow, in each pair, introduces the new system in each case.

3.7 The Radical Computational Vision Thesis

In this section, we introduce one of the main proposals of this book. In order to lead up to this proposal, it is worthwhile recapitulating some of the ideas of the chapter so far.

We began by contrasting Standard Computational Vision with Radical Computational Vision: Whereas Standard Computational Vision understands the task of perception to be the recovery of *surfaces*, Radical Computational Vision understands the task of perception to be the recovery of *causal interactions*. That is, according to Radical Computational Vision, the image on the retina is *memory* of past causal history.

This history is divided into two overall stages: Stage 1, environment formation, and Stage 2, image formation. Standard Computational Vision studies only Stage 2. Furthermore, it does not study it as the recovery of the past, i.e., as Radical Computational Vision would. It regards the environment as in the *present*. In contrast, we have argued that image formation (Stage 2) crucially involves a period of time: a period in which light interacts first with the environment and then with the retina.

The fact that, in image formation (Stage 2), causal interactions are taking place over time, makes our Asymmetry Principle the major controlling principle. This is because the causal interactions create asymmetries that end up in the retinal image and allow perception to recover time from the image. In fact, we propose:

RADICAL COMPUTATIONAL VISION THESIS. *(1) The image is best characterized as a collection of asymmetries. (2) All Computational Vision should be regarded as concerning the recovery of process-history from these asymmetries.*

In the book, up to the present chapter, we have in fact been trying to demonstrate this thesis with respect to Stage 1, environment formation. That is, we have been trying to demonstrate that the shapes of objects are assigned causal-histories *via the removal of asymmetries.* The purpose of the present chapter is to demonstrate the thesis with respect to Stage 2, image formation.

Standard Computational Vision concerns image formation. Therefore, in order to demonstrate the above thesis for image formation, we will go through some of the main research of Standard Computational Vision and argue that this research can be understood more deeply as concerning the *recovery of history from asymmetries in the image.* That is, our intention is to show that Standard Computational Vision should be re-interpreted as part of Radical Computational Vision. This chapter will show in detail how the former can be embedded in the latter. The embedding is illustrated in Fig 3.7. Under this re-interpretation, Standard Computational Vision concerns the recovery of that causal history in which the environmental structure is transferred to the retina.

Standard Computational Vision, being concerned with the image formation phase, takes, as its main problem, the recovery of surface structure; i.e., the recovery of the *shape* of the environment. As we saw (section 3.2), Standard Computational Vision argues that environmental shape can be recovered from five major cues in the image: shading, texture, contour, stereo, motion. Correspondingly, researchers speak of five major recovery problems: Shape-from-Shading, Shape-from-Texture, Shape-from-Contour, Shape-from-Stereo, Shape-from-Motion. We shall argue the following:

EMBEDDING PROPOSAL. *(1) Each of the five major image cues of Standard Computational Vision—shading, texture, contour, stereo, motion—*

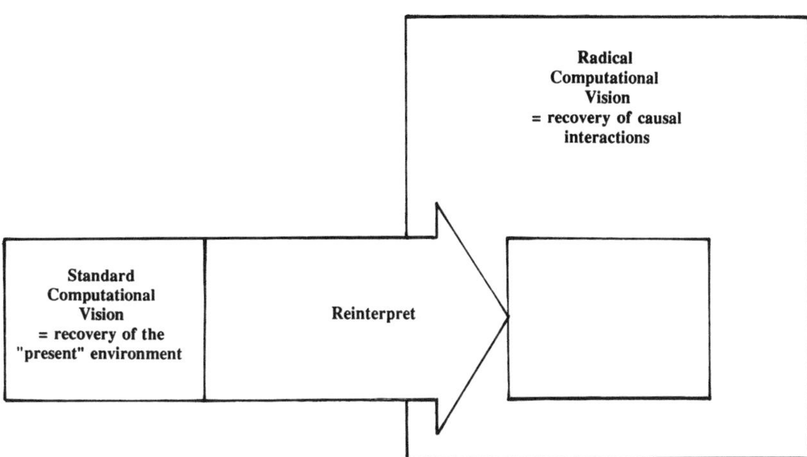

Figure 3.7
The embedding of Standard Computational Vision within Radical Computational Vision.

is a type of **asymmetry.** *(2) Each of the five major recovery procedures—Shape-from-Shading, Shape-from-Texture, Shape-from-Contour, Shape-from-Stereo, Shape-from-Motion—is the recovery of* **history from an asymmetry**; *i.e., an instantiation of the* **Asymmetry Principle**.

The two parts of the Embedding Proposal are determined by the two parts of the Radical Computational Vision Thesis (above). We shall now validate this proposal by going through the five major Shape-from-x problems in turn.

3.8 Shape-from-Shading

When human beings are presented with a black and white photograph of an object, they are usually able to infer the shape of the object without any difficulty. This means that human perception has the remarkable ability to infer shape merely from *variation in brightness*. Variation in brightness is called *shading*; and this ability to recover shape from brightness-variation alone is called *Shape-from-Shading*, and has been investigated, in computational vision, by Berthold Horn and his colleagues (Horn, 1977; Ikeuchi & Horn, 1981). We shall argue that the results of Shape-from-Shading corroborate our view that the perception of any environmental property is always the recovery of process-history from asymmetry.

3.8.1 Image-Formation

In order to show how human perception is able to recover shape from shading, Horn (1977) analyzes the geometry of the environment that allows the formation of the image on the retina, as follows:

As is generally stated in Computational Vision, an image is formed on the retina because two *causal interactions* have taken place:

> *Light has* **interacted** *with an object, and has subsequently* **interacted** with the retinal sensors.

This is illustrated in Fig 3.8. Observe that there are three separate systems:

(1) light,
(2) an object,
(3) the sensors.

These systems are involved in two successive interactions: system 1, the light, interacts with system 2, an object, and then with system 3, the sensors.

Horn analyzes the *geometric arrangement* of the light source, the object and the sensors, in order to understand how the shape of the object can be recovered from the brightness pattern on the image; i.e., how shape can be recovered from shading. However, we shall now argue that the recovery of shape from shading can be understood as the recovery of the *past*; in fact, the recovery of the *sequence of causal interactions* described above; i.e., the interactions of light, first with the object and then with the retina. That is, under our view, the recovery of shape from shading is the recovery of *process-history*; i.e., the recovery of *time*. That is, we replace Horn's

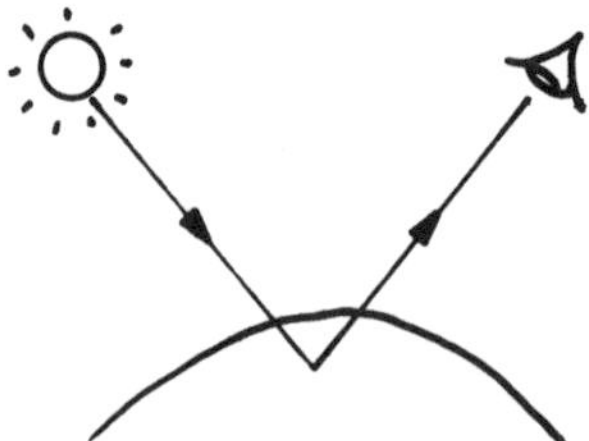

Figure 3.8
For an image to be formed in the eye, light from a source must have interacted first with an environmental surface and then with the retinal sensors.

configural analysis of image-formation with principles that are fundamentally temporal in character. In fact, we shall see that the recovery of shape from shading is based crucially on our Asymmetry Principle.

3.8.2 Orientation and Brightness

In this section, we recapitulate the central argument in the Shape-from-Shading literature. This argument relates the orientation of a surface to its brightness. The *shape* of an object can be regarded as the *variation in orientation* across the surface of the object. This variation in orientation produces *variation in brightness*. As we shall see, the relationship between orientation and brightness is the basis of the human capacity to recover shape, i.e., varying surface orientation, from shading, i.e., varying brightness.

Consider what happens when a flux of light-rays coming from a distant source hits a piece of planar surface of some specific size. If the plane faces the light source directly, as shown in Fig 3.9a, it will capture the maximum possible light rays. The more the plane is turned away from the light source, the fewer light rays will hit the surface, for reasons that can easily be illustrated thus: In Fig 3.9a, we show five rays hitting the surface patch when the latter is facing the source directly. If the patch is rotated a little, as shown in Fig 3.9b, the surface can capture only four of the light rays. This is because the surface patch remains the same size, but is angled obliquely to the light rays. If the patch is rotated still further, as shown in Fig 3.9c, the surface can capture only three light rays; and so on.

Observe that in each case, since there are fewer light rays spread over the same area, the light rays are *less dense* (more widely spaced) on the

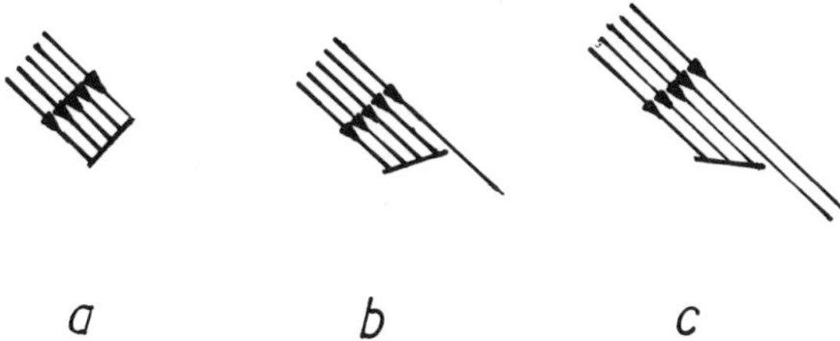

Figure 3.9
The same patch of flat surface at different orientations to a flux of light captures different amounts of light.

Figure 3.10
The different sides of a cube sitting in the sunlight have different levels of brightness.

surface. This means that the same surface patch receives less illumination, and will appear *less bright* to a viewer.

Thus we can see that there is a direct correspondence between *orientation* and *brightness*, because there is a direct correspondence between orientation and density of rays illuminating the surface. Therefore, given a specific planar surface, each orientation of the surface produces a particular brightness. This fact is easily recognized. For example, Fig 3.10 shows a cube sitting in the sunlight. Each face has a different orientation with respect to the light-source. Therefore, each has a different brightness. The more the surface is oriented away from the light-source, the darker it is.

Now consider an arbitrarily curved surface. The same correspondence between orientation and brightness exists. However, because the orientation is varying *continuously* across the surface, the brightness must be changing continuously. Nevertheless, any points with the same orientation have the same brightness. For example, consider the mountain terrain in Fig 3.11a. All points marked *A* have the same orientation. Thus they will have the same brightness as each other. Similarly, consider Fig 3.11b. All points marked *B* have the same orientation. Therefore they will have the same brightness as each other. Furthermore, they will have a different brightness from that of the points marked *A*. For example, if points *A* face directly at the sun, they will all be the brightest points on the mountains. Points *B* will be darker because they are oriented less towards the sun. In short, one will have variation in brightness across the mountains and this variation will be determined by two simple rules: (1) The more a point

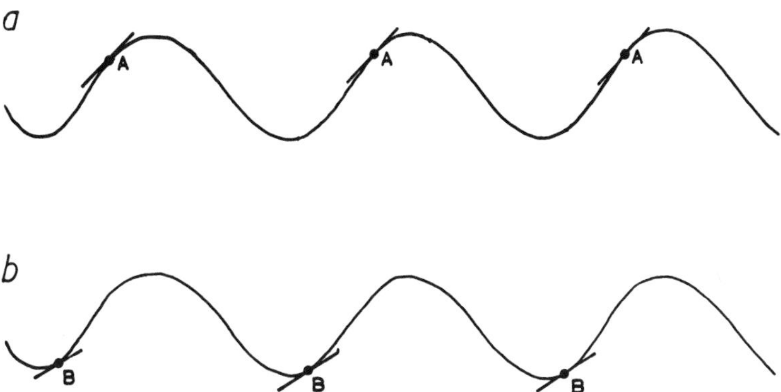

Figure 3.11
(a) All points *A* have the same level of brightness; (b) and so do all points *B*.

faces the sun, the brighter it will be. (2) All points with the same orientation will have the same level of brightness.

These simple rules are the basis of the human ability to recover shape, i.e., varying orientation, from shading, i.e., varying brightness.

3.8.3 Shading as Symmetry-Breaking

The use of the relation between orientation and brightness, in fact, relies on a crucial symmetry assumption: that the density of light rays is homogeneous before striking the surface, in the following way: Consider Fig 3.12a. It shows a flux of light rays, from a distant source, just before they have hit a surface. The crucial assumption is that, if one cuts the rays with a plane, the flux will be uniformly dense across the plane. That is, the density will be *symmetric*, i.e., indistinguishable, with respect to *translations* across the plane.

This assumption of uniformity is needed by an observer in order to be able to recover shape from shading. To understand why, consider what happens to a flux that is uniformly dense, when it hits a surface of varying orientation such as a mountain-range. As we said earlier, each different surface orientation will cut the flux at a different angle and will thus *capture* a different density of rays (recall Figs 3.9a, b, c). Thus, the uniform density of the initial flux will be converted into varying density across the

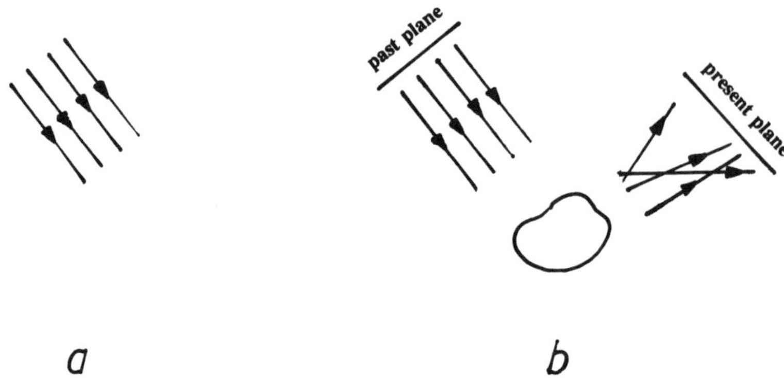

Figure 3.12
(a) A flux of rays before hitting a surface. (b) The visual system assumes that an object converted a past plane of uniform rays into the present plane of non-uniform rays on the retina.

different regions of the surface, because each region is at a different angle to the flux. Therefore, the density of light that is reflected out to the viewer will vary from region to region across the surface.

Consider now the problem faced by the viewer who examines the varying brightness in the image on the retina, and tries to infer the shape of the object out in the environment, i.e., infer the varying orientation of the object surface. In order to infer the environmental shape from the brightness variation across the retina, the viewer needs to make the uniformity assumption we have just described; i.e., the viewer needs to assume that the density of the light rays before hitting the environmental object was uniform. This assumption allows the viewer to attribute the variation in brightness reaching the eye to the variation in orientation (i.e., object shape) rather than to variation in the *initial* light density (i.e., the light prior to hitting the surface). If the light had a non-homogeneous density before hitting the object, then the viewer would not be able know which aspect of the varying density reaching the eye was due to the interaction of the light with the object, and which aspect was due to the prior inhomogeneity. The assumption that the prior light density was uniform therefore allows the viewer to be able to attribute the variation in brightness in the image purely to variation in surface orientation, and thus allows the viewer to infer the shape of the object from the varying pattern of brightness.

The crucial point we wish to make now is this: While the above uniformity assumption is standardly understood as a tool that specifically allows the recovery of shape from shading, it is actually a deep assumption about the nature of *time*. In fact, close examination reveals that it is an example of the use of our *Asymmetry Principle* which states that an asymmetry in the present originated from a symmetry in the past.

In the situation considered here, the asymmetry in the present is the varying light density across the retina, i.e., the image-plane. According to the Asymmetry Principle, this density must once have been homogeneous. This is exactly what is supposed under the above uniformity assumption, as is shown in Fig 3.12b. We can, in fact, imagine two planes cutting the light flux. One plane is the image-plane (retina) which cuts the light flux in the present. On this plane, one finds non-uniform density of the flux lines. The Asymmetry Principle implies that there is a point in time sufficiently far in the past at which, if *another* plane were to cut the light flux, one would find only uniformly spaced light-rays across the plane. Thus, time is understood as converting the uniform density on the past plane into the non-uniform density on the present one.

Observe that not only is the Asymmetry Principle fundamental to Shape-from-Shading, but so is the Second System Principle: According to the latter, the conversion of the past plane of uniform flux into the present plane of non-uniform flux must have been due to a causal interaction with some second system that asymmetrized the flux. This second system is, of course, the environmental shape. Thus according to our view, Shape-from-Shading is the recovery of the structure of a second system that can explain the asymmetrization of the flux-density on a plane.

3.8.4 Further History from Shading

The rule that corresponds the brightness of a surface patch with the orientation of the patch can allow one to recover the orientation (of the patch) *relative* to the light source, but does not allow one to recover the *absolute* orientation (of the patch). To understand this, consider a valley, as shown in Fig 3.13. Let us suppose that the sun is directly overhead, as also shown in the figure. Then each pair of opposite points of the valley will be equally bright; e.g., points *A* and *B*, as shown in Fig 3.13, will be equally bright. This is because the two points each have the same orientation

Figure 3.13
If the sun is directly overhead, points *A* and *B* have the same orientation relative to the incoming light rays.

relative to the incoming light and therefore capture the same amount of light.

Thus the brightness at *A* or *B* corresponds to the orientation at *A* or *B* *relative* to the light source. This means that if one knows the brightness of A or B (e.g., as one does when one sees a black-and-white photograph of the valley) one will be able to infer the orientation of *A* or *B* *relative* to the light source. However, the brightness will not be able to inform one of the *absolute* orientation of *A* or *B*; that is, whether *A* faces to the right or left, and whether *B* faces to the right or left.

However, it is clearly the case that, given a black-and-white photograph of the valley, a human being is easily able to identify the absolute orientation of each local patch such as *A* or *B*; i.e., whether a patch faces the left or right. This means that a human being uses an additional constraint to fully determine the absolute orientation. This additional constraint is easy to understand. Consider a photograph of the valley. The brightness will change smoothly down each side of the valley. The problem we are examining is why a human being does not see some of the orientations on a *single* side of the valley as facing left and others on the *same* side as facing right. The answer appears to be that, in visually hypothesizing a surface, a human being uses the following constraint: *The orientations on the surface should change as little as possible*. This will mean, for example, that all the orientations on one side of the valley will face the same way. The constraint that orientation has as little change, i.e., distinguishability, as possible is called the *smoothness constraint*, and it was proposed by Ikeuchi & Horn (1981).

Thus, the argument is that a human being uses two constraints in pinning down the absolute orientation of a surface patch. The first is the constraint we met earlier, that brightness corresponds to surface orientation (relative to the source). The second is the smoothness constraint. That is, the two constraints are:

BRIGHTNESS CONSTRAINT. There is a correspondence between the brightness of a surface patch and the orientation of the patch relative to the light-source.

SMOOTHNESS CONSTRAINT. The orientations on the surface change as little as possible (Ikeuchi & Horn, 1981).

The brightness constraint allows one to use brightness to pin down the relative orientation of the patch to the light source. The addition of the smoothness constraint allows one to convert this relative orientation into an absolute orientation.

Now let us return to the issue of process-history. We have argued that the first constraint is founded ultimately on the Asymmetry Principle; i.e., the non-homogeneous flux cutting the image plane was a homogeneous flux in the past.

We shall now argue that the second constraint, the smoothness constraint, is also founded on the Asymmetry Principle. Our argument will be elaborated in considerable detail later in this chapter. However, we give the argument here in a succinct form thus: The smoothness constraint requires that the orientational variation in the environmental surface is the smallest possible that will account for the brightness pattern in the retinal image. However, we shall see that, in Standard Computational Vision the smoothness constraint is realized by a particular trick: In order to compute the surface with the least amount of orientational variation, researchers have found that it is convenient to compute the surface to be that which has the least amount of *deforming history*. This trick is regarded merely as a computational convenience. Nevertheless, it is remarkable that, under this trick, orientational variation in the surface is viewed as the result of processes that deformed the surface; i.e., the non-flatness in the surface is the result of processes acting on a surface that was initially flat. In short, orientational distinguishability is *explained* by process-history.

Thus the two constraints correspond to two process-histories. (1) The brightness constraint explains the flux asymmetry on the retinal image as the result of a past uniform flux interacting with an object of varying surface orientation. (2) The smoothness constraint is regarded as explaining the varying surface orientation as the result of a flat surface interacting with deforming forces.

These two histories represent layers of time that are pealed off successively, as shown in the progression from left to right in Fig 3.14. That is,

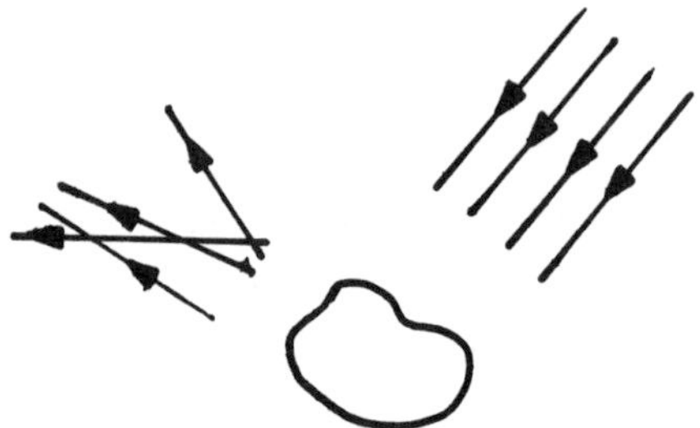

Figure 3.14
Backward in time, not only is asymmetry removed from the flux, but is also removed from the surface.

the two constraints correspond to two uses of the Asymmetry Principle, in order to remove the entire asymmetry backward in time. The first constraint removes the asymmetry from the light flux. The second constraint removes the asymmetry from the object surface.

In the previous section we showed precisely how the use of the brightness constraint in Standard Computational Vision is, in fact, based on an unstated use of the Asymmetry Principle. In order to see more clearly how the smoothness constraint is also ultimately based on an unstated use of the Asymmetry Principle, we need techniques that will be elaborated in section 3.12.

3.9 Shape-from-Texture

3.9.1 Projective Asymmetrization

Recall that an image is formed on the retina because two causal interactions have taken place: Light has interacted with an object and has subsequently interacted with the retina.

In Shape-from-Shading, one considers the first of these causal interactions: the interaction of the light-flux with the object and the resulting asymmetrization. We now consider the second interaction: that created when the light is projected from the object onto the retina.

The process of projection adds an extra layer of asymmetry. This layer is crucial to the recovery of both Shape-from-Texture and Shape-from-Contour. Thus, we will examine this asymmetrization process before examining, individually, these two recovery problems.

The layer of asymmetry, added by the projection process, has two separate components, which we will now discuss in turn.

3.9.2 Compression

Consider a plane consisting of a square grid, as shown in Fig 3.15. Now let us imagine that this grid is slanted with respect to the viewer, and that we are looking at the grid and the viewer from the side, as shown in Fig 3.16a. That is, the dots along the grid, on the right, show the positions of the horizontal grid-lines. The plane on the left represents the image-plane in the viewer.

Now suppose that the grid-lines are projected by parallel rays onto the image plane, as shown in Fig 3.16a. We can see, by comparing Fig 3.16a

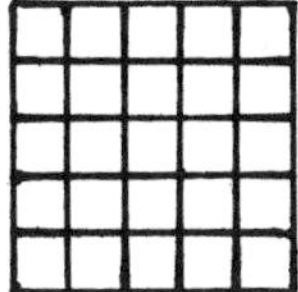

Figure 3.15
A square grid.

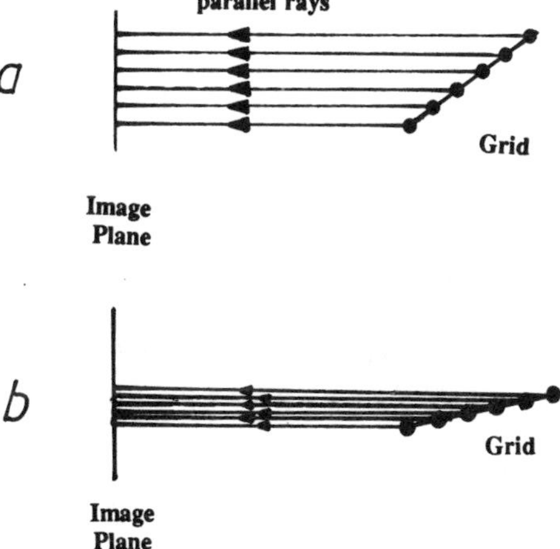

Figure 3.16
(a) A planar grid on the right is slanted with respect to the viewer on the left. (b) The grid is more slanted.

Figure 3.17
The two images resulting from the two situations in Fig 3.16.

with Fig 3.16b, that the greater the slant in the surface, the closer the parallel rays will be. Thus, greater slant causes the horizontal grid-lines to be more compressed together in the image. Fig 3.17a and 17b show how the grid appears in the image (to the viewer), in the two cases shown in Fig 3.16. That is, Fig 3.17a shows that the horizontal grid-lines appear compressed together due to the amount of slant in Fig 3.16a, and Fig 3.17b shows that the horizontal grid-lines appear still more compressed together when the slant is increased, as in Fig 3.16b.

Observe however that the horizontal spacings are not compressed in either Fig 3.17a or Fig 3.17b. This is because there is no slant along the horizontal direction.

We are therefore led to the following conclusion:

> *The slanting of a plane with respect to the retina causes the following* **asymmetry** *in the retinal image: Compression occurs in the direction of slant and no compression occurs in the perpendicular direction.*

3.9.3 Scaling

Fig 3.16a and 3.16b showed the process of projection as it occurs with parallel rays. In fact, in human perception, rays converge because projection takes place via a lens. The compression effect described above still occurs; but an additional asymmetry is introduced due to the converging process.

Consider Fig 3.18a. At the right of the figure, there is a pole sticking vertically out of the ground. The figure shows two rays, one from the top and the other from the bottom of the pole, traveling towards the left and converging there at the lens. The angle between the two rays is given by θ.

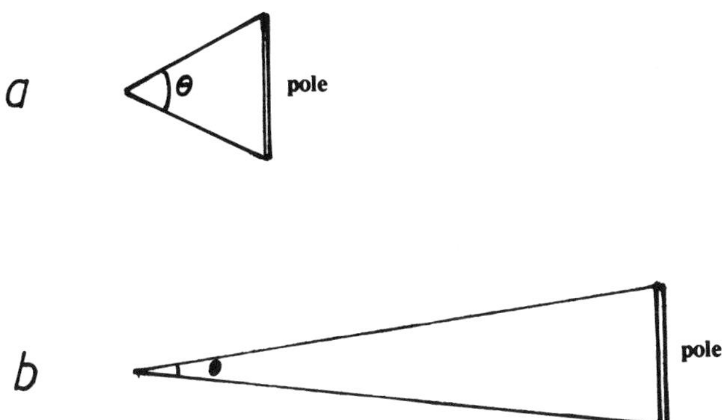

Figure 3.18
(a) A pole subtends some angle at the eye, and (b) a smaller angle when further away.

Now take the pole and stick it in the ground further away from the viewer, as shown in Fig 3.18b. The angle between the two rays will be much smaller. The consequence is that the pole will appear much smaller in the image (on the retina). This size-reduction is usually known as *scaling*.

We are therefore led to the following conclusion:

> *The projection process adds a second* **asymmetry** *to the image: Objects of equal size in the environment become unequal in size, in the image (in a manner that depends on their distance from the viewer).*

It is important to note that, unlike the *compression* effect, considered earlier, the *scaling* effect acts equally in all directions. That is, in the scaling effect, the pole is reduced in size by the same amount, whether it is vertical or horizontal. This is because, when the pole is moved further from the viewer, the angle of convergence is reduced by the same amount, whether the pole is vertical or horizontal.

3.9.4 The Total Asymmetry

The asymmetry created by the projection process is a sum of the asymmetry that constitutes the compression effect and the asymmetry that constitutes the scaling effect.

Figure 3.19
(a) The compression effect alone. (b) The compression effect together with the scaling effect.

To understand the total asymmetry, consider the grid again. Fig 3.19a illustrates, once more, the compressive effect due to the slant, i.e., the horizontal lines are brought closer together. Observe, however, that, because the plane is slanting away from the viewer, the top of the grid is further away than the bottom of the grid. Thus the scaling effect should also take place: That is, the elements at the top of the grid should appear smaller than those at the bottom. The total result is shown in Fig 3.19b. The sense of depth is very strong in Fig 3.19b, showing that the sum of these two asymmetries allows a strong recovery of the environmental structure.

The two asymmetries we have just discussed are the result of the process whereby the structure of a surface is projected onto the retina. These asymmetries are involved in both the recovery of Shape-from-Texture and the recovery of Shape-from-Contour. To understand this, consider again the grid example in Fig 3.19b. The repeated elements comprising the grid produce a *texture*. This texture has undergone both *compression* (in the vertical direction) and *scaling* (which increases towards the top). Alternatively, the boundary of the grid constitutes a *contour*. This contour again has undergone both *compression* (in the vertical direction) and *scaling* (which increases towards the top).

We will now consider Shape-from-Texture, and then consider Shape-from-Contour.

3.9.5 Texture and Uniformity

Texture is generally regarded as the repeated occurrence of some visual primitive over some region. The primitive is simply a collection of properties that occur in this repetitious way. For a square grid, the primitives are the individual square elements; or for the rough surface of a sheet of paper, the primitives are the tiny bumps.

Thus, a texture consists of a two-level structure:

(1) a primitive

(2) a placement structure; i.e., a structure that determines where the primitives are placed.

Each of these has its own organization. Each can be completely regular or statistically regular.[1] For example, at one extreme, a square grid has both a completely regular primitive and a completely regular placement structure in which the primitive repeats. At the other extreme, the rough surface of a sheet of paper has both a statistically regular primitive, and a statistically regular structure in which the primitive repeats. Thus although the grid-texture and the paper-texture are very different, what they have in common is that they have some kind of regularity, i.e., uniformity. It is the uniformity that allows one to see the surface as being a single texture.[2]

Now let us consider the problem of recovering Shape-from-Texture. When an environmental texture is projected onto the retina, the texture becomes compressed and scaled in the manner described in sections 3.9.2 and 3.9.3. The recovery of Shape-from-Texture consists of removing the compression and scaling from the image and returning the texture to its uniform environmental state. Thus, suppose the image on the retina consists of the structure shown in Fig 3.20a. A human being easily assumes that this structure corresponds to a square grid in the environment, i.e., Fig 3.20b. Thus, human perception removes scaling and compression from the lines and shapes in Fig 3.20a, and produces the *uniform* structure of

Figure 3.20
If (a) is the image, then (b) is the assumed environmental structure.

lines and shapes shown in Fig 3.20b. The same applies to the image of the rough sheet of paper that is slanted in depth, let's say, at the same orientation as the square grid in Fig 3.20a. The tiny bumps in the paper become more compressed together, and smaller, towards the top of the image. Again, human perception assumes that such an image corresponds to an environmental sheet in which the bumps are not compressed or scaled; i.e., where they are uniform.

It is important to understand why perception postulates the environmental surface in Fig 3.20a to be a *plane*. The plane is that *shape* which allows the environmental texture to be uniform; i.e., to be the square grid in Fig 3.20b. This accords with the general form of any method for the recovery of Shape-from-Texture. The hypothesized shape is that which allows the uniformity assumption to be fulfilled. This applies not just to simple shapes, such as planes, but to complex shapes, such as the American flag waving in the wind. *In the retinal image*, the texture, i.e., the stripes on the flag, are compressed and scaled at different points because the flag has varying slant and distance across it. However, the viewer assumes that the stripes, on the flag, *in the environment*, are uniform. Thus, the viewer postulates folds and waves in the flag, in the environment, in order to allow the stripes to gain a uniformity, in the environment, that they do not have in the image. Once again, shape is postulated so that the environmental surface has a uniformly repeated texture element.

3.9.6 Shape-from-Texture as the Recovery of Time

We shall now argue that these standard concepts from Computational Vision are, on a deeper level, examples of our theory of process inference.

To proceed: We noted above that, when one is presented with Fig 3.20a, one recovers Fig 3.20b. That is, one assumes that the non-square grid in the image, Fig 3.20a, corresponds to a square grid in the environment. However, this correspondence has actually been created by *time*. The projection process *started* with the square grid and *finished* with the non-square one. This means that, in going from the non-square grid (Fig 3.20a) to the square grid (Fig 3.20b), one is going backward in time; i.e., one is recovering the *past*.

Thus, in making the uniformity assumption, i.e., that the non-square grid arose from the square grid, one is making the assumption *that the past is more symmetric than the present*. Therefore, the uniformity assumption is an example of our *Asymmetry Principle* which states that an asymmetry in the present arose from a symmetry in the past.

Several different methods have been developed, in the literature, for recovering Shape-from-Texture. Examination of each of these methods, however, reveals that each is based on a uniformity assumption; i.e., an assumption that the texture was uniform prior to projection. Each method concentrates on a different structural aspect of the notion of texture and designates that aspect to be uniform prior to the projection process. Each method is therefore an example of the Asymmetry Principle, because each is an assumption that an asymmetry in the present arose from a symmetry in the past.

It is worth listing the several different kinds of symmetry, i.e., uniformity, that are assumed in the several Shape-from-Texture methods. In each case, we shall first state the symmetry assumed to exist in the texture in the environment, i.e., prior to projection, and then state the asymmetry assumed to be added in the projection process. Thus each example constitutes a complete instantiation of our Asymmetry Principle.

(1) Uniform Density (Gibson, 1950). We clearly see Fig 3.21 as a plane receding in depth. Gibson argued that this was because we make the assumption that, prior to projection, each unit area of the surface, contained approximately the same number of texture elements.

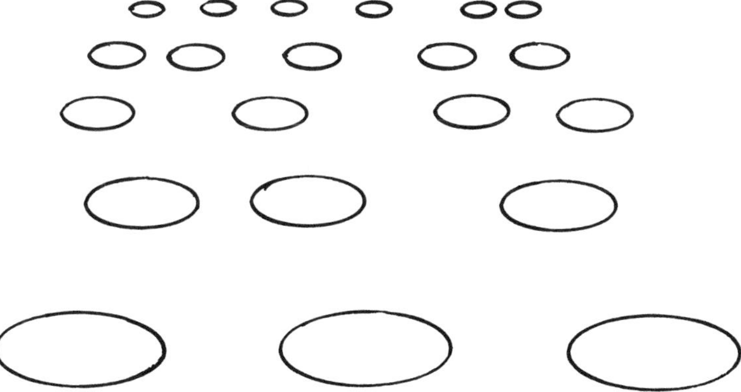

Figure 3.21
Gibson (1950) proposed that change in texture density in the image is interpreted as arising from uniformity in the environment.

> **Asymmetry added in the projection process:** After projection, the densities vary across the image, despite being uniform cross the original surface in the environment. For example, near the top of Fig 3.21, a unit area contains more elements than the same size area does at the bottom of the figure.

(2) Uniform Boundary-Lengths (Aloimonos, 1988). The assumption is that, prior to the process of projection, the boundaries of each texture element are all the same length.

> **Asymmetry added in the projection process:** After projection, the boundary-lengths vary across the image, despite being uniform across the original environmental surface.

(3) Uniform Spatial Frequency (Bajcsy & Lieberman, 1976). A spatial frequency across a region is an oscillation of a single repeated width that exists across the region. The assumption is that, prior to the process of projection, the spatial frequencies across the entire surface are the same.

> **Asymmetry added in the projection process:** After projection, the spatial frequencies vary across the image, despite being uniform across the original environmental surface.

(4) Uniform-Sized Texture Elements (Stevens, 1981). The assumption is that, prior to the process of projection, each texture element has the same size.

> **Asymmetry added in the projection process:** After projection, the sizes of the texture elements vary across the image, despite being uniform across the original environmental surface.

(5) Uniform Spacing (Kender, 1979). The usual assumption, here, is that, prior to the process of projection, the texture elements are arranged along a square grid.

> **Asymmetry added in the projection process:** After projection, the uniform spacing varies across the image, despite being uniform across the original environmental surface.

(6) Uniform Tangent Orientation (Witkin, 1981). Consider an individual texture element. Each point on the boundary of the element has a tangent that has a particular orientation. Witkin's assumption is that, prior to the process of projection, the tangents of each texture element are uniformly distributed across all possible orientations.

> **Asymmetry added in the projection process:** After projection, the distribution of tangents becomes non-uniform despite being uniform across the original environmental surface.

(7) Symmetric Texture Element (Ikeuchi, 1984). The assumption is that, prior to the process of projection, each texture element is symmetric.

> **Asymmetry added in the projection process:** After projection, the texture element becomes asymmetric, despite being symmetric in the original environmental surface.

In conclusion, we observe that, even though each of the above uniformity assumptions concerns a different structural factor, each is nevertheless an instantiation of the *Asymmetry Principle*; that is, each is the assumption that an asymmetry in the present originated from a symmetry in the past.

3.9.7 How Many Uniformity Assumptions?

There has been much controversy in Computational Vision as to which uniformity assumption is correct. Each uniformity assumption is usually introduced by a researcher who also argues that some previous uniformity assumption cannot be valid.

However, we shall now attempt to show that, in fact, there are logically only two possible uniformity assumptions, and the many uniformity assumptions, in the literature, are each examples, of only these two. This, in turn, will let us more fully understand the fundamental role of the Asymmetry Principle in Shape-from-Texture. The two types of uniformity assumption follow from the two types of effects that the projection process has on texture: *compression* and *scaling*. Let us consider these again:

Compression. As illustrated in Fig 3.22, compression causes squashing along one direction and leaves the perpendicular direction unchanged. Because compression alters the length-to-width ratio, it destroys any

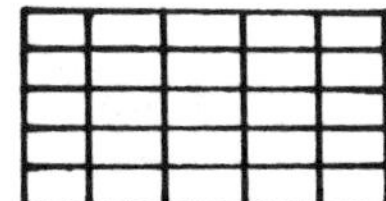

Figure 3.22
The compression effect destroys rotational symmetry.

rotational symmetry in the original environmental surface (recall that rotational symmetry means indistinguishability under rotations).

Thus, given an image, to deduce that it is the result of compression, one has to assume that a rotational symmetry existed in the original environmental texture, and that it had been destroyed in the projection process. That is, the recovery of the compression effect requires a particular *uniformity assumption* about the original surface: *rotational symmetry*.

Scaling. Scaling causes squashing in all directions, by the same amount. Thus its effect can be observed only if it occurs in different amounts at different positions across image. For example, in the grid in Fig 3.23, scaling has occurred in greater amounts towards the top. Thus, to have evidence that scaling took place, one has to translate an element at some position in the image (e.g., at the bottom) to another position (e.g., the top) and see whether it is the same size as the element at the second position. If it is not, then a *translational symmetry* does not exist in the image (recall that translational symmetry means indistinguishability under translation).

Thus, to hypothesize that scaling actually took place in the image, one has to assume that translational symmetry existed prior to the projection process, i.e., in the original surface, and that it had been destroyed in the projection. That is, the recovery of scaling requires a particular *uniformity assumption* about the original surface: *translational symmetry*.

Each of the seven uniformity assumptions listed in the previous section is an example of one of these two assumptions: rotational symmetry or translational symmetry. In fact, the first five listed are examples of translational symmetry, and the last two are examples of rotational symmetry. (Note that the last two are the statistical and deterministic versions, respectively, of the same assumption.)

We can now specify precisely the way in which Shape-from-Texture is ultimately based on our Asymmetry Principle, which states that any asymmetry in the present is understood to have arisen from a symmetry in

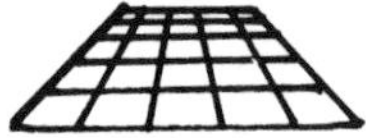

Figure 3.23
The addition of scaling destroys translational symmetry.

the past. The Asymmetry Principle is involved in two fundamental ways, corresponding to the two basic uniformity assumptions: (1) Rotational asymmetry in the image is explained as having arisen from rotational symmetry prior to the projection process; and (2) translational asymmetry in the image is explained as having arisen from translational symmetry prior to the projection process. That is, both asymmetries in the image are understood as *memory*; i.e., they are forms in which *time* has been locked into the present.

3.9.8 Asymmetry in the Environmental Texture

It is often the case that the texture of the *environmental* surface also contains some distortion, i.e. asymmetries, prior to the projection of the surface onto the retina. This means that the resulting image (on the retina) contains two forms of asymmetry: that contained in the original surface texture, and that added by the projection process.

The uniformity assumptions, discussed in the previous section, are therefore more generally stated as the attempt to remove one of these asymmetries, i.e., the projection asymmetry. The assumptions do so by trying to make the environmental texture as uniform as possible and still compatible with the image. Nevertheless, as we said, the environmental texture might still contain residual asymmetry. What is done with this latter asymmetry?

Zucker (1976) has argued that any environmental texture is itself describable as having originated from a completely regular texture—one that has undergone various deforming or randomizing transformations that changed it into its present environmental form. This seems to accord with human perception. For example, the grooves on a warped gramophone record are seen as having once been circular and having undergone the deforming process of melting.

Now observe that Zucker's claim conforms to the Asymmetry Principle. That is, the residual asymmetry in the environmental texture is itself explained as having originated from a past symmetry.

In conclusion, we see that the Asymmetry Principle is used successively twice to go backwards in time: First, given the image, the principle is used to hypothesize a preceding environmental texture from which the image arose by projection; and then, given the environmental texture, the principle is used to hypothesize a completely regular texture, which is still further back in the past, and from which the current environmental texture arose by deforming and randomizing processes.

Finally, observe that these two stages accord with the two-stage structure we gave in Fig 3.3. That is, the stages are, backwards in time, the *image-formation* stage and the *environment-formation* stage.

3.9.9 Externalization of Texture

It should be understood that, given an image, the two successive removals of asymmetry are accomplished via the inference of an *external* history. That is, in the first inference stage, the image texture is conjectured to have arisen from an environmental texture that is outside itself; i.e., not a substructure of the present configuration. Similarly, in the second inference stage, the environmental texture is conjectured to have arisen from a preceding texture that is outside itself; i.e., not a substructure. Observe, in contrast, that it would be possible to use only internal inference; i.e., one could describe the image-texture entirely as a trace (e.g., like the traced-out lines of a TV screen) and to remove its asymmetries by going back along the trace; i.e., inferring a past configuration that is always a *subset* of the present one (as is the case when one undoes a trace). However, human perception does not choose to do this. It chooses instead to describe the image texture as created by external history.

Recall now that there are two basic uniformity assumptions used in removing the first stage backwards in time, i.e., the projection process. One assumption is that, in the removal of compression from the image, one obtains as much *rotational* symmetry as possible; and the other assumption is that, in the removal of scaling from the image, one obtains as much *translational* symmetry as possible. Observe that both rotations and translations are Euclidean transformations.

The important thing to understand now is that these uniformity assumptions conform to our Externalization Principle which states that any external process is hypothesized in order to reduce the internal structure, as much as possible, to a *nested, repetitive, Euclidean hierarchy*. The requirement of a *hierarchy* is fulfilled because any texture has a two-level structure with a primitive on one level and the placement structure on the next level. The requirement of *nesting* is fulfilled because a primitive is embedded at each point of the placement structure. The *Euclidean* requirement is fulfilled because the uniformity assumptions produce rotational and translational symmetry backwards in time. Finally, the *repetitiveness* is fulfilled by the symmetry itself, because all symmetries are decomposed into repetitions, as we shall see later in the book.

The undoing of the second stage backwards in time, i.e., the environment-formation stage, removes any residual deforming or randomizing processes. This again conforms to the Externalization Principle, because the removal of randomization produces a purely smooth structure and the removal of deformation from this smooth structure produces a structure with no metric distortion, i.e., a structure with *Euclidean symmetry*.

In short, we see that, given a texture in the image, one performs two stages of external inference that conform to the Externalization Principle in that they both attempt to gain, as much as possible, an internal structure that is a nested repetitive Euclidean hierarchy.

3.10 Shape-from-Contour

In section 3.9, we saw that the process of projection adds two types of asymmetry to the image: (1) compression, due to the slant in the environment, and (2) scaling, due to varying distance in the environment. The successive addition of these two effects is illustrated in Fig 3.24. We saw also that the compression and scaling effects apply to the *texture* comprising the surface as well as the *edge* of the surface. Thus, the recovery of Shape-from-Texture often involves the same principles as the recovery of Shape-from-Edge; i.e., the removal of compression and scaling. The usual term used for Shape-from-Edge is *Shape-from-Contour*.

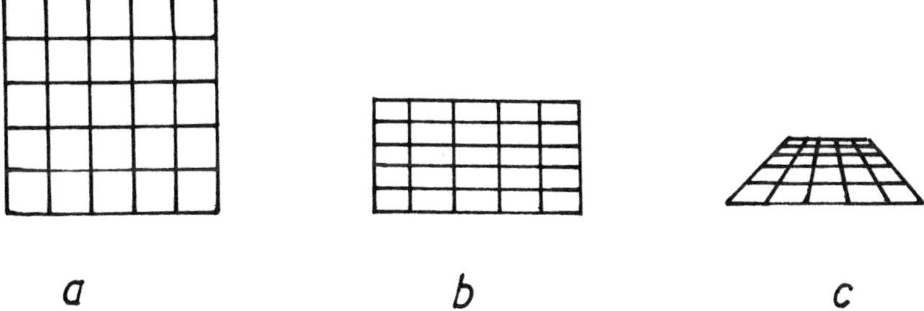

Figure 3.24
(a) An environmental texture. (b & c) The successive addition of compression and scaling respectively.

Figure 3.25
An irregular outline.

The recovery of Shape-from-Contour splits into two alternative cases. (1) The surface is flat (planar); or (2) the surface is curved. In the present section, we will consider the first case. The second case will be considered under our section on interpolation (section 3.12). The recovery of Shape-from-Contour in the planar case usually means the recovery of the *slant* of the plane from its outline. (The term *slant* is closely associated with the term *shape* because an arbitrary shape is an object with varying slant.)

In order to see that Shape-from-Contour is ultimately founded on our principles for the recovery of *time,* we shall consider prominent methods that have been developed for Shape-from-Contour.

We begin by considering one of the Shape-from-Contour methods that was developed also to handle Shape-from-Texture. This method, due to Witkin (1981) is applicable to highly irregular outlines such as Fig 3.25. In the case of Shape-from-Contour, such an outline could be the irregular coastline of an island. In the case of Shape-from-Texture, such an outline could be that of an irregular texture element such as a blotch on a blotchy surface.

Central to Witkin's method is a *uniformity assumption,* which can be understood in the following way. Consider a contour such as a coastline, *before* the projection of the coastline onto the retina has taken place, e.g., Fig 3.25. At each point on the outline, the curve has a tangent which has some direction. Witkin's uniformity assumption is that all the directions are equally likely. In fact, it is this equal likelihood that gives the coastline its irregularity: That is, at each point, the direction of the coastline can change arbitrarily; i.e., with equal likelihood for each direction. Since all directions are equal with respect to probability, the directions are *indistinguishable* with respect to probability. That is, probability is *symmetric* across all directions.

Now suppose that the coastline is drawn on a planar sheet of paper, and that the sheet is at a slant to the viewer. When the coastline is projected

Figure 3.26
(a) Before projection, the tangent arrows occur in several directions. (b) After projection, they align closer to the axis of rotation.

onto the viewer's retina, this uniformity or symmetry is lost. To understand why, consider Fig 3.26. Fig 3.26a shows part of the coastline *before* projection. The tangents arrange themselves in many directions. Now suppose that the plane, on which the coastline is drawn, is rotated around the horizontal, so that it becomes slanted with respect to the viewer. Then, after projection onto the retina, the coastline would look like Fig 3.26b. That is, the usual compression would have occurred in the vertical direction. However, as a result of this, the tangents would now be aligned more towards the horizontal; whereas initially they were equally likely to orient in all directions.

Given an arbitrary coastline in the retinal image, Witkin solves the problem of the recovery of the slant (i.e., recovery of Shape-from-Contour) by developing a method that he explicitly defines as having three parts:

(1) **Geometric Model.** This expresses the relationship between the surface in the environment and the surface in the image; i.e., compression. Witkin does not consider the scaling effect, because he uses *parallel* projection; i.e., the light rays are assumed to be parallel, as illustrated in Fig 3.16a.

(2) **Statistical Model.** This is the above uniformity assumption; i.e., the statement that the tangent directions in the environment are all equally likely.

(3) **Estimator.** This is an equation that estimates the orientation of the environmental plane, given the contour in the retinal image.

We shall now see that Witkin's three-part model is a historical model, and that crucial aspects of the model correspond, in a deep sense, to the principles of process-recovery defined in this book.

(1) **The Geometric Model.** Because the Geometric Model describes the relationship between the environmental surface and the image, it must ultimately involve *time*; i.e., be a model of the process of projecting the surface to the retina.

(2) **The Statistical Model.** This model, which is the uniformity assumption, is a statement about what the environment was like *prior* to the process of projection. Therefore it is an assumption about the *past*. Furthermore, it is a statement that the past was more symmetric than the present. This means that the uniformity assumption is an example of our Asymmetry Principle, which states that an asymmetry in the present grew out of a symmetry in the past.

(3) **The Estimator.** This is a means of going from the image to the environmental configuration that gave rise to the image. Therefore, it is a means of going *backward in time* from the image to the environment. Furthermore, it is a means of doing so in accord with the Asymmetry Principle, i.e., going from the asymmetric image to the symmetric environment.

We can see therefore that the components of Witkin's system, for the recovery of Shape-from-Contour, correspond to essential aspects of our theory of the recovery of time.

Let us now consider another prominent method for the recovery of Shape-from-Contour; that due to Kanade (1981). Kanade proposes the following rule called the Non-Accidentalness Rule:

> NON-ACCIDENTALNESS RULE. Regularities observable in the image are not by accident, but are some projection of real regularities (Kanade, 1981).[3]

We shall soon see that, by "regularities", Kanade means *symmetries*. Thus Kanade's rule states that a *symmetry* in the image arises from a *symmetry* in the environment.

Once again, however, closer examination reveals that this rule must actually be a rule about time. It can be reformulated as saying that, *in going backwards in time*, a symmetry in the image arose from a symmetry in the environment. Thus Kanade's rule is actually an example of our *Symmetry Principle* which states that a symmetry is preserved backwards in time.

One example that Kanade gives of this rule is a rule for parallel lines:

> PARALLEL LINES RULE. If two lines are parallel in the image, they depict parallel lines in the environment (Kanade, 1981, p423).

Since parallelism is simply an example of a type of symmetry (indistinguishability under translation), this rule can easily be seen to be an example of our Symmetry Principle; i.e., translational symmetry is preserved backwards in time.

A more complex rule that Kanade subsumes under the above Non-Accidentalness Rule is worth considering in detail.

Kanade develops a means for handling contours that have "skewed symmetries". Two examples are shown along the top in Fig 3.27. Each has the following property: Points along the contour are equidistant, about some central axis, to points on the other side of the axis. The axis is labeled *S* in each of the figures, as shown along the bottom of Fig 3.27. The crucial aspect of skewed symmetry is that the line, connecting symmetrical points, is not at right angles to the axis; e.g., in each figure, the line labeled *T* connects a pair of symmetrical (equidistant) points *A* and *B* on the contour. In a true reflectional symmetry, the transverse line *T* would be perpendicular to the symmetry axis *S*.

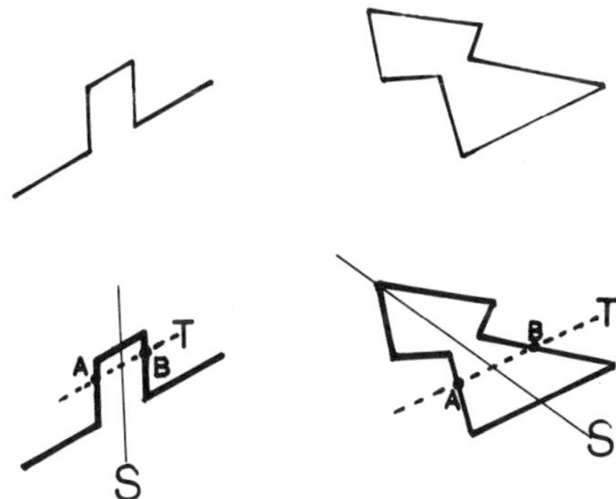

Figure 3.27
Examples of skewed symmetries.

Kanade proposes the following heuristic concerning skewed symmetries in the image:

> SKEWED SYMMETRY RULE. A skewed symmetry depicts a real symmetry viewed from some (unknown) view direction (Kanade, 1981, p424).

That is, in the *image*, a contour that contains a skewed symmetry (e.g., Fig 3.27) corresponds to a contour in the *environment* that has a true reflectional symmetry.

We shall now see that this rule is a consequence of both the Symmetry Principle and the Asymmetry Principle, as follows:

Symmetry Principle. Kanade's Skewed Symmetry Rule implies that a particular symmetry property of the image-contour, i.e., equidistance of points from a central axis, corresponds to the same property in the environmental contour, i.e., equidistance of points from a central axis. This, of course, is an example of our Symmetry Principle, which states that a symmetry in the present is preserved backwards in time.

Asymmetry Principle. Kanade's Skewed Symmetry Rule also implies that the *oblique* transverse line, T (Fig 3.27), of a skewed symmetry in the image, corresponds to a *perpendicular* transverse line, in the environment. A crucial aspect of this correspondence is revealed by considering Fig 3.28a. In the skewed symmetry in Fig 3.28a, the transverse line T makes two different angles, θ and ϕ to the symmetry axis S. However, in the case of a real symmetry, as shown in Fig 3.28b, the transverse line, T, makes

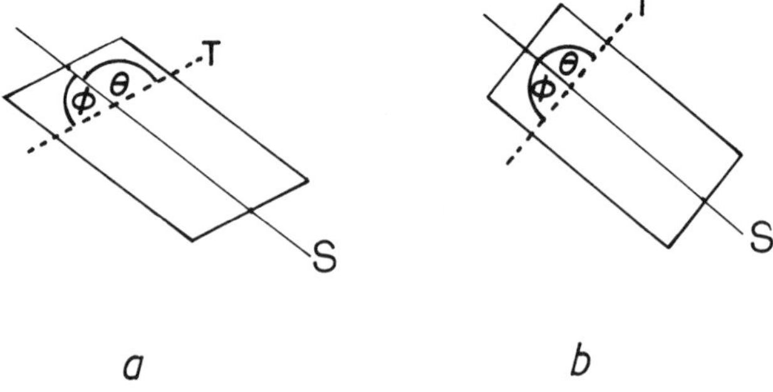

Figure 3.28
(a) A skewed symmetry. (b) The corresponding real symmetry.

only one angle, 90° to the symmetry axis. Thus, in going from a skewed symmetry to a real symmetry, the two *different* angles θ and ϕ become equal. That is, distinguishability is lost between the angles. However, observe now that, in going from the skewed symmetry in the image to the real symmetry in the environment, one is going backward in time. This means that, in going backwards in time, distinguishability between the angles θ and ϕ is lost. This accords with our Asymmetry Principle, which states that an asymmetry in the present originates from symmetry in the past.

Thus Kanade's Skewed Symmetry Rule follows from our Symmetry Principle and Asymmetry Principle, in the following way: It accords with the Symmetry Principle in that the *symmetry content* (equidistance of points) of the skewed symmetry is preserved backwards in time. It accords with the Asymmetry Principle in that the *asymmetry content* (distinguishable angles θ and ϕ) becomes symmetrized backwards in time.

Let us now consider a third method for recovering Shape-from-Contour, due to Brady & Yuille (1983). This method was developed to handle situations not handled by the first two methods we discussed, i.e., those by Witkin (1981) and Kanade (1981). Consider first Witkin's method. Recall that it was designed to deal with irregularly shaped outlines, such as coastlines (e.g., Fig 3.25). In irregular outlines, the tangent can change direction arbitrarily at each point along the outline. In contrast, Brady and Yuille developed a method that can handle situations in which the outline is regular; e.g., ellipses, parallelograms, etc. The importance of such cases can be understood by considering Fig 3.29. As can be seen from this figure, an ellipse is easily interpretable as a circle slanted in depth, and a parallelogram is easily interpretable as a square (or rectangle) slanted in depth.

Figure 3.29
Two regular shapes perceived as slanted in depth.

The ellipse example is worth considering in relation to Kanade's Skewed Symmetry Rule. Recall that the latter rule takes an image contour, in which the transverse axis is at an oblique angle to the symmetry axis, and produces an environmental contour in which the transverse axis is perpendicular to the symmetry axis. In an ellipse, however, the two axes are already perpendicular to each other. Therefore Kanade's rule would simply leave the ellipse as it is. This means that Kanade's rule would not be able to remove the effect of slant from the ellipse; i.e., recover the environmental circle that gave rise to the ellipse in the image. Thus, valuable as Kanade's rule is, it is silent about situations like the ellipse. The method by Brady and Yuille, which we now examine, was designed to handle such situations.

Brady and Yuille base their model on the observation, discussed earlier, that projection causes compression.[4] They observe that compression converts a radially symmetric object, e.g., a circle, into a radially asymmetric object, e.g., an ellipse. For example, under compression, the object that has a length-to-width ratio equal to one gains a length-to-width ratio that is greater than one. This, of course, is an example of our proposal, in section 3.9.7, that the general effect of compression is the destruction of *rotational* symmetry—whether this be radial symmetry in a regular shape, or statistical symmetry in the tangents of an irregular coastline.

Brady and Yuille use their observation as the basis of a rule: When radial asymmetry is seen in an image contour, perception tries to remove the asymmetry by hypothesizing that the environmental contour is radially symmetric and was made asymmetric by projection onto the retinal image. In fact, Brady and Yuille use the following simple but powerful measure to compute the hypothesized environmental contour, and they claim that the latter contour is that which maximizes the measure:

$$\frac{Area}{(Perimeter)^2}$$

That is, when the *environmental* contour is that which maximizes this measure, the *image* contour is explained as having arisen from an environmental contour that is the most symmetrical shape that could have given rise to the image. For example, under this measure, an ellipse in the image is explained by a circle in the environment, a parallelogram in the image is explained by a square in the environment, and a non equal-sided triangle in the image is explained by an equal-sided triangle in the environment.

Observe that, in each case, the resulting environmental shape has greater radial symmetry than the figure in the image.

We can see that the system, devised by Brady and Yuille to compute Shape-from-Contour, is an instantiation of our general theory of process-inference. According to Brady and Yuille, the radial asymmetry in the image is understood as having arisen from radial symmetry in the environment. Their rule can therefore be re-interpreted as a statement about the inference of the *past*. In this way, it conforms to our Asymmetry Principle which states that an asymmetry in the present is understood as having arisen from a symmetry in the past.

Finally, we close this section with the following conclusion: We have surveyed a number of very different methods of recovering Shape-from-Contour. These methods are different because they were designed to handle different situations; e.g., irregular coastlines, regular shapes, etc. What is remarkable, however, is that despite these differences, close examination reveals that the methods all accord with our Asymmetry Principle for the recovery of process-history.

3.11 Shape-from-Stereo

We noted in section 3.2 that there are five major Shape-from-x recovery problems: Shape-from-Shading, Shape-from-Texture, Shape-from-Contour, Shape-from-Stereo, Shape-from-Motion. It turns out that these five problems are divided into two classes:

CLASS I

Shape-from-Shading

Shape-from-Texture

Shape-from-Contour

CLASS II

Shape-from-Motion

Shape-from-Stereo

The members of the first class are characterized by the fact that they each try to recover shape from a *single* image. The members of the second class are characterized by the fact that they each try to recover shape from *two* images. In Shape-from-Stereo, the two images are from the two eyes;

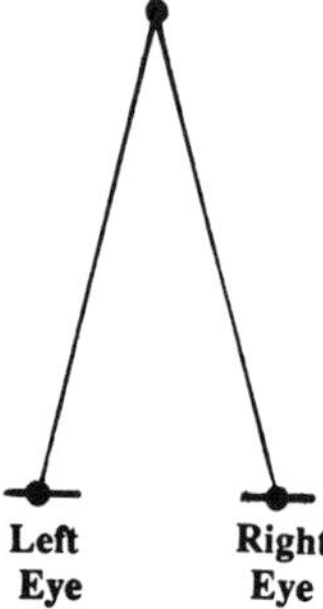

Figure 3.30
Two dots, one in each eye, perceived as arising from one dot in the environment.

i.e., two views separated in *space*. In Shape-from-Motion, the two images are from one eye, but are separated in *time*.

We have now completed our study of Class I, recovery from a single image; and have shown that each of the three methods in this class can be subsumed under our theory of the recovery of time. We now turn to the two methods in Class II, and show that these can also be subsumed under our theory.

Consider, first, recovery from *stereo*. The two eyes each have a view of some object. We will consider, for the moment, the simplest possible object, a dot. Thus the image in the left eye is a single dot and the image in the right eye is a single dot. The assumption made by the perceptual system is that these *two* dots, one in one eye and one in the other, arose from a *single* dot in the environment; as illustrated in Fig 3.30. This means that a distinguishability, two dots, is converted into an indistinguishability, one dot.

What is crucial for us to observe is that the removal of this distinguishability is the recovery of the *past*. That is, the two separate dots were once a single dot in the environment; i.e., the upward direction in Fig 3.30 represents the reversal of time. The conclusion therefore is that recovery from stereo conforms to our Asymmetry Principle which states that a distinguishability in the present, the separate image dots, arose from an indistinguishability in the past, the single environmental dot.

Of course, the same argument applies not just to dots but to complex objects and whole scenes. For example, a scene is represented by two images, one in each eye. The distinguishability between the two image

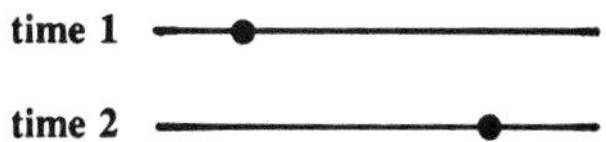

Figure 3.31
Two successive images in the same eye.

scenes is removed backwards in time to create a single scene—in accord with our Asymmetry Principle.

Now let us consider recovery from *motion*. In this case one has different images occurring in the same eye over time. We represent the fact that the images occur in the same eye by putting the images one below the other as shown in Fig 3.31. The downward direction in Fig 3.31 gives the successive images over time.

Let us again consider the simplest possible object, a dot. It has some position at Time 1, as shown in Fig 3.31. If motion takes place, then a different image is produced at a different time, Time 2, and the dot is in another position, as shown in Fig 3.31. In the perception of motion, one assumes that the second dot was prior to the first dot. This inference conforms to the Asymmetry Principle: A distinguishability, the two dots, is explained as having arisen from an indistinguishability, the single dot.

Thus, stereo perception and motion perception are fundamentally similar: The distinguishability between two images is removed backwards in time, in accord with our Asymmetry Principle. For example, two dots are understood as having arisen from one dot.

It is important to observe that the difference between stereo and motion perception is exactly our distinction between *external* and *internal* inference. Consider Fig 3.30 again. As illustrated in this figure, stereo takes two images and infers a past that is *outside* the two images. In fact, the past is literally *out* in the environment. Since the inferred past is not within the images, stereo perception is an example of what we call *external* inference. In contrast, consider Fig 3.31. As illustrated in this figure, motion perception takes two images and infers a past that is one of those images; i.e., one of the dots is assumed to have been an earlier state of the other. This means that the past is *contained within* the set of images. Therefore, motion perception is an example of *internal* inference.

Thus, stereo and motion perception are, respectively, *external* and *internal* versions of the removal of distinguishability between two images, backward in time.

Figure 3.32
A figure interpreted as an undulating shape.

The computation of stereo and motion each conform to several of our other principles of process-inference. To see this for stereo, we must move on to the study of *interpolation*, which we do in section 3.12. Motion will be studied individually in section 3.13.

3.12 Interpolation

When one looks at Fig 3.32, one immediately sees it as representing an *undulating surface*. However, Fig 3.32 gives only a small amount of data to support this percept. All that is given is a single closed line, in the two-dimensional plane of the page. In perceiving the figure, we do at least two things: (1) We see the different points on the line as having different *depths* (distances) from us. For example, the points along side *B* are assumed to be further away from us than the points along side *A*. (2) We see the entire line as the boundary of a single surface and we *interpolate* a particular surface between the boundary.

Observe that an infinity of different surfaces could be interpolated between the boundary. However, the surface that we perceptually choose is a particular one. It is *minimal* in this sense: Although we could choose a surface that has 50 or 100 bumps in it, we perceptually choose a surface that has only the bumps that can be deduced from the outline. Julesz (1959) and Attneave (1982) have argued that the process of choosing a minimal surface is like stretching a film of soap across a wire frame (the boundary) and letting the soap film settle, as soap films do, to the configuration with *minimal deformation*.

The study of surface interpolation is one of the major areas of Standard Computational Vision. The research essentially follows the soap bubble idea, but has examined in detail which types of elastic material—e.g., membranes, thin plates, etc.—best approximate the surface used by human perception. Any such material, stretched over a wire frame, settles to a shape of *minimal deformation*.

We shall now argue that the shape that human beings choose for the interpolating surface is that which contains the least amount of *time*. Quite simply: to choose a surface that has undergone the minimal deformation is to choose the surface that has the minimal process-history. Our conclusion therefore is that *interpolation is primarily concerned with the recapitulation of the past*.

Standard Computational Vision, however, is not concerned with the past. In fact, the usual description for the interpolating surface is that it has the least amount of *variation*, i.e., the least number of "bumps." However, we shall now summarize some of the main arguments in the contemporary research on interpolation and show that, in fact, the concept of *the past* underlies these arguments. We shall also see that several of our other principles are unstated assumptions in this research.

3.12.1 Sources of Depth Data

We observed, in our discussion of Fig 3.32, that the human visual system makes the assumption that the points along side *B* are *deeper* in space (from the viewer) than the points along side *A*. Thus, different depth values are assigned to different points along the boundary. The hypothesized surface must interpolate these depth values. That is, the surface must recede incrementally in depth from side *A* towards side *B*.

This example illustrates the central issue that is examined in the interpolation literature. One has a collection of *edges*. Along these edges, the perceptual system has assigned depth values. The additional task of the perceptual system is to hypothesize a surface that interpolates the depth values. That is, the depth values for edges act as *constraints* for the interpolation process.

Where does the visual system derive the depth values for the edges? One source can be stereo vision. The image of an edge in one eye is compared with the image of an edge in the other eye. The *disparity* between the images gives the depth of the edge from the viewer.

Depth data from stereo vision are necessarily confined to the edges. To see why, consider the perception of a perfectly uniform table. Figs 3.33a and b represent the image in the left and right eye respectively. It is clearly easy to match the edges of the left image with the edges of the right image. However, matches are not easy to make for points *within* the surface of the table; i.e., points not on the edges. For example, pick any point *within*

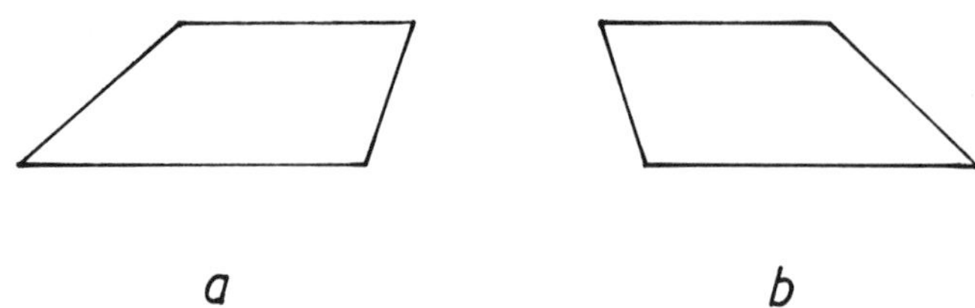

Figure 3.33
The image (a) in the left eye, and (b) in the right eye, of a perfectly uniform table.

the surface of the left image, and try to find the single point, within the right surface, that matches the left point. Clearly, the task is impossible if one does not refer to the edges. This is because, if one did not use the edges, any of the "white" points within the right surface could match the particular point in the left surface.

To summarize: The stereo process can easily obtain depth data at edges. To derive depth data elsewhere one has to *interpolate* the depth data at the edges. Thus, the input to the interpolation process is the depth data at the edges.

What we have said about stereo perception is true also of motion perception. In the latter case, matching is achieved across images over time. For example, Figs 3.33a and b could be two images of the table in the *same* eye at different times: i.e., the table has moved. By the argument given above, matching can be achieved at the edges, but not within the uniform regions. Thus, motion can be computed directly at the edges, but interpolation is needed to compute the motion of the points *within* the surface.

The conclusion is that the input to the interpolation process must be *edge* data; i.e., depth values at edges. This is because perceptual components such as stereo and motion perception can provide data only at the edges.

So far, we have been assuming that the edge data are produced only at the boundary between an object and the rest of the environment. But there are other sources of edge data. Edges are sudden changes of intensity. These occur at essentially three types of places.

(1) A boundary between two objects; e.g., between foreground and background.

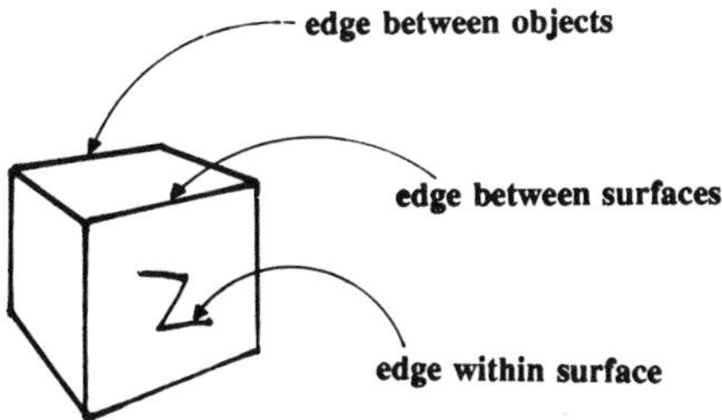

Figure 3.34
The three types of edges.

(2) A boundary between two smooth surfaces of the same object; e.g., creases.

(3) A boundary within a single smooth surface; e.g., scratches, paint marks, etc. We shall generally use the term *surface mark* for this type of edge.

These three types of edges are illustrated in Fig 3.34.

Stereo perception has no difficulty in matching any of these edges across pairs of images. Therefore, the *depth* values for any of these edges can be derived directly. Thus, all three types of edges can act as data to the interpolation process. That is, the hypothesized surface is constrained to interpolate these three kinds of data. Observe, however, that the surface must do different things at each of these three types:

(1) At object boundaries, the surface must stop.

(2) At creases, the object surface must continue but must suddenly change orientation.

(3) At scratches and other surface marks, the surface must continue smoothly through the marks.

These facts are again illustrated in Fig 3.34.

3.12.2 "No News is Good News"

The perceptual system must examine the two-dimensional *image* it has of the environment and infer, from this image, what the structure of the external environment is. At some points in the image, there are sudden changes of intensity; i.e., edges. The perceptual system can conclude that, in the environment, one of the above three types of boundary must exist (e.g., an object boundary, crease, or scratch) and that one of these types of boundary in the environment must have given rise to an edge in the image. Thus, at edges in the image, one can draw certain conclusions about the environmental structure.

What however is it reasonable for the perceptual system to conclude from areas of the two-dimensional image where there are no edges? The definitive answer to this problem was given by Grimson (1981). It is stated as the following principle

> NO-NEWS-IS-GOOD-NEWS PRINCIPLE. In an area of the image with no edges, the best hypothesis is that the corresponding area of the environmental surface has the least amount of variation (undulation). That is, where there is least variation in the image, one should conclude least variation in the environment (Grimson, 1981).

Grimson substantiates this principle by an analysis of the image-formation process; i.e., the process whereby light is reflected from a surface towards the eye. He shows that, under certain acceptable assumptions, the probability of an edge occurring in the two-dimensional image increases with the amount of variation (undulation) in the environmental surface. Therefore, if there is no edge in the image, the most likely environmental surface is one that has the least amount of variation (undulation).

One can regard the No-News-is-Good-News Principle as an example of a standard dictum in science: the Principle of Sufficient Reason. Essentially, the latter principle says that, if there is no bias in some empirical data, then one should assume that the phenomenon that gave rise to the data had a corresponding lack of bias. In other words, where the data is symmetrical, assume that the phenomenon that caused the data is correspondingly symmetrical.

What is important now to realize is that, under the view taken in this book, the Principle of Sufficient Reason is reformulated as a strongly *temporal* rule. In fact, the principle, as we use it, follows from our Symmetry Principle, which states that a symmetry in the present cannot be destroyed *backwards in time*. The temporal factor in the situation we are considering is the movement of light from the environmental object to the two-

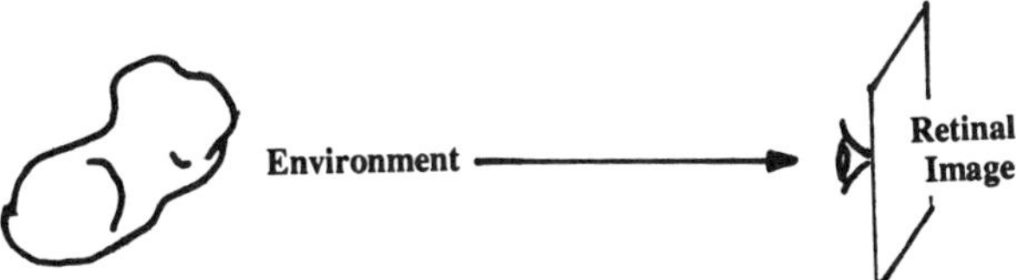

Figure 3.35
The forward direction of time from the environment to the eye.

dimensional image, as shown in Fig 3.35. The direction, environment-to-image, represents forward time. Thus, when one examines an image and tries to infer the corresponding environment, one is trying to move in the reverse direction, image-to-environment, and is therefore trying to move backwards in time. Thus, from the Symmetry Principle, which states that a symmetry in the present must be preserved backwards in time, one must conclude that a symmetry in the image implies a symmetry in the environment.

This means that Grimson's principle, that "no news is good news" (i.e., symmetry in the image means symmetry in the environment), can be regarded as a particular instantiation of our Symmetry Principle.

3.12.3 Energy and Interpolation

Consider the situation as we have discussed it so far: In Standard Computational Vision, it is argued that data from stereo perception or motion perception is necessarily sparse; i.e., confined only to edges such as object boundaries, surface scratches, etc. There are many surfaces that could fit this data. According to Grimson, the most likely surface is one that has the least amount of distinguishability (orientational variation).

Something remarkable now happens in Standard Computational Theory: The discovery is made that the amount of distinguishability in a surface equals the amount of *energy* the surface would have if one considered it to be a thin bent plate. More precisely, consider a thin plate that is perfectly flat when it is in equilibrium with no energy. Now bend the plate so that it conforms to the data points, e.g., so that it conforms to the boundary shown in Fig 3.32. The plate acquires potential energy as it bends because the internal stress that is created, in the plate, forces the plate to try to return to its flattened state; i.e., the plate acquires the potential to move. (Recall that energy is the capacity to create movement.) It turns out that the amount of potential energy acquired by the plate equals the amount of distinguishability in the plate.

The conversion of distinguishability into energy is the dominant metaphor in interpolation research. One should emphasize that researchers regard energy merely as a *metaphor* for distinguishability. For recall that the argument that makes distinguishability the central issue in visual interpolation has nothing to do with energy. That is, the argument is Grimson's which says that, given a set of edges in the two-dimensional image, the most probable environmental surface that created the image is the one that conforms to the edges and has the least amount of distinguishability (e.g., as in Fig 3.32). The reason is this: If the actual environmental surface had more than the minimal distinguishability, more edges would have probably appeared in the image.

Thus, the distinguishability issue has nothing to do with any actual energy.

Why then would interpolation theorists continually use energy as a metaphor for distinguishability? The answer, we argue, is given by our Energy-Asymmetry Principle, which states that, psychologically, asymmetry corresponds to energy. The asymmetry, in the present case, is undulation in the surface; in fact, surface curvature. The thin plate is used as a metaphor because the thin plate is that object whose potential energy is given by curvature, when one assumes that the plate has zero energy in its flat state.

Now, what is fundamentally important is that interpolation theorists regard the thin plate, with curvature, as having been *bent*. Why is this? The answer, we argue, is given by the Asymmetry Principle which states that any distinguishability is understood as the memory of a process that created it. In interpolation theory, the distinguishability is curvature (undulation) in a surface. It appears that interpolation theorists cannot help but see the curvature as having been caused by the process of *bending*. That is, although their theoretical interest is purely to regard the surface as having minimal distinguishability, they cannot help but see this minimal distinguishability as the *result* of process-history.

Furthermore, because interpolation theorists understand the distinguishability as having been produced by a process-history, and because they identify the distinguishability with energy, they are also conforming to our Energy-is-Memory Principle (section 2.17), which states that the energy of a system can be regarded as memory of the causal interactions that transferred the energy to the system. The researchers not only equate curvature with energy but they explicitly describe the energy as having been introduced by bending; i.e., in accord with our principle.

Finally, we note the following: We have seen that, although interpolation theorists are theoretically interested only in the *distinguishability* contained in the surface, and not in the *history* of the surface, the theorists very readily invoke the concept of history when discussing distinguishability. However, because they are not *theoretically* interested in the history, they regard the use of the history concept as a mere *metaphor*; in fact, a tool—one among many alternative tools—to help *compute* the surface. The reason for this is that the current goal of interpolation theory is to recover a "fixed" surface that exists in the present environment; i.e., the goal of *Standard* Computational Theory. In contrast, *Radical* Computational Theory regards the primary goal of vision to be the recovery of causal interaction. Under this latter view, the hypothesizing of a causal description, as in interpolation theory, is not merely an incidental device to help the researcher's computation, but is the fundamental goal of perceptual representation itself.

3.12.4 Edges as Memory

The use of a thin plate model, to interpolate data, involves not only a history of bending, but also histories of other processes, as follows:

Recall first, from section 3.12.1, that there are three types of situation in the environment that can produce edges in the image

(1) A boundary between two objects; e.g., between foreground and background.

(2) A boundary between two smooth surfaces of the same object; e.g., a crease.

(3) A boundary within a single smooth surface; e.g., a scratch or other surface mark.

Recall also that the surface has to do different things at each of these boundaries:

(1) At an object boundary, the surface must stop.

(2) At a crease, the object surface must continue but must suddenly change orientation.

(3) At a scratch or other surface mark, the object surface must continue smoothly through the mark.

In order to properly fit an interpolating surface, one has to identify which edges in the two-dimensional image belong to which type of boundary in the environment, because one has to make the surface do the appropriate things at the different types of boundary. For example, one should make the surface stop, rather than continue, through an object boundary.

Thus, at some level of perception, one needs to have some *method* of identifying what types of boundary one has in the image. The most well-known method, developed by Terzopoulos (1984) and Blake & Zisserman (1987), is the following: Consider an environmental scene in which there are examples of each of the three types of boundaries. Try to fit a single global plate to the entire scene. At each of the different edges, let the plate undergo one of three alternative processes: *fracturing*, *creasing*, or *continuing*. Of these three alternatives, the one that is chosen at an edge is that which leads to the lowest energy in the entire plate. One then defines the three types of edges by the three types of processes thus:

(1) An object boundary is an edge where the plate *fractures*.

(2) A boundary between two smooth surfaces (within the same object) is an edge where the plate *creases*.

(3) A surface mark is an edge where the plate smoothly *continues* under the mark.

It is crucial that we fully understand what is going on here: Interpolation theorists are trying to find a method simply of *identifying* the correct edges; i.e., they are purely interested in establishing *edge detectors*, so that the appropriate types of interpolation can take place between the edges. Any method would be satisfactory, that leads to a correct edge-detection. However, these researchers find that the most successful method is to define the *edges as memories of process-history*. That is, an edge between objects is a memory of *breaking*; an edge between two smooth surfaces (within an object) is a memory of *creasing*; and an edge within a single smooth surface is a memory of *layering*, i.e., marking that was imposed on a surface that smoothly *continues* under the marking.

It is important now to observe that edges are distinguishabilities, i.e., asymmetries (in fact, regions of rapid change or distinguishability). Thus, once again, we see that a major aspect of interpolation theory conforms to the Asymmetry Principle: i.e., distinguishability is explained by process-history.

Interpolation theorists do not use the term *distinguishability*. They, in fact, use the term *energy*. They say that, as the surface is broken or creased, it acquires energy. A greater number of breaks or creases means more energy. Thus more edges mean not only more process-history, but more energy. This, once again, conforms to our Energy-is-Memory Principle, which states that the energy of a system can be regarded as memory of the causal interactions that transferred the energy to the system.

Again, as we said above, more edges mean more distinguishability. Thus the amount of distinguishability corresponds to the amount of energy. Thus, standard interpolation theory conforms also to our Energy-Asymmetry Principle, which states that, psychologically, asymmetry corresponds to energy.

3.12.5 Noise as Memory

There is a final type of distinguishability considered in current interpolation theory: noise. In order to appreciate the relationship between noise and surface interpolation, consider the image of a face on a television screen. Across the image, there are a number of dots ("white snow") that are the result of noise entering the system. The viewer can assign two alternative interpretations to the dots: (1) The dots can be seen as part of the face, and thus make the face rough and blotchy, or (2) the dots can be separated from the face, and thus the face can be seen as a smooth surface with noise added somewhere between the viewer and the face. Clearly, human beings choose the second interpretation. Thus a crucial part of visual interpolation, as carried out by human beings, is the creation of a smooth surface despite noise that would otherwise make the surface rough.

Terzopolous (1984) developed a model of how this could be carried out by the visual system. Consider Fig 3.36a. Let the dots in Fig 3.36a represent the points at which one has data about the environment. If one interpolated a surface through those dots, one would get a surface as shown in Fig 3.36b. However, some of the dots might actually be due to noise

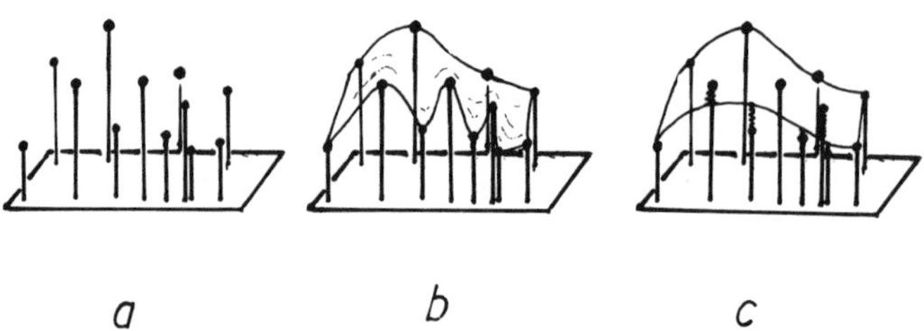

Figure 3.36
(a) The data points, (b) the smoothest surface exactly interpolating the data, and (c) a smoother surface resulting when springs are inserted between the data-points and the surface.

("white snow"). Thus some of the undulation in the surface, in Fig 3.36b, might not correctly be present; it would the result of trying to make the noise part of the surface. In order to produce a more correct surface, one inserts a spring between each data point and the surface. The surface then settles to a smoother shape as shown in Fig 3.36c, and the springs stretch to allow this. One can think of the smooth surface as the face on a television screen. The dots that are far away from the surface in Fig 3.36c constitute noise that has been visually *distinguished* from the face.

Observe, however, that the further a dot is away from the surface, the more it stretches its spring. However, the further a dot is away from the surface, the more *distinguishable* it is from the surface (i.e., greater distance corresponds to greater distinguishability). Thus the above model for noise conforms to our Asymmetry Principle, which states that distinguishability is explained by causal history. In this case, the distinguishability of a dot is explained by the stretching of a spring.

As any spring is stretched, it acquires more energy; i.e., a greater capacity to cause movement. Thus, a dot that is further from the surface not only has greater process-history, i.e., stretch, but it has greater acquired energy. This, once again, conforms to our Energy-is-Memory Principle, which states that the energy of a system can be regarded as memory of the causal interactions that transferred the energy to the system.

Again, as we said above, the further the dot is from the surface, the more distinguishable it is from the surface. Therefore, the amount of distinguishability in the dot corresponds to the amount of energy in the spring. Thus, the above model conforms to our Energy-Asymmetry Principle, which states that, psychologically, asymmetry corresponds to energy.

3.12.6 Five Processes

We have seen that *five* types of distinguishability are central to current interpolation theory. The first, considered in section 3.12.3, is curvature distinguishability. The next three are the three types of edges, as discussed in section 3.12.4. The final type, considered in the previous section, is noise.

These can simply be summed, to give the total distinguishability, thus:

Total distinguishability = curvature distinguishability
+ object-boundary distinguishability
+ surface-boundary distinguishability
+ mark distinguishability
+ noise distinguishability

The five types of distinguishability are illustrated in Fig 3.37.

In accord with the Asymmetry Principle, we have seen that the five distinguishability terms, in the above sum, correspond to five types of process-history, thus:

Total process-history = bending + breaking + creasing
+ marking + noise

The components in the distinguishability sum correspond, component for component, with those in the process sum.

Let us consider the first sum, the distinguishability sum. The technique of summing the distinguishabilities in the above way was proposed and precisely quantified in an elegant analysis by Blake & Zisserman (1987). In fact, they have four of the five components, since the marking compo-

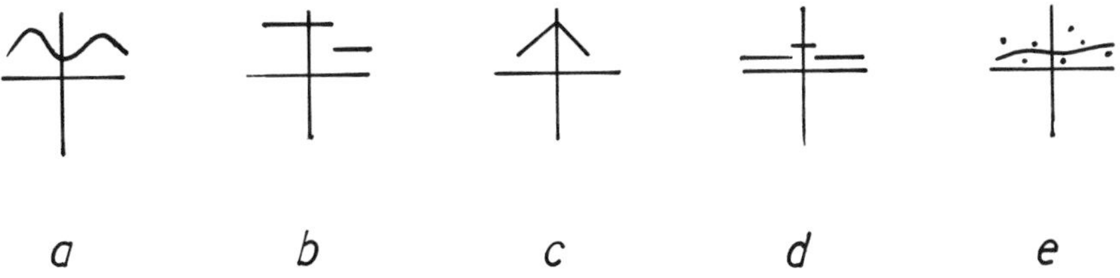

Figure 3.37
The five types of distinguishability in current interpolation theory.

nent, in the above sum, was added by us in the present book. The presence of this process is triggered when there is a sudden change in a factor such as intensity but no accompanying change in depth (distance from the viewer).[5] Note that the lower surface, in Fig 3.37d, represents what psychologists call *constancy*; i.e., what the main surface would be if the process represented by the upper surface, the mark, had not been present. In our system, the constancy is, of course, the past.

One should observe, however, that Blake and Zisserman do not talk about their components as representing *distinguishability*, but instead talk about them as representing *energy*. Nevertheless, examining the four mathematical terms that comprise their sum reveals that what each term actually measures is *distinguishability*. For example, their curvature term is a measure of the how much the gradients along the surface change (i.e., the gradient distinguishability); and their noise term is a measure of the distance (positional distinguishability) between the surface and the noise points. The mathematical reader can see this by closely examining their equation which is presented in Footnote 6 to this chapter.

Once again we argue that the Energy-Asymmetry Principle is responsible for the use of the term *energy* rather than *distinguishability*. Furthermore, the Asymmetry Principle is responsible for understanding the distinguishability as being a consequence of a causal history; i.e., that transferred energy to the surface.

3.12.7 Energy Minimization

Recall the problem that is being examined in interpolation theory. The data in the two-dimensional image is necessarily sparse; i.e., confined to edges. The visual system has to try to infer, from this sparse data, the most likely environmental surface that gave rise to the image (recall Fig 3.32). According to arguments such as that of Grimson (1981), the most likely environmental surface is that with the minimal distinguishability compatible with the data.

Recall now that the distinguishability is decomposed into five components, as given by the distinguishability sum at the beginning of the last section. However, because distinguishability is understood as energy, these five components are translated into five energy components, thus:

$$\begin{aligned}\text{Total energy} = {} & \text{energy due to bending}\\ & + \text{energy due to breaking}\\ & + \text{energy due to creasing}\\ & + \text{energy due to marking}\\ & + \text{energy due to noise}\end{aligned}$$

Interpolation theorists thus state the requirement that the hypothesized surface have minimal distinguishability as the requirement that this total energy be minimized. However, since each energy component represents a different *process-history*, the minimization of the total energy means that the total *process-history* must be minimized. The energy minimization requirement, in current interpolation theory, can thus be interpreted as an example of our History Minimization Principle; i.e., that the conjectured process-history be minimal.

3.12.8 Externalization

In interpolation theory, the interpolated surface is assigned an external history; i.e., the surface is understood as derived from a flat plane and this latter object is not a substructure of the interpolated surface itself. Because the past surface is conjectured to be flat, we see that interpolation theory conforms to our Externalization Principle which states that any external history is hypothesized in order to reduce the internal structure, as much as possible, to a *nested structure of repetitive, Euclidean processes*. The *Euclidean* requirement is fulfilled because the flat plane can be generated purely by translations, one along the x-axis and the other along the y-axis. The *repetitiveness* requirement is fulfilled because both of these translations are merely repeated in their respective directions. Current work by Zucker (1983, 1985) seems to suggest that, in the human visual system, the *nesting* requirement is also fulfilled because translations in one direction are acted on by translations in the other direction, thus creating a two-level structure (see Zucker's notion of Type 1 and Type 2 processes[7]).

3.13 Shape-from-Motion

As was noted before, there are five main Shape-from-x recovery problems in Standard Computational Vision: Shape-from-Shading, Shape-from-

Texture, Shape-from-Contour, Shape-from-Stereo, and Shape-from-Motion. We have placed our discussion of the fifth problem, *motion*, after our discussion of the other four problems because motion perception is the only one, out of the five, that is based on *internal* inference, the recovery of a *trace*. The other four cases are each an example of external inference. That is, in those cases, the viewer is given a single state, an image on the retina, and has to infer a past state that is unavailable—the shape in the environment that caused the retinal image. However, in motion perception, there is crucially another form of inference: the image is a sequence of states from which a *trace* must be compiled.

It is important first to remember the definition of the process-inference problem, the central problem examined in this book. This problem is defined as that of inferring the past, given *only* the present. That is, all that one has available is a single moment in time, a frozen snapshot, and from this one tries to infer the past. We have seen that the process-inference problem has two alternative forms. One of these two forms is internal inference, the inference of a trace or trajectory.

Now, because internal inference is an example of the process-inference problem, all that can be available, in internal inference, is a single frozen moment, the present. Our definition of internal inference is that this single frozen moment is regarded, by the perceiver, as containing records of several states of a process. For example, graffiti contains a record of each of the past states along the trace; i.e., the present *contains* stimuli that can be regarded as past states.

However, this does not, at first, appear to be the case in motion perception. If one takes a photograph of a single moment, one has only a single position of the object; e.g., the object suspended in mid-air. In contrast, a photograph of graffiti contains all the past states. Therefore, because internal inference can have access to only a single snapshot, it appears that internal inference cannot produce motion perception.

Thus, it seems that motion perception must involve a third type of process-inference: one where access to the present alone is not enough. That is, the organism must be able to go into the immediate past and retrieve snapshots from that past.

Adding a third form of process-inference to our list would not be a problem, except for one inescapable fact: *Organisms can experience only the present*. That is, all that is ever available is a single snapshot in time. It is this that makes the process-inference problem—i.e., recovery from

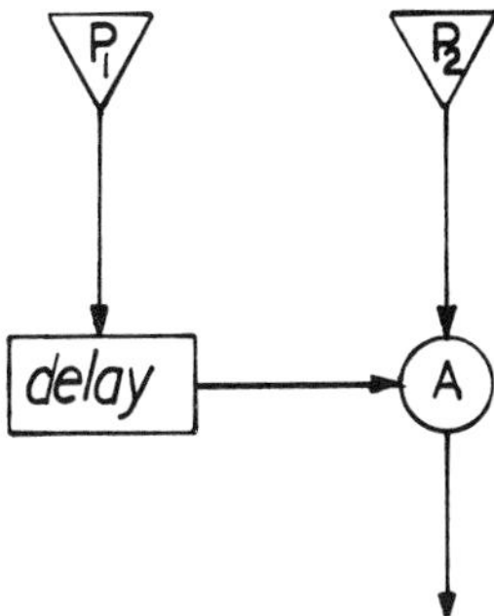

Figure 3.38
The motion detector of Hassenstein and Reichardt (1965; Reichardt, 1961).

a single snapshot—such a general one. One lives only in the present moment. Therefore the past is always inaccessible. The past has always disappeared.

It is because the past has disappeared that motion perception seems to violate our definition of internal inference. Closer examination however reveals that motion is in fact an example of internal inference. Motion perception occurs as a result of the *transfer* of one or more past states into the present. That is, the present must contain a record not only of the present state but also of one or more past states. The transfer of the past into the present is what is generally meant by the term *memory*. The use of memory, in motion perception, is an example of internal inference, as we shall now see.

The fact that motion perception involves memory is most clearly illustrated by examining the motion-detection mechanism that was proposed by Hassenstein & Reichardt (1956; Reichardt, 1961) as a model for detectors in the visual system of the fly. This model is depicted in Fig 3.38. At the top of the figure, two receptors are shown, one at position P_1 and the other at position P_2. The receptors are one unit distance apart. When a stimulus crosses a receptor, at the top of the figure, the receptor gives a response that is sent along the downward fiber from the receptor. Now consider a stimulus traveling from left to right across the two receptors, and suppose the stimulus is traveling at the speed of one unit distance per one unit time. The stimulus first passes P_1 which sends its response downward. However, this response is delayed, for one unit time, by the DELAY cell shown in the figure. Meanwhile, the moving stimulus continues along the top and passes over the second receptor P_2, which then sends its response down its own fiber. But because the response from the first receptor P_1 has been delayed by one unit time, the response from P_1 arrives at the cell labeled A at the same time as the response arrives from receptor P_2. The cell A fires in exactly this situation, i.e., when it receives a response *simultaneously* from both P_1 and P_2. This means that the cell A possesses, within the present moment, the present response of receptor P_2, and the past response of receptor P_1. That is, the past of receptor P_1 has been *transferred* into the present by the DELAY cell, and the present therefore contains two states of the movement process.

In conclusion, we can see that motion perception is made possible by giving the present a multi-state interpretation, in strict accordance with our definition of internal inference.

3.13.1 Two Motion Problems

In Computational Vision, the perception of motion is usually regarded as involving two problems:

(1) *The Two-Dimensional Problem.* The changing pattern of light *on* the retina must be organized into the movement of light *across* the retina. This movement of light is a two-dimensional movement in the plane of the image.

(2) *The Three-Dimensional Problem.* From the two-dimensional movement of light across the retina, the perceptual system must infer the structure of the three-dimensional object, out in the environment, that is creating the retinal movement. This second problem is usually called *Structure-from-Motion*, and will be referred to here as *Shape-from-Motion*, to give a uniformity of terminology.

These two problems will be discussed successively in the next two sections.

3.13.2 The Two-Dimensional Problem

In order for motion to be perceived, the changing pattern of light in the image must be *organized* into motion. However, a changing pattern of light can be given many different organizations, each corresponding to a different motion; i.e., the pattern is highly ambiguous. This ambiguity has at least two forms: (1) *Correspondence ambiguity*: Usually, the points in one image can each be corresponded to many alternative points in the next image. (2) *Intervening-path ambiguity*: There are an infinite number of positions that a point can take between its positions in two successive images.

Both of these ambiguity problems are solved, in the research literature, by invoking a minimization rule. In the case of correspondence ambiguity, the correspondence chosen is that which minimizes the amount of motion involved (e.g., Ullman, 1979). In the case of intervening-path ambiguity, the path that is chosen is, again, that which minimizes the amount of motion involved (e.g., Carlton & Shepard, 1990a, 1990b; see also Shepard & Cooper, 1982).

Clearly, minimizing the amount of motion means minimizing the amount of *history*. Thus we can see, quite directly, that motion perception conforms to the History Minimization Principle (section 2.14).

3.13.3 The Three-Dimensional Problem

Even if the changing intensity on the retina were organized into a two-dimensional flow of motion across the retina, an important problem would still remain: one would still have to recover, from this flow, a three-dimensional object in the environment. As we said, this latter recovery problem, usually referred to as *Structure-from-Motion*, will be called here *Shape-from-Motion.*

When a three-dimensional object moves in the environment, e.g., rotates, the two-dimensional image that it produces is often one of complex deformation. For example, when a head rotates in the environment, its two-dimensional silhouette in the image deforms such that different regions of the outline fluctuate in and out at different rates. It is remarkable that, from this flat deforming image, the perceptual system can recover a unique three-dimensional object.

The remarkable nature of this perceptual recovery was emphasized in a classical series of experiments by Wallach & O'Connell (1953), where an unfamiliar object was rotated behind a screen, and a shadow was cast on the screen by a distant light source. An observer, in front of the screen, was allowed to view only the flat deforming shadow on the screen. Nevertheless the observer automatically recovered the shape of the three-dimensional object that was casting the shadow.

What is important to recognize about this recovery is that there are many different three-dimensional objects, changing in many different ways, that could produce the same image. Nevertheless, for each image, the observers consistently choose a single three-dimensional interpretation.

How is this single interpretation chosen? The answer given by Wallach & O'Connell was that the perceptual system chooses the interpretation in which the object is moving the most rigidly; that is, although the distances between the points on the two-dimensional image might be changing to quite an extent, the distances between the corresponding points on the original three-dimensional shape are changing as little as possible. For example, the complex *deforming* image produced by a rotating head is converted into a three-dimensional head having a *fixed* shape in the environment. This rule of maximizing rigidity in the hypothesized environmental object has been investigated and thoroughly formalized by Ullman (1979). In fact, in Ullman (1984), the three-dimensional object is computed by considering the three-dimensional points to be connected by springs that try to prevent the points from changing their distance from each other.

It is important now to recognize that the maximizing-rigidity constraint conforms to our Asymmetry Principle, and does so in an interesting way. Up till now, we have seen that motion generally conforms to our Asymmetry Principle in that different positions of an object are explained as having arisen temporally from a single position, i.e., the different positions are understood as a trace starting from that initial position. Thus, we have so far used the Asymmetry Principle to make *internal* inference.

Now, the maximizing-rigidity constraint also conforms to the Asymmetry Principle, but does so via *external* inference. In the two-dimensional image, the distances between points change greatly over time, i.e., the image is deforming. However, under the maximizing-rigidity constraint, the corresponding points on the three-dimensional object, in the environment, change distance as little as possible over time. That is, distinguishability in the image is converted into indistinguishability in the environment. However, in moving from the image to the environment one is going backwards in time. Therefore distinguishability is being removed backwards in time. Thus the maximizing-rigidity constraint conforms to our Asymmetry Principle.

This means that, in the recovery of environmental shape from motion in the image, the Asymmetry Principle is being used in two ways: *internally*, in constructing a motion trace across the image, and *externally*, in taking the image trace and removing distinguishability from the trace itself, by moving the trace backward in time until it becomes the trace of the object in the environment.

3.14 Summary

The purpose of this chapter was to show that all vision is the recovery of the past. That is, rather than understanding the visual system as recovering the three-dimensional environment, the cohesive objects that constitute the environment, the spatial layout of the environment, etc., we have argued that vision serves only the single function of unlocking *time* from the image.

According to our view, the image on the retina is *memory* of past process history. This memory is in the form of asymmetries. In accord with the Asymmetry Principle, the visual system recovers the process-history by undoing these asymmetries backwards in time. The history is divided into three overall stages, corresponding to three layers of asymmetry in the retinal image. In accord with the Second System Principle, each layer of asymmetry is created by a causal interaction with an additional system. Going forward in time, the successive layers are: (1) object asymmetries, i.e., shapes, created by objects interacting, e.g., rock against rock, feet on snow, pen across paper, etc.; (2) illumination asymmetries, created by the interaction of the object shapes with an additional system, the light rays; (3) viewpoint asymmetries, created when the light rays, reflected from the objects, interact with an additional system, the viewer's retina.

Based on this approach, we can then embed Standard Computational Vision (the study of the recovery of static surface structure) into Radical Computational Vision (the study of the recovery of time), as follows: We observe that each of the five major image cues of Standard Computational Vision—shading, texture, contour, stereo, motion—is a type of *asymmetry*. We then claim that each of the five major recovery procedures—Shape-from-Shading, Shape-from-Texture, Shape-from-Contour, Shape-from-Stereo, Shape-from-Motion—is the recovery of *history from an asymmetry*; *i.e., an instantiation of the Asymmetry Principle.*

These claims were substantiated by examining each of the five Shape-from-x procedures in detail. (1) We found that Shape-from-Shading is the recovery of the structure of a second system (in fact, shape) that caused an asymmetrization in a translationally symmetric flux of light. (2) We found that, although each of the many procedures used to recover Shape-from-Texture rests on a different uniformity assumption, each assumption is again an instantiation of the Asymmetry Principle. In fact, the assumption is always the requirement that rotational or translational asymmetry in the image originated from rotational or translational symmetry in the environment. (3) We found that the Shape-from-Contour procedures rest on the same assumptions. Furthermore, we found that Kanade's Non-Accidentalness Rule, that symmetries in the image originated from symmetries in the environment, is an example of our Symmetry Principle, which states that a symmetry in the present is preserved backwards in time. (4 & 5) Again, in turning to stereo perception and motion perception, we found that each are examples of our Asymmetry Principle: Both remove the distinguishability between two images backwards in times. Stereo and motion perception are, respectively, *external* and *internal* versions of this removal.

In addition to examining the research on the Shape-from-x problems, we examined the research on surface interpolation. We found that, in computing the interpolating surface, researchers again presuppose the Asymmetry Principle, because the distinguishability in the surface is assumed to have arisen from a surface without that distinguishability.

4 Representation Is Explanation

4.1 The Representation-is-Explanation Principle

In this book, much of the discussion explicitly centers on perception. However, we shall, over the next few sections, propose that fundamental aspects of our view of perception generalize to all cognition, and we will require this generalization a number of times throughout the book. Our generalization proceeds as follows:

In the previous chapter, we have shown in considerable detail that visual perception is the recovery of causal history from the stimulus on the retina. That is, the visual system creates a representation of the retinal stimulus in terms of the history that produced the stimulus.

What we have said of visual perception can also be said of auditory perception, i.e., hearing.

To understand this, note first that, although an auditory stimulus occurs at the ear, one does not hear it as occurring there. Instead, one hears it as occurring out in the three-dimensional environment; e.g., one hears a twig snap some way off, in the forest. Thus, the situation with respect to hearing exactly corresponds to that of seeing, as follows: One does not see the visual stimuli that are on the retina; one sees instead *the three-dimensional environment that has been recovered* from the retinal stimuli. Similarly, one does not hear the sound stimuli that are at the ear; one hears instead *the three-dimensional environment that has been recovered* from the auditory stimuli. Indeed, Wildes & Richards (1988) consider hearing as allowing one to recover the material properties of objects *in the environment*; e.g., whether an object is made of wood or metal, whether it is solid or hollow, etc.

Now, in the previous chapter, we argued that visual perception is not, in fact, the recovery of the static three-dimensional structure of the environment. Rather, it is the recovery of *causal history*, i.e., causal interactions in the past of the retinal stimuli. Similarly, we now propose that auditory perception is not the recovery of the static three-dimensional structure of the environment. It too is the recovery of causal history, i.e., causal interactions in the past of the stimuli at the ear. For example, when one hears a twig snapping, some way off in the forest, one hears an event that has happened some time before the ear received the sound waves. Furthermore, what one hears is a *causal interaction*; i.e., a twig being broken.

It can therefore be concluded that both, the visual and auditory systems, represent the stimuli they receive in terms of the causal history of those stimuli.

We wish now to argue that what we have said about the representation of visual and auditory stimuli is true for any representation formed by the brain—e.g., any representation formed in thinking. That is, we are going to propose that our claims are true not just for *perception*, but for all of *cognition*.

What is cognition? Generally, cognition is regarded as the set of *knowledge* or *informational* functions of the brain. These functions include perception, categorization, language, planning, problem solving, etc. Cognitive activity is regarded as being carried out via *computations*. A computation proceeds by the manipulation of the contents of memory stores. Thus, given the centrality of memory to computation, we must try to understand what memory is. Our proposal concerning the nature of memory consists of two parts:

(1) We propose that memory is some physical object, in the present, that some observer interprets as holding information about the past. The object can exist in the external environment of the observer; e.g., it can be a squashed beer-can; or it can exist within the observer; e.g., it can be a piece of neuronal material. In either case, the interpretation of the object as memory belongs to the observer.

(2) We also propose that, when an object is memory, the past, about which the object is holding information, is the past *of the object itself*. In fact, an object becomes memory for an observer when the observer assigns to the object a history that causally explains how certain features of the object came about.

Therefore, our proposal concerning memory can be stated thus:

DEFINITION OF MEMORY. *A memory is a physical object (1) that is in the present of some observer and (2) that has features that the observer causally explains.*

Let us first check that this proposal underlies the phenomenon that people intuitively call memory. Consider the memory one has of a meeting that one attended last night. This memory must exist in one's head *in the present*. Otherwise, one could not access it. In the present, it is stored as a state of part of one's neuronal material. That is, it is a physical object with a certain set of features. These features were *caused* by the meeting last night. Thus, in order to remember that meeting, it is necessary for some observer, another piece of neuronal material in the brain, to *recover* the causal history of those features.

It is important now to note that the above definition of memory implies that a *stimulus* can be memory, thus: For the observer, a stimulus is a state of a receptor. This means that the stimulus is a physical *object*, the receptor, at the moment of stimulation. This allows the stimulus to become memory in the strict sense defined above; that is, it is a physical object that exists in the present of the observer, i.e., at the moment of stimulation, and that has features that the observer attempts to causally explain, e.g., as the visual system attempts to explains a retinal stimulus.

With these concepts, it is now possible to return to the topic of computation. A computation proceeds by the reading and writing of memory. Both, the reading and writing of memory, are the determination of the past. The reading of memory determines the past by inferring it from some object. The writing of memory alters an object in order to determine the past that will be inferred from it. In either case, the basis for the definition of the object as memory is the *reading* relationship, i.e., the inference of the past from the object. We will therefore concentrate exclusively on the reading relationship, since this relationship is co-extensive with the phenomenon of memory.

We argue that the reading relationship is given by the above definition. That is, the relationship involves two objects, one of which, the observer, defines the other as memory. In conventional discussions of computational systems, the two objects are often called "processors". Thus, consider some processor A that receives a stimulus S coming from another processor B, informing processor A of the memory content of B; for example, the stimulus informs A that a certain number is stored in B. Then, A has the role of the observer in the above definition of memory. When A receives stimulus S, the stimulus contains memory of a history divided into two phases, backward in time: (1) the transmission of energy from processor B to processor A; i.e., the transmission of energy that became stimulus S; (2) the history of processor B prior to the transmission, i.e., the formation of B as memory, e.g., the number stored in B is memory of the causal process that resulted in that number being entered into B.

It is important now to observe that the situation has exactly the same structure as that we described for visual perception, as follows. In this case, the visual system receives a stimulus from the three-dimensional environment. As we saw in Chapter 3, the stimulus arriving at the retina is memory of a history that is also divided into two phases. Going backward in time they are, respectively, (1) the image-formation phase, i.e., the causal process by which the light left the environment and arrived at the retina; (2) the prior environment-formation phase; i.e., the causal processes that formed the environment itself.

In both the case of two processors and the case of visual perception, we have an observer (processor *A* or the visual system) which receives a stimulus that is memory of a history that divides into two phases: (1) the movement of energy from some object (processor *B* or the external environment) to the observer, and (2) the formation of those aspects of the object that will be interpreted as memory.

Two important points emerge from this:

(1) In a computation, the relationship of a processor *A* to another processor *B*, from which *A* receives information, can be regarded as a perceptual one in this sense: (a) processor *B* is in the external environment of processor *A*, and (b) perception is, as we argue, the definition of the external environment as memory.

(2) Since perception defines the external environment as memory, the external environment is a memory component of the computational system defined by the mind, and should therefore be considered to be part of the mind. That is, the external environment has a computational role within the mind that is indistinguishable from the computational role of any part of the brain.

Let us now consider again some of the central components of our argument. We have said that cognition necessarily involves the interaction of objects as memory, and that a cognitive system can therefore include both internal objects such as pieces of neuronal material and external objects that are made memory by perception. This means that any object *X* can gain a cognitive role by virtue of being memory for some other component—which thereby becomes an example of what we call an *observer*. Furthermore, the communication from object *X* to the observer takes place via a stimulus which the observer converts into memory, i.e., memory of the object *X* from which the stimulus emanated.

We now propose that any *cognitive representation* is the result of the type of situation just described, thus: A cognitive representation is a causal history hypothesized to explain a stimulus. Note that, by defining the stimulus as memory, the observer can define any object, from which the stimulus arrives, as memory, and can thereby bring that other object within the cognitive system.

We are thus led to one of the central proposals of this book:

REPRESENTATION-IS-EXPLANATION PRINCIPLE. *Any cognitive representation is the description of a stimulus as a state (or set of states) in a history that causally explains the stimulus to the organism.*

That is, according to this principle, a stimulus has no cognitive existence independent of its causal explanation. Any role that the stimulus has in the cognitive system is by virtue of its causal explanation.

A crucial conclusion that follows from this proposal is that the notion of cognition is founded on the notion of *causality*. That is, the *content* of cognition is causality. In fact, the Representation-is-Explanation Principle can be regarded as *characterizing* cognition in this way:

COGNITIVE CHARACTERIZATION PRINCIPLE. *A cognitive system is one that creates and manipulates causal explanations.*

That is, given a system, what makes it a cognitive system is that it can be described, on some level, as a system in which the medium that it processes consists of causal explanations. We shall regard the Representation-is-Explanation Principle and the Cognitive Characterization Principle as making the same claim.

We began this book by observing that the present is the only moment with which one can have contact. This simple observation can be regarded as a key to the purpose of cognition. One can argue that cognition exists because of a single inviolable constraint: It is impossible to have contact with two moments in time simultaneously. Causal explanation is the means by which it becomes possible to define moments other than the present, while remaining imprisoned within the present. This issue will be examined more fully in section 4.8.

4.2 The Well-Definedness Principle

Most of this chapter will be devoted to developing a *structural characterization* of the Representation-is-Explanation Principle. We shall first present the characterization in a rough form, and then present it rigorously from section 4.6 onwards.

The assignment of a history is not simply the assignment of a starting and a finishing state, but the assignment of *each step* of the intervening history. A description of a process where each step of the process is well-defined is generally called an *effective procedure*. That is, an effective procedure defines, at each moment in time, exactly what is to happen next. Thus we will assume the following to hold for cognition:

WELL-DEFINEDNESS PRINCIPLE. *The histories defined in cognitive representations are well-defined, i.e., they are effective procedures.*

This proposal goes against currently held beliefs. For example, Pentland (1983) argues that human perception models parts of the visual environment in terms of statistically-generated processes that have the structure of fractals. We would argue that, while indeed the environment itself can be modeled in this way, human cognition nevertheless structures it in a completely well-defined way. Again, in music, the composer Xenakis produces compositions that have the structure of statistically generated processes like those of gases. The implication is that the human ear is able to hear the music as such. However, according to the above principle, this could not be the case.

4.3 The Representational Machine Analogy

The next stage in our initially rough characterization of the Representation-is-Explanation Principle is as follows:

The Representation-is-Explanation Principle states that any cognitive representation is a causal explanation. A causal explanation is the specification of a relationship between a *cause* and an *effect*. If the relationship is *well-defined*, the cause must completely determine the effect.

There is no concept that more clearly embodies the notion of well-defined causal interaction than *machine*. A machine is an object that behaves completely mechanically, i.e., without choice or creativity. Thus, if one applies an operation to a machine, its response is completely determined by this applied operation and the machine's state at the time. Therefore, one can regard the application of the operation to the machine's state as the *cause* and the machine's response as the *effect*. Thus the machine can be regarded as a causal relationship. The relationship is *well-defined* in the sense that, given the cause, one can predict the effect.

There is an important consequence of the well-defined nature of the causal interaction embodied in a machine. Suppose one designed a *program* (i.e., an ordered sequence of operations) to be applied to the machine. Suppose further that one had some object that could systematically apply the program to the machine. Then, if one started the machine at the beginning of the program, one could leave the machine, and it would undergo the succession of effects determined by the program sequence. Because the machine would be purely mechanistic, it would undergo a well-defined history, i.e., a history where, at each moment in time, the next step would be completely determined.

A close relationship between well-defined histories and machines was proposed by Turing (1936), and is called Turing's Thesis. Turing proposed that any history that is well-defined, i.e., described by an effective procedure, can be "imitated" by some machine with a program on a tape. In what follows, however, the term *program* will mean a sequence of environmental inputs, and not specifically a tape of symbolic inputs.[1]

Let us now consider the discussion so far in this chapter. The discussion has involved three proposals:

(1) Cognition represents each stimulus as a state, or set of states, in a causal history (Representation-is-Explanation Principle).

(2) Cognition makes such histories well defined (Well-Definedness Principle).

(3) Any well-defined history is structurally equivalent to some programmed machine.

These three proposals lead us to advance the following principle, which we will call the *Representational Machine Analogy*.

REPRESENTATIONAL MACHINE ANALOGY. *Any cognitive representation of a stimulus is the description of the stimulus as a state (or set of states) of a programmed machine. That is, cognition attempts to define any environment as a collection of machines.*[2]

Note that the machine might be represented as a component of a larger machine. The larger machine is a causal structure in the following way. It can be decomposed into a configuration of components, e.g., just as a clock mechanism can be decomposed into parts. Each component is itself a machine and is thus a well-defined relationship between causes and effects. Furthermore, the components are causally connected such that they affect each other in a specific set of well-defined ways, like the parts of a clock mechanism. This ensures that the configuration of components defines an overall machine, i.e., a specific overall relation between causes and effects.

The Representational Machine Analogy is a central principle of this book. It helps significantly toward *structurally* understanding the histories that are recovered in representing the environment as memory. The histories are machines and a structural understanding of machines is currently available for use. In fact, four of the main principles of this book, the Asymmetry Principle, the Symmetry Principle, the History Minimization Principle, and the Representational Machine Analogy are crucial because *together* they allow us to fully understand the structure of the recovered histories.

4.4 Two Machine Analogies

We have just proposed what we call the Representational Machine Analogy. Current cognitive psychology is founded upon a different machine analogy, which we can call the *Processing Machine Analogy*:

PROCESSING MACHINE ANALOGY. *A cognitive system can be represented as a machine.*

The two machine analogies, the Processing Machine Analogy and the Representational Machine Analogy, can be contrasted in the following way: Consider the following three objects: (1) a cognitive system, (2) a machine, and (3) a stimulus set. The two analogies define respectively two different relationships between the objects, as illustrated in Figure 4.1. In the upper diagram, the Processing Machine Analogy has been represented. Here, the cognitive system is regarded as equivalent to a machine; this is represented by the "=" sign between the machine (top left corner) and the cognitive system represented by the human head. An environmental stimulus set, the set of dots on the right, is input into the head just as if it were input into a machine. The head processes the stimuli in a way that is analogous to the action of a machine. Figure 4.1b, however, illustrates the Representa-

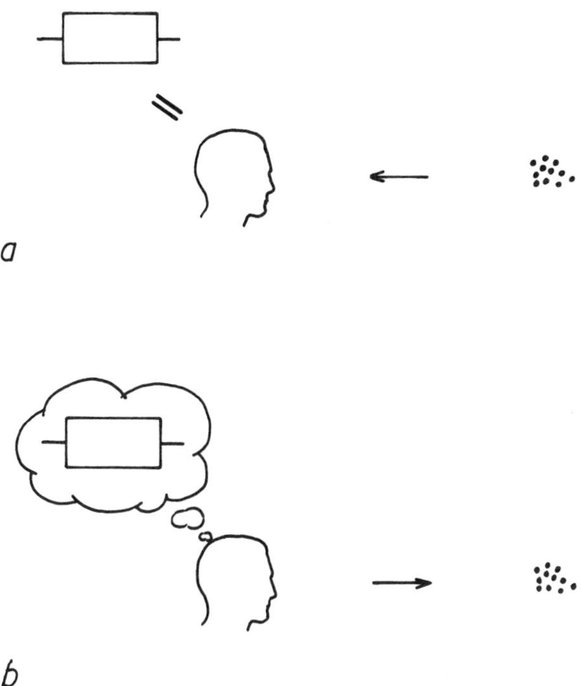

Figure 4.1
(a) The Processing Machine Analogy, and (b) the Representational Machine Analogy.

tional Machine Analogy. Here, the cognitive system examines the stimulus set, and tries to explain the set causally. That is, it sees the set as part of the trajectory of a well-defined environmental process, i.e., the environment becomes structurally a machine. This is why we cannot help but see graffiti as the outcome of a drawing process, or a dent as the outcome of a kick.[3]

4.5 Some Comments

4.5.1 Evidence from Contemporary Cognitive Science

The Representation-is-Explanation Principle states that cognitive representations are descriptions of stimuli in terms of causal histories (i.e., programmed machines) that explain the stimuli. In the present chapter, we have derived this proposal from the claim that cognitive activity necessarily involves the interaction of objects by virtue of their being memory. However, besides holding to this theoretical derivation of the proposal, we shall also attempt, throughout the book, to justify the proposal *empirically*.

In addition to the supporting theoretical argument, and the corroborating empirical data, there is a further way of substantiating the Representation-is-Explanation Principle: When one carries out a review of the conceptual analyses and theoretical constructs that are emerging in the various sub-disciplines of cognitive science, one finds that these analyses and constructs also support our proposal.

Let us carry out such a review now:

(1) Perception

We have already given an extensive review of Standard Computational Vision and shown that a number of its constructs—the analysis of image-formation, the use of energy as a construct in interpolation, etc.—lend themselves, quite naturally, to the view we are developing in this book. It is unnecessary to repeat our argument showing this, here.

(2) Categorization

The ability to categorize is an essential psychological function: It is the ability to group together several entities into a category, in order to see them all as examples of one type of object; e.g., the capacity to see several entities as examples of a *chair*. One advantage of this ability is that it allows the entities to be used interchangeably. For example, a person does not go through life being able to use only one specific chair; but can use many chairs interchangeably.

The modern study of categorization is an outgrowth of the seminal ideas of Eleanor Rosch (1975, 1977, 1978)—ideas that have had a major impact on a number of disciplines including psychology, anthropology, linguistics and philosophy. A central idea of Rosch's is that there is a level of abstraction at which people tend to use terms—a level she calls the *basic level*. For example, consider the term *furniture*. Less abstract versions of this term are *chair* and *table*. Still less abstract versions are *kitchen chair* and *kitchen table*. It turns out that people tend to use, most frequently, the terms in the middle level of abstraction, i.e., the terms *chair* and *table*; and only rarely use the more abstract term *furniture* or the less abstract terms *kitchen chair* and *kitchen table*. Rosch argued that an important reason for this is that the middle level of abstraction is the level at which people tend to *functionally interact* with the objects. For example, when a person enters a room, he or she does not require a piece of furniture, but requires specifically a chair or table. Furthermore, the person rarely requires anything more specific, i.e., a kitchen chair or kitchen table. Any chair or table tends to suffice.

Rosch thus argues that an important factor that groups objects into the same category is that they have the same function. Observe that a functional description of an object is a description of it in terms of one's *causal interaction* with it. Thus we conclude that an important factor that groups objects together is that they share a common causal description; i.e., causality is an essential aspect of their representation.

In relation to this idea, consider the branch of Artificial Intelligence that is mainly concerned with categorization, the discipline known as *machine learning* (Winston, Binford, Katz & Lowry, 1984; Mitchell, Keller, & Kedar-Cabelli, 1986; DeJong & Mooney, 1986). This discipline has investigated, among other things, the way in which a set of functions becomes mapped to the physical appearance of an object. It is the fact that this mapping can be made to the object that *defines* the object as a member of the category. For example, consider what are recognizably the main physical features of a cup. Each has a crucial function assigned to it, thus: The flat bottom is for standing the cup on a table; the handle is for grasping the cup when lifting; the inside is for containing the liquid; the rim is for supporting the cup against the lips when drinking. The assignment of functions, i.e., causal interactions, to features, *defines* the object as a cup. That is, in accord with Rosch, a cup is defined by the particular causal structure in which it is supposed to participate. A further expansion of the concept of causal embedding is given in recent work by Rosch and her colleagues (see Varela, Thompson & Rosch, 1991).

One of the developments that was influenced by Rosch's theory of categorization is Lakoff's metaphor-based analysis of categories. Lakoff (1987) argues that a category is cognitively understood in terms of some *model* that is usually a physical model. For example, he argues that people's concept of *anger* is understood in terms of a model of a *liquid in a container undergoing heating*. In the standard model that people have of a liquid undergoing heating, the liquid boils and rises up the container. Furthermore, the liquid could be restrained using a lid, but it might then explode, etc. Lakoff argues that this detailed *physical* model is the basis of our understanding of anger. Thus, we consider anger to be something that can "rise", "boil over", "be kept under a lid", "be about to explode", etc. The relevance of this view, for us here, is that we see once again that a cognitive representation of an entity is in terms of a causal structure in which it is embedded; i.e., in accord with our Representation-is-Explanation Principle.

It should be observed that the above model of anger is a machine; i.e., a well-defined causal structure. Furthermore, the machine is decomposed into a system of machines which are causally interlinked. The component machines include the container, the lid, and the liquid. The container and lid are each described as a machine in the same way that a cup was described as a machine in the above discussion; i.e., as causal structures that are mapped to the visual features. However, the liquid is also a machine. Hayes (1985) has argued, in considerable detail, that we understand liquids in terms of their causal structures and histories.

(3) Language

The most generally held view of grammar (following Chomsky, 1957, 1965, 1981) is that a sentence is *generated* from an underlying structure or from an alternative sentence that is more basic. Therefore, the sentence is, in effect, assigned a *history*. This historic explanation is central to the interpretation of the sentence. For example, part of the history is the trace of successive movements that a Noun Phrase makes through a sentence; and this trace allows one to establish the meaning of the Noun Phrase (e.g., its case role). Thus we see that the cognitive representation of a sentence conforms to our Representation-is-Explanation Principle. Furthermore, the representation conforms to our Representational Machine Analogy; that is, the sentence is described as a state in a well-defined history, i.e., a state of a programmed machine. These issues will be discussed in great detail in Chapter 7.

Whereas Chomsky's work gives an analysis of the grammar of individual sentences, the work of Schank & Abelson (1977) gives an analysis of connected discourse. It has been found that, in discourse—for example, in describing to a friend what happened at a restaurant—one leaves out a large amount of material that is necessary to make sense of what one is saying. Schank and Abelson propose that the reason that one can leave out so much important material is that one assumes that the listener has a "script" (in his or her head), which describes what standardly happens in a restaurant and which therefore gives a context for the few things that one actually chooses to mention. A script is a large *causal* or *process* structure. Thus, the few events that one describes are *explained* in terms of a causal structure in which they are embedded. Therefore, a script conforms to the Representation-is-Explanation Principle which states that the representation of any stimulus is a causal explanation that is assigned to the stimulus.

In fact, a script is an example of a machine that is decomposed into a set of component machines which are causally linked together. In a restaurant script, the component machines include the waiter, the menu, the table, etc. The speaker, who is relating a story about the events in a restaurant, is describing particular causal interactions between particular components, e.g., the waiter bringing the menu. The particular causal interactions are embedded within the larger framework of causal interactions that constitutes the whole script—in the same way that the particular interactions between any components of a clock are embedded within the full causal framework of the clock.

(4) Planning

Planning is the attempt to find a behavioral path from the present state to some goal in the future. Thus, it is clearly the creation of a causal sequence—that which will obtain the goal. One might think that planning is not an example of the history-recovery problem, as we have described it in the previous chapters, because the state being explained is in the future rather than in the present. In fact, close examination of planning reveals that it accords exactly with what we have called *internal inference*; i.e., the description of two states as two positions along the same trace. In this case, the trace is the plan. Furthermore, the internal inference problem requires that the two states be *represented* within the present. In the planning problem, the future goal state and the present state are *both* represented within the present moment. Thus, the planning problem is formally indistinguishable from the shape problem we examined in Chapter 2; e.g., of bridging two X-rays taken of the same object at two developmental stages.

These considerations lead us to conclude that planning is an example of our Representation-is-Explanation Principle, i.e., the goal state is causally explained by a history. Observe also that since the causal history is well-defined, it is the trajectory of a machine, and thus planning conforms to our Representational Machine Analogy.[4]

The discussion in this section seems to show that researchers in different areas of cognition are beginning to find that cognitive representations are causal explanations. This supports our Representation-is-Explanation Principle, which proposes that *any* cognitive representation is the assignment of a causal explanation. It simultaneously supports our Representational Machine Analogy because the causal explanations are structurally the same as machines.

4.5.2 What is the Future?

The phenomenon of planning leads us to consider, more generally, the status of the future.

The only actual temporal moment, with which a cognizer can have contact, is the present. The cognizer can never have contact with the future. Therefore, what the cognizer takes to be the future must exist in the present in the same way that what the cognizer takes to be the past exists in the present. In fact, the similarity is greater than one might first suspect. For example, as is made clear by research on planning, a representation of the future that is derived and manipulated in planning has reality for the cognizer only as a record in memory, i.e., as the state of a memory store. This means that the future is given by features of a physical object, the store. Furthermore, the cognizer takes the features as *referring* to the future. However, as we shall see in the next section, the referential relationship of an object to its referent is a *retrospective* one. Thus, to recover the referent of an object, the cognizer must recover the object's causal history. But, in the case being considered, the referent is an envisaged future state. We are therefore led to conclude that, in this sense, the future is an example of the past, and that the recovery of the future is yet another example of the recovery of the past.

4.5.3 Cognition and Symbols

Researchers have argued that a necessary property of cognition is that it proceeds by *symbol interpretation* and *manipulation* (Newell & Simon, 1976; Newell, 1980; Pylyshyn, 1984; Fodor & Pylyshyn, 1988). However, we shall now see that this view can be subsumed under the view being proposed in the present book, thus:

For an object to be a symbol, it must be taken as *referring* to some other object. For example, the symbol or name, *Aristotle*, is taken as referring to a certain person in ancient Greece. Now, the process by which a symbol is created proceeds forward in time from the referent to the symbol; e.g., first there was some individual in ancient Greece, and then various processes led to the creation of a name for this individual. Furthermore, a chain of communicating individuals across time led to the symbol being used, e.g., spoken, in the present moment. Thus, there was a forward progression in time from the initial object to the current use of the symbol referring to the object. This means that, when one is presented with a symbol and one attempts to *interpret* it—i.e., when one attempts to recover the referent of a symbol—what one is in fact doing is attempting to recover the *past*, i.e., recover the object from which the long chain of naming and communicating proceeded. We are therefore led to conclude that *the interpretation of a symbol is an example of the process-recovery problem, i.e., an attempt to reverse the effects of time*. Furthermore, symbol manipulation must be an example of the manipulation of the past; i.e., the organization and reorganization of the structure of time.

By our definition of memory, symbols are an example of memory, as follows: Our definition states that a memory is a physical object (1) that is in the present of some observer, and (2) that has features that the observer causally explains. Similarly, a symbol must be instantiated as a physical object (e.g., a name written on a page), and it fulfills the above two conditions thus: (1) It must exist in the *present* of an observer, for it to be interpreted, and (2) the observer explains it as originating from the referent.

4.5.4 Time Required to Construct Causal Explanations

We have argued that all cognition proceeds by the construction of causal explanations of the present. However, the causal explanation of the present takes time to construct. Therefore, by the time the explanation is constructed, the present is the past and the causal explanation is of the past and not the present. Note that this is another example of the two-phase history that we described for perception, i.e., a transmission phase preceded by an environment-formation phase: That is, the activity of constructing the causal explanation is an example of the transmission phase and the constructed causal explanation is an example of a description of the prior environment-formation phase, where, in this case, the environment is the previous present that is being explained.

Thus, any cognitive activity—being the reading and writing, i.e., construction, of causal explanations—must produce causal explanations that take into account the time taken for their construction.

4.6 How Machines Embody Causal Interactions

Our Representational Machine Analogy proposes that any cognitive representation is the description of a stimulus as a state (or states) of a machine. Thus we can gain a better understanding of cognitive representations if we understand in more detail what is meant by a machine and how machines can embody causal structure.

Any machine is defined by *five* components. The first component consists of actions that are applied to the machine. These are called the *input actions*, or simply *inputs*. They are represented as in Figure 4.2. The input actions are applied by some part of the machine's external environment. For example, the actions can be provided by a human being, or they can be actions of inanimate objects that collide with the machine, etc. To restate: the term *inputs* refers to the causal actions that can be *applied to* the machine *by the environment*.

As a result of applying the actions to the machine, the machine produces actions in response. These are called the *output* actions as shown in Figure 4.2. For example, suppose the machine is a radio. If the *input* action applied to the radio is *switching the radio on*, then the radio will respond with the *output* action of sending sound waves into the environment. Again, suppose the machine is a gun. If the *input* action applied to the gun is *pressing the trigger*, then the gun will respond with the *output* action of projecting a bullet into the environment. The point is that outputs are the resulting actions that the machine *performs on the environment*.

Thus a machine involves two sets of actions. (1) *Inputs*—the actions that the environment applies to the machine. (2) *Outputs*—the resulting

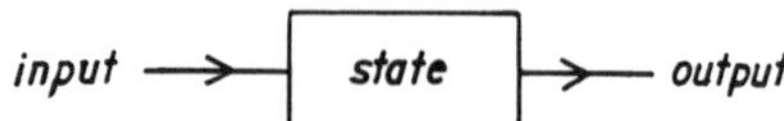

Figure 4.2
A machine.

actions that the machine applies to the environment. These two sets of actions are depicted in Figure 4.2.

In fact, *two* factors are responsible for the output actions. The first are the input actions, which we have just discussed. The second factor consists of certain properties that the machine has at the time of the input. For example, suppose that the machine is a gun. An input action is pulling the trigger and an output action is projecting a bullet. However, this output action occurs only if the gun is loaded. If the gun were empty, a different output would occur, e.g., the gun would just make a dull click. Thus, a property of the gun, *loaded* or *empty*, determines which output occurs. Any property of an object that determines which output occurs is called a *state*. Thus, the state of the gun can be either *empty* or *loaded*. When the input is applied, i.e., pressing the trigger, the state *empty*, will determine one output, whereas the state *loaded* will determine another. It is in this sense that the input together with the state determine the output.

To summarize so far: A machine consists of (1) a set of *input* actions that are applied by the environment to the machine, (2) a set of *output* actions that the machine applies back to the environment, and (3) a set of properties of the machine, called the machine's *states* that determine which output actions the machine will produce in response to its input actions. Notice that the relationship between the inputs and the states is that of a causal *interaction*. That is, the system represented by the inputs is acting on the system represented by the states. We shall now see that this causal interaction has two causal effects:

The first causal effect to arise from the interaction between the input and the state is the *output*. We will call this the first causal relationship:

FIRST CAUSAL RELATIONSHIP

Input × State → Output

Reading from left to right, the above line simply says that a particular **Input** acting on a particular **State** leads to a particular **Output**. This relationship is *causal* for two reasons: (1) The left of the arrow, **Input × State**, depicts a causal *interaction*; i.e., the input acting on the state. (2) The left of the arrow is the *cause* and the right of the arrow is the *effect*. That is, the cause is the causal interaction, **Input × State**, and the effect is the **Output**.

There is in fact a second effect that arises from the causal interaction of the input on the state. This second effect is a change in the state itself. For example, firing a gun might cause its state to change from loaded to empty. This type of causal relationship is specified thus:

SECOND CAUSAL RELATIONSHIP

Input $\times$ **State** $\rightarrow$ **Next State**

Reading from left to right, the above line simply says that a particular **Input** acting on a particular **State** leads to a **Next State**. This relationship is again *causal* for two reasons: (1) The left of the arrow, **Input** $\times$ **State**, depicts a *causal interaction*; i.e., the input acting on the state. (2) The left of the arrow is the *cause* and the right of the arrow is the *effect*. That is, the cause is the causal interaction, **Input** $\times$ **State**, and the effect is the **Next State**.

We have now gathered together the five entities that completely define a machine:

(1) INPUTS: actions that are applied by the environment on the machine.

(2) OUTPUTS: actions that are applied by the machine on the environment.

(3) STATES: properties of the machine that act as factors in determining the consequences of applying the inputs.

(4) FIRST CAUSAL RELATIONSHIP

Input $\times$ **State** $\rightarrow$ **Output**

(5) SECOND CAUSAL RELATIONSHIP

Input $\times$ **State** $\rightarrow$ **Next State**

We should emphasize that the first and second causal relationships each concern the same *causal interaction*, that of the inputs on the states. Thus the left of the arrow, in either case, is the same, i.e., **Input** $\times$ **State**. What differs in each case is the *effect*, the right of the arrow. In the first causal relationship, the effect is the *output* and in the second causal relationship, the effect is the *next state*. Note that the first effect, the output, is an action on the environment and is thus part of another, subsequent, causal interaction. But this latter interaction is not fully specified. For example, while, in this interaction, the actions applied by the machine on the environment are specified, the states of the environment are not specified, nor are the effects of this other interaction.

We therefore propose that the concept, *machine*, is the full definition of *a single causal interaction*. The definition specifies the two interacting systems to be represented by the *inputs* and the *states*. The definition specifies also the total set of effects, as follows: The total set splits into two subsets (1) the set of actions that the machine contributes to any resulting causal interactions with the environment, and (2) the set of changes in the machine itself.

We now have a clear picture of what a machine is and how it embodies the notion of causal interaction. Before we return to cognition, we need to define one other term: *program*. A program is an ordered sequence of input actions to be applied to a machine. For a gun, a program could be the following sequence of actions: (1) Take the gun from the holster, (2) aim the gun, (3) press the trigger, (4) press the trigger again, (5) press the trigger yet again, . . . etc.

Now let us return to our Representational Machine Analogy. This states that *any stimulus is represented as a state (or states) of some programmed machine*. We can now understand that this proposal means the following: Given a stimulus, the cognitive system does three things: (1) It defines the stimulus in terms of a fully specified causal interaction, that is, in terms of the full structure of a machine, i.e., all the five components listed above—inputs, outputs, states, first causal relationship, and second causal relationship. (2) The cognitive system represents the stimulus as one (or more) of the machine states, i.e., the stimulus is embedded within the space of machine states. (3) The cognitive system identifies a program—a sequence of input actions—as the actual history of the machine. Notice that, in response to the history of inputs, the machine goes through a history of states, i.e., a particular trajectory through the space of states. Thus the stimulus is embedded within a history of states, as shown in Figure 4.3.

One of the consequences of our proposal is that the history is not defined simply as a sequence of states but is given a *context*, the machine. This context is a sizable structure, in the following sense. It consists of (1) a space of *alternative* actions (inputs) that could have been applied, i.e., not just the particular inputs that were applied in the history; (2) a space of *alternative* states that could have resulted, i.e., not just the particular states that were used in the history; (3) a space of *alternative* outputs, not just the particular outputs that were used in the history; (4) a description of how *all* the possible causal interactions, between all the possible inputs and states, determine the outputs, i.e., a specification of the first causal rela-

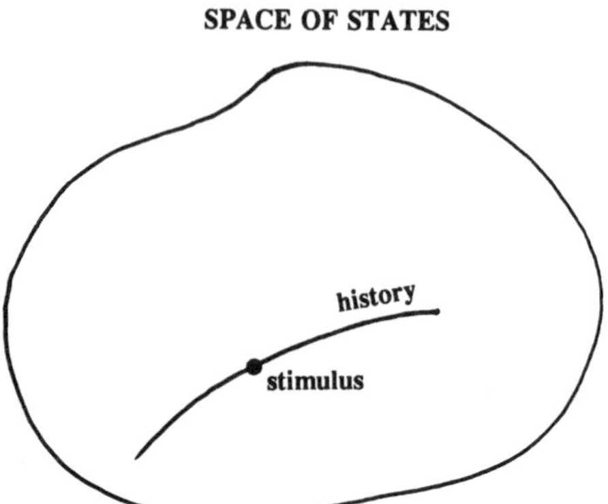

Figure 4.3
A stimulus embedded in a history embedded in a space of states.

tionship; and (5) a description of how all the possible causal interactions, between all the possible inputs and states, determine the changes of state, i.e., a specification of the second causal relationship.

Thus, although the stimulus is given a context or reference frame in terms of a history, this history is itself defined in terms of a reference frame of causal possibilities, i.e., the machine.

Most of this book will illustrate these concepts with respect to perception. Therefore, let us give an illustration here with respect to another area of cognition: categorization.

We saw that recent work on categorization indicates that what categorizes an object such as a cup is a causal framework for its use. Using the concepts described above, we argue that the *use* of a cup is specified in terms of actions applied to it, e.g., lifting, and in terms of the resulting actions that the cup applies back to the environment, e.g., conveying the liquid upward. This means that a cup is defined as a relationship between a space of input actions and a space of output actions—crucial aspects of a machine.

The other aspects of a machine are also involved in defining a cup. Observe that the concept of state is crucial. The state of the cup includes its shape. This does not alter during the cup's use. In fact, the use of the cup depends on this state not altering.

A moment's consideration will reveal that we have now defined a cup in terms of the five components specifying a machine:

(1) INPUTS: e.g., lifting

(2) OUTPUTS: e.g., conveying liquid

(3) STATES: e.g., the shape

(4) FIRST CAUSAL RELATIONSHIP

e.g., lifting (input) acts on shape (state) → conveying liquid (output)

(5) SECOND CAUSAL RELATIONSHIP

e.g., lifting (input) acts on shape (state) → same shape (next state)

We argue that the description of an object in terms of its *use* makes explicit the first causal relationship of a machine; i.e., the actions to be applied on the machine and the resulting actions that the machine applies to the environment. That is, *using* an object is understood as converting the input actions into the output actions, and this conversion is the first causal relationship. As we noted however, the other aspects defining a machine are also crucially present, although often left implicit. For example, with a cup, we saw that what happens to the state (the shape), over time, is essential in allowing the use. The evolution of the state is prescribed by the second causal relationship.

The Representational Machine Analogy specifies that a stimulus is described in terms of a *programmed* machine. Thus, in addition to defining a cup in terms of the five components of a machine, a cognitive description of a cup must contain a program or history for a cup. In fact, a cup involves two crucial histories, a past history defined with respect to the second causal relationship and a future history defined with respect to the first causal relationship. The past history is the history that created the shape of the cup. That is, one perceives the cup as having resulted from a certain formation history, e.g., the rounding action that created its concavity. The future history concerns the *use* of the cup. This is a history of input actions and resultant output actions. The history of inputs is a *plan*. It is usually the following sequence of actions: (1) stand the cup on the table, (2) fill the cup from a pot, (3) lift the cup to one's lips, (4) tip the cup against one's lips, (5) empty the cup by drinking, (6) return the cup to the table. Figure 4.4 shows how the plan is mapped onto the different parts of the cup. Observe that the numbering gives the order in which the plan uses the different parts of the cup.

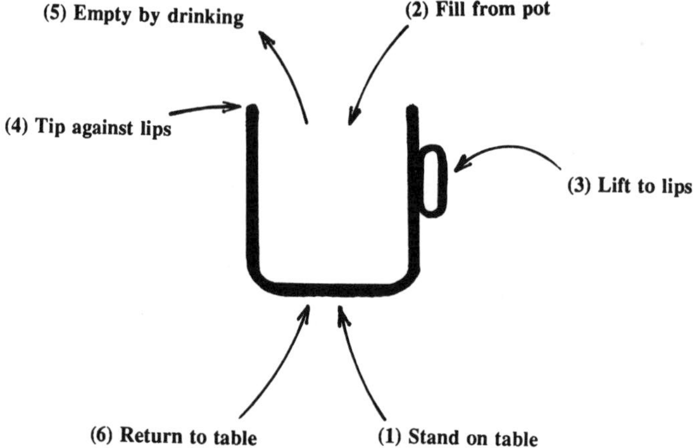

Figure 4.4
A plan for a cup mapped on to a cup.

The example of the cup illustrates the fact that the term *causal explanation*, in the Representation-is-Explanation Principle, should be understood as referring to both retrospective and prospective explanations. For example, explanations in terms of use are usually prospective. It can be easily seen that the retrospective history, based on the second causal relationship, and any prospective history of the object *interlock*. This is because the retrospective history is responsible for the current structure and the prospective history is dependent on the current structure. Principles that inter-relate retrospective history with prospective history will be called Interaction Principles. We shall propose such principles later in the book.

Finally, we should observe that, in accord with section 4.5.2, what the cognizer takes to be the prospective history sits in memory, and is therefore really an example of the past. Thus, when we use the terms "future" and "prospective", in our discussion, we use them merely as labels given by the cognizer to certain memories.

4.7 Machines as History

The description of an object in terms of its use is embodied in the first causal relationship of a machine. This is because the *use* of an object is the way in which it converts actions applied to it (input actions) into actions it applies back to its environment (output actions). In this book, we will only rarely discuss the actions that the object applies back to its environment, i.e., the output actions. Thus we will be only rarely concerned with the first causal relationship. Our concern will be with the second causal relationship. This relationship describes the effects that the input actions have on the object itself; that is, it describes the object as a historical record of the actions applied to it. The history could be labeled as in the past or in the future, but the point is that we consider the object as a set of *consequences* (the second causal relationship) rather than the transmitter of *further actions* (the first causal relationship).

Let us examine more deeply how the notion of *machine* contains exactly what we require to express the history-recovery problem. The importance for us of the concept of machine is two-fold:

(1) A machine embodies the notion of **causal interaction.**

(2) A machine embodies the notion of **history**.

The first of these two ideas was explained in section 4.6. We now explain the second.

The definition of machine involves five parts (1) inputs, (2) outputs, (3) states, (4) the first causal relationship, (5) the second causal relationship. It is important to observe now that the word *object* does not occur in any one of the five items. This might at first seem strange. After all, it is always some object that undergoes the input actions.

In fact, the notion of object is actually included as one of the five items. It is represented by the third item, the states. Consider the set of properties that define an object at a specific moment; e.g., that it is blue, horizontal, cold, etc. Some of these properties will change under the specified actions; some will not. The states are the properties of the object that are involved in defining the causal interaction. That is, the *states* represent the *object*, within the definition of machine.

It is important now to understand that the definition of machine implies a correspondence between the notion of *state* and that of *history*, as follows: The second causal relationship is this:

$$\textbf{Input} \times \textbf{State} \rightarrow \textbf{Next State}$$

This means that the **Next State** is different from the preceding **State** by only one factor, the **Input** action that was applied. This argument can be extrapolated backward in time as far as one wishes, e.g., to some designated starting state in the remote past. Therefore only the starting state and the sequence of subsequent input actions up to and including the current action determine the **Next State** of the machine. Thus, if we label the current action, **Input**, and we label the preceding history, consisting of the starting state and the sequence of subsequent inputs, **History**, then the **Next State** is determined thus:

$$\textbf{Input} \times \textbf{History} \rightarrow \textbf{Next State}$$

That is, the pair **Input** × **History** is the *causal interaction* that determines the **Next State**. However, the second causal relationship specifies that the causal interaction that determines the **Next State** is given thus:

$$\textbf{Input} \times \textbf{State} \rightarrow \textbf{Next State}$$

Therefore, we now have two pairs, **Input** × **History** and **Input** × **State**, representing the causal interaction that produces the next state. Comparing the two pairs, we see that **State** must correspond to **History**. In other words, the current state must be the history up to the present. This means that there is an equivalence between the notion of *state* and the notion of *history*.

Observe now however that a number of histories might be equivalent in that they determine the same subsequent behavior. For example, consider a watch with only a second hand, i.e., a hand that moves one notch per second and does a complete rotation in one minute. Suppose that the watch has been stopped at the 17th notch. Then we will know that the hand has traveled 17 seconds, but we will not know how many complete rotations the hand has traveled. That part of the history is lost. Thus, a whole set of equivalent histories is compatible with the state of the watch, i.e., any history that consists of a number of complete turns plus 17 seconds. We can see therefore that the state of the watch, notch 17, corresponds to an *equivalence class* of histories.

The discussion of the watch also emphasizes that a *state* is equivalent to *memory*. That is, notch 17 is a memory of 17 seconds but not of the number of complete rotations made by the watch.

The above discussion shows that the notion of machine involves a profound correspondence between four concepts: *object*, *state*, *history*, and *memory*. That is, we began this section by seeing that the concept of *object* is represented in the definition of machine by the concept of *state*. However, we also saw that the definition of machine is so economical that the concept of *state* is forced to be the same as that of *history*, or, more strictly, *equivalence class of histories*. Finally, we saw that the particular histories in the same equivalence class are those that produce the same state as the single memory of those different histories. Thus there is a correspondence between the notion of *state* and that of *memory*.

We see therefore that the concept of machine defines an object in terms of its history or memory. However, this view that an object is memory, a memory of history, is a view that is also assumed in the history-recovery problem. Thus the object-is-memory notion gives an important reason why we consider that cognition defines a stimulus in terms of a *machine*. Note that it is the very determinism of a machine, i.e., that it is causally well-defined, that allows one to understand its state, i.e., its object, as a historical entity.

We can also therefore conclude that the purpose of the history-recovery problem is to "extract" the machine from a stimulus.

4.8 The Construction of Temporality

4.8.1 Temporality from Atemporality

The ultimate constraint that determines the requirement of memory, or causal explanation, is this: No two distinct points in time can occur simultaneously. Therefore, the mind cannot *reach* two temporally distinct points simultaneously. It can reach one and only one point in time—a point we call the present.

Now, because no two distinct points can be reached simultaneously, the point that can be reached, the present, has no temporal extent. The present cannot be, as we perhaps experience it, a *neighborhood* containing a small fraction of the immediate past and immediate future. We cannot actually be in contact simultaneously with a small fraction of the immediate past and future, because this would mean being in contact with more temporal points than one—which is impossible. Thus our "neighborhood" experience of the present must be *constructed*, i.e., it is the *outcome* of a computational process, not a given for the cognitive system. The cognitive system must compute it from what is actually given, which is merely the present moment.

The present, being only one point and having no temporal extent, must be completely *atemporal*. Thus *every* aspect of our experience of time must be constructed. Nothing can be taken for granted. For example, we experience time as *continuous*. However, because we have available only a single punctate, dislocated, decontextualized moment—the present—the experience of continuity must be constructed. Again, we experience time as *flowing*, i.e., as *moving* from the past to the future. This experience also must be constructed, because we have only atemporality, which contains no movement, i.e., which is completely frozen. Again, we experience time as having *direction*. And this experience also must be constructed, because all we have available is a point, and to define direction one needs more than one point.

I wish therefore to propose that the mind is faced with the following computational problem:

CONSTRUCT TEMPORALITY FROM ATEMPORALITY

As we shall now see, one can choose to regard this as a *recovery* problem in the same way that one can choose to regard visual perception as a recovery problem. Recall that, according to Computational Vision, perception is given by the following structure: There is a three-dimensional physical environment in the outside world. This *projects* to a two-dimensional spatial array on the retina. In this projection, important information concerning the three-dimensional environment has been lost. The purpose of visual perception is to *recover* the three-dimensional environment from the two-dimensional retinal array; i.e., to put back the information that has been lost.

In a similar way, one can choose to regard the problem we are defining—the construction of temporality from atemporality—as a *recovery* problem, thus: Let us suppose that physical time actually exists and has certain properties such as continuity, movement from past to future, direction, etc. Now, in projecting the one dimension of time to what the observer actually has, the zero-dimensional point, the present moment, all time and its properties are lost. Thus, in constructing temporality from atemporality, the observer is attempting to *recover* time and its properties.

At first, one might say that, since physical time, i.e., time as it exists independently of our psychological experience of it, actually does have the properties listed above—continuity, movement from the past to future, direction—the task of constructing the experience of those properties is either not required or not difficult. But this would be similar to saying of visual perception that, because the world is actually three-dimensional, the task of constructing the three-dimensional experience of the environment in our heads, is either not required or not difficult. It is required because all that is available to the visual system is a two-dimensional array of stimuli on the retina. It is difficult because the retinal array, being two-dimensional, has lost one of the dimensions contained in the three-dimensional environment, and the loss of this dimension means a loss of such crucial information that the task of trying to put back that information requires a massive amount of computation. Similarly, with respect to time, since we have available only a single moment, we have lost a dimension, and all the information that it contains. How much greater the task is of putting back one dimension, in time, as opposed to visual perception, when the lost dimension of time is the entire content of time and the lost dimension in perception is only a third of the total to be reconstructed.

We take much for granted about our experience of time, in the same way that we take much for granted about our visual experience of the spatial world. In both cases, we are unaware of the immense amount of computation required to create that experience. We listed above a number of the properties that are taken for granted about time and must be constructed. An additional such property is the feeling of *familiarity* that many objects have for us. This feeling too must be constructed. For all that we have is the present, which consists of a set of objects whose past we cannot now reach. These objects are therefore, by definition, *new*. Thus, the sense of familiarity must arise from the *atemporal* relationship between the set of new objects that constitute the present.

The logic of this argument is exemplified, in its most simple form, in the motion detector discussed in section 3.13 and diagramed in Figure 3.38. That is, motion cannot be detected by the comparison between a past state and the present, because the past state no longer exists. Motion can be detected only if two *parts* of the present are taken as having being caused at two different points in time. The two parts are (1) a delay cell that is *currently* firing because it has delayed the output from a receptor P_1 that fired previously, and (2) a second receptor P_2 that is *currently* firing in response to the stimulus. Motion detection occurs only because both of these two cells, the delay cell and the second receptor, are firing *in the present*. That is, what is interpreted as motion is in fact a configuration of the present, i.e., the *shape* of the atemporal present. In particular, part of that shape—that part we call the delay cell (which is defined in the present by its location)—is interpreted as the past and is thus a very primitive example of what one can call "familiarity", because it is interpreted as designating something that one has experienced in the past. Nevertheless, the firing delay cell does not exist in the past; it exists in the present and its state is therefore *new*, and is not in any sense familiar.

What this example illustrates is that (1) anything that one takes to be familiar is in fact new, and (2) familiarity is a response to a particular shape that the new possesses; i.e., it is *shape* that is interpreted as time.

The same is true of the other properties that are taken for granted about temporality, properties such as those discussed above—the sense of continuity, the sense of temporal flow, the sense of temporal direction, etc. All these are responses to *shape* within the present.

What pushes cognitive activity forward, i.e., pushes it out of one present moment into the next, are the laws of physics that govern the physical substance of the brain and its evolution through physical time. These laws, together with the physical structure existing in the present, push the activity forward so that the present is followed by a present that is taken to be a causal explanation of the previous present. Thus the forward movement of physical time is responsible for the backward movement of psychological time. Thus the purpose of cognition is to compensate for the forward asymmetry of physical time with the backward asymmetry of psychological time.

Notice, finally, that this indicates that the classical dichotomy of mind and body rests ultimately on a difference between the two directions of time.

4.8.2 The Return to Perception

The fact that every property that one ascribes to time must be constructed emphasizes that the recovery of history is a vast problem. In the remainder of this chapter and for most of the lengthy Chapter 6, we will study some of the immense work that the cognitive system must carry out in order to *organize* time and therefore history. It is possible in this book to take only one domain and study it in the particularly close detail we shall attend to. The domain that we shall take is the *static image in visual perception*. When we return to another area of cognition, in Chapter 7, where we study language, the analysis will be much higher in level and more coarse grained.

However, our study of the static visual image can be regarded as a *paradigm* example, for the following reason: One can suppose that all domains must, at the level we study this one, have a certain amount of commonality with this one. The reason is that all domains must have, as input to the construction of history, the same thing: arrangement within the present, i.e., *atemporal shape*. First note that this must be true even for domains involving movement—because, as was argued earlier, motion cannot be perceived directly but must be derived from atemporal shape. Again, it must be true also for semantically meaningful domains such as language, because the input to processes such as language understanding is ultimately shape, e.g., phonetic shape; and even when this shape is presented sequentially, e.g., when a sentence is heard, it must conform to the same constraint as motion, e.g., its only existence as input to cognition must be as shape within the present. It is in this way that the static *visual* image exemplifies the general situation. It is the example we choose because it is the one concerning which there is the most knowledge available—at the level at which we wish to carry out our analysis.

Our hope therefore is that, by a detailed study of perception, we might gain a general sense, and perhaps even some specific sense, of the nature and extent of the work required by a system where all that is available to construct temporality is atemporality.

4.8.3 Temporal Construction as Program Construction

The Representational Machine Analogy states that cognition represents any environment as a collection of well-defined (i.e., machine) histories. A history is well-defined if it can be represented by a program. The advantage of understanding histories as machine programs is that we can start to understand the *structure* of histories; i.e., how *time* is organized by the mind. The branch of research that has carried out the most extensive analysis of programs is computer science. Computer science has established a number of principles concerning the way in which programs are best organized, and we shall argue that there are important reasons why these principles are relevant to understanding how the mind structures time.

4.8.4 Programs

Any computer program is a sequence of actions. An *action*, in a program, is an event brought about at a single moment in time, i.e., it is identified with only one moment in time. The program specifies the *order* in which the actions are to be executed.

We shall see later that within a *psychologically* inferred history or program there are collections of actions that are grouped together in a certain way. It will be natural to call these groupings *processes*. Thus in perception, there are three important levels to an inferred environmental history or program. At the most local level, there are the individual actions. (These will later be called *generators*). At the most global level, there is the entire program or history. At the intermediate level there is a division of the program into processes, particular groupings of actions.

Any computer program is executed in sequence, one action at the time. The program can be represented by a flowchart. One travels through a flowchart in sequence, executing one action after the other. Thus, at any point in time, one is at a single point in the flowchart. The chart *controls* how that point moves. Therefore, a program, or part of a program, is called a *control-structure*.

4.8.5 Structured Programming

In computer science, one can construct many alternative programs to perform the same task. An important property that distinguishes the different programs is their *complexity*. Some programs are simpler and others more complex. Clearly, it is much easier to deal with a simpler program than a more complex one. In fact, as Martin & McClure (1985) state, "The basic problem of computing is the mastering of complexity."

We can see that minimizing program complexity is exactly the requirement made in our History Minimization Principle. The principle states that perception chooses the history with the minimal amount of distinguishability. Complexity can be regarded as the quantification of distinguishability. Since each history is equivalent to a program, the History Minimization Principle shares with computer science the requirement that the program complexity be minimized.

In fact, we are going to argue that the type of complexity one attempts to minimize in an inferred history is *control complexity*. Recall that *control*, in computer science, refers to the way in which one is sent through a flowchart; i.e., one's position at any point in time is represented by a single point in the chart and the chart controls the way in which this point moves. One can regard the control structure as corresponding to the organization of the associated causal environment, in this sense: If the control structure is divided neatly into separately acting components, the environment is itself decomposed into well-separated causal factors. Therefore, minimizing control complexity corresponds to simplifying causal structure.

The branch of computer science that deals with the minimization of control complexity is called *structured programming*. We are therefore going to argue the following:

STRUCTURED PROGRAMMING HYPOTHESIS. *Perception is a structured programmer; i.e., it applies techniques of structured programming to construct the histories it assigns to explain the environment.*

In sections 4.8.5 to 4.11, we will take an initial simple look at the principles of structured programming and see how they are applied by perception to create histories. In Chapter 6, this examination will be carried out in much greater detail to explain many psychological phenomena.

The basic principles of structured programming are the following:

(1) Restricted set of standardized control structures. The program or history should be built out of a minimal vocabulary of standardized components used to structure any program or history.

(2) Decomposition. The program or history should be decomposed into a set of smaller simpler modules.

(3) Hierarchical form. A program or history should be stratified into levels. This is a basic means by which complexity is reduced in nature generally.

A discipline related to structured programming is *software engineering* which encompasses structured programming and concerns the process of developing and maintaining software generally. A basic principle of software engineering is the following:

(4) Localization. Place logically related items together.

This principle will also be seen to be central to the way perception constructs histories. Thus, more generally, we shall argue that perception is a software engineer.

Let us restate our position concerning perception. In order to do this, we begin with a quotation from a standard work on structured programming:

> Structured programming is a methodology that lends structure and discipline to program form, program design process, program coding, and program testing. It is a programming methodology for constructing hierarchically ordered, modular programs using standardized control constructs (Martin & McClure, 1985, p42).

Our argument is that this is an appropriate definition of *perception*. That is, since perception is the definition of physical history structure in the environment, it is primarily concerned with the design of programs (i.e., histories). Furthermore, it must have standardized techniques for creating process structures. That is, as in the above quotation, it must have a methodology that gives structure to program design, coding and testing. We shall mainly concentrate, in this book, on the second half of the above quotation, i.e., we shall argue that perception is a programming methodology for constructing hierarchically ordered modular programs using standardized control structures. It is to these standardized structures that we now turn.

4.8.6 Three Basic Control Structures

As noted above, structured programming uses a restricted vocabulary of control structures from which to build any program. In fact, Bohm & Jacopini (1966) proved that any program could be constructed out of a vocabulary of only three basic control structures, called *sequence*, *selection*, and *repetition*, which are described below. Note that these three components are called *control* structures because they control the order in which the program is executed. That is, in our terms, they determine how the history develops.

(1) Sequence. Sequence is the structure in which one program step follows another, as shown in Figure 4.5a. In this figure, one moves successively rightward executing one step after another.

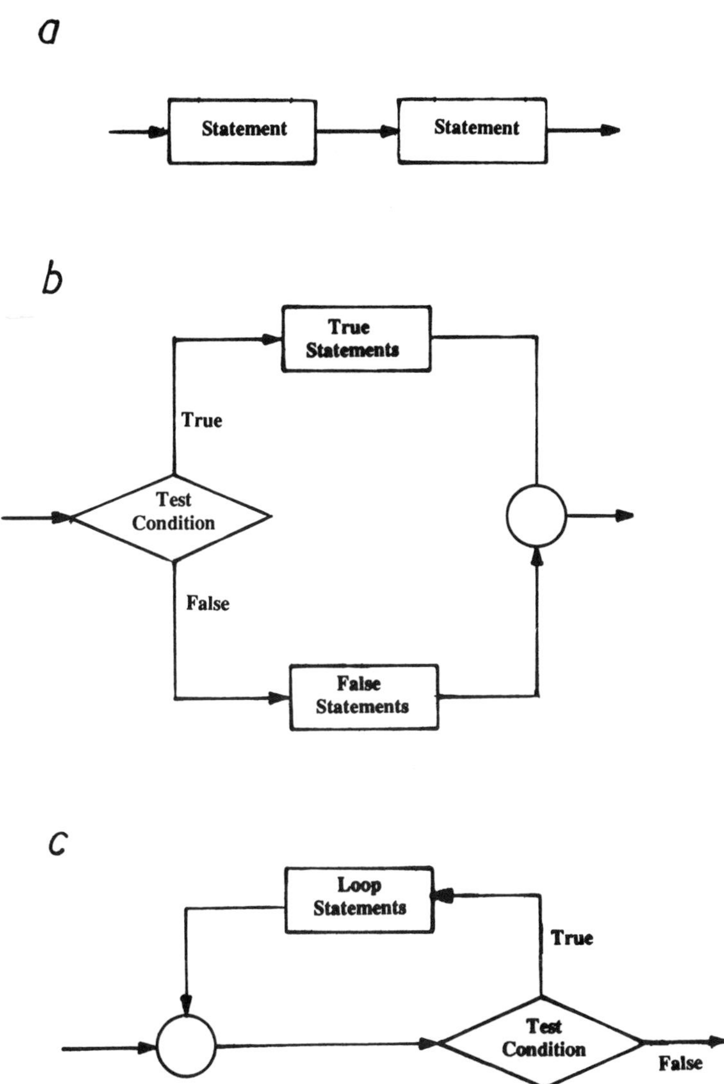

Figure 4.5
The three basic control structures: (a) sequence, (b) selection, and (c) repetition.

(2) Selection. Selection is the control structure in which some action is chosen if a certain condition is true and an alternative action is chosen if the condition is false. The structure is often described as the IF-THEN-ELSE structure. An example is this: IF today is Tuesday, THEN walk to work, ELSE take the bus. The flow diagram for the IF-THEN-ELSE structure is shown in Figure 4.5b. One enters the diagram from the left and first encounters the diamond-shaped test-condition box. If the condition mentioned in the box is true (e.g., it *is* Tuesday), then one takes the upper branch rightward. If the condition mentioned in the box is false, then one takes the lower branch rightward.

(3) Repetition. Repetition, usually called *iteration*, is a simple loop. The only necessary type of repetition is the DOWHILE loop, shown in Figure 4.5c. One enters the loop from the left and continues rightward until one encounters the diamond-shaped test-condition box. If the condition named in this box is true, one goes back around the loop. If the condition is false, one exits the loop by continuing rightward. Thus, in effect, one continues going around the loop *while* the test condition holds. Note the following important difference between the IF-THEN-ELSE structure in Figure 4.5b and the repetition structure in Figure 4.5c: While the former looks diagrammatically as if it is a loop, it is not. In Figure 4.5b, the program flows from left to right through that structure without being allowed to turn back at any point. In flowing through the diagram, the program will take either the top or the bottom branch but not both. The opposite is true of the repetition structure (Figure 4.5c). Here, the program can turn back and a complete loop can be made.

4.9 Modularization

As noted earlier, structured programming is a set of techniques used in computer science to lower the complexity of a program. Our interest in structured programming is that perception is the inference of well-defined histories, i.e., programs, and that our History Minimization Principle requires the inferred histories or programs to have minimal complexity. We listed the three main techniques, used in structured programming to lower complexity, as (1) use of a minimal set of standardized control structures, (2) decomposition of the program into modules, and (3) hierarchical organization of the program.

Having described the first item in the previous section, we now consider the second, and its use in perception.

A significant method of reducing complexity in a program is to decompose it into smaller, more manageable, components. Such components, called *modules*, should be simpler than the whole; they should be closed, discrete, and independent of each other.

We shall argue that perception decomposes histories into modules. These modules are, in fact, what are psychologically understood as *processes*—which we will soon characterize. That is, perception sees a history as decomposed into processes. Furthermore, we argue that the decomposition into processes follows from the attempt to reduce complexity by modularization.

The advantages of decomposing a history into processes are the same as the advantages in computer science of decomposing a program into modules, namely:

(1) The modules or processes are constructing blocks that can be used in building other programs or histories.

(2) The programs or histories can be more easily modified or revised (e.g., if new information is acquired) by modifying only the relevant module or process. That is, the effects of revision do not multiply through the whole program or history. In particular, errors in the history can be more easily isolated and changed.

(3) The programs or histories are more comprehensible because they can be understood in pieces, i.e., modules or processes.

As we said, in our system, a module is a single process. But we have not yet established what a single process is. In order to do so we return to computer science for more clues.

In computer science, a module is a set of logically related statements that performs a specific function. The idea of forming a module follows from the software-engineering Principle of Localization: Items that are logically related should be put together. Thus, in perception, a process should consist of items that are as logically related as possible. Now, suppose an inferred history contains a number of occurrences of the same action. Then, it would be natural to conclude that the occurrences are logically related by having the same causal origin. The Principle of Localization would then imply that they should be put into the same module. Let us apply the Localization Principle in the most extreme form possible by saying that a module should contain only the repeated occurrences of the same action. That is, we shall propose the following:

PSYCHOLOGICAL DEFINITION OF PROCESS. *Psychologically, a process is the repetition of the same action.*

We can understand this proposal further in terms of one of the main principles of module design, a principle called *Cohesion*. It states that a module should be as cohesive as possible, and in particular express only a single function. The single function in this case would be the single causal origin. By making a process express only this function, and, in accord with the Localization Principle, by putting together all the actions expressing this function, a module becomes simply a repetition of an action. In this sense, the notion of *repetition* realizes, for perception, the Principles of Localization and Cohesion in the most extreme form possible—even more extreme than is often possible in computer science.

We can now see that a process corresponds to the third type of control structure listed in section 4.8.6, i.e., *repetition* or loop. The loop is the control structure that most embodies the History Minimization Principle, because distinguishability between moments has been eradicated. Each action within the process is the same as every other action; that is, the loop is a history that is the most *symmetric* possible, i.e., under translations in time. The same cannot be said about the other two primitive control structures, *sequence* and *selection* (section 4.8.6). In the *sequence* structure, the successive actions must be different because, if they were not, they would be a repetition and hence be describable by a loop. In the *selection* structure, the actions must be different because, if they were not, there would be no reason to make a choice between them.

Our claims that a process is psychologically the repetition of a single action and that perception tries to decompose histories into such repetitions will be discussed and validated in considerable detail in this book.

4.10 Nested Control

A number of additional concepts can be introduced by examining, in detail, the flow diagram representing the history of a rotated parallelogram.

The history we shall represent is a fully externalized one. Recall that such a history is one in which sufficient external processes are hypothesized so that the remaining internal (trace) structure is a nested hierarchy of repetitive, Euclidean, actions. In the case of a rotated parallelogram, our standard externalized history is that in which, backward in time, we have externally removed rotation, shear, and stretch, leaving a square, i.e., a shape that has a trace structure that is a nested hierarchy of repetitive Euclidean actions. This trace structure consists of two levels, one that draws an individual side, and the other that shifts the "pen" from one side to the next around the square.

Going forward in time, the history is therefore the following:

1. Draw a side.
2. Shift from one side to the next, and repeat level 1 (thus creating a square).
3. Stretch (creating a rectangle).
4. Shear (creating a parallelogram).
5. Rotate (creating a rotated parallelogram).

Let us go through this history in order. First of all consider stage 1, drawing a side. Drawing a side is a repetitive action. One moves a point (the end of a pen) by a small amount, e.g., 1/1000 of an inch, repetitively, until the point has moved the full required distance which is, let us say, 1 inch.

Being the repetition of an action, this sequence is an example of what we call a *process*. Furthermore, being a repetition, it is represented in flowchart form by a loop—the third kind of control structure in section 4.8.6.

The loop required to draw a side is given in Figure 4.6. To understand the instructions comprising this structure, consider the following. First note that, in drawing the 1-inch side, a point should start at 0 inches and be moved till it reaches 1 inch. Therefore, we have an object, a point, which has a state, its *position*, which changes from 0 to 1. For our purposes, the state of any object will be represented in the computer (that runs the program) by a *memory*. Note that this accords with our view that an *object* equals a *state*, and the latter equals a *memory* (section 4.7). In the present case, the object, the state, or the memory, can be called "Point Position" and it is initially set to equal 0. Then one enters the loop in Figure 4.6. The first thing one encounters is the diamond-shaped box at the bottom

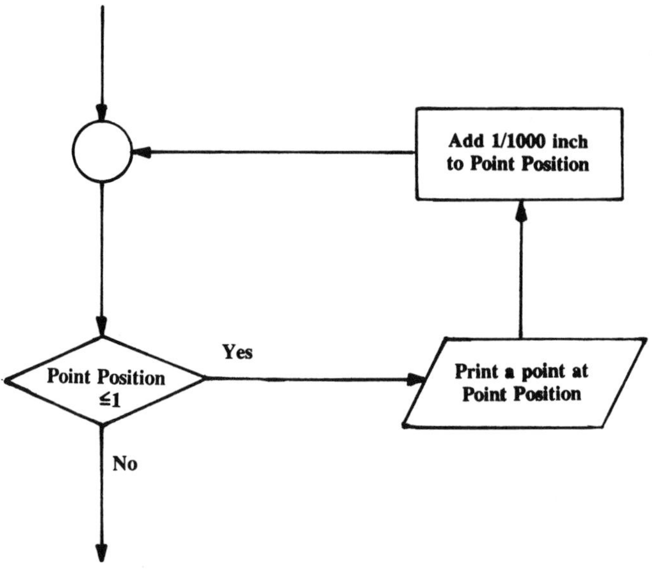

Figure 4.6
A flow-chart for drawing a side.

left. This tests to see whether the total amount the pen has traveled has exceeded 1 inch. If it has not, one continues around the loop. The next box that one encounters (bottom right) gives one the instruction to draw a dot. Finally, in the top right, one encounters the instruction to move the pen by 1/1000 inch. Thus, each time one goes around the loop, one prints a dot at the new "Point Position" and changes the "Point Position" by a 1/1000 inch. Finally, one stops going around the loop when a *trace* of length 1 inch has been drawn. The trace is that of the *point* which has been moved.

The relationship between the object, a point, and the computer memory is worth emphasizing. Effectively, the above flowchart takes the point-object and changes its position from 0 to 1 inch. As noted above, the object is given by a particular memory in the computer. This object changes its state under *actions* contained within the loop. Therefore, the actions in the loop are *interactions*, i.e., actions on the object represented in the memory. Thus, as in any interaction, there are two systems *A* and *B*: system *A* acts on system *B*. The computer memory is system *B*, the recipient of the action. The action itself, i.e., that described in the flowchart, is to be carried out by system *A* on system *B*. System *A* is the part of the computer that

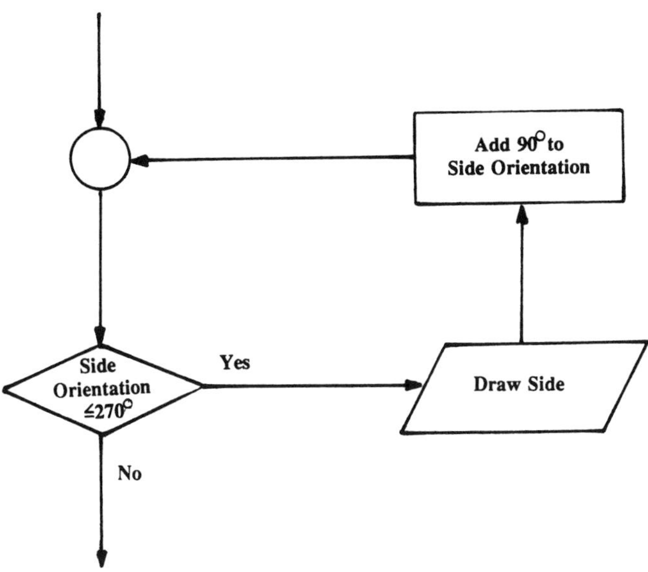

Figure 4.7
A flow-chart for drawing a square.

carries out the actions. Thus the computer, as a whole, represents *two* systems in interaction, *A* acting on *B*.

Now let us turn to stage 2, listed above, the shift from one side to the next. This is the repetition of a single action, that of rotating 90°. Being a repetition, the movement is again an example of what we call a *process*. Again, it can be represented by a loop, as shown in Figure 4.7. To understand this figure, observe that what the shift should do is to change the orientation of a side through the four successive values, 0°, 90°, 180°, and 270°. Therefore, one has an object, a side, which has a state, its *orientation*, which changes from 0° to 270°. This state is again represented by a *memory,* which we will call "Side Orientation". Initially, the state is set to 0°. Then one enters the loop shown in Figure 4.7.

The first thing one encounters, on entering the loop, is the diamond-shaped box at the bottom left. This tests to see whether the total amount one has shifted has exceeded 270°. If it has not, one continues around the loop. The next box one encounters (bottom right) gives one the instruction to draw a side. Finally, in the top right, one has the instruction to shift 90°. In short: each time one goes around the loop, one draws a side in the

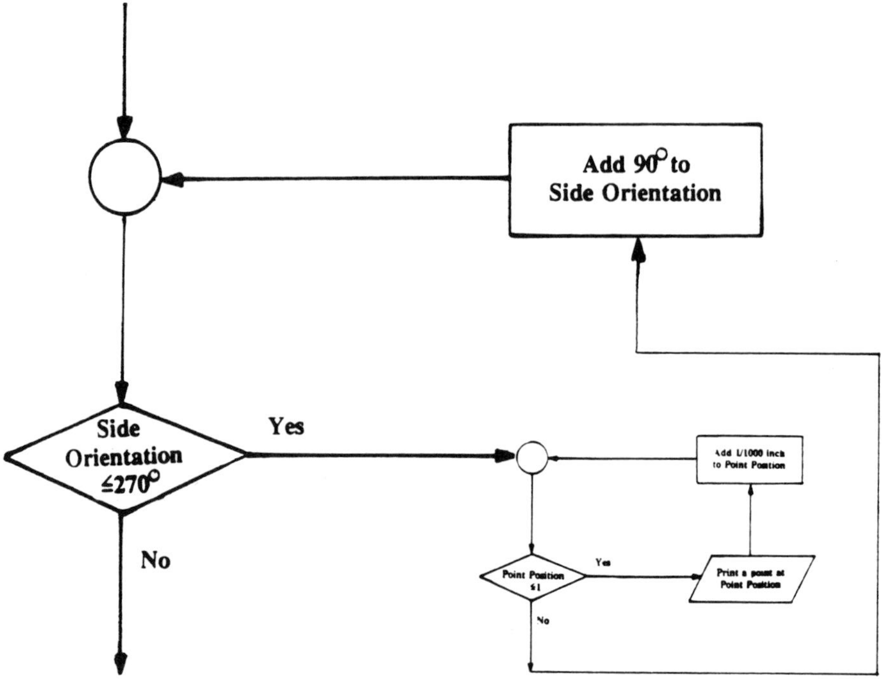

Figure 4.8
The flow-chart for drawing a side, nested within the flow-chart for drawing a square.

correct "Side Orientation", and one increases the "Side Orientation" by 90°. Finally, one stops going around the loop when one has shifted a total of 270°. This means that all four sides of the square have been drawn, and the successive sides are the *trace* of the shift movement.

Now observe that one of the instructions encountered in this loop is "Draw side". However, drawing a side is exactly what is accomplished by the first loop we described, i.e., that shown in Figure 4.6. This means that, at this position in the second loop (Figure 4.7), one has the first loop (Figure 4.6). That is, the first loop is nested within the second one. This nesting is shown in Figure 4.8.

The flowchart presented in Figure 4.8 embodies the claim we made in section 2.6, that, in a fully externalized history, the trace structure is a hierarchy of *nested*, *repetitive*, *Euclidean*, actions. That is, not only are the loops *nested*, but, being loops, they are *repetitive*; and furthermore, being translations and rotations, they are *Euclidean*.

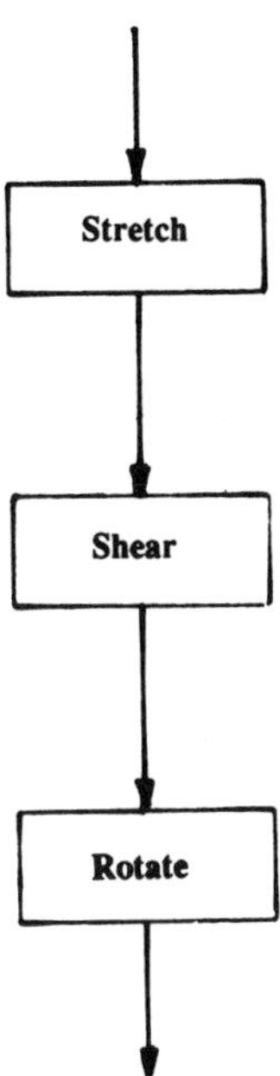

Figure 4.9
The sequence: stretch followed by shear followed by rotation.

We have considered the first two processes in the list of five (given near the beginning of this section) for creating a rotated parallelogram. The two create the trace, or internal, structure in this fully externalized history. The remaining three processes—stretch, shear, rotation—are external ones. Since these latter three arise in sequence, they represent an example of the first of the control structures listed in section 4.8.6, i.e., the *sequence* control structure. This means that the flowchart that represents this three-layered history is simply that shown in Figure 4.9.

However, we should observe that each of the three operations in Figure 4.9—stretch, shear, rotation—is a *process* in the strict sense of the term we have defined in section 4.9, i.e., the repetition of a single action. For example, the rotation phase is not assumed to consist only of the parallelogram first in its non-raised position and then in its raised position, but is assumed to consist of a sequence of intervening states given by the regular increase in the parallelogram's angle to the ground. That is, the rotation phase is the repetitive action that moves the parallelogram from its initial to final position. Thus the rotation phase must be given by a loop; i.e., it is a *process* in the strict sense we have defined. The same applies to the other two operations given in Figure 4.9. Therefore each box in Figure 4.9 is itself a loop, and Figure 4.9 should be understood as consisting of three successive loops.

Now, to create the entire five-process history, the flowchart in Figure 4.8 for the internal structure is simply added onto the top of that shown in Figure 4.9 for the external structure. In fact, the descending line at the bottom of Figure 4.8 becomes the descending line at the top of Figure 4.9.

The total flowchart consists of five loops, two for the internal structure and three for the external structure. As shown in Figure 4.8, the loops for the internal structure are nested one within the other. We shall call this type of nesting, *internal* nesting. When one process is nested internally within another, one goes through the former, in its entirety, at each moment of the latter. For example, in Figure 4.8, one goes through the entire side-drawing program (the inner loop), each time one does a single rotational shift (the outer loop).

The three successive *external* operations in Figure 4.9 are nested in a different sense: Each process acts on the entire results created by the previous stage. That is, one can consider each of the three successive processes in Figure 4.9 as a layer covering the previous process. We shall call this type of nesting, *external nesting*.

Both internal and external nesting are nestings, in this sense:

DEFINITION OF NESTING. *Control is said to be* **nested** *when the entire effects of one process are acted on by another process. The nesting is said to be* **internal** *when one goes through the former process, in its entirety, at every point in the latter process. The nesting is said to be* **external** *when one goes through the former process, in its entirety, only once and before the latter process begins*; *i.e., the latter is a "layer" over the former.*

Recall now that structured programming is the design of programs that use (1) standardized control constructs (i.e., sequence, selection, repetition); (2) modular decomposition; and (3) hierarchical structure. We see, in this section, each of these requirements put into effect, as follows: (1) A single standard control structure, the loop, is used to define each process. (2) The module, a process, is the loop itself. (3) The loops are in a nested hierarchy, either internally or externally.

One should note that, in structured programming, hierarchy tends to refer to levels of abstraction, whereas here it refers to the embedding of segments of time. Nevertheless, the structuring purpose is the same in both cases; i.e., simplification via stratification. Furthermore, it is certainly true that abstraction hierarchies are involved in our histories—e.g., each box in Figure 4.9 is an abstract representation of a loop. However, the abstraction hierarchy will be much less important to us than the nesting hierarchy. The latter represents the layering of segments of time. Remarkably, in the layering of time, the three structured programming techniques—(1) standard control constructs, (2) modularization, and (3) hierarchy—become one, as follows: The *standard control construct*, the loop, becomes the *module* we call a process, and the decomposition into modules becomes the *hierarchical organization* (nesting) in which processes act on each other.

4.11 Replacing Selection by Repetition

Recall that, according to structured programming, there are three basic control structures, (1) sequence, (2) selection, and (3) repetition (section 4.8.6). We have seen that the third of these is fundamental to the histories inferred by perception; i.e., repetition is the structure of the module. We have also seen that the first of the control structures, sequence, is used to order the modules in the external layering of processes, as illustrated in the previous section by the layering of stretch, shear, and rotation. However, we have not encountered a use of the second control structure, *selection.* To see why, let us consider an example where selection might be used.

Consider Figure 4.10, a square with a bump in one side. If this configuration were described purely as a trace (i.e., internal structure), then one would include a section involving selection, as follows: Recall that selection is the IF-THEN-ELSE structure; i.e., IF a condition is true, THEN one chooses the first alternative, ELSE one chooses the second alternative (recall section 4.8.6). That is, the structure is a selection between two

Figure 4.10
A square with a bump in the side.

alternatives. It is relevant in Figure 4.10 because one would need two alternative strategies for drawing a side—one for drawing any of the straight sides and the other for drawing the bent side.

To understand this more clearly, recall that, in drawing a square, one has two levels of control. The higher level is the shift that moves from one side to the next around the square. The lower level is the drawing of an individual side after each shift. Let us suppose that the shift is moving *clockwise* around the square, starting with the top side. Then, for the first, third, and fourth sides in Figure 4.10, the program can simply draw a straight side. However, for the second side (the right-hand side), the program has to use the alternative strategy of drawing a bent side. This means that, before drawing each side, the program has to test whether it is starting one of the former three sides or the latter side. That is, there has to be the selection structure shown in Figure 4.11, which first tests the amount of shift, and then has two branches, one for drawing a straight side if the total shift is not 90°, and the other for drawing a bent side if the total shift is 90°.

Recall now from Figure 4.8 that the program for drawing a square without the bump is simply two loops—an outer loop giving the repeated shifts, and the inner loop generating the side. In the case of a square with a bump, we need the selection structure in Figure 4.11, and this is inserted

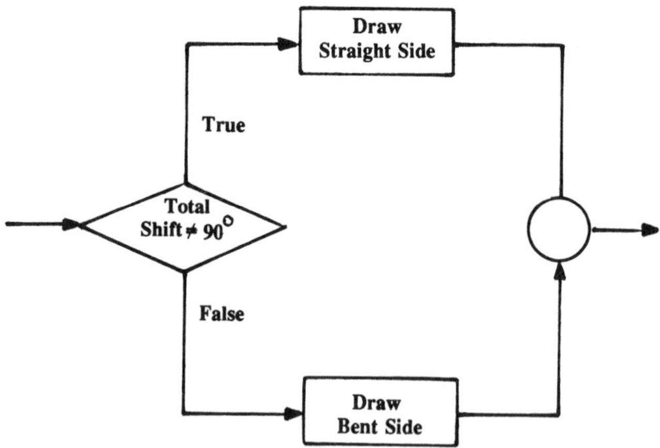

Figure 4.11
The use of a selection structure to generate the bent side purely by internal means.

into the outer shift loop. Thus the additional selection structure prevents the outer loop from being simply the repetition of an action, i.e., a process.

As shown in Figure 4.11, the inserted selection contains two branches, one for drawing a straight side and one for drawing a bent side. The former is in fact the inner loop from the simple structure in Figure 4.8. The other branch is more complex. In tracing out the side with the bump, the drawing movement on that side is initially straight, i.e., repetitive; then it must change to tracing around the bump; and finally it must become straight again, i.e., repetitive. This means the lower branch in Figure 4.11 must itself contain a selection structure—one that tests the position on the side—and it must choose repetitiveness in the initial and final segments, and a bump in between.

We can see from the above discussion that the attempt to draw the deformed square purely as a trace results in a complex control structure. In particular, the purely repetitive nature of the processes is lost; i.e., the outer loop ceases to be the pure repetition of a shift. The same result occurs in any situation where there is an *anomaly* or *irregularity*. Accounting for the anomaly purely within an internal (trace) structure means that a selection configuration has to be introduced where one branch accounts for the irregularity and the other accounts for the regularity.

The complexity that occurs when one tries to describe an anomaly purely as a trace, i.e., internal structure, can be reduced by *externalization*. For example, in the case of the square with a bump, one can understand the history to be a fully externalized one in which, going forward in time, one proceeds in two phases: (1) one first generates a perfect square, and (2) one then creates the bump by pushing out one of the sides. The bump is now given by an external process because, going backward in time, the deformed square points to a past configuration, the undeformed square, which is not contained in the deformed square. The two successive phases of the history are shown respectively in Figure 4.12a and b. That is, the internal structure is shown in Figure 4.12a and the external structure in Figure 4.12b. The complete program is the nested pair of pure loops given in Figure 4.8 for a perfect square, followed by a pure loop that then pushes out one of the sides.

Thus, the externalized history consists of pure loops in a simple arrangement.

We see therefore that externalization has caused the elimination of the selection (branching) component from the internal (trace) structure and

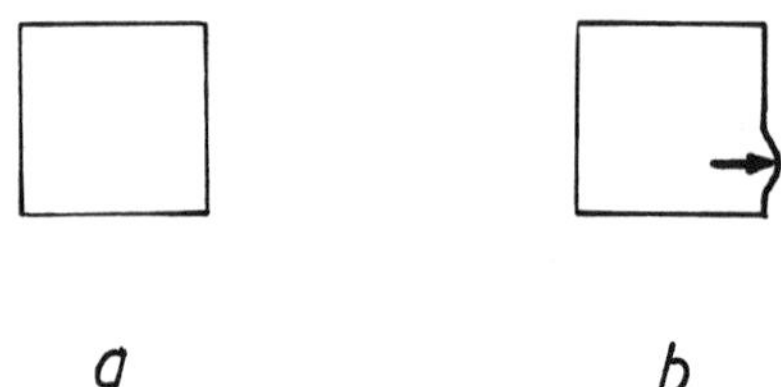

Figure 4.12
The externalization of the bent side.

the addition of an extra loop in the external structure. Furthermore, whereas the selection component had been situated in a complex arrangement with respect to the internal structure, the external loop has a simple relationship to the rest of the program.

The complex relationship that the selection structure had in relation to the internal structure was to break the purely repetitive nature of the internal loops. Externalization restored the simple, nested, repetitiveness. Thus we see here an illustration our Externalization Principle (in section 2.6) which states that the purpose of externalization is to reduce the internal structure to a nested hierarchy of pure repetitions.

4.12 Summary

In this chapter, we argued that, what we had claimed about perceptual representations, i.e., that they are representations of stimuli in terms of causal histories, is true of cognitive representations generally.

Our argument proceeded thus: We noted that cognitive activity, or computation, is generally regarded as an interaction between processors that hold memory. Furthermore, we also noted that the processors communicate via stimuli that are themselves memory; i.e., memory of the processors from which they emanate.

Given the centrality of memory to our discussion, it was necessary to give a strict definition of memory, thus: A memory is a physical object (1) that is in the present of some observer and (2) that has features which the observer has decided to causally explain.

In accordance with this definition, we noted that a stimulus is a physical *object* in the present of an observer, because the stimulus is the receptor at the moment of stimulation.

We then noted that, when a processor *A* receives a stimulus from another processor *B*, within a computation, the stimulus contains memory of a history divided into two phases—the transmission phase from processor *B* to processor *A*, and the prior phase in which processor *B* is formed as memory. This two-phase structure also defines any *perceptual* stimulus as memory, where processor *B* is the external environment.

Because cognition is the interaction of objects as memory, the cognitive system can therefore include both internal objects, such as pieces of neuronal material, and external objects, which are made memory by perception. Any object, within the system, that defines any other object as memory, brings that latter object within the cognitive system. In particular, a stimulus, as an object, has a cognitive role within the system only insofar as it is memory. Furthermore, any object from which the stimulus arrives can be made memory only if the stimulus is.

Thus we were led to one of the main proposals of the book, the Representation-is-Explanation Principle, which states that any cognitive representation is the description of a stimulus as a state (or set of states) in a history that causally explains the stimulus to the organism. According to this principle, a stimulus has no cognitive existence independent of its causal explanation. Thus, under this view, the notion of cognition is founded on the notion of causality. We regard this principle as *characterizing* cognition and being equivalent to our Characterization Principle, which states that a cognitive system is one that creates and manipulates causal explanations. That is, given a system, what makes it a cognitive system is that it can be described, on some level, as a system in which the medium that it processes consists of causal explanations.

We then proceeded to develop a structural characterization of the Representation-is-Explanation Principle. We began by proposing the Well-Definedness Principle, which states that the processes given in cognitive representations are well-defined, i.e., they are effective procedures. This claim led us to the Representational Machine Analogy, which states that any cognitive representation of a stimulus is the description of the stimulus as a state (or set of states) of a programmed machine.

The Representational Machine Analogy contrasts with the Processing Machine Analogy, which states that a cognitive system can itself be represented by a machine. However, the Processing Machine Analogy follows from the Representational Machine Analogy in two ways: (1) It is an instantiation of the latter principle, in that the cognitive system becomes an example of an object being represented as a machine. (2) However, in the sense that a cognitive system, i.e., a mind, includes anything it defines as memory, the mind, and its parts, are the only objects that are described as machines.

The Representational Machine Analogy is supported in three ways: (1) The theoretical argument leading to it, in this chapter; (2) the empirical data presented throughout the book; and (3) a review, carried out in this chapter, of the theoretical constructs emerging in the various branches of cognitive science.

Our review of planning led us to consider the nature of the future generally, and we concluded that since the future has existence, for the cognizer, only within some memory store, the future is always yet another example of the past.

We then went on to study how machines embody the notion of causality. We argued that a machine is equivalent to a single causal interaction. In the definition of machine, the causal interaction is represented by the action of the input on the state. Two effects are defined as coming from that interaction. One is given by the first causal relationship, which describes the effect that is the output, i.e., the resulting action of the machine on the environment. Functional definitions of objects are in terms of this relationship. The other type of effect is given by the second causal relationship, which describes the effect as the next state, i.e., in terms of the machine's own history. It is this second relationship that is of primary concern in the book. We observed that the definition of machine embodies the notion of object as memory of history; i.e., exactly the notion required by the history-recovery problem.

We then argued that the reason why memory, i.e., causal explanation, is required is that no two points in time can be simultaneously accessible. Because all that is accessible is a single moment, i.e., complete atemporality, the cognitive system must solve the following computational problem: construct temporality from atemporality. Every aspect of our sense of temporality must be constructed from complete atemporality. Nothing can be taken for granted. To create temporality, the cognitive system must therefore carry out an immense amount of organizational work. To illustrate the nature of some of the work, we devoted the rest of the chapter, and will devote most of the lengthy Chapter 6, to making clear some of the detail required in the visual perception of a static image. We claimed that this domain can be regarded as a paradigm domain because it takes as input, to the construction of time, the same as any other domain: atemporal shape.

Thus, for the remainder of the chapter, we returned to perception.

The Representational Machine Analogy states that cognition represents any environment as a collection of well-defined histories, i.e., machine programs. The theoretical advantage, for us, of understanding histories as machine programs is that we can start to understand the *structure* of histories; i.e., how *time* is organized by the mind. Thus we began, in this chapter, to look at program structure. In particular, we considered the issue of program complexity, and saw that the History Minimization Principle, in our system, shares with computer science the requirement that the complexity of a program or history be minimized.

In fact, we argued that the type of complexity one attempts to minimize in an inferred history is *control complexity*, because this type of complexity most corresponds to the causal structure of the environment. The branch of computer science that deals with minimization of control complexity is called *structured programming*, and we argued that perception is essentially a structured programmer.

In accord with this, we saw that perception lowers the complexity of histories by using the central techniques of structured programming, that is: (1) standardized control structures, (2) decomposition into modules, and (3) hierarchical organization. In fact, we found that, in perception, these techniques become one, as follows: The standard control construct, in perception, is repetition, which we claim is the psychological definition of a process. The repetition defines the module, into which the history is decomposed. The decomposition into modules is simultaneously hierarchical, having the structure of nesting.

5 Groups and Symmetry

5.1 Introduction

While the Representational Machine Analogy states that a stimulus is represented as a machine history, it does not state which of the alternative machine histories is chosen. Several such histories could be compatible with the presented stimulus. The particular choice of machine history is actually determined by those principles of ours that involve the concept of symmetry: the Asymmetry Principle, the Symmetry Principle, and the History Minimization Principle (the last of which we have examined in relation to structured programming). In order to understand the role of these three symmetry-based principles, we need to gain a much clearer *structural* understanding of symmetry. Thus, in this chapter we will give an analysis of what symmetry means structurally. Armed with this analysis, we will return in the next chapter to the relationship between symmetry and process-history.

In mathematics, symmetry is characterized by the concept of *group*. Fortunately, the concept of group is one of the simplest to understand in mathematics. One can think of a group simply as a list of the symmetries contained in an object. For example, a cube has a total of 24 symmetries, which are given by its reflectional and rotational axes. The group of a cube is therefore the list of those 24 symmetries. Any object has its own group which is a list of its symmetries. Because a group describes the symmetry of an object, the notion of group will be important to us in subsequent chapters, in understanding the role of our three symmetry-based principles.

The purpose of this chapter is to give the reader an introduction to groups and to the way they describe symmetries. If the reader has already had an introduction to the notion of group, then he or she can read the titles of the sections in this chapter and check that the topics are familiar. Any section with a familiar title can be omitted, since it is a standard exposition of the subject. The reader should look finally at Table 5.1, at the end of the chapter, which lists the *names* we shall use for the particular groups that will arise later in the book.

5.2 The Symmetry Group of a Triangle

Consider the equilateral triangle in Fig 5.1. It appears highly symmetrical to us. For example, it is reflectionally symmetric about the vertical axis.

Figure 5.1
An equilateral triangle.

Furthermore, it has other axes of symmetry. The *group* of the equilateral triangle is simply the total collection of symmetries that the figure possesses. By saying that the figure is reflectionally symmetric about the vertical axis, we have identified one of the members of the group. When we have listed all the other symmetries, we will have listed all the other members of the group. Let us therefore enumerate all the symmetries of the figure in order to gain a full specification of the group.

First recall that symmetry means *sameness* under some transformation. Therefore, when we say that the triangle in Fig 5.1 is reflectionally symmetric about the vertical axis, we are saying that the triangle is the same after applying the reflection about the vertical axis. That is, after applying the *transformation*, reflection, the triangle is brought into complete coincidence with itself.

Thus, to find all the symmetries of the triangle, one simply has to find all the transformations that bring the figure into coincidence with itself. This collection of transformations is the *group* of the triangle.

In order to keep track of the effects of the various transformations that we try out, we shall label the vertices of the triangle, A, B, C, as shown in Fig 5.2a. Notice that reflection about the vertical produces Fig 5.2b. In this transformation, the vertices B and C, at the bottom of the triangle, have exchanged positions, and the top vertex, A, has remained in the same position. We shall label this reflection transformation, t. That is, t will mean reflection about the vertical axis. In order to check whether t has happened, we need only check whether the letters have changed in the way shown in Fig 5.2b.

Now let us look at another type of transformation that makes the triangle coincide with itself. This transformation is clockwise rotation by 120°, shown in the transition between Fig 5.3a and 5.3b. In this transformation, the vertices have simply replaced each other in a simple cycle around the

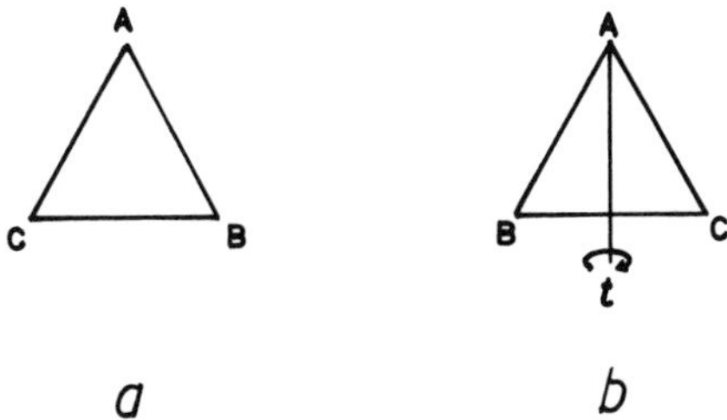

Figure 5.2
(a) An equilateral triangle, and (b) its vertical reflection.

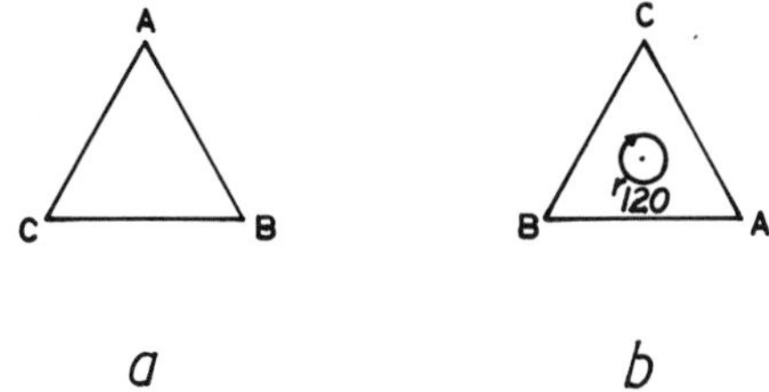

Figure 5.3
(a) An equilateral triangle, and (b) the triangle after rotation by 120°.

triangle; that is, vertex *A* has moved to *B*, vertex *B* to *C*, and vertex *C* to *A*. We shall label this rotation transformation, r_{120}. That is, r_{120} will mean rotation by 120° about the center of the triangle. In order to check whether r_{120} has happened, we need only check whether the letters have changed in the way shown in Fig 5.3b.

Now let us look at the next largest rotation that brings the triangle back into coincidence with itself. This transformation is clockwise rotation by 240°, shown in the transition between Fig 5.4a and Fig 5.4b. In this transformation, the vertices have replaced each other one further step in a simple cycle around the triangle. We shall label this rotation transformation, r_{240}. That is, r_{240} will mean rotation by 240° about the center of the triangle. Again, in order to check whether r_{240} has happened, we need only check whether the letters have changed in the way shown in Fig 5.4b.

Now let us see what happens when we try to obtain a still larger rotation bringing the triangle into coincidence with itself. This transformation is clockwise rotation by 360°. However, rotating the triangle by 360° is equivalent to applying no transformation at all. Thus, we can have a single

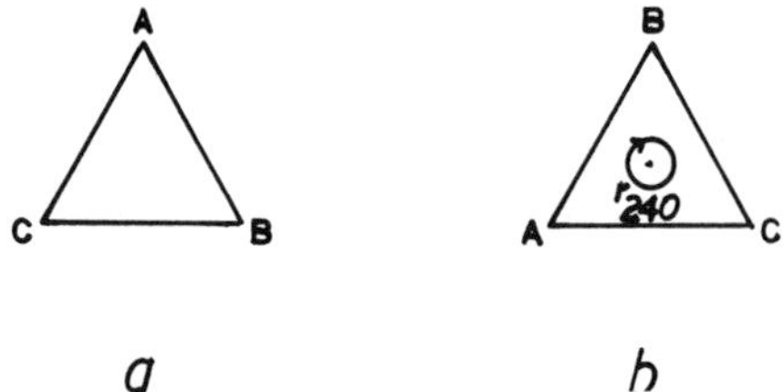

Figure 5.4
(a) An equilateral triangle, and (b) the triangle after rotation by 240°.

label, e for any transformation that has no ultimate effect on the vertices. For example, e means rotation by 0°; but it also means rotation by 360°. The null transformation e will be called the *identity element.*

To summarize so far, we have three rotations that bring the triangle into coincidence with itself:

e, the null transformation

r_{120}, rotation by 120°

r_{240}, rotation by 240°

Now let us try to increase the amount of rotation above 360°. The next larger rotation that brings the triangle into coincidence is rotation by 360° + 120°, which is 480°. However, observe that the effect of this rotation is exactly the same as rotation by 120°. This means that we do not have to include rotation by 480° in our list of symmetries if we already have rotation by 120°. Similarly, any higher amount of rotation will simply duplicate the rotations we have already. Thus, the above list of three transformations, e, r_{120}, and r_{240}, exhausts all the distinguishable rotations the triangle can undergo to be brought into coincidence with itself.

Having established all the distinguishable *rotations* that create coincidence, let us return to the possible *reflections*. We have already seen that reflection about the vertical axis brings the triangle into coincidence with itself. This reflection, which is called t, is shown in Fig 5.5a. There are two other reflections that also work. One is shown in Fig 5.5b, and is the reflection about the axis indicated there. The other is shown in Fig 5.5c, and is the reflection about the axis shown there.

It is important to observe however that the reflection in Fig 5.5b can be obtained by combining two of the transformations we already have. This

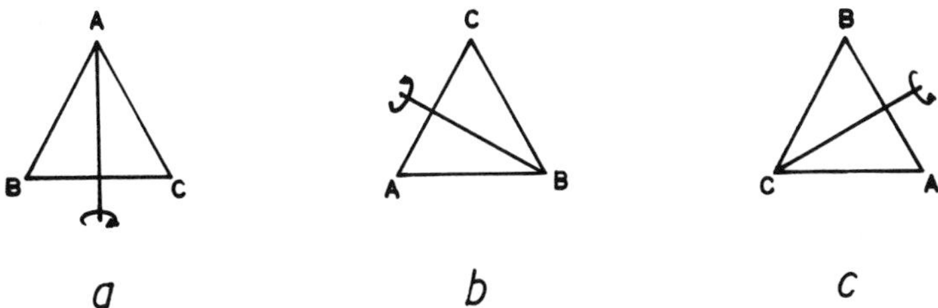

Figure 5.5
The three possible reflections of an equilateral triangle.

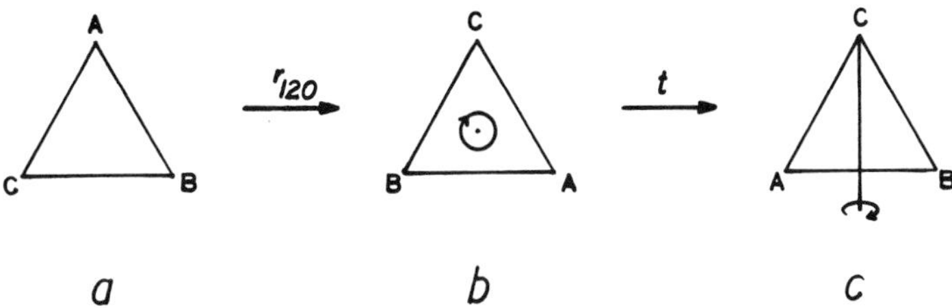

Figure 5.6
The reflection in Fig 5.5b can be generated in two successive stages.

is demonstrated in Fig 5.6. We start in Fig 5.6a with the triangle in its initial position. We then apply r_{120} obtaining Fig 5.6b. Then we apply t obtaining Fig 5.6c. However, the letters around the triangle in Fig 5.6c are in the same order and position as those in Fig 5.5b. Therefore the reflection in Fig 5.5b is equivalent to the application of r_{120} followed by t. This combination can be written as $r_{120}t$. Similarly, the reflection in Fig 5.5c is equivalent to the application of r_{240} followed by t, which we shall write as $r_{240}t$.

We have now enumerated all the rotations and all the reflections that bring the equilateral triangle into coincidence with itself. There are three rotations and three reflections, thus:

Rotations: e, r_{120}, r_{240}

Reflections: t, $r_{120}t$, $r_{240}t$

These are all the *symmetries* of the equilateral triangle. That is, the equilateral triangle has a total of six symmetries which we list between the parentheses on the next line:

$$\mathbf{D}_3 = \{e, r_{120}, r_{240}, t, r_{120}t, r_{240}t\}$$

The entire list is labelled $\mathbf{D}_3$ meaning "Dihedral group of rank three". The first three members of the list (in the parentheses) are the three rotations, and the last three are the three reflections.

$\mathbf{D}_3$ is an example of what is called a *group*. Any group is a set of elements that has four very simple properties, as follows:

(1) Closure

Any pair of members of a group can be combined to produce another member of the group. For example, we saw that the rotation r_{120} can be combined with the reflection t to produce the reflection $r_{120}t$. Again, the rotation r_{120} can be combined with the rotation r_{240} to obtain $r_{120}r_{240}$, which turns out to be another member of the group. To see which member this is, observe that $r_{120}r_{240}$ is rotation by 120° followed by rotation by 240°. This gives a total rotation of 360° which is the null rotation. Therefore, the combination $r_{120}r_{240}$ is the same as the transformation e. As another example, we can combine r_{120} with itself, obtaining $r_{120}r_{120}$, which is a rotation of 120°, applied *twice*. This combination is clearly r_{240}, which is also another member of the group.

In summary, therefore, we have illustrated the *closure* property of the group $\mathbf{D}_3$. The property is simply this: Combining any pair of members of the group produces another member of that same group.

The closure property assumes the fact that we have a *means* of combining any two members of the group. We will make this means explicit by using the symbol $*$ to denote combination. For example, we have

$$r_{120} * r_{240}$$

which denotes the combination of r_{120} and r_{240}. As we saw above, this combination is the same as the element e. That is,

$$r_{120} * r_{240} = e$$

Another example is the combination

$$r_{120} * r_{120}$$

which is, of course, r_{240}. That is, we have

$$r_{120} * r_{120} = r_{240}$$

The symbol $*$ will be called the *combination rule*. When the symbol occurs, it can easily be translated into English. For example:

$$r_{120} * r_{240}$$

means r_{120} *followed by* r_{240} Thus the combination rule $*$ means *followed by*.

We have seen above that the combination rule, $*$, in $\mathbf{D}_3$ obeys the property called *closure*. That is, if one takes any two members of $\mathbf{D}_3$ and combines them, one obtains another member of $\mathbf{D}_3$. The combination rule $*$ obeys another three simple properties, as follows:

(2) Associativity

Note first that the operation $*$ combines any *two* elements of a group. Suppose now that we wish to combine *three* elements. For example, suppose we wish to form the combination

$$r_{120} * r_{240} * r_{240}$$

Then we can do this by a simple trick: we can break the triple into pairs. For example, we can break the triple like this:

$$(r_{120} * r_{240}) * r_{240}$$

This trick allows us to combine the elements one pair at the time, as follows: First, we combine the two elements in the parentheses to produce one element, which in this case is e. Then, we combine e with the third element r_{240} to produce r_{240}. At any of these stages, we combined only two elements. The parentheses indicated how to do this.

Observe now that our initial triple,

$$r_{120} * r_{240} * r_{240}$$

can be broken into pairs, either like this:

$$(r_{120} * r_{240}) * r_{240}$$

or, like this:

$$r_{120} * (r_{240} * r_{240})$$

The *Associativity* property of a group guarantees that either method of breaking the triple into pairs yields the same final result. That is,

$$(r_{120} * r_{240}) * r_{240} \;=\; r_{120} * (r_{240} * r_{240})$$

In other words, we do not have to worry about which way we choose to break a triple into pairs. Both ways will yield the same result. In fact, the reader can check that both sides in the above equation reduce to r_{240}.

The Associativity property is this ability to break a string into pairs in any way. It can be simply stated by saying that, for any three elements x, y, z, the following holds:

$$(x * y) * z = x * (y * z)$$

(3) Identity Element

The group $\mathbf{D}_3$ contains a null element, e; that is, an element that has no effect. As we said, this element is called the *identity element*. Any group must contain an identity element, and must contain only one identity element. The way to characterize this element is to specify that it has no effect *when combined* with any other element of the group. For example,

$$r_{120} * e$$

gives the result r_{120}. Furthermore, the reverse order,

$$e * r_{120}$$

also gives r_{120}. That is, in general, we characterize the identity element by saying that, given any member x of the group, the following holds:

$$x * e = x = e * x$$

(4) Inverses

Close examination reveals that any member of $\mathbf{D}_3$ has a corresponding member that undoes the effect of that member. For example, consider the element, r_{120}, which is *clockwise* rotation by 120°. Clearly, the transformation that would undo the effect of r_{120} would be *anticlockwise* rotation by 120°. In fact, there is a member of $\mathbf{D}_3$ that has exactly this effect. It is r_{240}. Although the element r_{240} is clockwise rotation by 240°, its effect is equivalent to anticlockwise rotation by 120°. Now because the element r_{240} undoes the effect of r_{120}, it is called the *inverse of* r_{120}. In fact, the elements r_{240} and r_{120} are mutual inverses. That is, not only does r_{240} undo the effect of r_{120}, but r_{120} undoes the effect of r_{240}.

Let us consider the reflection t. The effect of any reflection is undone by performing the reflection again. For example, consider Fig 5.7a, the triangle in its initial position. Now apply t and thus obtain Fig 5.7b. In Fig 5.7b, the bottom vertices B and C have been reversed. Now apply t again and

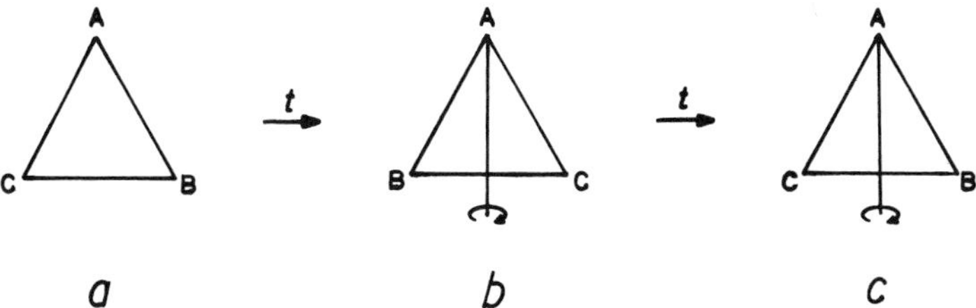

Figure 5.7
Application of t followed by t returns the figure to the original configuration.

obtain Fig 5.7c. Here the bottom vertices, B and C, have been reversed again, which means that they are in the same positions that they had in the initial configuration in Fig 5.7a. Thus, the second application of t undoes the first application of t. This means that t must be *its own* inverse. The same argument applies to each of the other reflections $r_{120}t$ and $r_{120}t$. Each is its own inverse.

The inverse of any element x will be denoted by x^{-1}. In a group, every element of the group has its own unique inverse. We can characterize this property by saying that, given any member x of the group, there is a unique member x^{-1} in the group, such that x combined with x^{-1} has a total of no effect; that is, the total effect of the combination is equivalent to the identity element, e. That is,

$$x * x^{-1} = e = x^{-1} * x$$

5.3 Four Basic Properties

Let us look back over the discussion in the previous section, in order to extract the general conclusions. We have been considering the symmetries of an equilateral triangle. On systematically elaborating all these symmetries, we found that there are a total of six; that is, three rotations and three reflections. The collection of six symmetries is given the label $\mathbf{D}_3$. The collection, $\mathbf{D}_3$, is a group for the following reasons. First, quite simply, it is a set of elements—in this case, it is a set of six transformations. Second, there is a rule, $*$, for combining any pair of elements. Finally, $\mathbf{D}_3$ obeys four basic properties, as follows. (1) *Closure*: When any pair of elements in $\mathbf{D}_3$ are combined, the resulting element is also in $\mathbf{D}_3$. (2) *Associativity*: The combination of three elements in $\mathbf{D}_3$ can be found by breaking the triple into pairs in either possible way. (3) *Identity Element*:

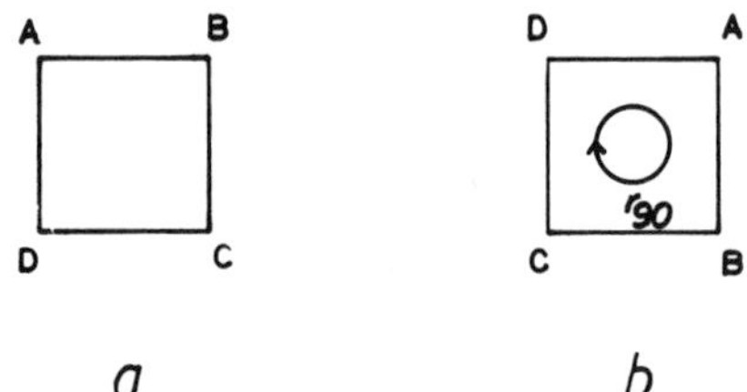

Figure 5.8
(a) A square. (b) A square after the application of rotation by 90°.

There is an element in $\mathbf{D}_3$ that has no effect when combined with any other element in $\mathbf{D}_3$. (4) *Inverses*: Each element, x, in $\mathbf{D}_3$ has a corresponding element, x^{-1}, that undoes the effect of x.

Thus we can state the definition of a *group*: A group is a set that has a rule for combining any pair of elements in the set, and that obeys the properties of *Closure*, *Associativity*, *Identity Element*, and *Inverses*.

5.4 The Symmetry Group of a Square

The symmetry structure of an equilateral triangle is expressed by the group $\mathbf{D}_3$. In exactly the same way, the symmetry structure of a *square* is captured by the group $\mathbf{D}_4$. Because an equilateral triangle has three equal sides, its group $\mathbf{D}_3$ consists of three rotations and three reflections. Similarly, because a square has four equal sides, its group $\mathbf{D}_4$ consists of four rotations and four reflections. These transformations can be systematically elaborated as follows.

In order to keep track of the effects of the various transformations that we try out, we shall label the vertices of the square, A, B, C, D, as shown in Fig 5.8a.

The smallest clockwise rotation that brings the square into coincidence with itself is rotation by 90°, shown in the transition between Fig 5.8a and 5.8b. In this transformation, the vertices have simply replaced each other in a simple cycle around the square; that is, vertex A has moved to B, vertex B to C, vertex C to D, and vertex D to A. We shall label this transformation, r_{90}. In order to check whether r_{90} has happened, we need only check whether the letters have changed in the way shown in Fig 5.8b.

Now let us look at the next largest rotation that brings the square back into coincidence with itself. This transformation is clockwise rotation by

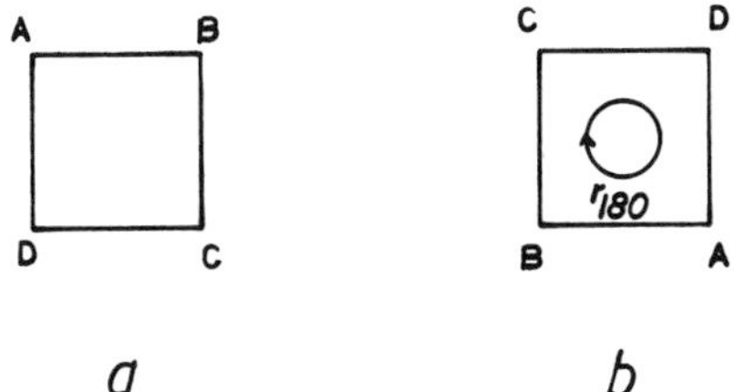

Figure 5.9
(a) A square. (b) A square after the application of rotation by 180°.

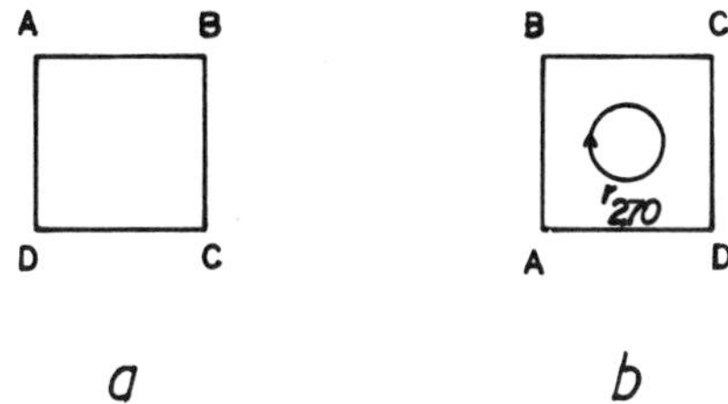

Figure 5.10
(a) A square. (b) A square after the application of rotation by 270°.

180°, shown in the transition between Fig 5.9a and 9b. In this transformation, the vertices have replaced each other one further step in a simple cycle around the square. We shall label this transformation, r_{180}. Again, in order to check whether r_{180} has happened, we need only check whether the letters have changed in the way shown in Fig 5.9b.

Now let us look at the next largest rotation that brings the square back into coincidence with itself. This transformation is clockwise rotation by 270°, shown in the transition between Fig 5.10a and 10b. In this transformation, the vertices have replaced each other one further step in a simple cycle around the square. We shall label this transformation, r_{270}. Again, in order to check whether r_{270} has happened, we need only check whether the letters have changed in the way shown in Fig 5.10b.

Now let us look at the next largest rotation that brings the square into coincidence with itself. This transformation is clockwise rotation by 360°. However, rotating the square by 360° is equivalent to applying no transformation at all. Thus, we can have a single label, e, for any transformation that has no ultimate effect on the vertices. For example, e means

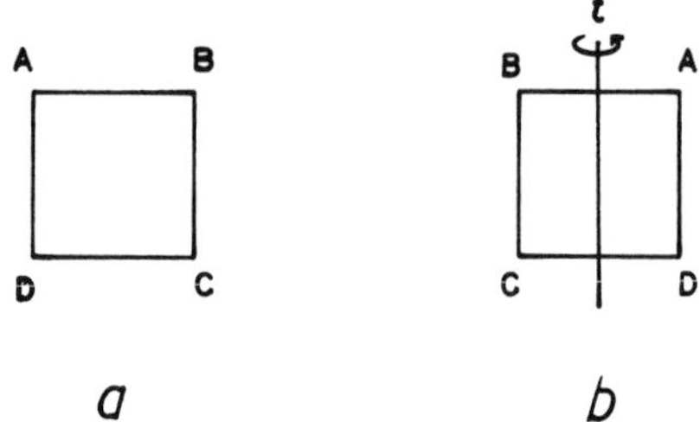

Figure 5.11
(a) A square. (b) A square after the application of the reflection t.

rotation by 0°; but it also means rotation by 360°. The null transformation e is, of course, the *identity element* of the group $\mathbf{D}_4$.

To summarize so far, we have four rotations that bring the square into coincidence with itself:

e, the null transformation

r_{90}, rotation by 90°

r_{180}, rotation by 180°

r_{270}, rotation by 270°

Now let us try to increase the amount of rotation above 360°. The next possible rotation, above 360°, that brings the square into coincidence, is rotation by 360° + 90°, which is 450°. However, observe that the effect of this rotation is exactly the same as rotation by 90°. This means that we do not have to include rotation by 450° in our list of symmetries if we already have rotation by 90°. Similarly, any higher amount of rotation will simply duplicate the rotations we have already. Thus, the above list of four transformations, e, r_{90}, r_{180}, and r_{270}, exhausts all the distinguishable rotations the square can undergo and be brought into coincidence with itself.

Having established all the distinguishable *rotations*, let us look at the possible *reflections*. Clearly, reflection about the vertical axis brings the square into coincidence with itself. This is the reflection that is shown in the transition from Fig 5.11a to Fig 5.11b. Observe that what has happened is that the top pair of vertices have been reversed, and the bottom pair of vertices have been reversed. This reflection will be labelled t.

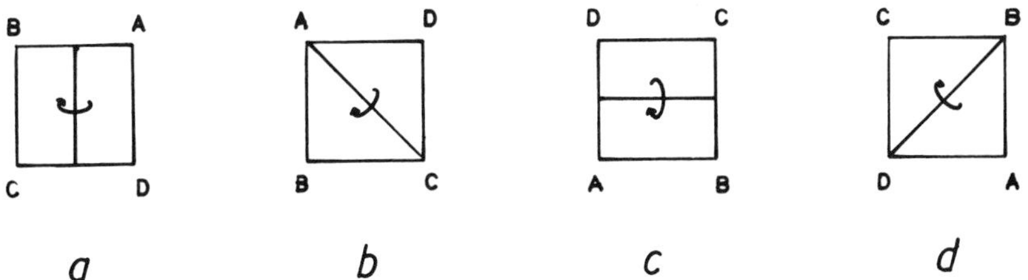

Figure 5.12
The four possible reflections of a square.

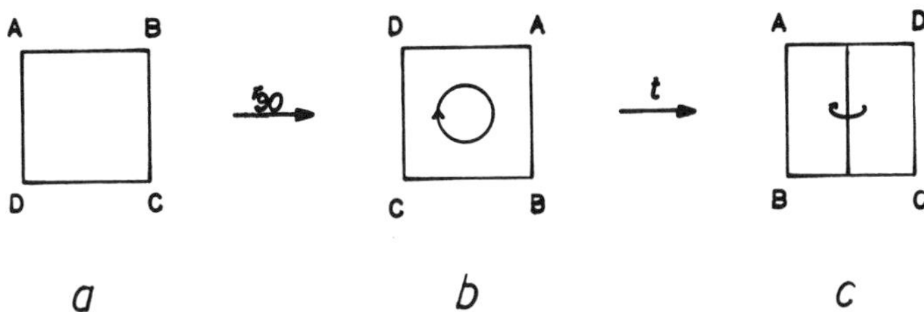

Figure 5.13
The reflection in Fig 5.12b can be generated in two stages.

There are a total of four distinct reflections that bring a square into coincidence with itself. The effect of reflection about the vertical, which we have just considered, is shown in Fig 5.12a. The three other reflections are shown in Fig 5.12b, c, and d, with their respective axes of reflection.

It is important to observe however that the reflection in Fig 5.12b can be obtained by combining two of the transformations we already have. This is shown in Fig 5.13. We start in Fig 5.13a with the square in its initial position. We then apply r_{90} obtaining Fig 5.13b. Then we apply t obtaining Fig 5.13c. However, the letters around the square in Fig 5.13c are in the same order and position as those in Fig 5.12b. Therefore the reflection in Fig 5.12b is equivalent to the application of r_{90} followed by t. This combination can be written as $r_{90}t$. Similarly, the reflection in Fig 5.12c is equivalent to the application of r_{180} followed by t, which we shall write as $r_{180}t$. Finally, the reflection in Fig 5.12d is equivalent to the application of r_{270} followed by t, which we shall write as $r_{270}t$.

We have now enumerated all the rotations and all the reflections that bring the square into coincidence with itself. There are four rotations and four reflections.

$$\textit{Rotations}:\quad e,\ r_{90},\ r_{180},\ r_{270}$$

$$\textit{Reflections}:\quad t,\ r_{90}t,\ r_{180}t,\ r_{270}t$$

These are all the *symmetries* of the square. That is, the square has a total of eight symmetries which we list between the parentheses on the next line:

$$\mathbf{D}_4 = \{e,\ r_{90},\ r_{180},\ r_{270},\ t,\ r_{90}t,\ r_{180}t,\ r_{270}t\}$$

As we have said, this entire list is labelled $\mathbf{D}_4$ meaning "Dihedral group of rank four". The first four members of the list (within the parentheses) are the four rotations and the last four are the four reflections.

It can easily be shown that $\mathbf{D}_4$ is a group. This is done by showing that $\mathbf{D}_4$ conforms to the four properties defining a group, as follows:

(1) Closure

Any pair of members of $\mathbf{D}_4$ can be combined to produce another member of the group. For example, the rotation r_{90} can be combined with the rotation r_{180} to obtain $r_{90} * r_{180}$, which turns out to be another member of the group. To see which member this is, observe that $r_{90} * r_{180}$ is rotation by 90° followed by rotation by 180°. This gives a total rotation of 270°. Therefore, the combination $r_{90} * r_{180}$ is the same as r_{270}. As another example, we can combine r_{90} with itself, obtaining $r_{90} * r_{90}$, which is a rotation of 90°, applied *twice*. This combination is clearly r_{180}, which is also another member of the group.

(2) Associativity

As noted earlier, the operation $*$ combines any *two* elements of a group. The Associativity property ensures that any *three* elements can be combined with consistent results. For example, suppose we wish to form the combination

$$r_{90} * r_{270} * r_{90}$$

Then the Associativity property assures us that we can break the triple into pairs, in any way. That is, either as

$$(r_{90} * r_{270}) * r_{90}$$

or as

$$r_{90} * (r_{270} * r_{90})$$

In either case, one will obtain the same result, r_{90}.

(3) Identity Element

The group $\mathbf{D}_4$ contains a null element, e; that is, an element that has no effect. As we said, this element is called the *identity element* of the group. When it is combined with any other element of the group, the total effect is the same as the latter element on its own. For example,

$$r_{90} * e$$

is the same as r_{90}.

(4) Inverses

Any member of $\mathbf{D}_4$ has a corresponding member that undoes the effect of that member. For example, consider the element, r_{90}, which is *clockwise* rotation by 90°. Clearly, the transformation that would undo the effect of r_{90} would be *anticlockwise* rotation by 90°. In fact, there is a member of $\mathbf{D}_4$ that has exactly this effect. It is r_{270}. The element r_{270} is clockwise rotation by 270°, and this is equivalent to anticlockwise rotation by 90°. Because the element r_{270} undoes the effect of r_{90}, it is the *inverse of* r_{90}.

Now consider the reflections t, $r_{90}t$, $r_{180}t$, $r_{270}t$. As noted earlier, the effect of any reflection is undone by performing the reflection again. Therefore, each of these reflections is its own inverse. Therefore, quite trivially, each reflection in $\mathbf{D}_4$ has an inverse in $\mathbf{D}_4$.

Let us finally look back over the discussion in this section. We have been considering the symmetries of a square. On systematically elaborating these symmetries, we found that there are a total of eight; that is, four rotations and four reflections. This collection of eight symmetries is given the label $\mathbf{D}_4$. The collection, $\mathbf{D}_4$, is a group for the following reasons. First of all, it is a set of elements—in this case it is a set of eight transformations. Second, there is a rule, $*$, for combining any pair of elements. Finally, $\mathbf{D}_4$ obeys four basic properties, as follows. (1) *Closure*: When any pair of elements in $\mathbf{D}_4$ are combined, the resulting element is also in $\mathbf{D}_4$. (2) *Associativity*: The combination of three elements in $\mathbf{D}_4$ can be found by breaking the triple into pairs in either possible way. (3) *Identity Element*: There is an element in $\mathbf{D}_4$ that has no effect when combined with any other element in $\mathbf{D}_4$. (4) *Inverses*: Each element, x, in $\mathbf{D}_4$ has a corresponding element, x^{-1}, that undoes the effect of x.

5.5 The Symmetry Group of a Regular Planar Polygon

Let us continue looking at polygons in the plane.

The symmetry structure of an equilateral triangle is expressed by the group $\mathbf{D}_3$. The symmetry structure of a square is expressed by the group $\mathbf{D}_4$. Correspondingly, the symmetry structure of any regular polygon of n sides is given by the group $\mathbf{D}_n$. For example, the symmetry group of a pentagon is $\mathbf{D}_5$, the symmetry group of a hexagon is $\mathbf{D}_6$, and so on.

For each regular polygon of n sides, the associated group $\mathbf{D}_n$ has the same general structure. The group consists of n rotations, and n reflections. That is, the group consists of

Rotations: $e, r, r^2, r^3, \ldots, r^{n-1}$

Reflections: $t, rt, r^2t, r^3t, \ldots, r^{n-1}t$

where r is the smallest rotation that brings the polygon into coincidence with itself. Notice that the rotation list consists of successively increasing multiples of r. The highest multiple is r^{n-1} because the next highest multiple, r^n, is simply rotation by 360°, which is e. Notice also that the reflections are exactly the same rotations each multiplied by t.

5.6 Cyclic Groups

Consider again the symmetry group $\mathbf{D}_3$ of an equilateral triangle. We have seen that $\mathbf{D}_3$ splits into two halves:

Rotations: e, r_{120}, r_{240}

Reflections: $t, r_{120}t, r_{240}t$

The three *rotations*, in fact, together form a group. That is, the set

Rotations: e, r_{120}, r_{240}

is a group in its own right. This can easily be checked by showing that this set satisfies the four properties of a group: (1) *Closure*: When any pair of the rotations is combined, the resulting transformation is also one of the rotations. (2) *Associativity*: The combination of three of the rotations (e.g., including possible repetitions) can be found by breaking the triple into pairs in either possible way. (3) *Identity Element*: One of the rotations, e, has no effect when combined with any of the other rotations. (4) *Inverses*: To each of the three rotations there is another rotation (amongst the three) that undoes the effect of the rotation.

Thus the three rotations form a group. Now observe the following: $\mathbf{D}_3$ is, as we have seen, a group. However, it splits into two halves, its set of rotations and its set of reflections. The set of rotations forms a group in its own right. Generally speaking, if a subset of a group is itself a group, then it is called a *subgroup*. Therefore, the set of rotations is a subgroup of $\mathbf{D}_3$.

The set of rotations we have been considering, e, r_{120}, r_{240}, is a particularly simple group. It is a *cycle*, as follows. Applying r_{120} three successive times moves one successively through the three rotations. That is, applying r_{120} once is simply r_{120} (rotation by 120°); applying r_{120} twice yields r_{240} (rotation by 240°); applying r_{120} three times is equivalent to no rotation, that is, e. Now if one continues to apply r_{120}, one will again cycle through the three rotations, and so on.

Because the above group of rotations forms a cycle of size three, we shall call the group, **Cycle(3).** That is we have:

$$\textbf{Cycle(3)} = \{e, r_{120}, r_{240}\}$$

We have seen therefore that the symmetry group $\mathbf{D}_3$ of an equilateral triangle splits into two halves, its set of rotations and its set of reflections. The set of rotations forms a group **Cycle(3)**.

Consider now the symmetry group, $\mathbf{D}_4$, of a square. We have seen, earlier, that $\mathbf{D}_4$ also splits into two halves.

Rotations: $e, r_{90}, r_{180}, r_{270}$

Reflections: $t, r_{90}t, r_{180}t, r_{270}t$

The four rotations, in fact, together form a group. This can easily be checked by showing that together, they satisfy the four properties of a group: Closure, Associativity, Identity Element, and Inverses.

Therefore, the four rotations e, r_{90}, r_{180}, r_{270}, form a *subgroup* of $\mathbf{D}_4$. This subgroup is again a *cycle*. That is, applying r_{90} four successive times moves one successively through the four rotations. Therefore, the group formed by the four rotations will be called **Cycle(4)**. That is we have:

$$\textbf{Cycle(4)} = \{e, r_{90}, r_{180}, r_{270}\}$$

Thus, we have seen that the symmetry group $\mathbf{D}_4$ of a square splits into two halves, its set of rotations and its set of reflections. The set of rotations forms a group **Cycle(4)**.

Recall now that, given a regular polygon, of any number of sides n, its symmetry group $\mathbf{D}_n$ splits into two halves, thus:

$$\textit{Rotations}: \quad e,\ r,\ r^2,\ r^3,\ \ldots,\ r^{n-1}$$

$$\textit{Reflections}: \quad t,\ rt,\ r^2t,\ r^3t,\ \ldots,\ r^{n-1}t$$

The *rotations* together form a group. Furthermore, the group is simply a cycle of size n. That is, applying r, successively, moves one successively through all the rotations, in a cyclic fashion. This group of rotations will therefore be labeled **Cycle(*n*)**. That is,

$$\mathbf{Cycle}(\boldsymbol{n}) = \{e,\ r,\ r^2,\ r^3, \ldots,\ r^{n-1}t\}$$

5.7 The Reflection Group

As we noted $\mathbf{D}_3$ has three reflections, t, $r_{120}t$, and $r_{240}t$. Let us consider t. We noted that applying t twice has the same effect as doing nothing; that is, it has the same effect as applying e. Now, applying t a third time produces t. Furthermore, applying t a fourth time produces e. Thus, continuing to apply t simply cycles one backwards and forwards between e and t. This means that the set consisting purely of e and t forms a group. The group is, of course, **Cycle(2)**. That is

$$\mathbf{Cycle(2)} = \{e,\ t\}$$

The same argument applies to the reflection $r_{120}t$. This means that repetition of $r_{120}t$ produces a cycle of size two. That is, applying $r_{120}t$ once produces $r_{120}t$; applying $r_{120}t$ twice produces e; and continuing to apply $r_{120}t$ simply cycles one backwards and forwards between e and $r_{120}t$. Thus the set consisting purely of e and $r_{120}t$ forms a group. The group is, again, an example of **Cycle(2)**.

Exactly the same argument applies when we take the third reflection $r_{240}t$ in $\mathbf{D}_3$. This reflection too forms a group together with e. The group is, again, an example of **Cycle(2)**.

Now consider $\mathbf{D}_n$, that is, the symmetry group of the n-sided polygon. We saw that, besides its n rotations, it contains n reflections. We can now see that each reflection can be paired with the identity e to form a group of size 2, the group **Cycle(2)**.

Often, throughout the book, we will use an alternative label for the group **Cycle(2)**. We shall label it **Reflections**.

One final point is worth making. Let us return again to $\mathbf{D}_3$, the symmetry group of an equilateral triangle. We have often considered $\mathbf{D}_3$ split into two halves: the three rotations and the three reflections. We have also observed that the three rotations *together* form a group.

The reader might now ask whether the three reflections also, *together*, form a group; that is, whether the set consisting purely of the reflections, t, $r_{120}t$, $r_{240}t$, is a group. The answer is negative. Nevertheless, we have now seen that each *individual* reflection, paired with the identity element, e, forms a group, **Cycle(2)**.

These considerations are of course general. The symmetry group, $\mathbf{D}_n$, of an n-sided polygon, contains n rotations and n reflections. However, whereas the n rotations *together* form a group, the n reflections do not. Nevertheless, each *individual* reflection, together with e forms a group of size 2, the group **Cycle(2)**.

5.8 The Continuous Rotation Group

We have seen that the group $\mathbf{D}_n$ of any regular n-sided polygon contains a group of rotations called **Cycle(n)**. The rotations are n equally spaced rotations. Thus they can be represented by n equally spaced notches around a circular clock. For example, Fig 5.14 shows the particular notches on a clock that mark each successive quarter of an hour (i.e., the 12, 3, 6, and 9 o'clock positions). There are four such notches, and the hand rotates 90° between each successive notch. Since the hand goes

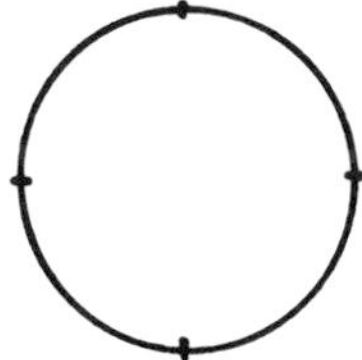

Figure 5.14
A clock with the quarter hour positions marked.

through four such successive rotations to get back to the starting position, the hand goes through the group **Cycle(4)**. This is exactly the same group as the rotations group of a square; i.e., the rotation group within $\mathbf{D}_4$.

Again consider those twelve notches that mark the *hours* on a clock. The hand rotates 30° between each successive notch. Since the hand goes through twelve such successive rotations to get back to the starting position, the hand must go through the group **Cycle(12)**. This group is of course the same group as the group of rotations of a 12-sided polygon. Again, if we increase the number of equally spaced notches to 1000, we will have the group **Cycle(1000)**.

Now let us suppose that we allow *all* the points on the circle to become notches; i.e., the entire continuum of points around the circle. That is, we allow rotations between *any* points on the circle. The entire set of rotations forms a group. For example, consider Fig 5.15a and Fig 5.15b. They show rotations to two arbitrary points on the circle. These two rotations can be added as shown in Fig 5.15c. The combination is of course equivalent to another rotation on the circle.

Observe also that the group forms a cyclic group. This is because rotating 360° past the zero point will cause one to "wrap around" and start again.

The main difference between any of the previous cyclic groups, **Cycle(*n*)**, and the new cyclic group, is that the former were *discrete*, in this sense: In a previous **Cycle(*n*)** group, the rotations could increase only by discrete steps; e.g., in **Cycle(4)**, the rotations increased by 90° discrete steps. In the new group, however, the amount of rotation can increase *continuously*, rather than be confined to discrete steps. That is, there is a *continuum*

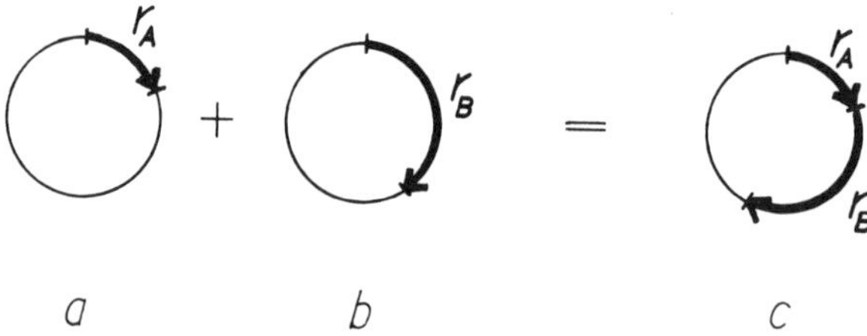

Figure 5.15
The addition of rotations produces another rotation.

of rotations, one rotation for each point on the circular continuum around the clock.

We shall label the new group simply **Rotations**. It is the largest possible group of rotations. Any other rotation group is an example of **Cycle(*n*)**. Furthermore, any **Cycle(*n*)** is a subgroup of the group **Rotations**. For example, **Cycle(4)**, the rotation group of a square, has selected the four 90° rotations from the overall group **Rotations**. Again, the twelve-hour clock, **Cycle(12)**, has selected the twelve 30° rotations from the overall group **Rotations**.

5.9 Translations Along a Line

The group **Rotations** is a continuous group. That is, rotations can increase continuously around the circle rather than be confined to discrete steps like any **Cycle(*n*)**. The group **Rotations** has a very simple structure. However, it is not the simplest possible continuous group. There is a continuous group that is even simpler. This is the group of translations along a line.

Imagine an infinite straight line. Now consider translations, or movements, along that line. Notice that any pair of translations can be combined to form another translation. For example, Fig 5.16a represents a translation by 5 inches and Fig 5.16b represents a translation by 2 inches. These can be added together, as shown in Fig 5.16c, to yield a total translation of 7 inches. Consider all possible translations along the line, in both the negative and positive directions. The entire collection forms a group that we will simply call **Translations**.

One can easily show that **Translations** forms a group, and we will show this, in order to remind the reader of this procedure: First observe that, if we combine any pair of translations, as illustrated in Fig 5.16, we obtain another translation. This means that the set of translations is closed under combination.

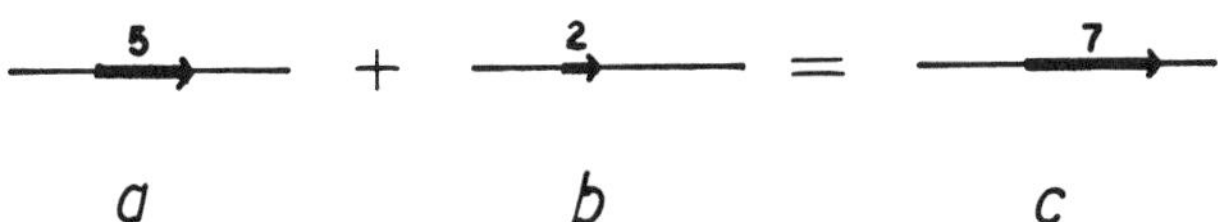

Figure 5.16
The addition of translations produces a translation.

Again, observe that a combination of three translations, for example,

$$5 \text{ inches} + 2 \text{ inches} + 3 \text{ inches}$$

can be partitioned into pairs in either way. That is,

$$(5 \text{ inches} + 2 \text{ inches}) + 3 \text{ inches} = 5 \text{ inches} + (2 \text{ inches} + 3 \text{ inches})$$

Thus translations conform to the Associativity property.

Again, observe that there is a translation with no effect: The translation by 0 inches. This acts as an identity element. For example,

$$5 \text{ inches} + 0 \text{ inches} = 5 \text{ inches}$$

Finally, observe that, for each translation, there is a translation that undoes its effect. The latter is simply the former applied in the opposite direction. For example, the translation, 5 inches, is undone by applying the translation, −5 inches. Thus, generally, each translation has an *inverse*.

We conclude therefore that **Translations** is a *group*, i.e., it conforms to the four requirements for a group: *Closure*, *Associativity*, *Identity Element*, and *Inverses*.

Recall now the relationship between the groups **Cycle(*n*)** (for all sizes *n*) and the group **Rotations**. The former groups are all discrete; that is, the amount of rotation in a **Cycle(*n*)** can increase through the successive members of **Cycle(*n*)** by only discrete steps (e.g., by intervals of 90° in **Cycle(4)**). In contrast the amount of rotation in **Rotations** can increase by continuous amounts. Each **Cycle(*n*)** is therefore a *discrete subgroup* of the continuous group **Rotations**.

Now let us return to **Translations**. This is a continuous group. That is, one can increase the amount of translation continuously through the group, i.e., along the straight line. However, just as **Rotations** has discrete subgroups, **Translations** has discrete subgroups. Each of the discrete subgroups in **Translations** is simply the successive translation by an equal interval. For example, one such subgroup is the group of translations by the following amounts:

$$\{\ldots, -3, -2, -1, 0, 1, 2, 3, \ldots\}$$

This group is called **Numbers**. Thus, whereas the complete group **Translations** allows movement to any point along the line shown in Fig 5.17a, the subgroup **Numbers** is confined to movements to only the equally spaced dots in Fig 5.17b.

Figure 5.17
Translations allows movements to all points; (b) **Numbers** to equally spaced points.

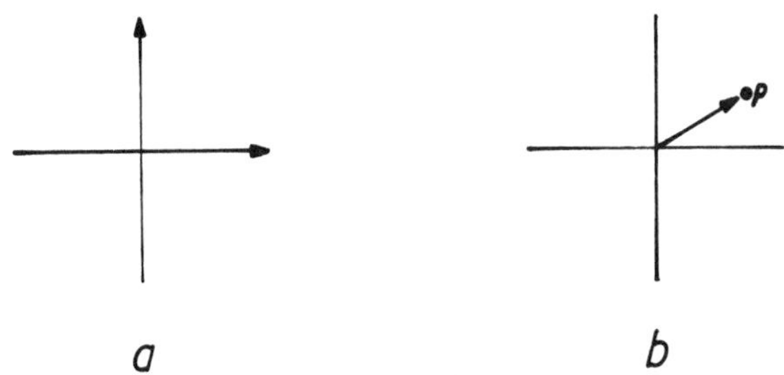

Figure 5.18
(a) Horizontal and vertical translation. (b) Translation to a point p.

5.10 Translations in the Plane

In the previous section, we considered translations along a line. Now let us consider translations in a plane. The plane is depicted in Fig 5.18a with a horizontal and a vertical axis. Consider any translation in this plane, for example, the translation shown in Fig 5.18b from the origin to point p. Clearly, this translation can be accomplished by combining a translation in the horizontal direction, and a translation in the vertical direction. However, the set of movements along the horizontal direction is simply a copy of the group **Translations**, defined in the previous section (i.e., translations along a line). Similarly, the set of movements along the vertical direction is a copy of the group **Translations** (i.e., translations along a line). This means that the entire set of translations in the plane is produced by a combination of two copies of the group **Translations**. This combination will be written simply as

Translations $\times$ **Translations**

The set of translations in the plane, **Translations** $\times$ **Translations**, is itself a group. That is, any pair of planar translations can be combined to give another planar translation (i.e., the Closure property holds). A triple of

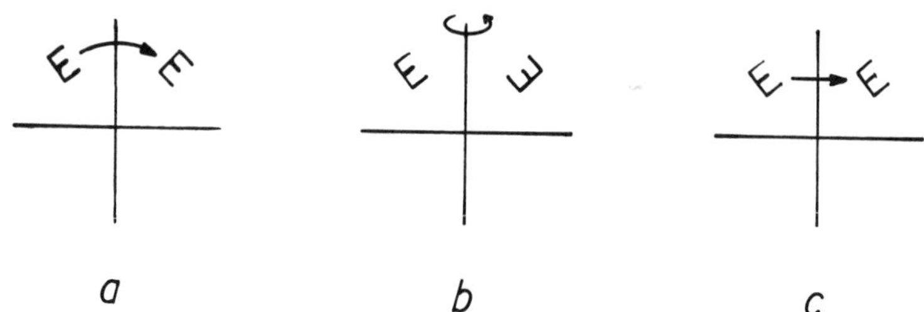

Figure 5.19
The Euclidean motions of (a) rotation, (b) reflection, and (c) translation.

successive planar translations can be broken into pairs in either way (i.e., the Associativity property holds). There is a planar translation that has no effect, i.e., the zero translation, and this serves as the identity element. Finally, any planar translation has its inverse—the translation by the same distance in the opposite direction.

5.11 Euclidean Motions

Although we have so far mentioned several examples of groups, the transformations that have made up the groups have been of three types: rotations, reflections, and translations. There exist groups with very different types of transformations, as we shall see. However, rotations, reflections and translations share a common property. *They all preserve size and shape*. For example, consider the letter E on the upper left of Fig 5.19a. After it has been rotated, as shown in Fig 5.19a, it has the same size and shape as it had before rotation. Fig 5.19b shows it undergoing a reflection. Again, although it has changed sidedness, its size and shape are the same. Finally, in Fig 5.19c, the letter has undergone a translation. Again, its size and shape have been preserved.

A transformation that preserves size and shape is called a *Euclidean transformation*. Any Euclidean transformation is a combination of a rotation, a reflection and a translation. For example, Fig 5.20a shows a letter E that has undergone a Euclidean transformation, i.e., its size and shape have remained the same. Fig 5.20b shows that this transformation can be reconstructed using a translation followed by a rotation.

The set of all Euclidean transformations, i.e., size-and-shape preserving

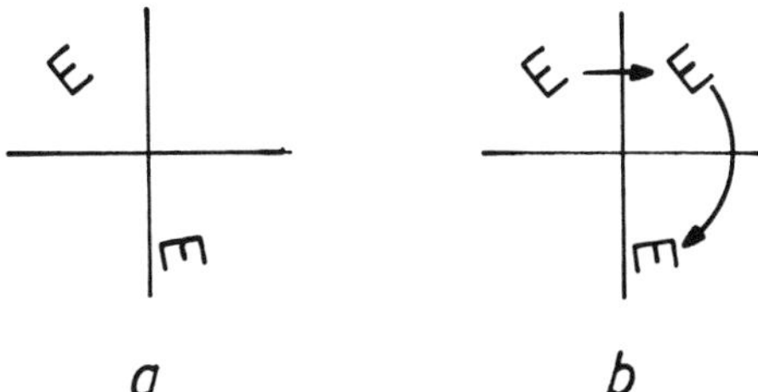

Figure 5.20
(a) A Euclidean motion, (b) decomposed into a translation followed by a rotation.

transformations, forms a group. This is easy to check, as follows: The combination of any pair of size-and-shape preserving transformations is a transformation that also preserves size and shape, and thus is in the same group; i.e., the Closure property is satisfied. A triple of such transformations can be broken into pairs in either possible way; i.e., the Associativity property is satisfied. There is a null transformation that preserves size and shape, i.e., the transformation that does nothing at all; i.e., there exists a unique identity element. And finally, to each transformation that preserves size and shape, there is a transformation that undoes its effect while preserving size and shape; i.e., the existence of inverses.

The group of size-and-shape preserving transformations will be called **Euclidean**. It consists of all the rotations, reflections and translations, as well as all their combinations.

All the groups we considered in the previous sections were subgroups of the Euclidean group. For example, the group $\mathbf{D}_n$, the symmetry group of the n-sided polygon, consists of n rotations and n reflections; i.e., Euclidean transformations. Recall also that the n rotations form a group called **Cycle(*n*)**. Observe that both $\mathbf{D}_n$ and **Cycle(*n*)** have a finite number of elements. $\mathbf{D}_n$ consists of $2n$ elements, and **Cycle(*n*)** consists of n elements. The groups $\mathbf{D}_n$ and **Cycle(*n*)** have a special significance in the group **Euclidean**: They are the *only* finite subgroups of **Euclidean**.

Recall now that three of the groups we examined in the previous sections were

Rotations: the continuous group of rotations

Reflections: the group of two elements $\{e, t\}$

Translation × **Translation**: the group of translations of the plane

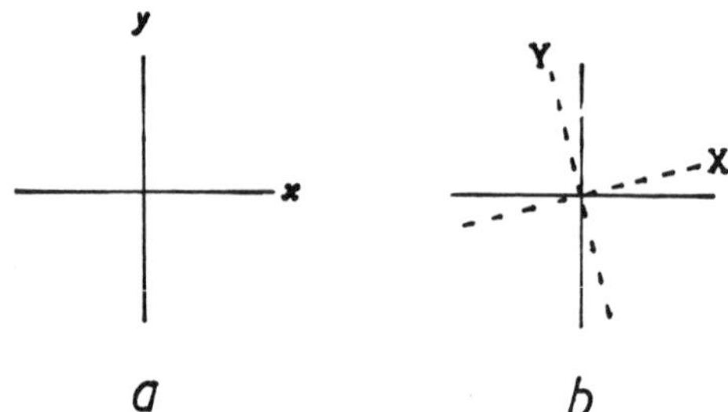

Figure 5.21
A rotation preserves straightness.

Each of these groups is a subgroup of the Euclidean group. In fact, the Euclidean group is simply a combination of these three groups. That is, any Euclidean transformation can be written as a combination of transformations from these three groups.

5.12 Linear Transformations

Euclidean transformations preserve shape and size. We shall now consider a slightly more general class of transformations: those that preserve *straightness*; i.e., do not involve bending. Such transformations are called *linear transformations*. In fact, to be more precise, linear transformations preserve straightness *and* the position of the origin.

We can see that any rotation is a linear transformation. For example, consider the rotation shown in going from Fig 5.21a to Fig 5.21b. The x-axis in Fig 5.21a is a straight line. After it is rotated, it becomes the axis labeled X in Fig 5.21b. This is also a straight line. Thus, although the x-axis has been moved, its straightness is preserved. The same is true of the y-axis.

The fact that a rotation preserves straightness is due to the fact that rotation preserves shape; i.e., straightness is a shape-like property. However, there are some transformations that preserve straightness but do not preserve shape. An example we have encountered in this book is in the assignment of a history from a rotated parallelogram back to a square. The transformation between a square and a rotated parallelogram does not preserve shape, but nevertheless preserves straightness. Fig 5.22 shows the transformation involved. Consider, for example, the x-axis and y-axis

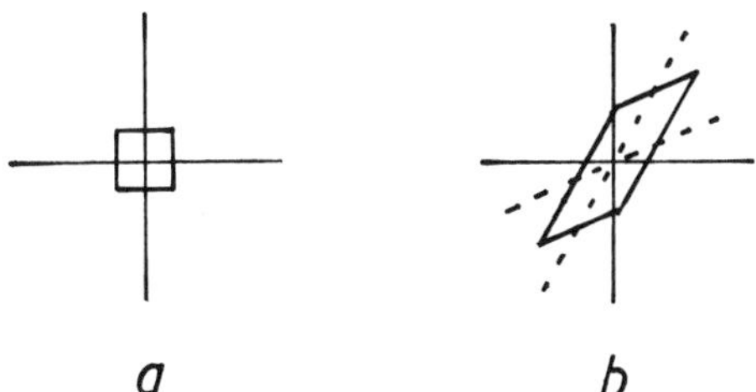

Figure 5.22
An example of a linear transformation that preserves straightness but not shape.

in Fig 5.22a. They are both straight lines. Under the transformation, the axes become the dashed lines shown in Fig 5.22b. These latter lines are also straight. Thus, as a result of the transformation, the axes in Fig 5.22a change position and their relation to one another, but nevertheless remain straight. In fact, any of the straight lines in Fig 5.22a remain straight lines in Fig 5.22b. Observe also another property of this transformation: the origin has remained in the same position.

Generally, a linear transformation is one that preserves straightness and preserves the position of the origin. Note that, in a larger dimensional space, a linear transformation preserves not only the straightness of lines, but also the straightness of planes, and any higher-dimensional straight slices. For example, a linear transformation of a three-dimensional space preserves the straightness of lines and of planes. From now on, unless otherwise stated, we shall assume that the term "linear transformation" refers to a linear transformation acting on a two-dimensional space.

The set of all linear transformations forms a *group*. This is not difficult to see. The combination of two transformations that preserve straightness must be another transformation that preserves straightness. Thus, the Closure property holds. Again, a combination of three transformations that preserve straightness can be broken down into pairs, in either way possible. Thus the Associativity property holds. The identity element is simply the transformation that does nothing. Clearly, this transformation preserves straightness. Finally, the inverse of a transformation that preserves straightness must be a transformation that preserves straightness. Thus, the four properties that define a group—Closure, Associativity, Identity Element, and Inverses—are true of the group of linear transformations. We shall call the group of linear transformations, **Linear**.

5.13 The Affine Group

The group **Linear** consists of all transformations that preserve straightness and the position of the origin. What about translations? Clearly, translations preserve straightness; i.e., straight lines and planes do not bend under translations. But observe that translations shift the position of the origin. Therefore, they are not members of **Linear**.

If we wish to have a group that includes *all* the transformations that preserve straightness, we simply add the translations to the linear transformations. That is, we add the group, **Translations** × **Translations**, to the group, **Linear**. The resulting, larger group is called **Affine**. That is:

Affine = Linear + (Translations × Translations)

Any transformation in **Affine**, i.e., any transformation that preserves straightness, can be decomposed into a transformation from **Linear** and a transformation from **Translations** × **Translations**.

5.14 Summary

The concept of group is used to describe the phenomenon of symmetry precisely. A group is a set of elements, which we will always regard as a set of transformations. With respect to this set, there is a rule, $*$, for combining any pair of elements. This combination rule satisfies the four following properties: (1) *Closure*: When any pair of elements in the group are combined, the resulting element is also in the group. (2) *Associativity*: The combination of three elements in the group can be found by breaking the triple into pairs in either possible way. (3) *Identity Element*: There is an element in the group that has no effect when combined with any other element in the group. (4) *Inverses*: Each element in the group has a corresponding element that undoes its effect.

In this chapter we studied a number of particular groups that will be used later in the book. These are listed and described in Table 5.1

TABLE 5.1 THE GROUPS DESCRIBED IN CHAPTER 5.

$\mathbf{D_3}$, the symmetry group of a triangle—three rotations and three reflections.

$\mathbf{D_4}$, the symmetry group of a square—four rotations and four reflections.

$\mathbf{D_n}$, the symmetry group of a n-sided regular polygon—*n* rotations and *n* reflections.

Cycle(3), the cyclic group of 3 elements. The group can be understood as the rotation group of a clock with 3 equally spaced numbers.

Cycle(4), the cyclic group of 4 elements. The group can be understood as the rotation group of clock with 4 equally spaced numbers.

Cycle(n), the cyclic group of *n* elements. The group can be understood as the rotation group of clock with *n* equally spaced numbers.

Reflections, the group of two elements: *e*, the null transformation, and *t*, which can be interpreted as a reflection. The group can also be understood as **Cycle(2)**, the cyclic group of size 2.

Rotations, the continuous group of rotations.

Translations, the continuous group of translations along an infinite line.

Numbers, the additive group of positive and negative whole numbers (sometimes called the group of integers).

Translations × **Translations**, the continuous group of translations of the plane.

Euclidean, the group of rotations, reflections, and translations, together with all their combinations.

Linear, the group of linear transformations; i.e., transformations preserving straightness and the position of the origin.

Affine, the group consisting of all linear transformations and translations, together with all their combinations.

6 Domain-Independent Rules

6.1 Introduction

This chapter is entirely devoted to perception. The problem we will examine is introduced as follows: The perceptual assignment of process-history to a shape clearly depends on which physical material the shape is assumed to be made from. For example, a sheet of paper is seen to have a different type of history from a sheet of rubber or from a sheet of cotton. That is, the same *shape* in each case is seen to have been produced by different causal interactions or forces; and, conversely, the same forces in each case would be understood to produce different shapes for the different materials.

Given the dependency of history on physical materials, it appears that, in order to gain any understanding of the recovery of history from shape, one first needs to consider the particular physical materials that are understood to be involved. Thus, it appears that the process-recovery problem is a collection of many separate process-recovery problems each determined by a different material domain.[1]

Following from this conclusion, it might seem that the appropriate research strategy, in understanding perception, is to work out the process-inference rules used for each material domain *independently*. This would be done by studying the particular dynamics of each domain and trying to discover the process-inference rules appropriate to that domain. However, while the study of the individual dynamics of each domain must be carried out to fully investigate the process-recovery problem, a fundamental proposal we wish to make in this book is that *there are rules that are valid irrespective of the material domain under consideration*. We shall call such rules *domain-independent rules*. The goal of the present chapter is to elaborate a number of such rules. We have actually proposed some already in this book. In fact, the most important rule of the book, the Asymmetry Principle, is a domain-independent rule. That is, we propose that, no matter which material domain one is investigating, asymmetries in that domain are regarded as arising backwards in time from symmetries.

In fact, the Asymmetry Principle holds across all domains despite the fact that the asymmetries are defined differently in each domain. To illustrate, let us consider again the problem, discussed in section 3.12, of interpolating a surface between a boundary in the image; e.g., between the boundary in Fig 6.1. As we noted, one chooses the surface to be that with the least distinguishability. However, observe that there are many alterna-

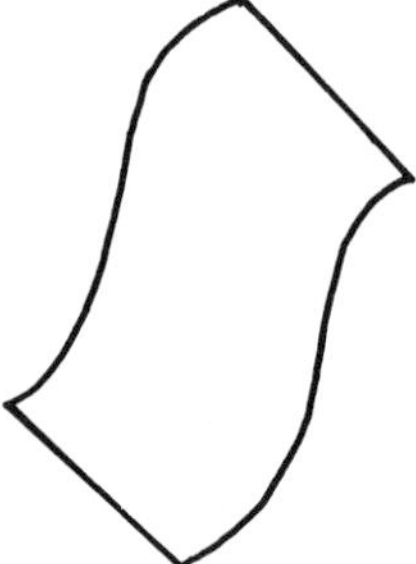

Figure 6.1
A boundary can be interpolated with alternative materials.

tive materials that could be conjectured for the surface. For example, the surface could be a membrane or a thin plate. It turns out that, if the surface is a membrane, then the distinguishability is taken to be based on the rates of change across the surface (i.e., the first derivatives). In contrast, if the surface is a thin plate, then the distinguishability is taken to be based on the rates of rates of change across the surface (i.e., the second derivatives). In both cases, the surface is understood as derivable from a flat surface; i.e., a surface that has no distinguishability relative to the distinguishability measure.[2] Thus, although the two types of surface are in two different material domains and each type of surface is evaluated by a measure that is particular to the domain, the measure is nevertheless a measure of distinguishability in that domain. Furthermore, although the two material domains necessarily involve two different types of dynamics, both types of dynamics consist of the generation of distinguishability from non-distinguishability, i.e., the surface is seen as deformed from a flat surface.

The membrane and thin-plate models also illustrate other domain-independent principles we have proposed in this book. One is the Externalization Principle, which states that the removal of external history always reduces the internal structure to purely Euclidean processes. Despite the very different dynamics of the membrane and thin plate, the removal of the deformational histories reduces both models to flat planes, which are defined by purely Euclidean processes, i.e., translations.

Again, the membrane and thin-plate models illustrate the Energy-Asymmetry Principle (Chapter 2) which is also a domain-independent rule. The principle states that asymmetry is taken, psychologically, to correspond to energy transferred to an object. Despite the fact that the membrane and thin-plate models are in different domains and have different measures of asymmetry, both *asymmetry* measures are regarded as measures of *energy* in the surface, in fact, energy transferred to the surface by deformation. Note that, in both cases, the energy is *potential* energy in accord with our proposal that the asymmetry used in external inference corresponds to potential energy.

6.2 The Structure of Histories

The central issue of this book is the recovery of process-history. We are concerned in this chapter with exploring domain-independent principles for the *structure* of recovered histories.

Clearly, the Representational Machine Analogy is a fundamental domain-independent rule because it claims that the mind structures any history so that it is well-defined; i.e., the history is represented as a machine undergoing some program of actions. However, while the Representational Machine Analogy states that a stimulus is represented as a machine history, it does not state which of the alternative machine histories is chosen. Several such histories could be compatible with the presented stimulus.

The particular choice of machine history is actually determined by those principles of ours that involve the concept of symmetry: the Asymmetry Principle, the Symmetry Principle, and the History Minimization Principle. The Asymmetry Principle states that the chosen history is one that reduces asymmetry backwards in time; the Symmetry Principle states that the chosen history must preserve symmetries backwards in time; and the History Minimization Principle states that the chosen history must be as symmetrical as possible with respect to translations across time, and, in particular, that any process is simply a repetition, i.e., a process is completely symmetrical with respect to temporal translations.

To gain a full understanding of the structure of an inferred history, we have to create a bridge between the Representational Machine Analogy, which concerns the concept of machine, and the three symmetry-based principles.

To begin, recall that a machine consists of five components:

(1) INPUTS: actions that are applied by the environment on the machine.

(2) OUTPUTS: actions that are applied by the machine on the environment.

(3) STATES: properties of the machine that act as factors in determining consequences of applying the inputs.

(4) FIRST CAUSAL RELATIONSHIP

Input × State → Output

(4) SECOND CAUSAL RELATIONSHIP

Input × State → Next State

Recall also that the first causal relationship defines the machine in terms of its *use* while the second causal relationship defines the machine in terms of its *history*. Our concern in this book is mainly with the second causal relationship. We shall be concerned with the first causal relationship in a further book.

The second causal relationship describes the changes in state, under the action of the inputs. These changes or transitions in state can be represented by a diagram called the *state-transition diagram*.

To illustrate such a diagram, let us look at a simple example. Fig 6.2 shows the state-transition diagram of the simplest possible machine, the light switch. The diagram can be explained as follows. In any state-transition diagram, one simply draws a box for each state, and draws an arrow between each box to represent the change in state caused by each input action. The action labels the arrow. To illustrate: In a light-switch, there are two states, ON and OFF. Therefore, there are two boxes in the state-transition diagram shown in Fig 6.2. Furthermore, at any time, one of two

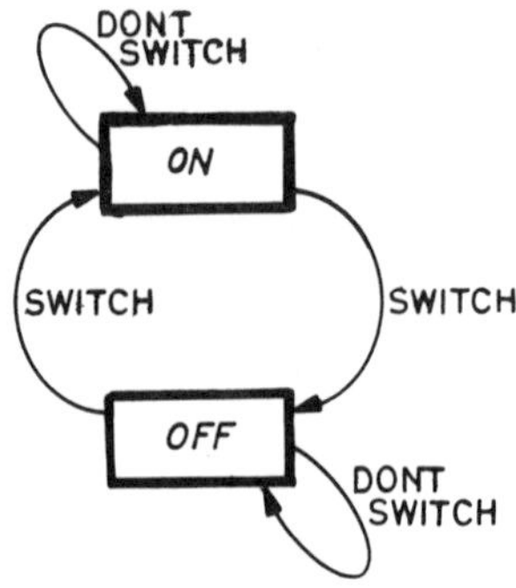

Figure 6.2
The state-transition diagram of a switch.

input actions can be applied—SWITCH or DONT SWITCH. Therefore, for each state, there are two arrows leading from the state, one for each input action. Each arrow describes the effect of one of the inputs. Thus suppose that the switch were in the upper state marked ON in Fig 6.2. If one applied the input action SWITCH, one would follow the right-hand arrow marked SWITCH emerging from ON, and this would lead to the state marked OFF. This sequence represents the simple fact that flicking a switch, while it is on, turns it off. However, suppose that the switch were in the state marked ON, but one applied the input DONT SWITCH. One would follow the top-left arrow marked DONT SWITCH emerging from ON but this would lead back to the state ON. This sequence represents the simple fact that *not* flicking a switch, while it is on, leaves it on. Conversely, suppose that the switch were in the lower state marked OFF in Fig 6.2. If one applied the input action SWITCH, one would follow the left-hand arrow marked SWITCH emerging from OFF and this would lead to the state marked ON. This sequence represents the simple fact that flicking a switch, while it is off, turns it on. Finally, suppose that the switch were in the state marked OFF but one applied the input DONT SWITCH. One would follow the lower-right arrow marked DONT SWITCH emerging from OFF, but this would lead back to the state marked OFF. This sequence represents the simple fact that *not* flicking a switch, while it is off, leaves it off.

Now let us return to the Representational Machine Analogy. It claims that any stimulus is represented as a *state, or set of states*, of a machine. This means that the stimulus is embedded within a state-transition diagram of a machine *because the diagram contains the states*. The particular history becomes a trajectory through that diagram. We shall see both, the inferred history and the inferred state-transition diagram, have a highly particular structure that is determined by the three symmetry-based principles.[3]

In order to see this, we shall begin, in an extremely simple way, by going back to our example of a rotated parallelogram. We will concentrate on this example for a considerable time because several principles will emerge that will turn out to be entirely general. In fact, these principles will become very important later when we progress to *complex shape*. Recall that the fully externalized history of a rotated parallelogram divides into two phases: (1) the external history, in which the rotated parallelogram is referenced backwards through time, via a non-rotated parallelogram and a rectangle, to a square, and (2) the internal history of the square itself, i.e., the representation of the square as a trace. We shall start by examining the internal history.

6.3 The Internal History of a Square

The analysis we shall give of the trace-structure of a square is important because it is generalizable to any regular polygon. The significance of regular polygons is that they are nested hierarchies of repetitive Euclidean actions, and we should recall the claim made in section 2.7 that any shape prototype has this structure. Thus the analysis we shall now give of a square is generalizable to any shape prototype. Again, the detail we shall give may, for a while, seem rather excessive. However, as we said, this detail will allow principles to emerge that will be of quite general importance later.

Recall that one infers a square to be the past state of a rotated parallelogram because one is using the Asymmetry Principle, i.e., the present is referred to a past state that is symmetrical. The symmetrical structure of a square is given by the group $\mathbf{D}_4$, which consists of

Rotations: $e,\ r_{90},\ r_{180},\ r_{270}$

Reflections: $t,\ r_{90}t,\ r_{180}t,\ r_{270}t$

That is, there are a total of eight elements, four of them are rotations and four of them are reflections. The *rotations* increase by intervals of 90°, starting with the zero rotation which is represented by e, the identity element. The four *reflections* are simply the four rotations each multiplied by t, which is the reflection about the vertical axis. Section 5.4 described this group in detail.

However, the square not only has a role as a symmetrical state in the past history of the rotated parallelogram, but has an additional role as the figure that is assigned an internal (trace) structure in that history. This latter role is also given to the square by virtue of its symmetries, i.e., by virtue of the group $\mathbf{D}_4$. The role can be put into effect by virtue of the Symmetry-to-Trace Conversion Principle (section 2.5) which we re-state here:

SYMMETRY-TO-TRACE CONVERSION PRINCIPLE. *Any symmetry can be re-described as a trace. The transformations defining the symmetry generate the trace.*

To re-iterate: the group of symmetries $\mathbf{D}_4$ of a square has two roles in the history of the rotated parallelogram: (1) the group allows the square to be the past state of the rotated parallelogram, via the Asymmetry Principle, and (2) the group provides the trace structure in the history, by virtue of the Symmetry-to-Trace Conversion Principle.

Let us now use the Symmetry-to-Trace Conversion Principle to produce a trace structure from $\mathbf{D}_4$. Any trace has three properties, as follows:

TRACE PROPERTIES.

(1) A trace consists of transformations that go between parts of the present moment.

(2) One part of the present moment is a distinguished starting state.

(3) The other parts of the present moment follow in sequence.

We shall argue that the Symmetry-to-Trace Conversion Principle is psychologically put into effect by realizing these three properties in three successive stages, respectively.

Stage 1: Internalization

The first thing to observe is that the symmetries comprising $\mathbf{D}_4$, i.e., the transformations in $\mathbf{D}_4$, act on the *entire* square. For example, when one applies the rotation r_{90}, one is rotating the entire square by 90°. This means that 90° is acting *externally* on the square. As an external transformation, such a rotation would be part of the rotation that produced the rotated parallelogram from the square. However, having arrived at the square from the rotated parallelogram, we have removed all of that rotation. We are now left with only internal structure. In an internal structure, or trace, the transformations do not act on the whole figure, they act between the parts of the figure, i.e., successively tracing out one part after another (recall section 2.1).

Thus we have a problem. The Symmetry-to-Trace Conversion Principle states that the symmetry transformations of the square become the transformations that generate the square as a trace; but the symmetry transformations act on the entire square, and the trace transformations act on individual parts of the square; i.e., in accord with Trace Property (1) above. This means that to realize the Symmetry-to-Trace Conversion Principle we require a stage that we shall call *internalization*.

By *internalization* we shall mean that the external transformations acting on the figure are converted into internal transformations acting

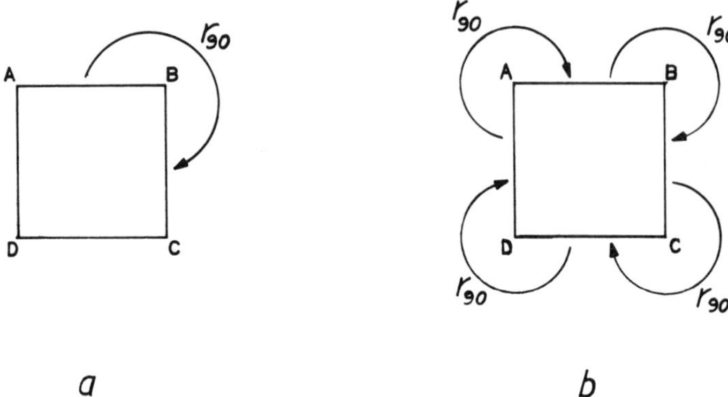

Figure 6.3
The r_{90} relationships between sides of a square.

Figure 6.4
The r_{180} relationships between sides of a square.

between parts of the figure. For example, whereas we previously considered r_{90} as rotating the square as a whole, we will now consider it as acting only on an individual side, sending the side to the side next to it in the clockwise direction. For instance, in Fig 6.3a, the operator r_{90} is sending side AB to side BC. Fig 6.3b shows all possible uses of r_{90} in this way. That is, consider any side. There is an arrow labeled r_{90}, emerging from that side and going to the next side in the clockwise direction. Again, the arrow indicates that the two sides are related to each other in this way.

As another example, consider the rotation r_{180}. This rotation always sends a side to the second side away from it in the clockwise direction. For instance, in Fig 6.4a, the operator r_{180} is sending side AB to side CD. Fig 6.4b shows all possible uses of r_{180} in this way. That is, consider any

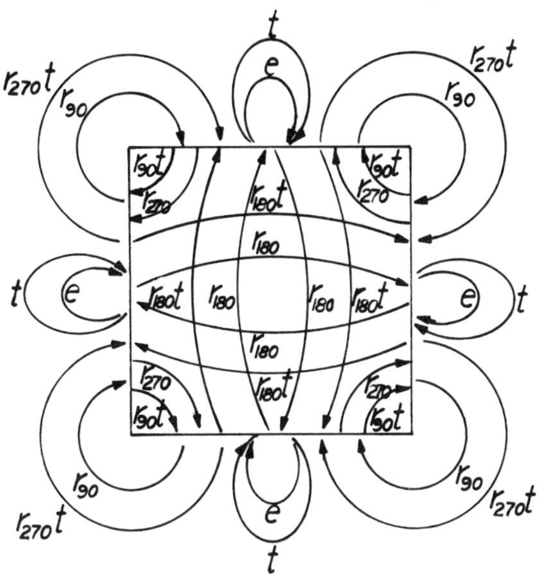

Figure 6.5
The entire application of $\mathbf{D}_4$ to each of the sides of a square.

side. There is an arrow labeled r_{180}, emerging from that side and going to the second side from it in the clockwise direction. The arrow indicates that the two sides are related to each other in this way.

Fig 6.3b shows r_{90} applied to each of the sides. Fig 6.4b shows r_{180} applied to each of the sides. Similarly, we could go through the rest of the eight transformations, comprising $\mathbf{D}_4$, and apply them to each of the sides. Fig 6.5 shows the application of the entire $\mathbf{D}_4$ to each side. That is, choose any side in Fig 6.5. One finds that there are eight arrows emerging from that side. This is because there are eight operations in $\mathbf{D}_4$, and the arrows show how the side changes under the eight operations. Similarly, choosing any other side, one finds that there are eight arrows emerging from this other side, showing how this side changes under the eight operations. Because there are four sides, there must be a total of 32 arrows in Fig 6.5, because there are eight arrows per side.

Now recall that $\mathbf{D}_4$ consists of the operations

Rotations: $e, r_{90}, r_{180}, r_{270}$

Reflections: $t, r_{90}t, r_{180}t, r_{270}t$

Thus, given any side, four of the arrows emerging from it are rotations and the remaining four arrows are reflections. Footnote 4 gives details concerning the labeling of the reflections.

Fig 6.5 should not be regarded as some abstract piece of formalism. It should be regarded as a *psychological* statement about how a square is perceived. The figure in fact documents all the perceptual relationships between the sides of a square. The relationships are actually contained within the symmetries of the square. For example, one of the symmetries of the square is that it can be rotated *as a whole* by 90° and look the same. This, of course implies that any individual side of the square is related to some other side by a 90° rotation. Thus, the fact that one sees the square as symmetrical means that one sees all the relationships shown in Fig 6.5.

Fig 6.5 is important to us for another reason. Closer examination of the figure reveals that it is the *state-transition diagram of a machine*. Recall that a state-transition diagram consists of a set of states and a set of input actions that are applied to each state and are represented by arrows going between the states. For example, the state-transition diagram of a light switch, shown in Fig 6.2, consists of two states, ON and OFF, and a set of input actions, SWITCH and DONT SWITCH, that are applied to each of those states. These actions label the arrows that go between the states because they show how the states change.

Fig 6.5 is exactly analogous to Fig 6.2. We can consider the four sides as four *states*. The four states are, in fact, the four different *positions* that a single side can take in a square. That is, according to Fig 6.5, a square contains only *one* side, and this side is moved to different possible positions. The side is moved by the different input actions, i.e., the eight members of $\mathbf{D}_4$; that is, the four rotations and four reflections. These actions label the arrows shown in Fig 6.5. The arrows indicate how the states, i.e., positions, change under the input actions. Thus, Fig 6.5 represents the square, *internally*, as a precisely defined state-transition diagram.

In fact, $\mathbf{D}_4$, viewed as the symmetry group of a square, i.e., as part of the external structure, also defines a square in terms of the state-transition diagram of a machine. The transformations are again the input actions but they are understood as acting on the square as a whole. They simply happen to leave the square invariant, i.e., they happen to define the symmetries of the square. The process of *internalization* (i.e., converting external transformations into internal ones) has converted this external state-transition diagram into the internal one shown in Fig 6.5.

The consequence is that the square now conforms *internally* to the Representational Machine Analogy, which claims that any stimulus is cognitively represented as a machine. Recall furthermore that a machine involves two causal relationships. We are concentrating in this book, on the second causal relationship, that embodied in the state-transition diagram of the machine, as in Fig 6.5.

To summarize: We have been looking at Stage 1 in the process of realizing the Symmetry-to-Trace Conversion Principle. This stage is the realization of Trace Property (1), i.e., that a trace consists of transformations that go between parts of the present moment. Stage 1 realizes this property by converting the symmetries that act externally on a figure into transformations that act internally between the parts of the figure. We have seen that this process turns part of the *state-transition diagram* that defines the figure externally (i.e., that part of the diagram which defines the symmetries), into a *state-transition diagram* that defines the figure internally.

Stage 2: Initialization

In Stage 1, the transformations defining *symmetries* become transformations between parts. However, the transformations are still acting *symmetrically*. That is, in Fig 6.5, every side is related to any other side by mutually inverse transformations. This means that there is no reason why one side should precede the other as is required in Trace Properties (2) and (3) above. To satisfy properties (2) and (3), one must produce an *asymmetric* structure.

We now consider property (2), that of having a distinguished starting state. Observe that, because the state-transition diagram in Fig 6.5 is symmetric, there is no distinguished starting state. Any side has a set of eight arrows leading from it to the other sides in structurally the same way that any other side has a set of eight arrows leading to the other sides. That is, each side has a copy of $\mathbf{D}_4$ that radiates from it to the other sides. This means that each side could equally be a starting state for the histories leading to other sides.

Thus to fulfill Trace Property (2) above, we must choose one of the sides as the starting state, i.e., the unique side from which the others are generated. This means that the system of eight arrows, $\mathbf{D}_4$, will be allowed to emerge from only one of the sides. There is psychological evidence that, in any regular polygon, human beings choose either the top or bottom side as the starting one (Leyton, 1986c, p269); in fact, it was found that, in any

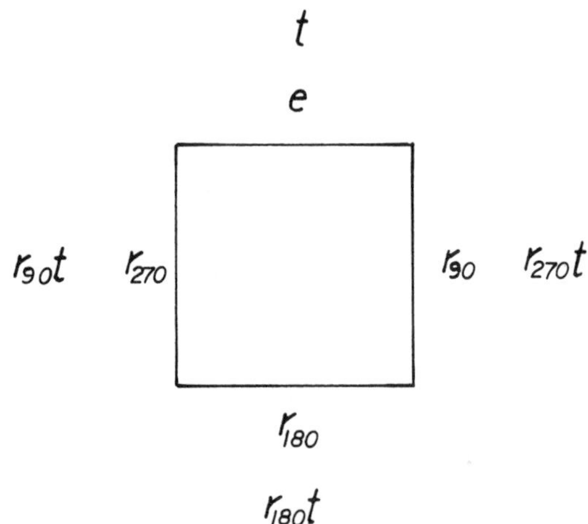

Figure 6.6
The internal structure when the top side is chosen as the starting state.

polygon, the sides are seen as uni-directionally related to the top or bottom side.

We can therefore assume the starting state to be the top side. This means that the other sides are obtained from that side by the transformations in $\mathbf{D}_4$, *but not vice versa.* The resulting structure is shown in Fig 6.6. This figure labels each side with the transformations that were used to obtain the side from the top side. For example, one of the labels of the right side is r_{90} because this side was obtained from the top side by rotating the latter 90° clockwise. The other label of the right side is $r_{270}t$, because the side was obtained from the top side also by a 270° clockwise rotation followed by a reflection about the perpendicular axis of the top side.

Now consider the labels of the top side itself. Observe that this side has been assigned both e and t, for the following reasons: It is labeled by e because it is obtained from the top side by doing nothing, i.e., applying the null transformation e. However, it is also labeled by t because it is obtained from the top side by reflecting that side about its bisecting axis.

Observe now that there are a total of eight labels around the square in Fig 6.6. These are the eight members of $\mathbf{D}_4$. They have been distributed, two per side, always a rotation together with a reflection. This is because any side can be obtained from the top side either by a rotation or by a reflection.

Thus we see that, in Fig 6.6, the eight members of $\mathbf{D}_4$ have been *mapped* onto the four sides of the square. This is a consequence of choosing one side as the initial side. By making that choice, the other sides can be *characterized* by the input actions that produce those sides from the starting side. The characterization allows the sides to be labeled, as in Fig 6.6, by the actions that produced them, i.e., the actions can be mapped directly onto the sides.

The choice of one state as the starting state, for all the actions, is an example of what we shall call *initialization*. The important consequence of initialization is that all the other states can be characterized by the actions that produced them.

We see that, as a result of initialization, each side is given an *explanation*, in accordance with the Representation-is-Explanation Principle. That is, each side is *explained* by an action that produced it. Furthermore, these explaining actions *label* or *characterize* the sides in Fig 6.6. Thus, in accord with the Representation-is-Explanation Principle, each side is *represented* by its explanation.

Let us take stock: Although Stage 1 converted the external transformations defining the symmetries into internal transformations between the parts, Stage 1 still produced a symmetric arrangement of arrows; i.e., the state-transition diagram in Fig 6.5 was still symmetric. In particular, there was no distinguished starting state. In Stage 2, we took the state-transition diagram of Fig 6.5, and *initialized* it; i.e., picked a distinguished starting state. The consequence was Fig 6.6, which shows the structure that is obtained by choosing the top side as the starting side. In Fig 6.6 each side is labeled by the transformations that obtained it from the top side. In other words, the input actions in $\mathbf{D}_4$ are mapped onto the sides themselves.

Stage 3: Subsumption

However, Fig 6.6, our newly derived figure, omits an important piece of information. The third of the Trace Properties is that the parts of the present moment are ordered in a sequence. Thus the sides in Fig 6.6 should be ordered, in some sequence. However, although we know from Fig 6.6 that various actions produced the various sides, we do not know the order of these actions. In fact, we do not know from Fig 6.6 whether the sides were created *independently*. That is, we do not know whether, after creating a side, the machine returned to the top side before going on to creating another side. For example, the machine might have first moved from the top side clockwise by 90°, created the right side, returned to the top side; then moved clockwise by 180°, created the bottom side, returned to the top side; then moved clockwise by 270°, created the left side, etc. Worse still, a "create and return" strategy allows one to create the sides in any order, because any order "costs" as much time and energy as any other order.

The "create and return" strategy produces each side independently. This is a perfectly valid means of creating a square. The crucial fault is that it violates the History Minimization Principle. Many more actions are taken than need be taken. A much shorter history is possible, and it is based on the following strategy:

INCORPORATION STRATEGY. *When a required action X can be accomplished as part of accomplishing another required action Y, then let it be accomplished in this way.*

Thus, the assumption, in seeing the square as a trace, is that the sides were not created independently, but that each side was created *on the way* to creating another side.

The Incorporation Strategy realizes the History Minimization Principle by removing unnecessary steps. However, the strategy takes a particular form in perception that causes a still more radical removal of temporal distinguishability, and has profound consequences for perceptual structure.

In order to see this, we will examine how the Incorporation Strategy is contained in the structure of a *group* and is an important reason why groups embody for us the symmetry involved in the History Minimization Principle.

To see how the notion of incorporation is embodied in a group, first recall some simple facts about $\mathbf{D}_4$. Recall that $\mathbf{D}_4$ consists of the following elements:

Rotations: $e, r_{90}, r_{180}, r_{270}$

Reflections: $t, r_{90}t, r_{180}t, r_{270}t$

Recall also that all the rotations can be generated by applying r_{90} four successive times. And finally, recall that the reflections are obtained from the rotations by taking the rotations and multiplying them by t.

This means that one can generate all the elements of $\mathbf{D}_4$ from only two of the elements: r_{90} and t. The element r_{90} will generate all the rotations, and the element t will generate all the reflections from those rotations. Thus any element of $\mathbf{D}_4$ can be described as a number of repetitions of r_{90} multiplied by a number of repetitions of t (the number of repetitions might be zero in either case). The elements r_{90} and t are called *generators* of $\mathbf{D}_4$. In fact, r_{90} and t form a *minimal* set of generators for $\mathbf{D}_4$, in the sense that a smaller set of generators will not generate all of $\mathbf{D}_4$.

Now recall the problem we discussed concerning Fig 6.6. In Fig 6.6, one does not know the order in which the actions produced the sides and one does not know whether the actions were carried out independently, i.e., whether the machine created a side and returned to the top before moving to another side and creating that side. The problem is resolved of course by using the Incorporation Strategy. For example, a side such as r_{90}, that could be created *on the way* to creating the side r_{180}, should be created thus.

However, observe that the incorporation structure for creating a square is given by the generators of $\mathbf{D}_4$, as follows: (1) Consider first the inner ring of labels in Fig 6.6. These are the rotations. This ring can be created by repeating the generator r_{90}. That is, any member of the inner ring is *incorporated* into a repetition of the generator r_{90}. (2) Now consider the outer ring of labels in Fig 6.6. The outer ring is obtained from the inner ring by applying the generator t to the inner ring. Thus the outer ring can be regarded as the reflection of the inner ring of rotations. That is, the outer ring can be regarded as the ring of rotations in the reverse direction. This means that any member of the outer ring can be incorporated into a repetition of the generator r_{90} and an application of the generator t.

The generator structure realizes the Incorporation Strategy because, if a larger action and a smaller action are both described as the repetition of the same generator, then one can simply make the smaller action an initial segment of the repetition that creates the larger one. That is, one realizes the Incorporation Strategy by using the following:

SUBSUMPTION STRATEGY *Incorporate actions under repetitions of generators. Make the number of generators minimal.*

Generally, we can see that the Subsumption Strategy lets a larger set of distinguishable actions be re-described in terms of a smaller set of distinguishable actions; i.e., generators. For example, the eight distinguishable actions of $\mathbf{D}_4$ are re-described in terms of only two distinguishable actions r_{90} and t. We shall say that the larger set is *subsumed* under the smaller, because the members of the larger set are re-described as instances of the same thing: repetitions of members from the smaller set. Subsuming the larger set under the smaller causes a reduction in distinguishability. Thus, the Subsumption Strategy is a means of realizing the History Minimization Principle.

Recall now that our over-arching concern, in this section, is the Symmetry-to-Trace Conversion Principle, and in particular the conversion of the symmetries of a square into a trace structure. We are realizing this conversion in three successive stages. Stage 3 can now be defined as the description of the square in terms of its subsumption structure; i.e., in terms of its generators.

It will be useful now to develop a diagram that describes precisely the perceptual representation that is produced by Stage 3. Recall that Fig 6.5 gave the representation produced by Stage 1. Fig 6.5 was the description of the square internally as the state-transition diagram of a machine. Recall secondly that Fig 6.6 gave the representation produced by Stage 2. Fig 6.6 initialized the state-transition diagram produced in Fig 6.5. We now wish to create a third diagram that describes the subsumption structure of a square.

We will argue that this third diagram, is one that comes from group theory, and is called the *color graph* of a group. The color graph of the group $\mathbf{D}_4$ is shown in Fig 6.7, and is understood as follows. There are eight

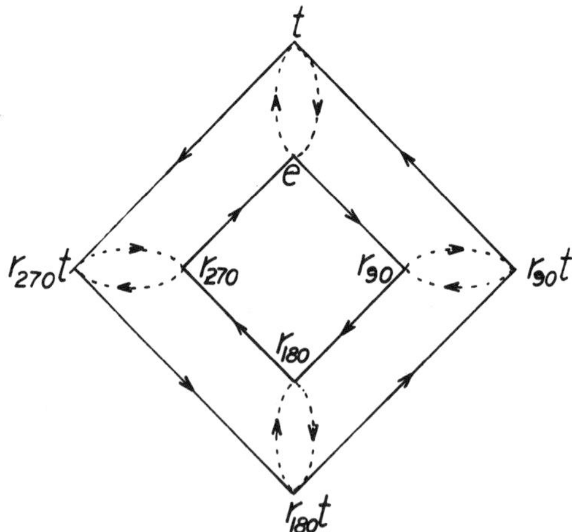

Figure 6.7
The color graph of the group $\mathbf{D}_4$.

vertices in Fig 6.7, and these are labeled by the eight elements of $\mathbf{D}_4$. However, the real substance of Fig 6.7 is in the *links* that connect the vertices. The links represent the subsumption structure of $\mathbf{D}_4$. In fact, each individual link corresponds to a generator, either r_{90} or t. One can infer which generator is being represented in a link by looking at the "color" of the link. A link has one of two colors, "solid" or "dotted": a solid link represents r_{90}, and a dotted link represents t.

The importance of the color graph is that it shows how to express any element of the group as a multiplication of only generators. Thus, for example, suppose that we wish to express the element $r_{180}t$, at the bottom vertex of Fig 6.7, in terms of generators. Then we start at e, at the upper vertex of the inner square, and take a path that leads to the vertex labeled $r_{180}t$. Such a path is shown in Fig 6.8. This path consists of two solid links followed by a dotted link. Since any solid link represents the generator r_{90}, and any dotted link represents the generator t, the path consists of r_{90} followed by r_{90} followed by t. Thus the bottom vertex $r_{180}t$ has been expressed as the succession of only generators.

Thus the color graph of a group shows the *subsumption* structure of the group; that is, it shows how any element can be subsumed under a succession of generators.

It is important now to understand that the color graph represents the *group* $\mathbf{D}_4$, not the *square* on the sheet of paper. To assign the subsumption

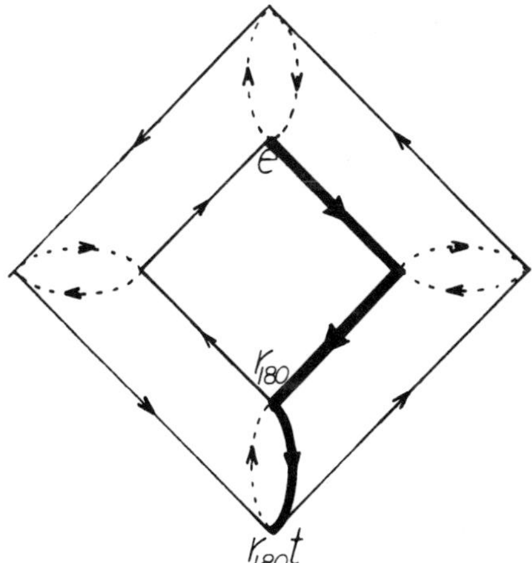

Figure 6.8
A path generating $r_{180}t$.

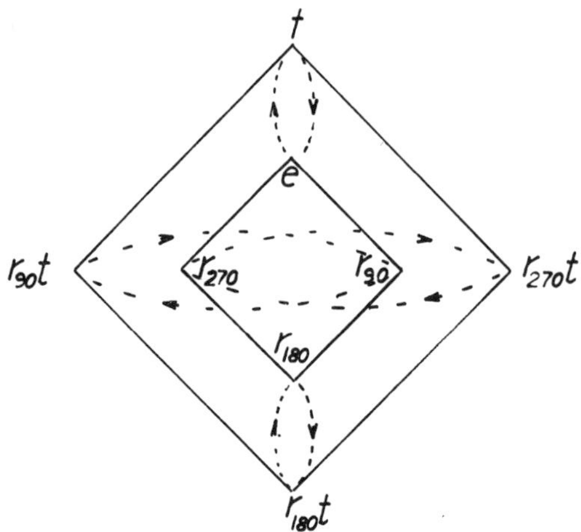

Figure 6.9
In laying the color graph of $\mathbf{D}_4$ over the square, the outer ring has been twisted around the inner ring.

structure to the square itself, one has to map the color graph onto the square. This is rather like laying a grid over a globe of the world. The grid exhibits various relationships between the points on the globe. Similarly, we must lay the color graph over the square, to exhibit the subsumption relations between various parts of the square.

In order to lay the color graph over the square, we do the following. Observe that Fig 6.6 shows how the elements of $\mathbf{D}_4$ distribute around the sides of the square. These elements are the vertices in Fig 6.7. Thus the *vertices* in Fig 6.7 must be assigned to the *sides* of the square (Fig 6.6). Consequently, the links in Fig 6.7, i.e., the generators, must go between the sides. (Note that, in laying the color graph, Fig 6.7, over the square, Fig 6.6, the outer ring of the graph has to be twisted around the inner ring, as shown in Fig 6.9. The reason is given in Footnote 5 to this chapter).

Summarizing Stages 1, 2, and 3

Our purpose in this section was to describe what is needed to realize the Symmetry-to-Trace Conversion Principle. One is converting the symmetries of a figure into a trace structure, and any trace must have three properties: (1) The trace must consist of transformations that go between *parts* of the present moment; (2) one part of the present moment must be a distinguished starting state; and (3) the other parts of the present moment must follow in sequence. We realize these properties in three separate stages, respectively: Stage 1 converts the symmetries that act externally on a figure into transformations that act internally between the parts of the figure. This stage therefore converts a section of the *state-transition diagram* that defines the figure externally (i.e., that section of the diagram which defines the symmetries), into a *state-transition diagram* that defines the figure internally. Fig 6.5 represents the outcome of Stage 1. Stage 2 then takes the state-transition diagram in Fig 6.5, and *initializes* it; i.e., picks a distinguished starting state. Fig 6.6 shows the outcome of Stage 2 (where the top side was chosen as the starting side). Observe that, in Fig 6.6, each side is labeled by the *actions* that obtained it from the top side. Finally, Stage 3 ensures that any action is incorporated under another action, whenever this is possible, rather than let the actions be carried out independently. This incorporation structure is fully realized by the subsumption of actions under repetitions of a minimal set of generators. The subsumption structure is given by Fig 6.7, the color graph of the group.

6.4 Color Graphs and Processes

The importance of the color graph is that it gives the process structure, as follows. Consider Fig 6.7 again. Choose any link, i.e., generator. Observe that the generator occurs within a *cycle*—the cycle created by the repetition of the link. For example, if one picks a solid link, it occurs within either the inner cycle of solid links or the outer cycle of solid links. Again, if one picks a dotted link, it occurs within a cycle of two dotted links, one link going in one direction and the other going in the reverse direction.

Recall now that we claim that a *process* is psychologically defined as a repeating action. Therefore the cycle in which the generator sits is in fact a *process*. Thus the cycles correspond to the processes.

In some groups, such a cycle might be infinite. For example, the group **Numbers**

$$\{\ldots, -3, -2, -1, 0, 1, 2, 3, \ldots\}$$

is an infinite cycle. The infinite aspect does not matter. The point is that any such cycle is a *repeating action*, and psychologically a repeating action is a *process*.

The color graph gives the configuration of such cycles or repetitions, and thus gives the configuration of processes. It is the *colors* of the color graph that give the processes; i.e., any cycle of one *color* corresponds to a process. Since a color graph represents the group as a structure of colored cycles, the graph represents the group as a structure of processes. That is, we can re-express the Subsumption Strategy in the following form:

SUBSUMPTION STRATEGY. *Incorporate actions into* **processes.** *Make the number of processes minimal.*

We conclude therefore that the notion of *process* exists psychologically in order to provide subsumption of environmental actions.

6.5 Processes as Cyclic Groups

We have characterized processes as cycles of a single color in the color graph. However, we now also argue that there is an equivalent characterization of processes as a particular kind of group. Recall that a cycle of a single color is the repetition of a single generator. Such a repetition creates a type of group we met previously in section 5.6, a cyclic group, **Cycle(*n*)**.

For example consider the inner ring of one color in Fig 6.7. Its vertices are the set of rotations in $\mathbf{D_4}$

$$\textit{Rotations}: \quad e,\ r_{90},\ r_{180},\ r_{270}$$

We saw, in section 5.6, that this set of four elements is an example of the group, **Cycle(4)**, the cyclic group of size 4. Recall that one can obtain all the four elements of this group merely by repeating r_{90}. This means that the group is generated by the single element r_{90}.

Any cyclic group, **Cycle(*n*)**, is generated by one element. As another example, observe that, in going around a cycle consisting of only dotted links, in Fig 6.7, one is repeating the generator *t*. This cycle is of size 2, and is equivalent to the group **Cycle(2)**.

Because we argue that a process is psychologically defined as a repeating action, we argue that a process is psychologically defined as a cyclic group. Again, the cyclic group might be infinite, in which case it has the same structure as **Numbers**

$$\{\ldots, -3, -2, -1, 0, 1, 2, 3, \ldots\}$$

Because a cyclic group can be represented by the repetition of one generator, it corresponds to a cycle of one color in a color graph.[6] This means that we have a correspondence between three notions thus:

process ↔ cyclic group ↔ cycle of one color

The first notion, *process*, is a psychological concept. The second notion, *cyclic group*, is a notion from group theory. The third notion, *cycle of one color*, is a notion in the domain of color graphs. Our claim, that a process is psychologically understood as a repeating action, allows us to go freely between these three notions.

6.6 The Group and the Program

The Representational Machine Analogy claims that any stimulus is represented as a state (or set of states) of a programmed machine. The program is the history that the machine undergoes; i.e., the history that is used to explain the stimulus. Let us now consider whether Stages 1, 2, and 3, of section 6.3, realize this claim.

Stage 1 describes a square as the state-transition diagram of a machine (Fig 6.5); Stage 2 then *initializes* this machine, thus giving a starting point to the input actions that created the various states (Fig 6.6); and finally Stage 3 ensures that these actions are incorporated under other actions, wherever they can be, rather than letting the actions be carried out independently. This incorporation structure is given by the color graph of the group; i.e., the *subsumption* of the group elements entirely under the repetition of a minimal set of generators.

The final structure, the color graph (Fig 6.7) should be the structure that realizes the Representational Machine Analogy, which requires that a stimulus be represented in terms of a programmed machine. However, in our discussion of the color graph, no mention was made of the word, *program*. How, therefore, does the color graph involve the notion of program?

To answer this, let us return to the actual program that we gave in section 4.10 for the creation of a square. The program consisted of two levels:

Level 1: Draw an individual side.

Level 2: Shift from one side to the next.

Let us concentrate, for a while, on Level 2. The program for this level is given in Fig 6.10. It is a simple loop structure. Every time one goes around

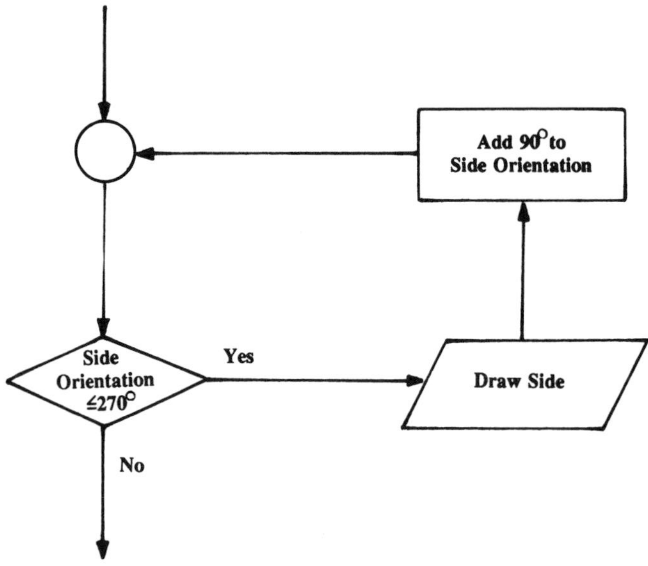

Figure 6.10
The flow-diagram for drawing the successive sides.

this loop, one draws a side (bottom right-hand box, Fig 6.10) and does a 90° rotation (top right-hand box).

The crucial feature to observe here is that the 90° rotation, occurring in the loop, is the generator r_{90} in the group $\mathbf{D}_4$. Equivalently, it corresponds to the solid link in the color graph of $\mathbf{D}_4$. This means that, in going through the repetition of the loop, one is repeating the generator. However, repeating a generator means that one is traveling along a cycle of a *single* color in the color graph. To summarize: A program loop corresponds to a group generator. Repeating a loop corresponds to tracing a cycle of one color.

The important point we wish to make here is that *the structure of the program is determined by the group*. That is: (1) the program is divided into modules where each module corresponds to a generator from a group, and (2) the *structure* of an individual module, i.e., *repetition*, corresponds to a cycle of a single color in the graph; i.e., a cyclic subgroup. Therefore the purpose of the module is to implement a cycle in the color graph of the group.

The reason why the group determines the program is that its color graph allows one to optimally realize the History Minimization Principle. This is because the only distinguishability in the color graph is between the generators themselves, and these are reduced to a minimal set. No distinguishability exists within a cyclic group in that any step along a cyclic group is like the previous step, i.e., the application of a generator. Thus, the color graph allows one to construct a history where distinguishability has been reduced to a minimum.[7]

6.7 The Internal History of a Side

Recall that the perceived history that traces a square has two levels:

Level 1: Draw an individual side.

Level 2: Shift from one side to the next.

In sections 6.3 to 6.6, we examined Level 2 of the history. Let us now examine Level 1, the tracing of an individual side. We shall now see that the principles we gave in the previous sections apply to Level 1, as they did to Level 2. That is, a trace structure is provided for Level 1 by using the Symmetry-to-Trace Conversion Principle. This is realized by the three stages described in the above sections: Stage 1, internalization of the symmetry transformations; Stage 2, initialization of the transformations; Stage 3, subsumption of the transformations, using the color graph. Finally, a generator from the color graph is taken for the loop that constitutes the program defining the history. We will now go through these consecutive stages. They will be presented in as much detail as before, to leave no doubt that exactly the same procedure is involved, and also to carefully illustrate the procedure once again for any reader who did not fully understand it the first time.

To begin, let us assume, as we did in section 4.10, that a side consists of a line of dots 1/1000 inch apart. We could equally assume that the dots form a continuum, and will discuss this alternative assumption later.

Let us also suppose, for the moment, that the line is infinitely long. This supposition will automatically disappear later, as part of the procedure we are describing.

Observe now that, if the dots are 1/1000 inch apart, the side is indistinguishable under a translation by 1/1000 inch. That is, a translation of the line by 1/1000 inch will map the line to itself. Thus the line is *symmetric* (indistinguishable) under a translation by 1/1000 inch.

In fact, the line is symmetric under translations by *any* multiple of a 1/1000 inch. For example, if one translates the line by 2/1000 inch, or by 3/1000, the line will map to itself; i.e., be indistinguishable from itself.

Thus the line is symmetric under the following set of translations:

$$\textbf{Thousandths} = \{\ldots, -3/1000, -2/1000, -1/1000, 0, 1/1000, 2/1000, 3/1000, \ldots\}$$

This set forms a group which we shall call **Thousandths**.

The group of translational symmetries of the side is the group **Thousandths**. This group can be converted into a *trace* structure for a side, using the Symmetry-to-Trace Conversion Principle. However, any trace has three properties: (1) The trace must consist of transformations that go between *parts* of the present moment; (2) one part of the present moment must be a distinguished starting state; and (3) the other parts of the present moment must follow in sequence. Therefore, in order to convert the symmetry group, **Thousandths**, into a trace structure, we must realize these three properties. We do so by carrying out the three successive stages defined in section 6.3. With respect to a side, the three stages look like this:

Stage 1: Internalization

The first thing to observe is that the symmetries comprising **Thousandths**, i.e., the transformations in **Thousandths**, act on the *entire* side. For example, when we applied the translation, 1/1000 inch, in the above discussion, we moved the entire set of dots by 1/1000 inch. This means that the translation was acting *externally* to the side.

The Symmetry-to-Trace Conversion Principle implies that the symmetry transformations of the side can become the transformations that generate the side as a trace. However, the symmetry transformations must act on the entire side, and the trace transformations act on individual points (or parts) of the side, in accord with the first Trace Property. This means that, to realize the Symmetry-to-Trace Conversion Principle, we must *internalize* the symmetry transformations. That is, the symmetry transformations must be converted from being external transformations, into internal transformations. For example, consider the *symmetry* that translates the whole side by 1/1000 inch. *Internally*, this translation will send any point to the next *individual* point rightwards, as shown in Fig 6.11a. Fig 6.11b shows all possible internal uses of this translation. That is, given

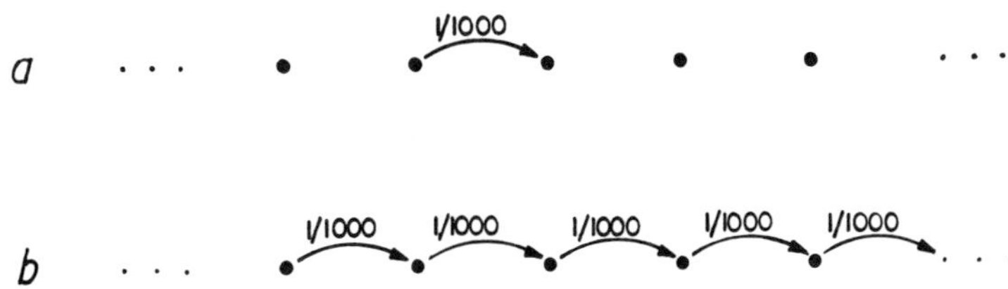

Figure 6.11
The internalization of the symmetry "translation by 1/1000".

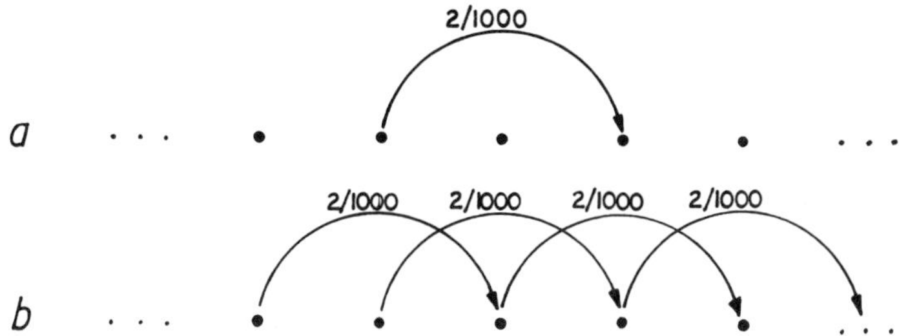

Figure 6.12
The internalization of the symmetry "translation by 2/1000".

any point, in Fig 6.11b, there is an arrow labeled 1/1000, emerging from that point, and going to the next point.

As another example, consider the symmetry defined by translation by 2/1000. Internally, this translation sends any individual point to the second point away from it to the right, as shown in Fig 6.12a. Fig 6.12b shows all possible internal uses of this translation. That is, given *any* point, in Fig 6.12b, there is an arrow labeled 2/1000, emerging from that point and going to the second point from it.

Similarly, we could go through the rest of the translations in **Thousandths** and apply them to each of the points individually. Fig 6.13 is meant to represent the application of the entire group **Thousandths** to each point. However, the group **Thousandths** is infinite, and we can show only a finite number of arrows. Furthermore, we have shown only the 1/1000, 2/1000, $-1/1000$ and $-2/1000$ arrows. In the complete picture, one would be able to choose any point in Fig 6.13, and find that there are all the arrows in **Thousandths** emerging from that point. From now on, when we refer to Fig 6.13, we shall mean this complete picture.

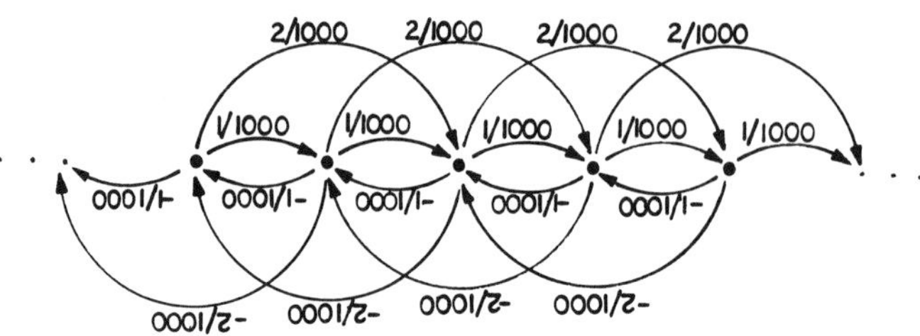

Figure 6.13
The internalization of the group **Thousandths**.

Fig 6.13 should not be regarded as some abstract piece of formalism. It should be regarded as a *psychological* statement about how a side is perceived. The figure is meant to document all the perceptual relationships between the points of a side.

Let us now observe that Fig 6.13 is the *state-transition diagram of a machine*. Recall that such a diagram consists of a set of states and a set of input actions that are applied to each state, where the input actions are represented by arrows going between the states. Fig 6.13 has exactly this structure. We can consider the points as *states*. The states are, in fact, the different *positions* that a point can take in a side. That is, according to Fig 6.13, a side contains *only one point*, and this point is moved to different possible positions. The movements are created by the different input actions, i.e., the translations in the group **Thousandths**. These label the arrows shown in Fig 6.13. In this way, Fig 6.13 represents the side, *internally*, as a precisely defined state-transition diagram.

In fact, **Thousandths** viewed as the symmetry group of a *side*, i.e., as part of the *external* structure, also defines a side in terms of the state-transition diagram of a machine. The transformations are again the input actions, but they are understood as acting on the side as a whole. They simply happen to leave the side invariant, i.e., they happen to define the symmetries of the side. The process of internalization has converted this external state-transition diagram into an internal one.

The consequence is that the side now conforms to the Representational Machine Analogy *internally*.

Stage 2: Initialization

In Stage 1, the transformations defining *symmetries* become transformations between points. However, the transformations are still acting *sym-*

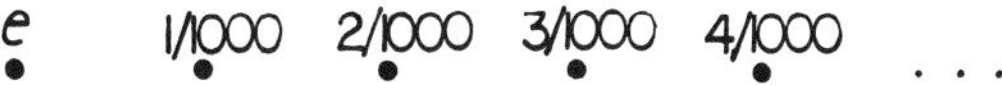

Figure 6.14
Choosing the left-hand endpoint as the initial state.

metrically. That is, in Fig 6.13, any point is related to every other point by mutually inverse transformations. This means that there is no reason why one point should precede the other as is required in Trace Properties (2) and (3) given earlier. To satisfy properties (2) and (3), one must produce an *asymmetric* structure.

We now consider Trace Property (2), that of having a distinguished starting state. Observe that, because the state-transition diagram in Fig 6.13 is symmetric, there is no such starting state. Any point has the set of arrows **Thousandths** leading from it to the other points in structurally the same way that any other point has the set of arrows **Thousandths** leading to the other points. This means that all the points could equally be starting states for the histories leading to other points.

Thus to fulfill Trace Property (2), we must allow the set of arrows to emerge from only one of the points. There is psychological evidence that, in any regular polygon, human beings choose the starting point to be either one of the end-points, or the mid-point, of the side (Leyton, 1986c, p269). We shall assume, without loss of generality, that the starting point is the left-hand end-point.

By choosing one point as the starting point, the other points are obtained *uni-directionally* from that point by the translations in the group **Thousandths**; i.e., the selected point cannot be obtained from the others. The resulting structure is shown in Fig 6.14. In this figure, each point is labeled with the translation that was used to obtain it from the chosen initial point. Note that the initial point is labeled by *e* because it was obtained from itself by applying the null transformation *e*. Then, the first point to the right of *e* is labeled by 1/1000 because it was obtained from the point *e* by a right-ward translation of 1/1000 inch. And so on, for the other points.

We had begun our discussion of the trace structure of a side, by assuming first that the side is of infinite length. Any side is, of course, of finite length. We shall now see that this is not a problem. Let us consider Fig 6.14. Examination of this figure reveals that it has a particular structure: it is the result of *mapping* the elements of the group **Thousandths** to the points of the side; i.e., individual elements of the group are assigned to individual points.

If the side is infinite, then *every* element of the group is mapped to a corresponding point on the side. However, if the side is finite, only some of the elements of the group find corresponding points on the side. The rest of the group members are uninstantiated in the side. Nevertheless, we argue that, *psychologically*, the entire group is involved. The group represents the *complete* space of actions that could be applied to an individual point, and the finite side is understood as part of that complete space; i.e., it is embedded in that complete space. We shall argue later that this embedding within a complete group is the basis of the Gestalt phenomenon of *closure*, i.e., understanding an incomplete figure to be *part* of a complete figure. Thus the group acts as a reference frame into which a stimulus is embedded. The group is however the space of input actions of the machine. Thus the machine is acting as the reference frame for the stimulus. Much more will be said on this later.

Stage 3: Subsumption

In Fig 6.14, each point is given an explanation, in accordance with the Representation-is-Explanation Principle. That is, the point is *explained* by an action that produced it. Furthermore, these explaining actions *label* or *characterize* the points in Fig 6.14. Thus, in accord with the Representation-is-Explanation Principle, each point is *represented* by its explanation.

However, Fig 6.14 omits an important piece of information. The third of our Trace Properties is that the parts of the present moment are ordered in a sequence. Although the transformations in Fig 6.14 look as if they are in sequence, they are not. They could have been created *independently*. That is, we do not know whether, after creating an individual point, the machine returned to the first point before going on to creating another point. As noted previously, the "create and return" strategy, allows one to create the points in any order, because any order "costs" as much time and energy as any other order.

The "create and return" strategy produces each point independently. This is a perfectly valid means of creating a side. The crucial problem is that it violates the History Minimization Principle. Many more actions are taken then need be taken. A much shorter history is possible, that based on the Incorporation Strategy. The strategy implies that, if a point A can be created on the way to creating point B, then A should be created in this way.

The incorporation structure is optimally given by the subsumption structure, i.e., the generator structure, of the group **Thousandths**. Furthermore, for **Thousandths**, this structure is extremely simple: One can generate all the elements of **Thousandths** by repeating the element 1/1000. Therefore 1/1000 is a generator of **Thousandths**, and is the only generator required for that group.

The subsumption structure gives the order in which the actions produced the points. Any element that is created by a certain number of repetitions of the generator is incorporated under any element that is created by a larger number of repetitions, and also incorporates any element that is created by a smaller number of repetitions. With this structure, the order is not one in which the points are created independently, but is the linear order in which the points occur along the side.

Recall now that the subsumption structure of any group is given by the group's color graph. The color graph of **Thousandths** is shown in Fig 6.15. Because there is only one generator, 1/1000, the links all have the same color; i.e., all the links represent the same generator, 1/1000.

It is important now to understand that the color graph represents the *group* **Thousandths** not the side drawn on the sheet of paper. To assign the subsumption structure to the side itself, one maps the color graph onto the side. As we noted before, this is rather like laying a grid over a globe of the world. The grid exhibits various relationships between the points on the globe. Similarly, we lay the color graph over the side to exhibit the subsumption relations between various points of the side.

Recall now that the color graph gives the structure of the *processes*, involved. A process is a repeating action. The action that is repeated is therefore a generator of the process. Any generator occurs within a *cycle* in its color graph—the cycle created by its repetition. Furthermore, the cycle must be of one color, because it is the repetition of the same generator. In the color graph for **Thousandths** (Fig 6.15) there is, in fact, only one cycle of a single color: It is the entire color graph. The cycle is an infinite one, and is therefore represented by an infinite line. Being the repetition of a single generator, it therefore corresponds to a single process.

Figure 6.15
The color-graph of **Thousandths**.

Now recall that the repetition of a single generator creates a cyclic *group*. However, since a process is the repetition of a generator, a process must be a cyclic group. This means that, since the group **Thousandths** is created by the repetition of only one generator, the entire group must be a single cyclic group, and must correspond to a single process.

Finally, let us return to the Representational Machine Analogy, which claims that any stimulus is represented as a state (or set of states) of a *programmed* machine. The program is the *history* that the machine undergoes; i.e., the history that is used to explain the stimulus. Now, Fig 6.16 presents the program we had given in section 4.10 for generating an individual side. The program is a simple loop structure. Every time one goes around this loop, one draws a point (bottom right-hand box, Fig 6.16) and carries out a translation of 1/1000 inch (top right-hand box). The crucial feature to observe here is that the 1/1000 inch translation, in the loop, is the generator in the group **Thousandths**. That is, this component of the loop corresponds to the solid link in the color graph of **Thousandths**. This means that, in going through the repetition of the loop, one is repeating

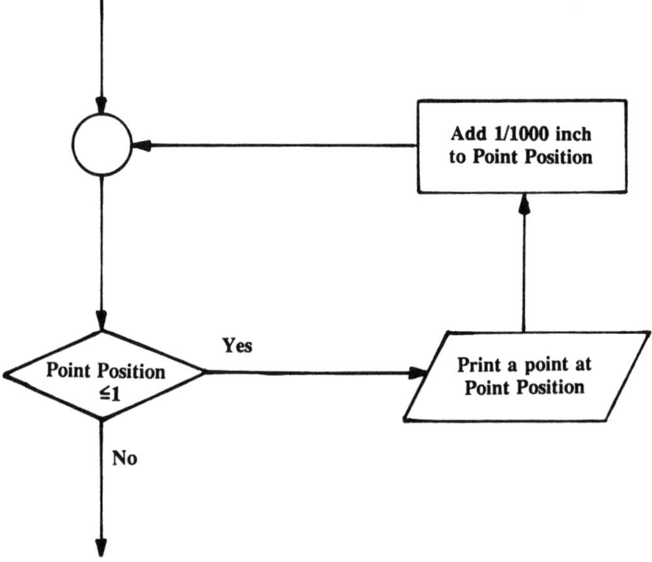

Figure 6.16
The program for generating a side.

the generator. Therefore, the program shown in Fig 6.16 corresponds to a cycle of a *single* color in the color graph, or, equivalently, a cyclic group.

6.8 The Two-Level History of a Square

Let us now return to the full trace history that is assigned to a square. We had observed that the history is equivalent to a program that is structured on two levels (shown again in Fig 6.17). We can now understand from where this full program, or history, is derived. A square has two levels of *symmetry*: the symmetries that map the entire figure to itself, and the symmetries that map each side to itself. The former level is given by the group $\mathbf{D}_4$, and the latter is given by the group **Thousandths**. This symmetry structure determines the program as follows:

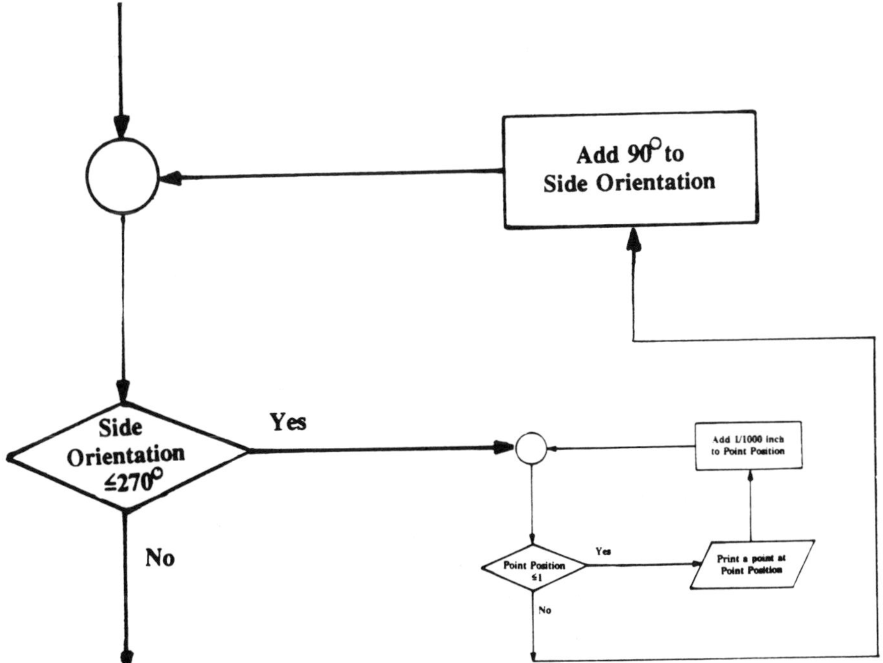

Figure 6.17
The two-level loop structure for drawing a square.

Each level of symmetry determines its level in the program via the Symmetry-to-Trace Conversion Principle. As we have seen, this requires three stages: internalization of the symmetry transformations, initialization of these transformations, and subsumption of the transformations under repetitions of generators. Each loop in the program is then defined by a generator in its associated color graph.

Thus we see that the program, or history, is determined by a structure of color graphs, which is determined, in turn, by the structure of symmetries in the figure.

6.9 The Trace Structure of Regular Polygons

We have shown, in detail, how the trace structure of a square can be extracted from a stimulus that is a square. The value of this analysis is that it can be generalized to the trace structure of any regular polygon. As we said earlier, regular polygons are important because they are nested, repetitive, Euclidean hierarchies, and any shape prototype seems to be a hierarchy of this type. That is, such hierarchies are psychologically salient as *past states* of any object irrespective of the physical material involved.

The only polygon we have investigated so far with respect to its inferred trace structure is a square. To illustrate how the rules we defined are applicable to any other polygon, we will apply them here, in a summary form, to a hexagon. We should note that the rules will be applied in a later section to understanding prototypes that are not regular polygons, but are nevertheless nested repetitive Euclidean hierarchies. For example, we shall later examine the cylinder, which constitutes the most famous prototype used for biological limbs.

Just as a square has two levels of action, a hexagon has two levels of action: Level 1, the drawing of an individual side; and Level 2, the shift from one side to the next. The two levels come from two levels of *symmetry* in a hexagon. Let us deal first with Level 2.

The symmetries of the hexagon, as a whole, are given by $\mathbf{D}_6$, which consists of

Rotations: $e, r_{60}, r_{120}, r_{180}, r_{240}, r_{300}$

Reflections: $t, r_{60}t, r_{120}t, r_{180}t, t_{240}t, r_{300}t$

That is, there are six rotations and six reflections. The rotations increase by increments of 60°, starting with zero rotation which is represented by e, the indentity element. The reflections are simply the six rotations each multiplied by t, which is the reflection about the vertical axis.

As symmetries of a hexagon, these transformations are external ones, acting on the entire figure. According to the Symmetry-to-Trace Conversion Principle, these transformations can become the transformations that define the hexagon as a trace. In order to realize this conversion, we carry out the three stages defined earlier, in a summarized form, as follows:

In Stage 1, the symmetry transformations are internalized so that they act between the individual sides. The internalization of three of the transformations, e, r_{60}, and r_{300}, is shown in Fig 6.18. Rather than having only three arrows emerging from each side, as in Fig 6.18, there should be twelve arrows, i.e., all of $\mathbf{D}_6$, emerging from each side. The complete figure defines a hexagon, internally, as the state-transition diagram of a machine. Under this description, the hexagon is reduced to a *single side* that is pushed into six alternative states; i.e., positions.

Stage 2 initializes the state-transition diagram. Psychological studies (Leyton, 1986c) seem to show that human beings regard the top or bottom side as the starting state. Here we shall assume the top side to be the starting state. Fig 6.19 shows this initialization of the state-transition diagram. The transformations of $\mathbf{D}_6$ are now mapped directly onto the sides, two per side. This defines the sides as derived from the top side.

Stage 3 subsumes these transformations under *processes*, i.e., repetitions of generators. Two generators are sufficient: r_{60} and t. The corresponding subsumption structure is given by the color graph of $\mathbf{D}_6$, which is shown in

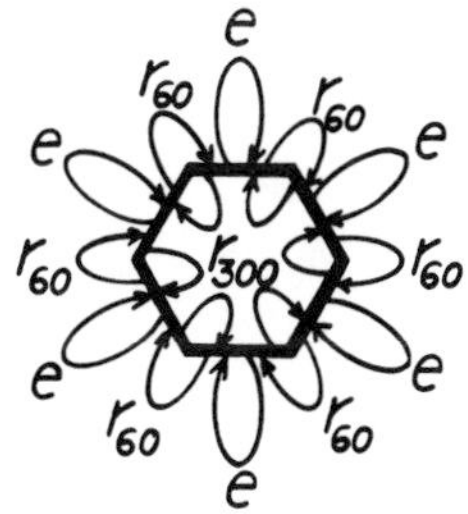

Figure 6.18
The internalization of three symmetries from $\mathbf{D}_6$.

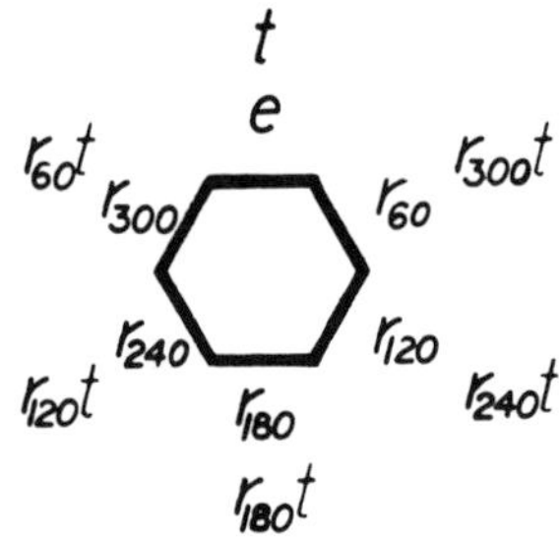

Figure 6.19
Choosing the top side as the starting side.

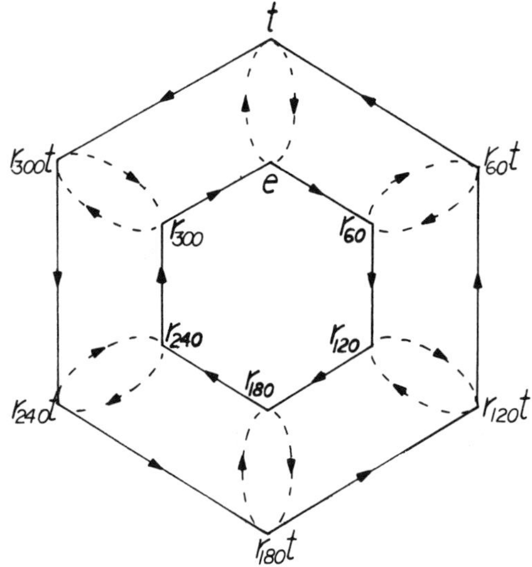

Figure 6.20
The color graph of $\mathbf{D}_6$.

Fig 6.20. Each solid link represents the generator r_{60}, and each dotted link represents the generator t. Now, given an actual hexagon, as in Fig 6.19, its subsumption structure is defined by mapping the color graph in Fig 6.20 onto the hexagon in Fig 6.19. The vertices in Fig 6.20 map to the sides in Fig 6.19.

This concludes the procedure used to define Level 2 of the trace structure of a hexagon; i.e., the shifts between successive sides. Level 1, the tracing of an individual side, is exactly identical to that used for the individual side of the square. That is, in Stage 1, the translational symmetries of a side are internalized to become the state-transition diagram defining the side as the states of a single point (recall Fig 6.13). In Stage 2, this state-transition diagram is initialized, choosing the end-point of the side as the starting state. The translations are then laid out along the points of the side, defining each point as a particular translation from the starting point (recall Fig 6.14). Finally, in Stage 3, the translations are subsumed under a single process, the repetition of a generator. This subsumption structure is again given by the color graph of the translation group (recall Fig 6.15). The structure organizes a side by mapping the graph onto the side.

6.10 The External History of a Square

In sections 6.3 to 6.8, we examined the trace history of a square. We are interested in a square firstly because it is representative of trace structures that are nested Euclidean hierarchies, and such structures are psychologically the most salient past states. However, we are interested in a square also because it is the past state in a particular history that we are examining for a while—the history assigned to a rotated parallelogram. This history divides into two phases: (1) the external history, in which the rotated parallelogram is referenced backward through time, via a nonrotated parallelogram and a rectangle, to a square, and (2) the internal history of the square itself. We now examine the external history.

As was pointed out in Chapter 1, we can assume that the external history is constructed in this way: The perceptual system examines the presented stimulus, the rotated parallelogram, and finds the following distinguishabilities: (1) the orientation of the parallelogram is different from the orientation of the environment (a condition that will be discussed in more detail later); (2) the two pairs of opposite angles of a parallelogram have different sizes; and (3) the two pairs of opposite sides have different lengths. The *history* that is inferred is a successive removal of these three perceived distinguishabilities.

While this removal of distinguishabilities appears to be different from that we described in the preceding sections, for internal structure, the constructed histories have some significant commonalities that are important to understand. In order to gain access to these commonalities, it is necessary first to observe that, in defining the external history, the perceptual system has once again defined a *machine*, as follows:

The four shapes involved—the rotated parallelogram, non-rotated parallelogram, rectangle, and square—are related to each other by certain actions—rotation, shear, and stretch. These actions are *linear transformations*. Recall (section 5.12) that a linear transformation is a transformation that preserves *straightness* (and the position of the origin[8]). Thus, in removing the distinguishabilities, the perceptual system hypothesizes transformations that are minimal in the sense that they preserve straightness.

The collection of linear transformations forms a group that we call **Linear**. Using **Linear**, we can now define a *machine*, as follows: Given any shape, one can apply to it the set of actions in **Linear**. This will produce an infinite number of other shapes, because there are an infinite number of actions in **Linear**. Furthermore, given any shape in this entire collection, if one applies **Linear** to this shape, one will obtain the rest of the shapes in the collection. That is, each shape in the collection is related to all the other shapes by the actions in **Linear**.

Observe now that, if any one of the four shapes—rotated parallelogram, non-rotated parallelogram, rectangle, or square—is in the collection, then so must the other three be in the collection. This is because the four shapes are related by members from **Linear**.

We have therefore defined the state-transition diagram of a machine. The states are an infinite number of shapes, four of which are the four we are interested in. The arrows between the states are the actions of **Linear**. Given any shape, there are an infinite number of arrows emerging from it, i.e., the infinite number of actions in **Linear**, leading from the shape to the rest of the collection.

Part of this state-transition diagram is shown in Fig 6.21. The shapes given in Fig 6.21 are only a small subset of the infinite number of shapes that constitute the states. Furthermore, the arrows shown there are only a

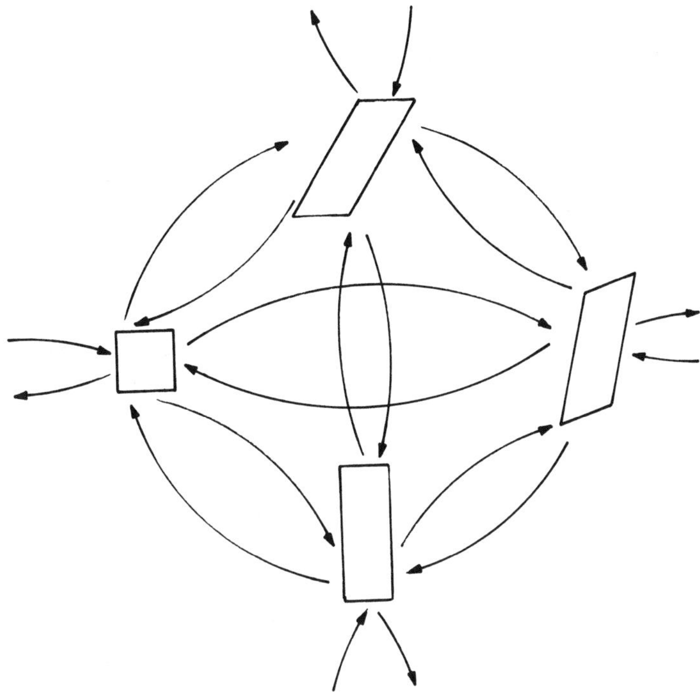

Figure 6.21
Part of the state-transition diagram for **Linear**.

small subset of the infinite collection of arrows, the entire group of arrows, **Linear**, that emerges from each shape.

The situation we now have conforms to the Representational Machine Analogy in that the rotated parallelogram is described as a state of a machine. However, the Representational Machine Analogy requires the machine to be a programmed one; i.e., to have a well-defined history. Any program is given by (1) a specific starting state, and (2) an ordered sequence of actions. Observe, however, that these two properties occur respectively in the second and third Trace Properties (section 6.3).

This means that, to satisfy the Representational Machine Analogy, and produce a well-defined history, we must satisfy the second and third Trace Properties, insofar that they demand a distinguished starting state and an ordering on states. But recall that these two properties were realized by Stage 2 (initialization) and Stage 3 (subsumption) in the procedure of the previous sections. We can gain some insight into external histories by examining these two stages in relation to the history hypothesized for a rotated parallelogram.

Initialization

The state-transition diagram described above (Fig 6.21) must be initialized; i.e., one of the shapes must be selected as the starting shape. The state that is chosen is the one that is most symmetric; i.e., the square. Now observe the following: The symmetry of any shape is defined by external transformations that send the shape *to itself*. In the state-transition diagram, these transformations are the arrows that emerge from a shape and return to the same shape. It therefore becomes easy to find the shape that is most symmetric in a state-transition diagram: It is that shape (state) which has the most number of arrows leading from the shape back to itself. In the particular state-transition diagram we are considering, the state with the most self-returning arrows is a square. These arrows are certain rotations and reflections in the group **Linear**. This set of arrows forms the group $\mathbf{D}_4$. There are eight transformations in $\mathbf{D}_4$, i.e., eight self-returning arrows on the square. No other shape (state) in the state-transition diagram has as many as eight self-returning arrows. The shapes with the second largest number of self-returning arrows are the rectangles, which only have half the number of self-returning arrows—four.

A second criterion is also involved in determining the starting state. This criterion will be discussed in greater detail later when we develop the tools to fully understand it. For completeness, however, we briefly describe the criterion here although the reader should not worry if he or she does not grasp it: Observe that there are several squares in the state-transition diagram, all differing by the rotations not in $\mathbf{D}_4$; in fact by rotations less than 90°. Because each of these squares is apparently symmetric under $\mathbf{D}_4$, each apparently has a copy of $\mathbf{D}_4$ emerging from and returning to it. We shall see later, however, that the operation, t, in $\mathbf{D}_4$, is actually reflection about the *gravitational* vertical in the environment. This means that the only copy of $\mathbf{D}_4$ that really exists in the state-transition diagram is that in which t is gravitationally aligned. Only one square in the state-transition diagram carries this copy of $\mathbf{D}_4$. It is the gravitationally aligned square. The apparent $\mathbf{D}_4$ structure on any other square is actually described as a rotation of the $\mathbf{D}_4$ structure of the gravitationally aligned square. This means that the gravitational square is selected as the starting state. To summarize: The gravitationally aligned square is chosen as the starting state because it has the maximal number of self-returning arrows, and this set of arrows includes the gravitationally aligned reflectional arrow.

Subsumption

The initialization of the state-transition diagram to the square makes the arrows of **Linear** emerge only from a square (going out to the other shapes and to the square itself).

In the history between a square and a rotated parallelogram there are many possible intervening shapes. However, a remarkable phenomenon is evident in the history that is actually chosen by perception. Perception makes use of a deep mathematical property of **Linear**, as follows:

Before we can state this property, we must first observe that **Linear** is composed of two halves, called the *positive* and *negative* halves. These halves are labeled, respectively, **Linear**$^+$ and **Linear**$^-$. The positive half, **Linear**$^+$, consists of all the linear transformations that do not involve a reflection. Such transformations are called *positive* linear transformations. The other half, **Linear**$^-$, consists of all the linear transformations that do involve a reflection. These transformations are called *negative* linear transformations.

We can see that the history that is psychologically chosen between the rotated parallelogram and the square does not involve reflection. Therefore the history must be confined to transformations in **Linear**$^+$. Let us therefore, temporarily, throw away the negative half of **Linear**, and concentrate on **Linear**$^+$.

We can now state the deep mathematical property which perception seems to use in constructing the history of the presented parallelogram. It is this: Any member of **Linear**$^+$, that is, any positive linear transformation, can be decomposed into three types of linear transformation: *stretch*, *shear*, and *rotation*. It is clearly the case that perception is using this decomposition in constructing the history from the rotated parallelogram back to the square.

The decomposition of **Linear**$^+$ can be represented as in Fig 6.22. The diagram shows the actual group, **Linear**$^+$. Movement along the *stretch* dimension, from the origin, increases the amount of stretch; movement along the *shear* dimension, from the origin, increases the amount of shear; and movement along the *rotation* dimension, from the origin, increases the amount of rotation. Any point in the space represents a member of

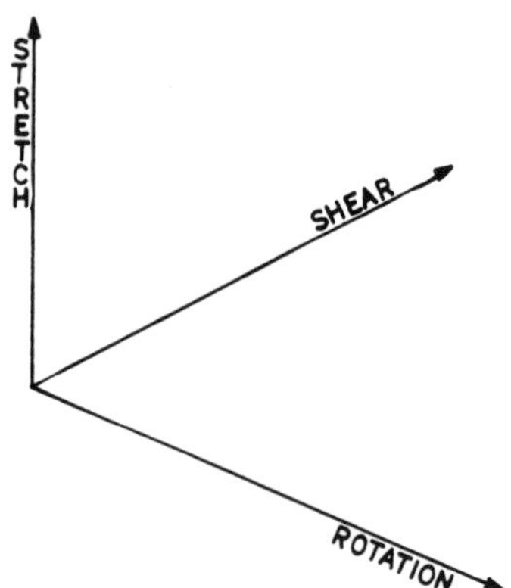

Figure 6.22
The decomposition of **Special-Linear** into stretch, shear and rotation.

Linear$^+$, that is, a positive linear transformation (a linear transformation without a reflection). Any point has a value on the stretch dimension, a value on the shear dimension, and a value on the rotation dimension. These values are respectively the amount of stretch, the amount of shear, and the amount of rotation that make up the transformation represented by the point.

As is shown in Fig 6.22, there is only one rotation dimension in **Linear**$^+$. This dimension corresponds to increasing *angle* of rotation. Furthermore, as is shown in Fig 6.22, there is only one shear dimension in **Linear**$^+$. This dimension corresponds to increasing *angle* of shear. However, although we have shown stretch in Fig 6.22 to be given by one dimension, it should actually be given by two: one for stretching a square horizontally, and the other for stretching it vertically. Nevertheless, we have carried out pilot psychological experiments that seem to indicate that subjects use only a one-dimensional version of stretch, in the hypothesized history. This is because they seem to use either of two alternative strategies in the stretch phase of the history; i.e., in going from a rectangle to a square: (1) Subjects reduce only the height of the rectangle, not the width; or (2) subjects try to preserve the area of the rectangle. Each strategy involves only a single stretch dimension. We shall assume, without loss of generality to the ensuing discussion, that all the subjects employ the second strategy.

The use of only one stretch dimension means that subjects perform one further reduction on **Linear**. Besides throwing out the negative half of **Linear**, thus obtaining **Linear**$^+$, they throw out one of the two stretch dimensions in **Linear**$^+$. This final reduced version is precisely the three-dimensional space shown in Fig 6.22. Furthermore, this version forms a group, which we shall call **Special–Linear**. One can regard Fig 6.22 as showing **Special–Linear** with its decomposition into stretch, shear, and rotation.[9]

What we have seen so far is that the history that connects the presented parallelogram to a square is a linear transformation. That is, the history takes place in **Linear**. However, we have also seen that this history is confined to a particular section of **Linear**. This section is the subgroup **Special–Linear**. Finally, we have also seen that, mathematically, this latter subgroup has a specific decomposition: into stretch, shear and rotation.

The main point of this section can now be stated. It is this: Each dimension of the decomposition, shown in Fig 6.22, is the *repetition of a generator*. That is, each dimension is a *process*. Thus Fig 6.22 represents a decomposition into *processes*.

This vividly illustrates the point we made in section 4.8.5, that, in constructing a history, cognition uses the structured programming principle of decomposition into modules, and, furthermore, that each module is a process, i.e., a repeating action.

Let us now plot, in Fig 6.22, the history that is taken by cognition. Recall that Fig 6.22 represents the group **Special–Linear**. The history is a trajectory of points through **Special–Linear** because each such point represents a transformation used in the history.

The initial rotated parallelogram can be given by some point in Fig 6.22 away from the origin, because the shape is understood as having non-zero stretch, non-zero shear, and non-zero rotation. In Fig 6.23, we have shown **Special–Linear** again, and have indicated this starting point as point A in the figure; that is, point A represents the rotated parallelogram, or, more precisely, the entire transformation that was used to obtain the parallelogram from the square. Now, the first stage in the backward history is the removal of rotation. This segment of history is given by the bold line from point A to point B. The line is in the direction of decreasing rotation. It terminates at point B because, at point B there is zero rotation. Thus point B represents the non-rotated parallelogram. Since only rotation has been removed from the shape, the line from A to B is parallel to the rotation axis. Furthermore, since it is parallel to the rotation axis, it is equivalent to the repetition of the rotation generator.

The second stage in the backward history is the removal of shear. This segment of history is given by the bold line from point *B* to point *C*. The line is in the direction of decreasing shear. It terminates at point *C* because, at point *C* there is zero shear. Thus point *C* represents the rectangle. Since, during this stage, only shear has been removed from the shape, the line from *B* to *C* is parallel to the shear axis. Therefore, it is equivalent to the repetition of the shear generator.

Finally, the third stage in the backward history is the removal of stretch. This segment of history is given by the bold line from point *C* to point *D*. The line is in the direction of decreasing stretch. It terminates at point *D* because, at point *D* there is zero stretch. Thus point *D* represents the square. Since, during this stage, only stretch has been removed from the shape, the line from *C* to *D* is parallel to (in fact, coincident with) the stretch axis. Therefore, it is equivalent to the repetition of the stretch generator.

Thus the entire history, inferred by cognition, is the bold trajectory shown in Fig 6.23. This consists of the three successive straight lines shown. Each line is the repetition of a generator, i.e., a process. Thus the trajectory shown is a succession of processes.

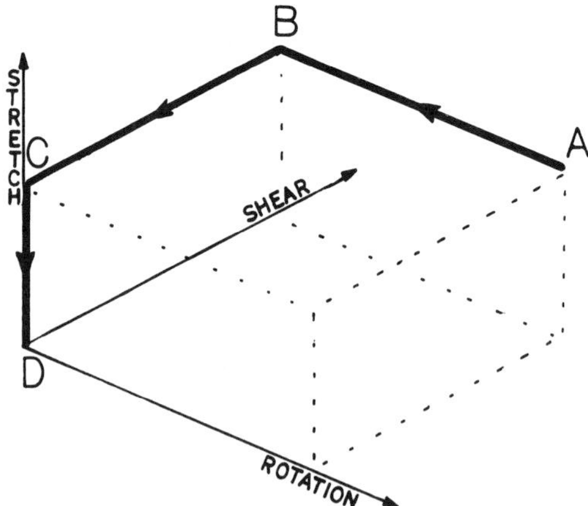

Figure 6.23
The group **Special-Linear** with the trajectory inferred by cognition.

The cognitive system clearly carries out the decomposition into processes as part of the *subsumption strategy*, as follows: The intervening history between a square and a rotated parallelogram could contain many alternative types of shapes created by arbitrary members of **Linear**. Furthermore, each of these intervening shapes could be produced by the create-and-return strategy. However, we have seen that arbitrary members of **Linear** are not used, and neither is the create-and-return strategy. The history is decomposed into processes, i.e., is subsumed under the repetition of generators.

We can therefore see that, in an external history, subsumption has an additional role besides that which it has in an internal history. In the latter history, the intervening states are actually presented in the visible stimulus, and subsumption has the role of ordering the states. However, in an external history, the intervening states are not presented in the visible stimulus, and thus subsumption has to define them as well as order them. It accomplishes both these tasks by defining the intervening states to be those that are repetitions of generators.

6.11 Continuous Generators

In the previous section, we concluded that the decomposition of **Special–Linear**, into stretch, shear, and rotation, allows the rotated parallelogram to be described as the outcome of processes, i.e., repetitions of generators. The repetition of one generator creates the stretch process, the repetition of another generator creates the shear process, and the repetition of the remaining generator creates the rotation process.

It should be noted that these three processes are all *continuous*. For example, rotation is understood as increasing continuously from the nonrotated parallelogram to the rotated one. This contrasts with the situation in which we first met generators, in discussing the group $\mathbf{D}_4$. Recall that the four rotations in $\mathbf{D}_4$—that is, e, r_{90}, r_{180}, r_{270}—are generated by the repetition of r_{90}. That is, the rotations increase in discrete steps of r_{90}. Thus the generator r_{90} is the size of the smallest discrete step between the rotations. Generally speaking, the generator of any discrete process is the size of its *shortest repeated element*.

What about a continuous process? Because the process is a continuum, the smallest step is infinitesimally small; i.e., one can think of two "adjacent" points on a continuum as being infinitesimally close. Thus the generator that bridges two adjacent points must be infinitesimally small. This infinitesimally small generator will be repeated in one direction to generate the continuum.[10]

Recall now that the repetition of a discrete generator produces a cyclic group. For example, the repetition of r_{90} produces the group **Cycle(4)**, which is discrete because rotations can increase only discretely in that group. Correspondingly, the repetition of an infinitesimally small generator produces a *continuous cyclic group* (the cycle might, of course, be infinitely large). In the decomposition of **Special–Linear**, shown in Fig 6.22, each of the dimensions—stretch, shear, rotation—is a cyclic subgroup, and each is continuous. The standard term for a continuous cyclic subgroup is *one-parameter subgroup*. A continuous process is therefore a one-parameter subgroup.

Recall again that, in the color graph of a group, a cyclic subgroup is represented by a cycle of one color. Using this concept, we can see how Fig 6.22 yields the color graph of **Special–Linear**. We allow each axis to have a different color. Furthermore, we draw, for each axis, the entire set of lines parallel to the axis and give, to the lines, the color of their associated axis. Fig 6.24 shows some of these lines. We will take Fig 6.24 as representing the complete color graph, although it represents only a small section of the graph. (The three different thicknesses of lines, in Fig 6.24, represent the

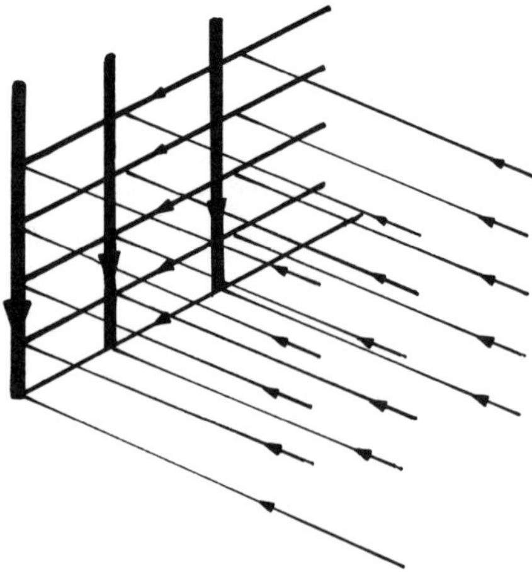

Figure 6.24
Part of the color graph of **Special Linear**.

three different colors.) Given this, we can now see that the trajectory in Fig 6.23 was based on the color graph. The three straight lines comprising the trajectory in Fig 6.23 are each parallel to an axis. The history travels down these successive lines. Thus the history travels along three successive lines in the color graph.

6.12 Mapping the Group and the Color Graph

In sections 6.3 to 6.8, we examined the *internal* component of the fully externalized history of a rotated parallelogram (i.e., the trace structure of a square), and in sections 6.10 to 6.11, we examined the *external* component of that history (i.e., that between the square and the rotated parallelogram).

The inference of internal history necessarily involves different rules from the inference of external history, because, in the former case, the rules apply to the present as a set of states in which the asymmetries are between the states, whereas, in the latter case, the rules take the present to be a single state in which the asymmetries are endpoints.

However, we have discovered something remarkable: Although the process inference rules are necessarily different, they result in the same type of representational structure, i.e., one involving a color graph. We shall now examine this structure more closely.

Let us begin by returning to Fig 6.22. It is important first to understand that this figure represents a group of actions, not a space of shapes. Each *point* in Fig 6.22 represents an action. The *collection* of points comprising Fig 6.22 is the group **Special–Linear**.

Although each individual point in Fig 6.22 represents an individual action, each point becomes *equated* with an individual shape, in the following way. First, because the square is regarded as the initial state, it is equated with the identity element of the group, i.e., the origin in Fig 6.22. Once this is done, the rest of the actions in Fig 6.22 generate the rest of the shapes. To illustrate: Let us consider, in Fig 6.22, the plane that goes through the shear axis and the stretch axis. This plane is shown in Fig 6.25a. Each point in this plane is equated with a shape. Thus the plane is a space of shapes, as shown in Fig 6.25b. As we said, the origin is the identity element, and this corresponds to the square. Moving upward along the stretch axis, in Fig 6.25b, means greater and greater stretch, which creates longer and longer rectangles. Now, if one takes any of these rectangles,

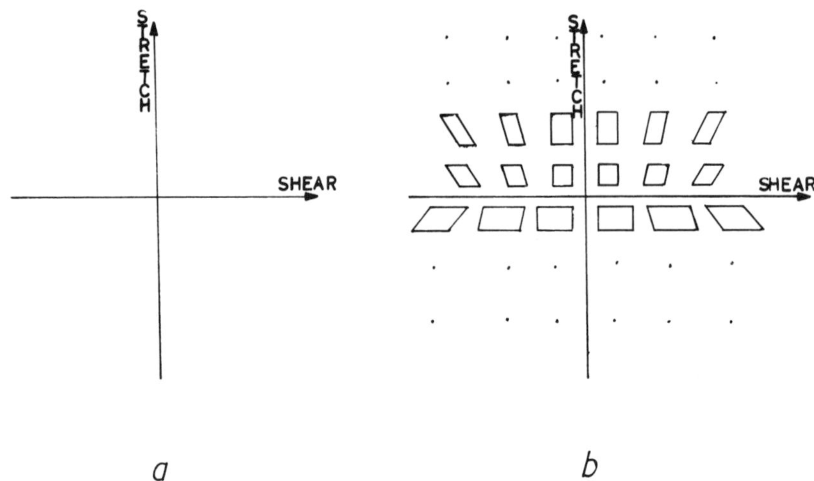

Figure 6.25
(a) The stretch-shear plane within the group **Special-Linear**. (b) The shapes positioned in the plane when the square is at the origin.

and one moves in the direction of greater shear, i.e., horizontally in Fig 6.25b, one will obtain more and more slanted versions of that rectangle, i.e., parallelograms. Since one can do this for each of the rectangles, one can fill the whole space with shapes, as shown in Fig 6.25b. There is one shape for each point in the space.

Fig 6.25b gives a single slice of Fig 6.22; that is, the plane through the stretch and shear axes. To obtain the remainder of the shapes assigned to Fig 6.22, one moves out along the rotation axis (in Fig 6.22). This has the effect of making rotated versions of the shapes in the stretch-shear plane. As a consequence, each point in Fig 6.22 now has its own particular shape.

It is important to recognize why each point in Fig 6.22 has its own shape. Each point is a specific *action*; i.e., a specific member of the group **Special–Linear**. At the origin of the group, we have placed the square. Thus each action corresponds to the shape that is obtained by applying that action to the square. Therefore, it is because a shape has been assigned to the origin point of the group, that each other point of the group, i.e., each action, is assigned a shape.

Observe now that the square is assigned to the origin in the *initialization* phase of the procedure we defined. Thus a shape is assigned to each action in this phase.

Again, observe that, by assigning to each action a particular shape, one is essentially mapping the space of actions onto a space of shapes, each action going onto a particular shape. The space of shapes is a particular (but infinite collection) of rectangles, parallelograms, rhombi, rotated rectangles, rotated parallelograms, etc. It is *initialization* that has mapped **Special–Linear** onto this collection.

We have met this kind of situation before—when we were examining how the group $\mathbf{D}_4$ defines the trace structure of a square. $\mathbf{D}_4$ consists of eight elements:

Rotations: $e, r_{90}, r_{180}, r_{270}$

Reflections: $t, r_{90}t, r_{180}t, r_{270}t$

We saw that these eight individual actions are mapped onto the four sides of the square, two per side, thus creating Fig 6.6. The map is produced in the *initialization* phase. This is because, in defining the starting state to be the top side, i.e., assigning the identity element to the top side, one lets the other actions in the group be assigned to the particular sides that they create from the top side.

Thus we see that the group **Special–Linear** is *mapped onto* a space of shapes in the same way that $\mathbf{D}_4$ is mapped onto the set of sides of a square. The difference is that $\mathbf{D}_4$ is mapped onto parts of a single shape; i.e., the sides, of a single shape, because it describes internal history; whereas **Special–Linear** is mapped onto a collection of shapes because it represents external history. In the former case, the sides of the square are the states of a machine, and mapping $\mathbf{D}_4$ onto the sides shows how the machine's input actions define the sides with respect to the initial side. In the latter case, the space of shapes is the space of states of a machine, and mapping the group **Special–Linear** onto the shapes shows how the machine's input actions define the shapes in terms of the starting shape.

The crucial point to emerge from this is that *mapping a group of actions onto a space of states causes the individual states to be characterized as transformations of the state that receives the identity element.*

Now, consider the rotated parallelogram. This stimulus is characterized as a transformation of a square. However, there is no intrinsic reason why the parallelogram should be seen as derived from another object. It could be seen as a non-historic static shape, independent of any other object. It is the mapping of **Special–Linear** onto the space of shapes that causes the presented parallelogram to be assigned a transformation and hence to be characterized as the application of that transformation to the shape that receives the identity element.

The corresponding effect occurs with the map of $\mathbf{D}_4$ onto the square. The sides become *defined* as transformations of the top side.

Observe now that, in characterizing the stimuli as transformations of the stimulus that receives the identity element, the map defines each of the stimuli as having the identity element as a *past* state. Thus it is *initialization* that introduces temporal asymmetry, i.e., distinguishability between the past and the present.

Observe, however, that, in mapping a transformation onto a stimulus, the transformation defines the *overall* change that has occurred between the past state (the identity element) and the present one. The transformation does not define the particular historical path that was taken from the past state to the present. It is the *subsumption structure* that allows this to be done, i.e., the decomposition into repetitions of minimal generators. Under subsumption, the color graph is mapped onto the stimulus elements. Again, we saw that such a map occurs in both the *internal* structure of the square and the *external* structure. Thus, despite the fact that the inference rules are necessarily different in the internal and external situations, the structure that results in both procedures is a map from the group's color graph onto the stimulus set.

In mapping the color graph onto the stimulus elements, it is the vertices of the color graph that map onto those elements. Thus, in the case of $\mathbf{D}_4$, the vertices of the graph map onto the sides of the square, and in the case of **Special–Linear**, the vertices of the graph (the individual points in Fig 6.24) are mapped onto the individual shapes in the space of shapes.

Therefore, each of the two stages, *initialization* and *subsumption* involves its own mapping:

Initialization: maps the group onto the stimulus elements.

Subsumption: maps the group's color graph onto the stimulus elements.

We can combine the effects of the two maps in the following simple way: Observe first that, in the color graph of $\mathbf{D}_4$, as shown in Fig 6.7, we had chosen a particular vertex as the identity element and consequently generated the labels for the other vertices. That is, we had chosen an initialization of the vertices of the color graph. Observe also that, in the color graph of **Special–Linear** as shown in Fig 6.24, we had chosen a particular vertex as the identity element and consequently generated the labels for the other vertices. That is, we had chosen, again, an initialization of the vertices of the color graph. The standard definition of color graph does not involve the choice of some vertex as the identity element. This is due to the fact that, because a color graph is always symmetric, any vertex can be chosen as the identity element. However, since an initialization occurs in a map of the group to the stimulus elements (i.e., in the initialization phase), we can induce this initialization on the vertices of the color graph; i.e., we can choose one of the vertices to be the identity element. We shall call a color graph, in which such a choice has been made, an *initialized color graph.* Fig 6.7 and Fig 6.24 each show *initialized* color graphs. The notion of an initialized color graph allows us to combine the map induced in the initialization phase, with the map induced in the subsumption stage, into a single map. We shall say that the combined effects of the initialization and subsumption phases is a map from the initialized color graph onto the stimulus set. Observe that we have seen that both internal and external inferences result in such a map.

6.13 Reference Frames

The term *reference frame* is used in psychological studies of perception in a large variety of perceptual situations to denote quite different structures. However, despite the variety of denoted structures, there is, nevertheless, one overall usage that the term *reference frame* serves: The term is used to explain the different descriptions, given by subjects, of the *same* set of stimuli in different circumstances. Researchers argue that the stimuli are described differently because they are assigned to different "frames". The frames are therefore considered, in these situations, to be crucial in determining the descriptions. Let us review some examples.

(1) Simple Shape. A number of researchers have shown that the perception of simple shape can be fundamentally different if the figure is simply given different orientations. The classic example, due to Mach (1897), is unfortunately one of the weakest to see: It is the phenomenological difference between a square in its horizontal orientation as shown in Fig 6.26a, and a square in a 45° orientation, as shown in Fig 6.26b, where it ceases to be square-like, and becomes diamond-like. The psycho-

Figure 6.26
The square- diamond effect.

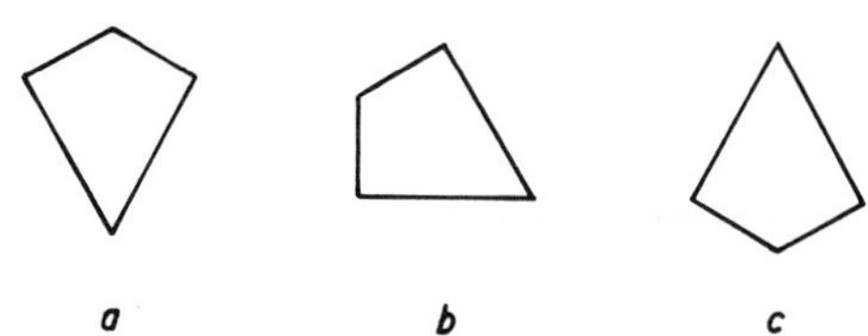

Figure 6.27
Goldmeier showed that subjects see the central figure as a different shape from the outer two figures.

logical difference between a square and a diamond is that the former must have only 90° angles, whereas the latter need not. Much more psychologically salient examples are due to Goldmeier (1936/1972), who demonstrated that the phenomenon is a theoretically substantial one underlying the very nature of perceptual representation. One of Goldmeier's examples is shown in Fig 6.27. Here, the central shape is perceived by subjects to be a completely different figure from the outer ones (Goldmeier, 1972, p88). Nevertheless, the three figures are actually identical; i.e., they can all be superimposed on each other. Rock (1973) has extensively analyzed the relationship between the various frames which are involved in the perception of shape; i.e., the Cartesian frame defined at the retinal level, the gravitational or ecological frame, the bodily frame, etc. What these studies show is that there is a crucial effect of *orientation* on the perceived form of an object. The studies are therefore concerned with what is called the "orientation and form" problem. One should note that the importance of the orientation and form studies is that they show that perception is a process of *describing* stimuli, rather than merely reproducing the outside world, an issue that we will attempt to make rigorous in this book.

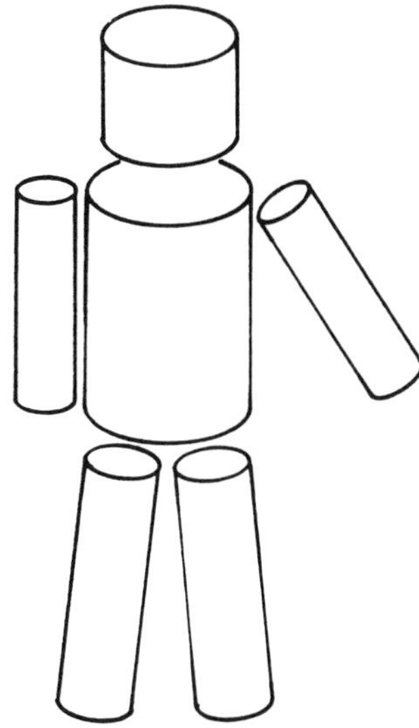

Figure 6.28
Describing a man in terms of cylinders, in accord with Binford (1971) and Marr & Nishihara (1978).

(2) Complex Shape. Binford (1971) and Marr & Nishihara (1978) have argued that complex three-dimensional shapes are perceived as the concatenation of approximately cylindrical modules, with the particular relative widths and lengths specified (Fig 6.28). For example, in the shape of an animal, the cylinders correspond to the limbs. Any such a description is achieved by assigning a collection of local two-dimensional Cartesian frames to regions of the stimulus configuration. For example, each limb of an animal receives a single frame, where one of the two axes of the frame defines the axis of the limb, and the other defines the cross-section. The limb-axis then becomes the axis of rotation defining the cylinder. Thus the assignment of the cylinders depends on how the frames are assigned to the stimuli.

(3) Induced Motion. Duncker (1929) has demonstrated, in several subtle experiments, that the perceived speed of a moving object depends on the assigned reference frame. For example, Fig 6.29 shows a rectangular frame moved relative to an observer, while a point inside the frame is kept still relative to the observer. However, Duncker found that the point is

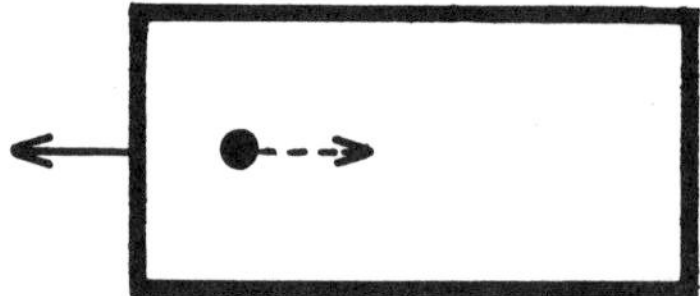

Figure 6.29
A rectangular frame moving and a point still, with respect to the observer, are seen as the opposite.

Figure 6.30
Two points of light moving as in (a), are seen moving as in (b).

perceived to be moving and the frame is perceived to be at rest. Thus the rectangle is said to provide a reference frame for the moving point.

(4) The Johansson (1950) Motion Phenomenon. Consider Fig 6.30a, two points of light moving perpendicularly, to and from each other, and in phase. Johansson (1950) showed that, if only one is presented, it is seen as moving along its path of motion; i.e., horizontally or vertically. However, he showed that if both are presented, they are seen as moving *diagonally* to and from each other, while together moving in the opposite diagonal direction as shown in Fig 6.30b. This means that the configuration of stimuli form a reference frame that defines the stimuli in terms of each other; e.g., the presence of a second stimulus defines the first as moving in relation to the second, and vice versa. Furthermore, the frame is seen as moving in relation to the observer.

From the above examples it appears that the term reference frame possesses a number of quite different specific meanings: In shape studies, it appears to mean a global or local Cartesian coordinate system; with respect to the Duncker motion phenomenon, it is a rectangular surround; in the Johansson motion phenomenon, it is a configuration of stimuli in relative motion.

However, during the course of this book, we will try to show that the above phenomena, and several other perceptual phenomena, are examples of exactly the same kind of structure. The purpose of the rest of this section is to define the structure.

Before starting, we should note the following: Assuming that a reference frame can be regarded, at least, as a set of objects, the term "reference" is usually used in one of two senses: (1) as the act of embedding an object into the set, i.e., referencing an object into the frame, or (2) as the act of judging or defining the objects within the frame with respect to each other, i.e., referencing the objects to each other. It will be a consequence of our study of perception to provide a theory of reference frames such that, using either sense of reference, the term *reference frame* denotes the same cognitive structure. That is, one references (embeds) objects into structures where the objects are referenced (judged with respect to) each other.

Our central claim concerning reference frames is the following:

REFERENCE FRAME HYPOTHESIS. *A reference frame is a programmed machine. The frame is imposed on a stimulus set by defining the stimuli as states of the machine; i.e., the reference frame is a causal explanation of the stimulus set.*

With respect to perception, we argue that reference frames have the following type of structure:

PERCEPTUAL REFERENCE FRAME HYPOTHESIS. *A perceptual reference frame is an initialized color graph of the input space of a machine. The reference frame is imposed on a set of stimuli by mapping vertices of the color graph onto the stimuli.*

To fully understand what is captured by this latter claim, we extract the following aspects from it:

(1) A reference frame is a machine that causally explains the stimulus set.

(2) Expressing the machine in terms of a color graph, forces the machine to be decomposed into processes, i.e., repetitions of minimal generators, each cycle of one color corresponding to a process. An individual process can be regarded as an individual causal factor. Thus we have:

A reference frame is a causal framework (machine) that is decomposed into causal factors (processes).

(3) The color graph is defined on the space of *inputs* of the machine. This is because the vertices of the color graph represent *actions*, and the inputs of a machine are the actions that are applied to the machine.

(4) Imposing a reference frame on a stimulus set means mapping the input actions (structured by the color graph) onto the stimuli. This means that individual actions are mapped onto individual stimuli. The consequence of this mapping is that the individual stimuli are characterized *historically* as derived from that stimulus that receives the identity element.

(5) Whereas the space of input actions is a complete group of elements, the set of stimuli onto which it is mapped might be incomplete with respect to that group. For example, in mapping **Special–Linear** onto the stimulus, a rotated parallelogram, only one member of **Special–Linear** (the point A in Fig 6.23) is mapped onto the stimulus. This situation is a general one for *external* inference: The presented stimulus is labeled by one point in the machine input space. In the case of internal inference, the trace might be incomplete with respect to the actions. For example, a side might be missing from a hexagon, as shown in Fig 6.31a, in which case, some of the members of $\mathbf{D}_6$ are not mapped to the stimuli. Again, points might

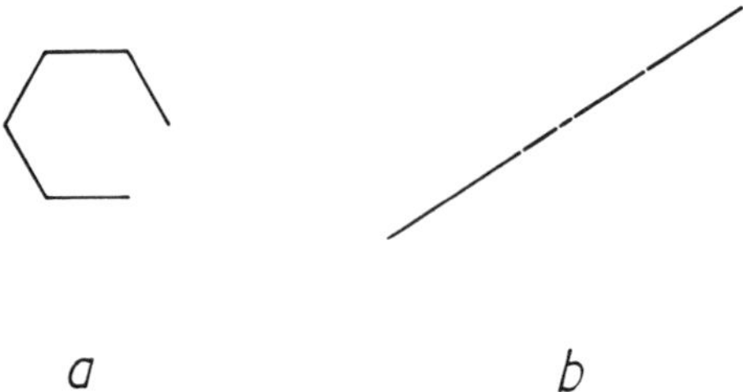

Figure 6.31
Both the two stimuli shown are seen as incomplete relative to a full group of actions.

be missing from a line, as shown in Fig 6.31b, indicating the absence of members of the group **Translations**.

We shall express this situation by saying that, in mapping the input space onto the stimulus set, not all the input actions are *instantiated* in the set. The crucial point is that the reference frame represents a complete process structure, i.e., a complete group, and the stimulus might instantiate only part of that structure. Nevertheless, we argue that *psychologically* the entire process structure is involved. It is this, we argue, that gives the sense that the stimulus sets in Fig 6.31 are incomplete. We believe this to be the basis of the Gestalt law of *closure*.

We call the map of the group onto the stimulus set, the *explanation map*. The subset of stimuli that are instantiated in the group will be called the *instantiated component* of the explanation map. Technically, one can represent the situation by defining the map from the group G to the stimulus set S as having the form $G \rightarrow S \cup \{\varnothing\}$ where $\varnothing$ is the empty set. The empty set receives those elements in G that are not instantiated in the stimulus set.

6.14 Full Nesting

We have been describing a psychologically salient history assigned to a rotated parallelogram. The history is constructed using five layers of processes:

1. Draw a side.
2. Shift from one side to the next, etc. (creating a square).
3. Stretch (creating a rectangle).
4. Shear (creating a parallelogram).
5. Rotate (creating a rotated parallelogram).

Each layer corresponds to a group. The groups are respectively:

1. **Thousandths**
2. $\mathbf{D_4}$

3. **Stretches**
4. **Shears**
5. **Rotations**

We shall make an amendment to the first item on the list. Recall that the group **Thousandths** successively translates a point at repeated intervals of 1/1000 inch, to generate a side, and is therefore a discrete group. The minimal generator for the group is a translation by 1/1000 inch. We will now make the group continuous by shrinking the minimal generator to an infinitesimally small one.[11] Thus the group becomes the continuous group of translations along a line, which we call **Translations** (discussed in section 5.9).

The successive application of the five groups will be represented thus:

Translations.$\mathbf{D_4}$.Stretches.Shears.Rotations

That is, reading from left to right, along this sequence, one obtains the five levels of processes listed above. That is, **Translations** creates the side, $\mathbf{D_4}$ creates the square, **Stretches** creates the rectangle, **Shears** creates the parallelogram, and **Rotations** creates the rotated parallelogram.

In fact, we will precede any such sequence, on the far left, by an additional layer. The production of a single point, which will be denoted simply by **Point**. Thus the successive layers of processes are given by the following left-to-right sequence:

Points.Translations.$\mathbf{D_4}$.Stretches.Shears.Rotations

The inclusion of this initial, almost trivial layer, **Point**, is required because the role of the next layer, **Translations**, is to move the point to generate a side. The group **Point** is simply the group consisting of the identity element, and it labels the first point to be drawn on a side.

The order from left to right, in the above six-fold sequence, represents the temporal order of the processes, in this sense: the left to right order is the order in which non-zero elements of the respective processes first occur in the history. However, the six successive group layers do not represent six successive *layers* of time. They in fact represent successive layers of *control* in the program that defines the history. To understand this, consider the relationship between **Translations** which generates a side, and the next layer $\mathbf{D_4}$ which produces the shifts from one side to the next. One creates a side by **Translations** and one then uses a shift, taken from $\mathbf{D_4}$ (e.g., rotation by 90°), to start the next side. However, one then uses **Translations** again to draw this next side. Then one shifts again, and then translates again, etc. So one oscillates backwards and forwards between the layers, **Translations** and $\mathbf{D_4}$. Therefore $\mathbf{D_4}$ does not *historically* follow, or lie over, the **Translations** layer. The $\mathbf{D_4}$ layer is above the **Translations** layer in another sense that we described precisely in section 4.10: That is, for each individual operator carried out in the $\mathbf{D_4}$ layer, *all* the operations in the **Translations** layer are carried out; i.e., for each shift taken from $\mathbf{D_4}$, an entire side is drawn. In the program, this is represented by embedding the side-drawing loop within the shifting loop (as discussed in section 4.10). We called this kind of structure, *internal nesting*. In fact, we defined two types of nesting, and we restate the definitions here for convenience:

DEFINITION OF NESTING. *Control is said to be* **nested** *when the entire effects of one process are acted on by another process. The nesting is said to be* **internal** *when one goes through the former process, in its entirety, at every point in the latter process. The nesting is said to be* **external** *when one goes through the former process, in its entirety, only once and before the latter process begins; i.e., the latter is a "layer" over the former.*

We can now see that, in our sequence,

$$\mathbf{Point.Translations.D_4.Stretches.Shears.Rotations}$$

the first three layers are internally nested and the last three are externally nested. We shall indicate this as follows:

$$[[\mathbf{Point.Translations.D_4}]_I\mathbf{.Stretches.Shears.Rotations}]_E$$

That is, we put the first three groups within a pair of brackets and index the right-hand member of the pair with the symbol I, indicating that each of these successive groups is *internally nested within the next*. Then we put the internal sequence and all the external groups together within a pair of brackets and index the right-hand member of the pair with the symbol E, indicating that each successive component is *externally nested within the next component*.

In the general case, where there might be several internal and external subsequences that are dispersed among each other, we bracket the subsequences, like clauses in linguistic syntax, and index them with I and E indicating that the internal and external interpretations respectively are required in reading those segments of the sequence. Later, in this chapter we shall meet a third type of bracketing, that indicating when processes are to be interpreted as parallel.

The phase we called *initialization* is equivalent to mapping this group-sequence onto the stimulus set. The corresponding sequence for the *subsumption* phase is the sequence of color graphs. This can be specified in the following way:

The color graph of an individual group can be very compactly specified without actually drawing the graph, but by using instead what is called a *presentation* of the group. A group presentation is simply a list of the generators, followed by a list of those sequences of generators that form the minimal cycles out of which the graph is built like a mosaic. For example, a presentation of $\mathbf{D}_4$ is simply

$$\{r_{90},\quad t:\quad r_{90}r_{90}r_{90}r_{90},\quad tt,\quad r_{90}tr_{90}t\}$$

The two elements before the colon are, respectively, the rotation generator, r_{90}, and the reflection generator, t. The strings after the colon specify the three minimal cycles out of which the graph is constructed as a mosaic. These are, respectively: (1) a cycle formed by repeating r_{90} four times, i.e., a solid cycle in Fig 6.7; (2) a cycle formed by repeating t twice, i.e., a dotted cycle in Fig 6.7; and (3) a cycle formed by applying r_{90} and t alternately, i.e., a cycle that is sandwiched between the inner and outer bold cycles in Fig 6.7. The presentation—i.e., the above list of generators and minimal cycles—completely specifies the color graph.

Therefore, to specify the sequence of color graphs corresponding to the sequence of groups

$$[[\mathbf{Point.Translations.D_4}]_I\mathbf{.Stretches.Shears.Rotations}]_E$$

one simply substitutes, for each group in the sequence, the presentation of the group, i.e., one obtains a sequence of group presentations. It is this latter sequence that is assigned to the stimulus set in the subsumption phase.

Finally, observe that the program, which defines the history, is a collection of repeat loops in which each loop takes a generator from one of the presentations in the sequence.

6.15 Nested Control and Memory

We now begin to examine what will be one of the central notions of our process theory: the concept of *nested control.* We shall see that this notion is profoundly linked to the notion of memory. We begin with some very simple considerations—so simple that they do not hint at the profundity of the phenomenon that lies behind them.

If a stimulus is assigned a process that explains it, it is not only assigned the process but the *amount* of the process that took place in creating the stimulus. For example, an object is not merely seen as having been rotated but as having been rotated *by a certain amount.* We shall say that the amount of a process used in creating the stimulus is the *coordinate* of the stimulus in that process.

It is important to observe that, since several processes could have been used to create a stimulus—e.g., the several processes used to create a rotated parallelogram—the stimulus can have several coordinates, one in each process. We shall call these coordinates the *process coordinates* of the stimulus. In this and the following sections, we will begin to study process coordinates and find that they have important psychological roles.

Let us consider first the internal structure of a square. Similar considerations will, of course, apply to any prototype. We have seen that there are two levels of process creating a square, given by the succession of groups

$$\mathbf{Translations}.\mathbf{D}_4$$

The first group defines a side and the second group defines the shifts between sides. We saw that this pair of groups is *mapped onto* the square. The mapping actually defines the *process coordinates* for each point on the square, as follows:

Translations is mapped onto the points of each side. As shown in Fig 6.32a, the zero translation, *e*, is mapped to the first point on the side, and each other point receives a translation corresponding to its distance from the initial point. For example, Fig 6.32a shows that the particular translation *g* has been mapped to some point on the side. This is because, as shown in Fig 6.32b, *g* is the distance that the initial point has been translated to obtain that point. Thus we say that *g* is the *coordinate* of the point in the process **Translations**.

Consider now the other internal group $\mathbf{D}_4$. We saw that this group is mapped onto the sides of the square, as shown in Fig 6.6. Thus each side

Figure 6.32
The translation is mapped to the point shown in (a), because the point is obtained by the translation shown in (b).

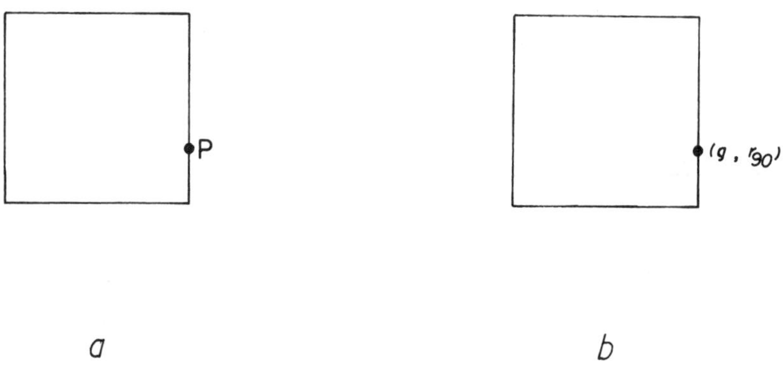

Figure 6.33
Point P shown in (a) has two coordinates as shown in (b).

has *coordinates* in $\mathbf{D}_4$. These show the amount of process that was used in creating the side. To make the discussion simpler, we shall consider only the inner ring of rotations in Fig 6.6.[12] Thus each side is labeled by the amount of rotation used to create that side, i.e., its rotation *coordinate*.

Observe now that any *point*, for example the point P in Fig 6.33a, therefore has two coordinates, because two processes were used to obtain the point: (1) the translation process that moved the "pen" along the side to the point, and (2) the rotation process that oriented the trajectory of the pen. That is, a point has a coordinate in **Translations** and a coordinate in $\mathbf{D}_4$. The pair of coordinates for point P are shown in Fig 6.33b. The first coordinate, g, specifies the amount of translation used to obtain the point and the second coordinate, r_{90}, specifies the amount of rotation used to obtain the side. That is, the pair of coordinates (g, r_{90}) comes from the pair of groups

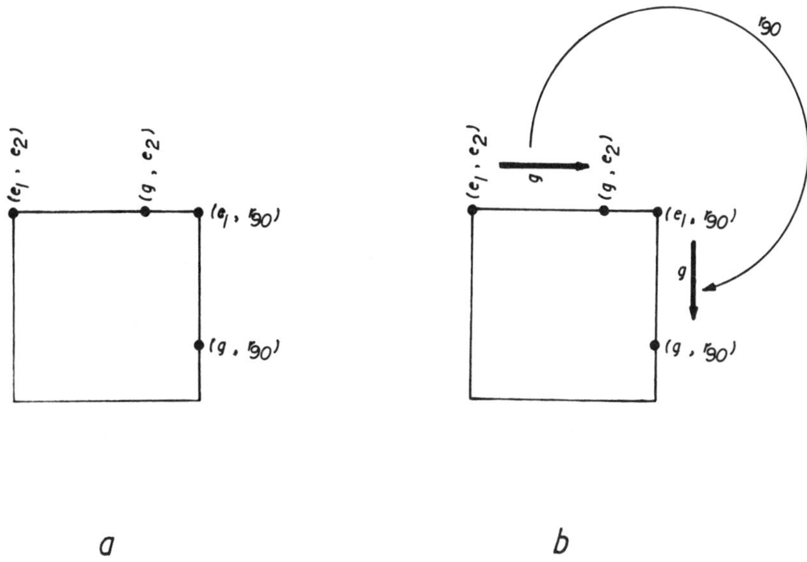

Figure 6.34
(a) The coordinates of four points on the square. (b) The nested structure of those coordinates.

Translations.$\mathbf{D}_4$

In order to describe the phenomenon we are going to investigate, it is necessary to fill in some other coordinates on the square. Consider the top left-hand corner point. We have assumed that the entire history starts here. Therefore the amount of translation here is zero—i.e., the point is at the identity element e_1 of **Translations**. Furthermore, the amount of rotation shift that has been applied is also zero—i.e., the side on which the point sits is at the identity element e_2 of $\mathbf{D}_4$. Therefore, as shown in Fig 6.34a, the top left corner-point has the pair of coordinates (e_1, e_2) in the pair of groups **Translations.**$\mathbf{D}_4$. Now consider the point we previously called g on the top side. It has involved the translation g from **Translations**. However, the side on which the point sits has not yet involved any rotation shift from $\mathbf{D}_4$ and therefore remains at the identity element e_2 in $\mathbf{D}_4$. Thus this second point has the pair of coordinates (g, e_2) in the pair of groups **Translations.**$\mathbf{D}_4$. Now consider the top right-hand corner point, and consider its description as the first point on the right-hand side. As the first point on this second side, it has undergone no translation along that side, and is therefore at the identity element e_1 of **Translations**. However, the point must have coordinate r_{90} in $\mathbf{D_4}$ because the pen direction has been rotated by 90° in order to create the right-hand side. Thus the point has the pair of coordinates (e_1, r_{90}) in the pair of groups **Translations**.$\mathbf{D_4}$. Finally, as we saw earlier, the lower labeled point on the right-hand side has coordinates (g, r_{90}) as shown in Fig. 6.34a.

We now observe that the *order* of the coordinates corresponds to the *nesting* of the processes, as follows: First observe that the action of $\mathbf{D_4}$ takes the process that generates one side (i.e., **Translations**) and images it onto the process that generates each other side. Thus, for example, consider the following two points on the top side: The first point (e_1, e_2) and an arbitrary point (g, e_2) on that side. The relationship between the two points is a translation g given by the top horizontal arrow shown in Fig 6.34b. This pair of points is then mapped via the rotation r_{90}, from $\mathbf{D_4}$, onto the pair of points (e_1, r_{90}) and (g, r_{90}) on the right side. But these latter two points are also related by the translation g in **Translations**. Thus, the translation, g on the top side, is *transferred* via the operation r_{90} in $\mathbf{D_4}$ to the translation g on the right-hand side as shown in Fig. 6.34b.

Now let us turn to the external history. Each shape in that history is a member of **Special–Linear** and thus has a stretch coordinate, a shear coordinate, and a rotation coordinate. Going temporally forward in that history, the first shape, the square, is seen as having zero stretch, zero shear, and zero rotation. Therefore it is at the identity element of each of these three groups. If we denote the identity elements of the stretch, shear, and rotation groups, respectively, by e_3, e_4, and e_5, then the square is given by

$$[\mathbf{Point.Translations.D_4}].e_3.e_4.e_5$$

The first half of this sequence, that is, **Point.Translations.**$\mathbf{D_4}$, is the internal structure of the square. The second half, that is, $e_3.e_4.e_5$, gives the three coordinates in the three groups **Stretches.Shears.Rotations**.

Now, as the history progresses forward from the square to the rectangle to the parallelogram to the rotated parallelogram, the identity elements e_3, e_4, and e_5, successively change and become non-zero values in the successive groups **Stretches.Shears.Rotations**. That is, going from the square to the rectangle, the stretch coordinate takes on a non-zero value, s, and thus the coordinates of the rectangle are

$$[\mathbf{Point.Translations.D_4}].s.e_4.e_5$$

Again, going from the rectangle to the parallelogram, the shear coordinate takes on a non-zero value, *h*, and thus the coordinates of the parallelogram are

$$[\mathbf{Point.Translations.D_4}].s.h.e_5$$

Again, going from the parallelogram to the rotated parallelogram, the rotation coordinate takes on a non-zero value, *r*, and thus the coordinates of the rotated parallelogram are

$$[\mathbf{Point.Translation.D_4}].s.h.r$$

Since the first half of each of these strings is **Point.Translations.D**$_4$, and the second half gives the three coordinates in the three groups **Stretches.Shears.Rotations**, the entire six-fold string, in each case, is taken from the six-fold process-hierarchy we defined in section 6.14:

$$[[\mathbf{Point.Translations.D_4}]_I.\mathbf{Stretches.Shears.Rotations}]_E$$

Now let us return us to the issue of the *nested* organization of the processes. For each *shape*, the first half of the sequence, i.e., **Point.Translations.D**$_4$, is the internal structure of the *square*, and the second half defines *what has been done to the square*. This implies that each shape is described as having been derived from the square; i.e., *the square is a past state of that shape*. Furthermore, it also implies the following deeper phenomenon: The structure of the square is mapped, or transferred, onto the structure of the derived shape. For example, consider the coordinates given in Fig 6.34a to four particular points of the square. On the rotated parallelogram, these four points *become* the four points shown in Fig 6.35a. That is, the four points we show on the rotated parallelogram are the *memory* of the four points shown on the square. In fact, not only does the parallelogram contain the memory of the points on the square, but it retains the memory of the process-structure of the square. Thus we saw, in Fig 6.34b, that, in a square, the translation process creating the top side is mapped onto the translation process creating the right side, via the rotation process; i.e., translation is nested within rotation. This nested structure is mapped, or transferred, onto the rotated parallelogram, as shown in Fig 6.35b. That is, the rotated parallelogram contains the memory of that structure.

Now observe that this relationship between the square and the rotated parallelogram is simply an additional layer of *nesting*. That is, just as, in a

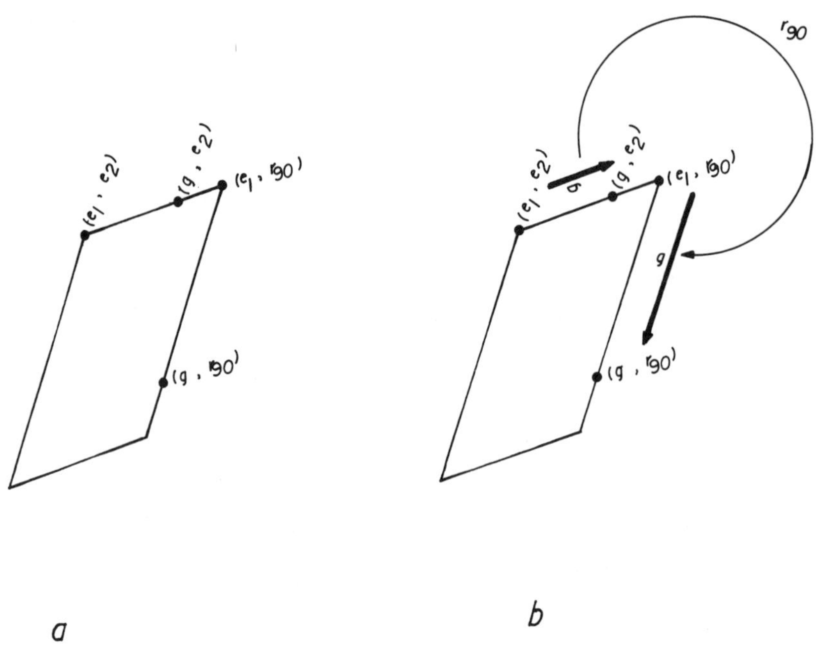

Figure 6.35
The four points labeled in Fig 6.34a become the four points labeled in Fig 6.35a. The nested structure in Fig 6.34b becomes the nested structure in Fig 6.35b.

square, nesting transferred the process structure of one side onto the process structure of another side, nesting now transfers the nested structure of a square onto the nested structure of a rotated parallelogram.

We can see therefore that the phenomenon of memory is closely linked to the phenomenon of nesting. It is nesting that permits the past to be imaged, or transferred, into the present, i.e., the past is seen within the present via nesting. The centrality of nesting to perception was captured by the title of the first paper in which we described this phenomenon: "Perceptual Organization as Nested Control" (Leyton, 1984).

It is now possible to see that the principles we have extracted concerning the fully externalized history of the rotated parallelogram apply to alternative histories that are not fully externalized. In fact, consider a fully internalized history of a rotated parallelogram; i.e., one that is purely a trace. That is, in this history, the rotated parallelogram is drawn, starting at one corner, say corner A in Fig 6.36a, and moving the pen so that it traces out the successive sides of the parallelogram directly.

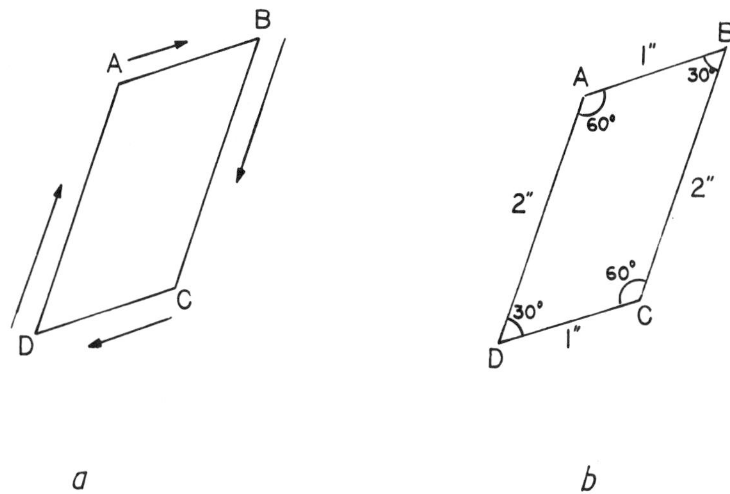

Figure 6.36
(a) The rotated parallelogram traced from point *A*. (b) The tracing requires several successive changes according to the sizes encountered.

It appears that, at each corner, a different decision has to be made from that made at the previous corner. For example, as shown in Fig 6.36b, the pen first has to draw a side of 1 inch, then change orientation by 30°, then draw a side of 2 inches, then change orientation by 60°, then draw a side of 1 inch, then change orientation by 30°, etc. This contrasts with the fully externalized history. In the latter history, the trace structure is carried out at the level of the square, and, at each corner, the same decision is involved: draw a side of 1 inch and change orientation by 90°.

It seems therefore that, in going to the fully *internalized* history, much greater distinguishability has been introduced; i.e., successively changing decisions at each corner. However, this need not be the case, as follows:

The clue to understanding this issue comes from the *nested* structure of the fully *externalized* history. Recall, for example, that, in the latter history, the external factors map the trace structure of the square shown in Fig 6.34b onto the rotated parallelogram, as shown in Fig 6.35b. We can now use this to construct a minimal purely internal history of the rotated parallelogram. All we do is apply the external factors *before* tracing, rather than afterwards, as follows:

Imagine that the pen is at the origin of a rectangular Cartesian coordinate system. In fact, in the human hand, the pen is fitted into such a system

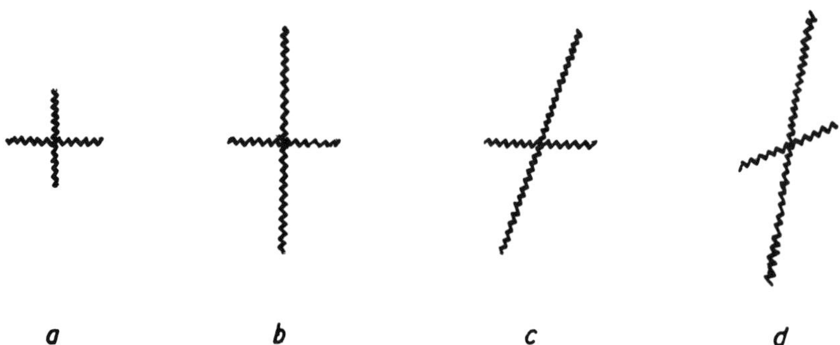

Figure 6.37
(a) The hand, as a pair of perpendicular springs, as proposed by Hollebach, 1981. (b, c & d) Successive changes to those springs.

thus: Hollerbach (1981) has given compelling evidence that, in handwriting, the hand is structured as if it were a pair of perpendicular springs holding the pen—an arrangement which we represent in Fig 6.37a, with the dot in the center representing the pen. We will identify this system of springs with the Cartesian coordinate system. Now, prior to drawing, i.e., moving the pen, let us apply some actions to the system of springs, i.e., to the coordinate system. Let us first stretch the spring system as shown in Fig 6.37b, then shear it, as shown in Fig 6.37c, and finally rotate it, as shown in Fig 6.37d. Having altered the spring coordinate system to that shown in Fig 6.37d, we then give the instruction simply to draw a *square*. A square drawn in that coordinate system, will be a rotated parallelogram. This means that, because we changed the coordinate system of springs, the *program* can be the same purely repetitive loop structure—draw a side of 1 inch and rotate by 90°—used in the fully externalized history. That is, we have removed the necessity of having a different decision at each successive corner.

We see therefore that, to create a minimal fully internal history of a rotated parallelogram, we simply use the same internal and external components that were used for the fully externalized history and reverse the order of application of those components.

The crucial point is that the sequence

$$[[\mathbf{Point.Translations.D_4}]_I\mathbf{.Stretches.Shears.Rotations}]_E$$

should be understood not as representing the order of processes, but as representing the hierarchy of *nesting* in the processes. That is, each level,

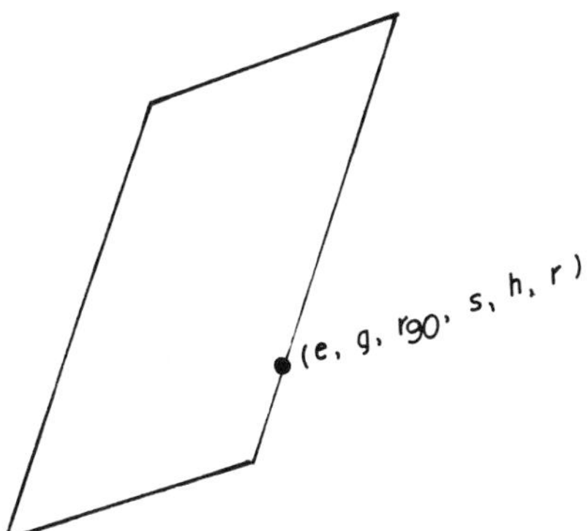

Figure 6.38
A point on a rotated parallelogram receives six coordinates.

right-ward in the sequence, maps the structure, defined by the preceding sequence of levels, onto other configurations. For example, $\mathbf{D_4}$ maps **Translations** from the top side onto the right side, as shown in Fig 6.34b. Again, on the next level **Stretches** transfers the **Translations.**$\mathbf{D_4}$ trace structure from a square onto a rectangle, etc. The entire sequence, given above, is a nested hierarchy of mappings, in that *each mapping is a mapping of the previous mappings, and so on.* For example, as we illustrated in Fig 6.34b the r_{90} mapping of **Translations**, from the top side to the right side, is itself mapped onto the rotated parallelogram in Fig 6.35b.

To emphasize: In the fully externalized history, the sequence

$$[[\mathbf{Points.Translation.D_4}]_I.\mathbf{Stretches.Shears.Rotations}]_E$$

does represent the temporal order in the particular sense that it is the order in which non-zero elements of processes are first applied. In the internalized history, the sequence does not represent temporal order. Nevertheless, in both the internalized and externalized histories, the sequence represents the *same hierarchy of mappings.*

Finally, we observe that any point on the rotated parallelogram, such as the point shown in Fig 6.38, is given six coordinates, corresponding to the six process levels in the above sequence. For example, the coordinates for the point shown in Fig 6.38 are

$$(e, g, r_{90}, s, h, r)$$

The first three coordinates are the internal coordinates that the point receives from the trace structure of the square, and the last three are the amounts of stretch, shear and rotation that were applied in the square.

These six coordinates represent the amount of *memory* contained in that point, i.e., they are a record of the processes that produced that point in that position. Furthermore, the order of the coordinates represents the nested structure in which the processes were organized.

6.16 Nesting as Transfer

In the previous section, we saw that histories necessarily involve a nested organization, i.e., one in which a process-structure is mapped or transferred from one stimulus set to another. In our discussion, we used the terms "mapping" and "transfer" interchangeably.

In section 2.3, we used the term *transfer* in the sense defined by the Transfer Principle, which states that, if an asymmetry in an object was not created in the history of that object, then it was transferred from the history of another object. An example is a printed word on some sheet of paper: The word was not traced out on the paper itself, but was transferred to the paper from a stamp on which the word was traced out. We shall call this type of transfer, *outcome transfer*. That is, outcome transfer is the transfer of the results of a history that has already taken place.

The notion of nesting allows transfer to be understood in the sense just given, and also in another sense that we will call *program transfer*. In the latter type, a program is transferred from one moment to another but has to be carried out after it is transferred. We can illustrate both types of transfer thus:

Consider the nested structure of the square, as shown in Fig 6.34b. A typical scenario that uses *outcome transfer* is this: A stamp created the top side (by stamping), and was then rotated to create the right side (by stamping). What is crucial here is that the right side was not, itself, traced out. The right side is the image of the trace structure that occurred even prior to the creation of the top side.

In contrast, under *program transfer*, the right side was directly traced out. However, the program used for this tracing was the program used for directly tracing the top side. That is, the *program* used in the history of the top side was transferred for use again in creating the right side. This type of transfer is embodied in the flow chart we studied in previous sections

(Fig 6.17), where the inner loop that traces out a single side is used again and again for each of the sides. In contrast, outcome transfer would use this loop only once, in the creation of the stamp.

As another illustration, consider the transfer of the structure of the square in Fig 6.34b onto the rotated parallelogram in Fig 6.35b. Under outcome transfer, the square would be traced out once, and then deformed to create the parallelogram, i.e., the previous outcome would be transferred from the past to the present. This is the type of transfer we used in the fully externalized history of the rotated parallelogram. In contrast, our fully internalized history could be considered to use program transfer: That is, the program for drawing a square is transferred from the equal-sided, perpendicular, gravitational frame, shown in Fig 6.37a, to the stretched, sheared, rotated frame, in Fig 6.37d, and is *then* carried out.

Thus the difference between outcome transfer and program transfer is that, in the former, the outcome of a realized history, is transferred to another moment and therefore does not have to be carried out again; whereas in the latter, the program for a history is transferred to another moment and does have to be carried out to produce an outcome.

A nested hierarchy, i.e., a hierarchy of mappings, yields either interpretation, and we shall not usually need to distinguish between them because the same principles are applicable in both. Since nesting allows either interpretation of transfer, we shall equate the notion of nesting with that of transfer.

6.17 Theory of Grouping

As argued by the Gestalt psychologists, a crucial determinant of a percept is the way in which the stimuli on the retina are *grouped* together by the perceptual system. An almost infinite set of alternative groupings are possible of the same collection of retinal stimuli. Each alternative grouping would lead to an alternative percept. The perceptual system chooses only one of the alternative ways to group the stimuli. In this section we examine what it is that the perceptual system chooses when it chooses a grouping, and in the next section we examine how the choice is made.

Despite the considerable research that has been devoted to the phenomenon of grouping, no one has proposed a *substantive* definition of what a grouping is. This is strange. For, if a substantive definition could be offered and validated, then it would be the core concept that would provide coherence to the many diverse data and ideas concerning grouping. Our intention, in this section, is to develop a substantive definition of grouping.

We argue that grouping is a *historical* construct. Therefore, we claim that grouping can be understood only within the view of perception as the recovery of history. In fact, we argue that grouping is *the* historical construct. That is, we propose:

A grouping is a history.

In order to analyze this claim, we need the group-theoretic notion of *coset*. One can illustrate the notion of coset using the group $\mathbf{D}_4$. As noted earlier, this group divides into two halves, its four rotations and its four reflections:

Rotations: $e, r_{90}, r_{180}, r_{270}$

Reflections: $t, r_{90}t, r_{180}t, r_{270}t$

The reflections are obtained by multiplying the rotations by the element t. This fact makes the set of reflections an example of what is called a *coset*. More generally, if one has a subgroup, $\mathbf{H}$, then a set obtained by multiplying $\mathbf{H}$ by some element g is called a *coset* of $\mathbf{H}$. The coset is simply notated by

$$\mathbf{H}.g$$

The element g is called a *coset leader*. In $\mathbf{D}_4$, the four rotations form a subgroup. The reflections are a coset of that subgroup, because they are the result of multiplying the rotations by t. The coset leader is t.

In what follows, we are going to use the term coset not just for groups but for sequences of groups. Technical details concerning this use are given in Footnote 13 to this chapter.

Now let us return to our claim that *a grouping is a history*. We saw, in the previous section, that a history is not just in the object which was first created by the history but is transferred onto other objects; e.g., from the top side of a square to the right side. The transfer is carried out by an operation on the next higher level of the nesting hierarchy. We shall now see that the history and the transferred history both have the structure of cosets.

Consider for example the top side of the square. It is generated by creating a point and translating it; that is, it is generated by the group sequence

Point.Translations

Now, to create the right side, one rotates this process-structure by 90°. That is, one applies the operation r_{90} to the above group sequence. This results in the sequence:

$$\mathbf{Point.Translations.}r_{90}$$

The latter sequence has the structure of a coset. That is, it has the form

$$\mathbf{H}.g$$

where **Point.Translations** corresponds to **H**, and the element r_{90} corresponds to the coset leader g.

The top side also has the structure of a coset, as follows: We said that this side is generated by the group sequence **Point.Translations**. However, the side is the initial state in the rotation process of $\mathbf{D}_4$, and is thus at the identity element e_2 of $\mathbf{D}_4$. This means that the side has the structure **Point.Translations**, acted on by e_2. Therefore, the top side is given by the sequence

$$\mathbf{Point.Translations.}e_2$$

This, again, has the structure of a coset $\mathbf{H}.g$. The sequence **Point.Translations** corresponds to **H**, and the element e_2 corresponds to the coset leader g.

Now recall that given a nested sequence such as

$$[[\mathbf{Point.Translations.D_4}]_I\mathbf{.Stretches.Shears.Rotations}]_E$$

the left-to-right order, in the sequence, gives the order of *transfer*. That is, given any level in the group sequence, the operations on that level *transfer* the groups on the left of the level (in the sequence), from one stimulus set to another. For example, consider again the sequence defining the right side of the square, that is:

$$\mathbf{Point.Translations.}r_{90}$$

The coset leader r_{90} has *transferred* the **Point.Translations** structure from the top side to the right side of the square. However, **Point.Translations** is to the left of r_{90} in the above group sequence.

Generally speaking, given a nested group sequence

$$\mathbf{G}_1.\mathbf{G}_2 \ldots \mathbf{G}_n$$

we shall say that a subsequence that starts from $\mathbf{G}_1$, and includes all the groups up to some level $\mathbf{G}_i$, is a *left-subsequence*. That is, any left-subsequence has the form

$$\mathbf{G}_1.\mathbf{G}_2 \ldots \mathbf{G}_i$$

Furthermore, given any level $\mathbf{G}_{i+1}$ in the sequence, we shall say that the entire sequence to the left of $\mathbf{G}_{i+1}$, that is, the sequence, $\mathbf{G}_1.\mathbf{G}_2 \ldots \mathbf{G}_i$, is the *left-subsequence of* $\mathbf{G}_{i+1}$. Because of the nested aspect of the hierarchy, the operations in $\mathbf{G}_{i+1}$ have the role of *transferring* the left-subsequence, $\mathbf{G}_1.\mathbf{G}_2 \ldots \mathbf{G}_i$, from one stimulus set onto another.

To illustrate, let us return to the sequence

$$[[\textbf{Point.Translations.D}_4]_I.\textbf{Stretches.Shears.Rotations}]_E$$

Consider the level $\mathbf{D}_4$. Its left-subsequence is **Point.Translations**. Furthermore, any operation in $\mathbf{D}_4$ has the role of transferring the left-subsequence **Point.Translations** from one stimulus set to another. For example, as noted above, the operation r_{90} (in $\mathbf{D}_4$) transfers **Point.Translations** from the top side to the right side of the square. Furthermore, the transferred process-structure is given by the coset

$$\textbf{Point.Translations.}r_{90}$$

which we can now see is a coset of a *left-subsequence*.

Now let us draw together the ideas presented in this section so far. The fundamental proposal we began with is that a *grouping is a history*. However, a history does not necessarily generate an object directly but can be transferred from another object. The crucial point is this: Since any transfer is carried out by an operation on the next higher level of the nesting hierarchy, the history is a coset of a left-subsequence. Before the history is transferred, it is such a coset, but one in which the coset leader is the identity element of the level above the left-subsequence. For a transferred history, the coset leader is a non-identity element of that next level. Since we propose that a grouping is a history, we are therefore led to this conclusion:

GROUPING PRINCIPLE. *A grouping is a coset of a left-subsequence of the nested group sequence.*[14]

Before continuing however, we should observe that, in the literature, a *grouping* is generally considered to be a *part*. This is because a figure is *partitioned* into its groupings. The above Grouping Principle, in fact, handles this issue in a particularly strong way. There is a theorem in group theory, the Cayley Theorem, which states that, if one is given a subgroup **H** of a group **G**, and one elaborates all the cosets **H**.g of **H**, one will find that the entire set of cosets completely partitions the overall group **G**. Correspondingly, in our case, we will find that the entire set of cosets of a left-subsequence completely partitions the full group sequence. Thus the groupings will be the parts determined in a manner corresponding to Cayley's Theorem. This concept will become easy to understand when we illustrate it shortly.

We shall now corroborate the Grouping Principle (and hence the original proposal that a grouping is a history), by considering a variety of examples.

Let us first consider a square. There are three types of groupings or parts that people commonly identify or refer to in a square: These are the *points*, the *sides*, and the *whole* square itself. People do not choose other subsets of the square as groupings, e.g., a third of one side. Our Grouping Principle explains the choices that are made by people. According to the proposal, the people will see the cosets of left-subsequences as groupings. Let us therefore enumerate all the cosets of left-subsequences, and see what we obtain.

The square is given by the process-sequence:

$$\mathbf{Point.Translations.D_4}$$

We first list all the left-subsequences. They are, in order of length,

Point

Point.Translations

$\mathbf{Point.Translations.D_4}$

We now take each of these left-subsequences, in turn, and elaborate all their cosets.

(1) The cosets of **Point.**

The cosets of **Point** in the sequence

$$\mathbf{Point.Translations.D_4}$$

must be of the form

$$\mathbf{Point}.g_1.g_2$$

where g_1 comes from **Translations** and g_2 comes from $\mathbf{D}_4$. The pair, $g_1.g_2$, is the coset leader.

What is such a coset perceptually? Recall, from Fig 6.34a, that a pair of the form $g_1.g_2$, from the group pair **Translations.D**$_4$, is simply the coordinate pair defining an arbitrary point in the square. For example, four such pairs label the four points shown in Fig 6.34a. Thus the set of cosets **Point.**$g_1.g_2$ is the set of points of a square. Therefore the decomposition of the group sequence **Point.Translations.D**$_4$, into the set of cosets of the left-subsequences of **Point**, is equivalent to decomposing the square into its set of points.

(2) The cosets of **Point.Translations**
The cosets of **Point.Translations** in the sequence

Point.Translations.D$_4$

must be of the form

Point.Translations.g_2

where the element g_2 comes from $\mathbf{D}_4$. This element is the coset leader.

What is such a coset perceptually? Recall, from Fig 6.34a that a coordinate g_2, from the group $\mathbf{D}_4$, is simply the coordinate defining an arbitrary side in the square. The eight such coordinates are shown in Fig 6.6. Thus the set of cosets **Point.Translations.**g_2 is the set of sides of a square. Therefore the decomposition of the group sequence **Point.Translations.D**$_4$, into the set of cosets of the left-subsequences of **Point.Translations**, is equivalent to decomposing the square into its set of sides.

(3) The cosets of **Point.Translations.D**$_4$
There is only one coset of **Point.Translations.D**$_4$ in the sequence

Point.Translations.D$_4$

It is the sequence **Point.Translations.D**$_4$ itself.

This coset is of course the whole square. Thus a consequence of our Grouping Principle is that the whole is a grouping itself—which accords with perceptual experience.

We conclude therefore that elaborating the cosets of the left-subsequences of the process-structure of the square gives the groupings or parts of the square to be the points, the sides and the whole, i.e., the psychologically salient groupings or parts of the square.

This example appears at first to be almost trivial. However, one can in fact show that it crucially involves the *transfer* structure in a non-trivial way. To do this, let us reverse the order of the components **Translations** and $\mathbf{D_4}$, in the group sequence, and show that one does not obtain parts of a square.

Thus, suppose that the square has $\mathbf{Point.D_4.Translations}$ as its group sequence (i.e., reversing the second and third factors). The left-subsequences are

Point

$\mathbf{Point.D_4}$

$\mathbf{Point.D_4.Translations}$

The first left-subsequence, **Point**, is the starting point because it precedes the application of the other operations. We have generally taken the starting point to be the top left corner-point of the square. Therefore, the left-subsequence **Point** represents the top left corner-point.

Now consider the second left-subsequence in the above list, $\mathbf{Point.D_4}$. Because $\mathbf{D_4}$ describes the polygon operations (rotations and reflections) the left-subsequence $\mathbf{Point.D_4}$ must be the four corner points, as shown in Fig 6.39a.

Now consider the cosets of this left-subsequence. Any coset must be of the form $\mathbf{Point.D_4}.g$. This is a translation g of each of the corner points by some distance g along each side; i.e., the configuration of four points shown in Fig 6.39b.

It is true that one can see these points as visually equivalent. This is due to the group-theoretic properties of the process-hierarchy. However,

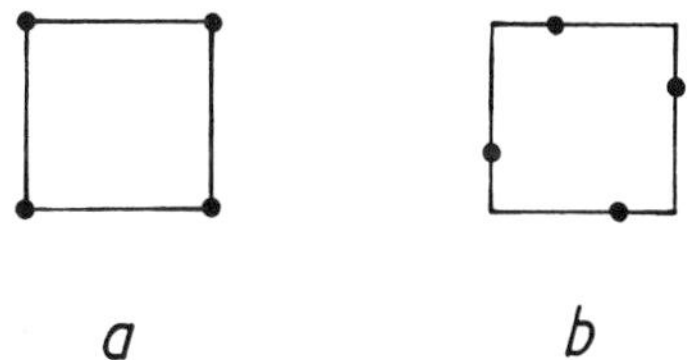

Figure 6.39
Groupings produced when the group-sequence is in the wrong order.

people never mention such a set as a "part" of a square. As was noted above, conventionally, the parts are regarded as the points, the sides, and the whole; that is, the sets elaborated earlier in the *non-reversed* group sequence. In conclusion, therefore, one finds that reversing the order leads to an empirically incorrect partition.

Observe that this situation is essentially due to the *transfer* structure. In the above reversed sequence, we would have to see **Translations** as *transferring* $\mathbf{Point.D_4}$ subsets like Fig 6.39a, onto other $\mathbf{Point.D_4}$ subsets like Fig 6.39b. Instead, we find it much easier to see $\mathbf{D_4}$ as transferring the sides, **Point.Translations**, onto each other. To re-emphasize: The grouping structure is a structure of observer-assigned *mappings*, which transfer the process-structure of subsets of stimuli onto other subsets of stimuli. It is the coset leader in each coset that does the transferring action. In the next section, we will describe the criteria that perception uses to determine the transfer order.

Now consider the rotated parallelogram. If we let the stretch, shear and rotation coordinates be s, h, and r, respectively, then the rotated parallelogram is given by the sequence

$$[\mathbf{Point.Translations.D_4}].s.h.r$$

This sequence is a coset of a left-subsequence. The triple $s.h.r$ is the coset leader, and $\mathbf{Point.Translations.D_4}$ is the subsequence on which it acts. Therefore, according to the Grouping Principle, the sequence must be a grouping. This, of course, accords with perception. A rotated parallelogram is a grouping.

The left-subsequence $\mathbf{Point.Translations.D_4}$ is the largest left-subsequence in the sequence $[\mathbf{Point.Translations.D_4}].s.h.r$ defining a rotated parallelogram. The next largest left-subsequence is **Point.Translations**. Its cosets are of the form

$$[\mathbf{Point.Translations}].g_2.s.h.r$$

where g_2 comes from $\mathbf{D_4}$. The coset leader is the quadruple $g_2.s.h.r$. Any coset of the form just given is a *side* of the rotated parallelogram. Since a side is a grouping in the rotated parallelogram, we see again that a grouping is a coset of a left-subsequence.

Notice how the above coset describes the *history* of a side. The left-subsequence is **Point.Translations**. This is the top side of a *square*. The coset leader $g_2.s.h.r$ is the following succession of operations that were applied in the subsequent history of that side. The operation g_2 rotated (or reflected) the top side to become one of the other sides of the square; and then the operations s, h, and r, that is, a successive stretch, shear, and rotation, were applied producing the side that appears on the rotated parallelogram. Thus the coset, or equivalently, the grouping, describes the *history* of the side, and this history is one of *transfer* where the structure of a side is transferred from one side of the square to one side of the rotated parallelogram.

Another way of showing that the perception of a grouping depends crucially on the arrangement of the transfer structure, or equivalently on the left-subsequences, is by causing the omission of one of the levels in the sequence. To illustrate this, let us consider a simple effect due to Wertheimer (cited in Rock, 1975, p273). In the sequence of dots in Fig 6.40, the separation between pairs *ab*, *cd*, *ef*, and *gh* is only slightly less than between *bc*, *de*, *fg*, and *hi*. Depending on whether the observer sees this difference in spacing, a different percept arises. The dots are perceived either as a line partitioned into pairs, or as a homogeneous line. According to the theory being offered, the phenomenological difference is determined by differences in the respective internal group decompositions.

Let us look at the non-homogeneous case first. We want to obtain the transfer structure of this case. In order to do this, we should first recall our claim that the internal structure (of a fully externalized history) comes from the symmetry structure of the configuration. There are three levels of symmetry in the non-homogeneous configuration. These are given by the following three groups:

Point, which maps a single point to itself, and describes the indistinguishability of the point from itself.

Reflections, which describes the fact that an individual pair of points is reflectionally symmetric; i.e., each point in the pair is mapped on to the other point in the pair, via **Reflections**.

Numbers, the group of whole numbers $\{\ldots, -3, -2, -1, 0, 1, 2, 3, \ldots\}$. This group describes the fact that the pairs of points

Figure 6.40
An example due to Wertheimer where two alternative groupings are equally likely.

are indistinguishable from each other under translations by whole-number amounts. The group **Numbers** was discussed in section 5.9.

Observe that these symmetries are arranged in a hierarchical transfer structure; that is, any level transfers the structure of the previous level from one set onto another. That is, **Reflections** transfers **Point** from one point to another point in a pair, and **Numbers** transfers **Reflections** from one pair to the next. Thus, the full structure is the following nested sequence:

Point.Reflections.Numbers

According to the Grouping Principle, the perceptual groupings must be the cosets of left-subsequences in the above sequence. To obtain these, we first list all the left-subsequences. They are, in order of length,

Point

Point.Reflections

Point.Reflections.Numbers

We now take each of these left-subsequences, in turn, and elaborate all their left cosets.

(1) The cosets of **Point**
The cosets of **Point** in the sequence

Point.Reflections.Numbers

must be of the form

Point.$g_1.g_2$

where g_1 comes from **Reflections** and g_2 comes from **Numbers**. The pair, $g_1.g_2$, is the coset leader.

Each such pair simply gives the coordinates defining an arbitrary point in the line. Thus the set of cosets **Point**.$g_1.g_2$ is the set of points. Therefore the decomposition of the group sequence **Point.Reflections. Numbers** into the set of cosets of the left-subsequences of **Point** is equivalent to decomposing the line into its set of points.

(2) The cosets of **Point.Reflections**
The cosets of **Point.Reflections** in the sequence

Point.Reflections.Numbers

must be of the form

Point.Reflections.g_2

where the element g_2 comes from **Numbers**. This element is the coset leader.

What is such a coset perceptually? To answer this, observe that the left-subsequence **Point.Reflections** generates one of the pairs of points. Therefore, the coset leader g_2, from the group **Numbers**, must move the pair by some whole-number step. This movement results in one of the other pairs. Thus the set of cosets **Point.Reflections.**g_2 is the set of pairs in the line. Therefore the decomposition of the group sequence **Point.Reflections.Numbers** into the set of cosets of the left-subsequences of **Point.Reflections** is equivalent to a decomposition of the line into its set of pairs.

(3) The cosets of **Point.Reflections.Numbers**
There is only one coset of **Point.Reflections.Numbers** in the sequence

Point.Reflections.Numbers

It is the sequence **Point.Reflections.Numbers** itself. This coset is of course the whole line. Thus a consequence of our Grouping Principle is that the whole is a grouping itself—which accords with perceptual experience.

The reader can easily check that any subsets of the group hierarchy, other than the cosets of left-subsequences, do not produce perceptual parts. This is because any such subsets break out of the transfer structure; i.e., do not accord with the way points are perceptually *mapped* onto each other.

To clarify this further, contrast the above with the homogeneous case. As in the non-homogeneous case, we consider first the levels of (internalized) symmetry: There are two such levels:

Point, which maps a single point to itself, and describes the indistinguishability of the point from itself.

Numbers, which describes the fact that the points are indistinguishable from each other under translations by whole-number amounts.

These symmetries are arranged in a hierarchical structure, thus:

Point.Numbers

Now let us look at the cosets of left-subsequences. The first left-subsequence is **Point**. Its cosets are all of the form **Point**.*g*. The coset leaders *g* are the translations, by whole-number amounts, that act on **Point**. Therefore, factorization of the full group sequence, into these cosets, yields a partition of the percept into points.

However, the next left-subsequence is **Point.Numbers**, the whole. Thus there is no intermediate transfer structure, *because each point is mapped onto any other point by* **Numbers**. Observe that this was not true in the non-homogeneous case. In that case, a *right-hand* point in a pair was not mapped directly by **Numbers** onto a *left-hand* point in the pair, or the left-hand point in any other pair. The mapping between right-hand and left-hand points was mediated by **Reflections**. In the homogeneous case, however, the mapping by **Numbers** is direct; i.e., there is no intermediate mapping or transfer level between **Point** and **Numbers**. Thus, according to the theory, there is no set of parts occurring between the lowest level, the points, and the highest level, the whole. This captures the homogeneity of the percept.

We will continue our study of grouping in the next section.

6.18 Grouping: Evaluation Theory

The retinal image consists of a large array of independent stimulus points, and the perceptual system chooses to group together some of these points as opposed to others. A major concern in the research on grouping has been why the perceptual system forms the groupings that it does as opposed to other groupings it could form in the same stimulus set. We call this problem the *evaluation* problem: That is, how does the perceptual system *evaluate* one grouping as better than its alternatives and thus reject the alternatives?

The evaluation problem faced us in the last section. We argued that a grouping is a coset of a left-subsequence of the process-group hierarchy. However, we still must decide how the hierarchy itself is chosen. For example, we saw that a square seems to be given by the hierarchy

$$\mathbf{Point.Translations.D_4}$$

and that elaborating all the cosets of left-subsequences in this hierarchy does indeed yield the perceptual groupings. However, given a square, an alternative order for the groups is possible—one in which the second and third groups are reversed. This ordering does not seem to be chosen by the perceptual system. The question therefore remains as to how the perceptual system *evaluates* one group-sequence as better than another. Thus, in our system, the problem of how groupings are evaluated is expressed in the problem of how alternative group-sequences are evaluated.

Most of the work done on the evaluation of grouping was done by the Gestalt school. This school provided substantial evidence that the perceptual system decides between alternative groupings of the same stimulus set by using a number of structural criteria. For example, one such criterion is that of *symmetry*: If the perceptual system has available two alternative groupings, one of which is more symmetrical than the other, then the perceptual system will choose the more symmetrical grouping.

Surveying the literature, one finds that the Gestalt school put forward at least eleven such criteria: (1) Closure, (2) Good Continuation, (3) Regularity, (4) Internal Consistency, (5) Inclusiveness, (6) Simplicity, (7) Symmetry, (8) Proximity, (9) Nearness to Singularity, (10) Stability, (11) Similarity (e.g., Schumann, 1900, 1902, 1904; Rubin, 1921; Wertheimer, 1923; Koffka, 1935; Kohler, 1947).

While it is clear, from various perceptual examples, that there is a certain amount of overlap between some of these criteria, the latter are regarded, on the whole, as *separately acting factors*. Indeed, the difficulty in establishing a theory of grouping is often understood as resulting from the independence of, and therefore competition between, the criteria (see Rock, 1975, p270).

In contrast to the structural approach taken by the Gestalt school, the researchers in Standard Computational Vision have taken a functional approach (e.g., Lowe, 1985). They observe that the groupings that human perception chooses, in a two-dimensional image, tend to correspond to objects in the three-dimensional environment. This leads the researchers to propose that the function of grouping in the image is the identification of objects in the environment. Observe, however, that, in contrast to our approach, this theory does not explain why an edge would be a grouping. In our theory, an edge is a process-history, e.g., cutting, and the function of grouping is to identify histories rather than objects.

Now, the Standard Computational view has the following evaluation theory: A grouping is more likely to be chosen in the image if it is more likely to correspond to an object in the three-dimensional environment. The researchers are then faced with the problem that a perceptual system does not have access to the three-dimensional environment, and thus can evaluate the groupings in the two-dimensional image only by using structural criteria applied to the image. The researchers then argue that the appropriate criteria must be the Gestalt criteria. Their argument is that groupings that are more symmetrical, regular, etc., are more likely to correspond to configurations that are symmetrical, regular, etc., in the environment. We reviewed this argument in section 3.10, and showed that it was an example of our Symmetry Principle.

Thus, despite the functional approach of Standard Computational Vision, the latter field is led back to a complete dependency on the Gestalt criteria. The field must apparently work out how to apply the eleven criteria to the image. Some initial work on this problem has been done by Lowe (1985) and Jacobs (1988).

The problem with using the Gestalt criteria is that (1) there seem to be too many of them to control, (2) they are ill-defined, (3) they seem to partially overlap and yet act independently, and (4) they seem to have different priorities over each other in different situations, as we shall shortly illustrate. We shall now argue that, within our theory of perception as process-inference, there is no need for eleven structural principles. They can all be encompassed by a single principle.

In order to see this, recall that we proposed that a *grouping is a history*. Thus the evaluation of one grouping in comparison with another is the evaluation of one history in comparison with another. In our system, the evaluation of history is carried out by a single principle: the History Minimization Principle. We will argue that this single principle replaces the eleven Gestalt criteria. What is important in making this replacement is that the History Minimization Principle focuses considerations on one aspect of the stimulus: the *historical* aspect.

Recall now that the History Minimization Principle splits into two halves: (1) the Subsumption Component, which requires the history to consist of a minimal set of processes, and (2) the Inertial Component, which requires the actions in a history to involve the least amount of change. We shall now argue that the eleven Gestalt criteria split into two classes, where the criteria in one class are simply manifestations of the

Subsumption Component and the criteria in the other class are simply manifestations of the Inertial Component. Let us consider the first class of criteria:

GESTALT CRITERIA IMPLIED BY THE SUBSUMPTION COMPONENT. *(1) Closure, (2) Good Continuation, (3) Regularity, (4) Internal Consistency, (5) Inclusiveness, (6) Simplicity, (7) Symmetry.*

Data is often put forward that these seven criteria perceptually compete with each other and have different priorities over each other in different situations. In contrast, we argue that there is only one criterion behind these seven, and that there can therefore be no competition or varying priority. In order to illustrate this, let us consider a standard set of data where several competing criteria have been invoked by researchers. These data are shown in Figs 6.41, 6.42, 6.43, and they come from Dinnerstein & Wertheimer (1957).

Consider first Fig 6.41. Subjects are presented with the top configuration in Fig 6.41, and are asked to choose between alternatives a, b, and c, on the bottom of Fig 6.41, as the perceived interpretation of the top configuration. Subjects tend to choose alternative a, that is, the interpretation

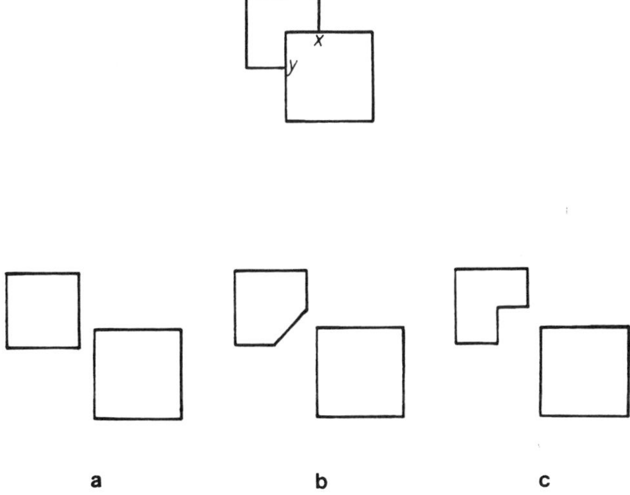

Figure 6.41
Dinnerstein & Wertheimer (1957) found that the top figure is interpreted as alternative (a) at the bottom.

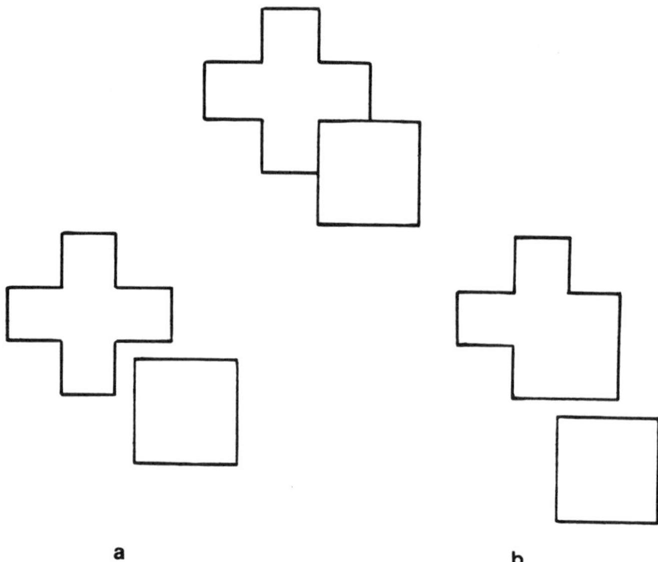

Figure 6.42
Dinnerstein & Wertheimer (1957) found that the top figure is interpreted as alternative (a) at the bottom.

with two complete squares. Notice that they reject interpretation c, which involves a corner "bitten out" of one of the shapes. The standard Gestalt argument is that the "bitten out" corner accords less with the Gestalt criterion of *good continuation* than interpretation a. That is, in the presented stimulus, at the top of Fig 6.41, one has to decide how to *continue* the line through point x to point y. If one continued the line around the bitten-out corner, as in interpretation c, the line would make more changes of direction than if it were continued as shown in interpretation a.

Now consider Fig 6.42. The subjects are presented with the top configuration in that figure and they have to choose between interpretation a or b below. They choose interpretation a rather than b. However, in this case, they are choosing the interpretation in which the corner is bitten out, and rejecting the interpretation in which the corner is not bitten out. This means that the subjects are violating the *good continuation* criterion. The continued line in interpretation a makes more changes than the continued line in interpretation b. Nevertheless, the former is chosen. The Gestalt criterion that is supposed to be used here is *regularity*. A bitten-out corner

Figure 6.43
Dinnerstein & Wertheimer (1957) found that the top figure is interpreted as alternative (a) at the bottom.

has occurred *regularly* around the cross-like shape and is thus predicted as occurring a further time in interpretation a. Therefore, it is argued, e.g., by Buffart, Leeuwenberg, & Restle (1981), that the Gestalt criterion of good continuation competes with the criterion of regularity and, in the present case, the latter criterion has priority.

Now consider Fig 6.43. The configuration at the top is the stimulus that is presented to the subjects. The interpretation that is chosen by the subjects is interpretation a. Notice that, in interpretation a, the shape nearest the square, i.e., the shape with the bitten-out corner, has itself no other bitten-out corner. Thus the criterion of regularity is not being used on that shape. Furthermore, the criterion of good continuation is not being used. This is particularly troublesome as follows: Consider again the top configuration in Fig 6.43. Observe that it contains the "two overlapping squares" configuration that occurs also at the top of Fig 6.41. However, in

Fig 6.41, interpretation a was chosen, and interpretation c was rejected. Nevertheless, this rejected interpretation was exactly what was chosen in the case of Fig 6.43, i.e., interpretation a in the latter figure.

Let us introduce some terms: Observe first that each of the top configurations (in Figs 6.41, 6.42, and 6.43) consists of two components, a lower square and the rest of the configuration which we will call the *remainder*. We shall say that the perceptual task is to *complete* the remainder, e.g., insert a line between x and y in the top configuration of Fig 6.41. The completion, or interpretation, chosen by the subjects is always shown by us, in part a of each figure. Now let us recapitulate the results with respect to the Gestalt criteria: In Fig 6.41a, the completion of the remainder is determined by good continuation; in Fig 6.42a, the completion of the remainder is determined by regularity in opposition to good continuation; and in Fig 6.43a, the completion of the remainder is determined by neither good continuation nor regularity.

We shall now argue that rather than invoking a number of apparently distinct Gestalt criteria, with different priority levels in different circumstances, one needs to apply only a single principle: the Subsumption Component of the History Minimization Principle. This component, given as the Subsumption Strategy in section 6.3, states (1) that actions must be subsumed under processes, and (2) that the number of processes must be minimal.

Now when we examine each of the chosen interpretations (i.e., interpretation a, in each case), we find that the completed remainder in each case is a repetition of the generator r_{90}; i.e., a *process* generated by r_{90}. None of the other interpretations can be described in this way. Thus, we see that the chosen interpretation in each case is determined by the phenomenon of *process*.

The Subsumption Strategy not only requires incorporation into processes, but requires that a *minimal* number of processes should be used. With respect to completing the remainder, the minimal solution would be that the process-structure evident in the remainder is simply continued without stop. This solution would add nothing new to the internal structure other than that in the remainder.

It is important to understand that minimality does not necessarily mean the continuation of *individual* processes evident in the non-completed remainder, but rather the continuation of the *process-structure*. As we have seen, a process-structure consists of a nested hierarchy, and it is this hierarchy that is continued. To illustrate, let us return to Figs 6.41, 6.42, and 6.43.

The generator, r_{90}, in each case, creates the top level of control in the internal structure of the remainder; i.e., it is the shift level in this structure. Below this level, in the control hierarchy, is a *translation* level that moves a point along an individual side. What is evident in the uncompleted remainder (i.e., in the top stimulus, in each case) is that this translation process starts and stops one or more times, but does so in places that are all made equivalent by the r_{90} generator. That is, in the uncompleted remainder, r_{90} *transfers* the translation structure from one shift position to the next. The *completed* remainder (interpretation a), in each case, merely continues this repeating transfer, i.e., it preserves the nested hierarchy.

We can therefore understand the confusion in the use of the Gestalt principles as follows: In the standard literature, the "good continuation" in Fig 6.42b means continuation of the translation level. However, according to our system, the actual continuation should be the continuation of the nested control structure. This is violated in Fig 6.42b, because, in the completion, the translation process starts and stops in places that are different from those places determined by the next level of control (r_{90}) in the non-completed stimulus. That is, in the non-completed stimulus, the r_{90} generator repeatedly transfers the start-and-stop structure of the translation level, and *continues to do so* in interpretation a. Thus, interpretation a actually fulfills good continuation, but it is good continuation of the nested hierarchy.

Observe now that each of the seven Gestalt criteria listed above are simply manifestations of this strategy. For example, consider closure. We have argued (e.g., section 6.5) that any process is a cycle. When the cycle is infinite, continuing the process closes it at infinity. When the cycle is finite, continuing the process closes it within a finite area. Thus, closure is simply a by-product of continuing a process-structure. The same is true of the remainder of the seven Gestalt criteria listed above.

Now let us turn to those of the eleven Gestalt criteria not listed in the above seven. We claim that the remaining four are simply manifestations of the Inertial Component of the History Minimization Principle. That is:

GESTALT CRITERIA IMPLIED BY THE INERTIAL COMPONENT. *(1) Proximity, (2) Nearness to Singularity, (3) Similarity, (4) Stability.*

The Inertial Component of the History Minimization Principle requires that the actions in a history involve the least amount of change. We shall see, a number of times in this book, that this component seems to take the following specific form:

INERTIAL STRATEGY. *At each point in a history choose, from the available actions, the action that involves the least amount of change.*

That is, our claim is that the Inertial Strategy is used to realize the Inertial Component, just as the Subsumption Strategy is used to realize the Subsumption Component.

Consider the standard Gestalt example shown in Fig 6.44. In both Fig 6.44a and 6.44b, there is a rectangular grid of 6 dots by 6 dots. However, the former is perceptually grouped into rows and the latter is grouped into columns. If one applies the Inertial Strategy, one can see why this is the case, as follows: Both sets of stimuli can be generated as the trace of a dot. Given a starting dot, e.g., the top left-corner dot, the Inertial Strategy requires the dot to be moved to the position that involves the least amount of change. In Fig 6.44a, this means that the dot is moved horizontally. When this strategy is reapplied repeatedly, it generates a row. Notice that the row has been produced by a single *process*, i.e., the repetition of a single generator. Having created a row, one need provide only one further process to create the remainder of the stimuli: the process generated by applying, to the row, the smallest vertical movement consistent with the stimulus. Therefore, the row is moved successively downward in the vertical direction.

There are thus two processes involved in generating Fig 6.44a: the first is **Numbers** (the group of positive and negative whole numbers) applied in

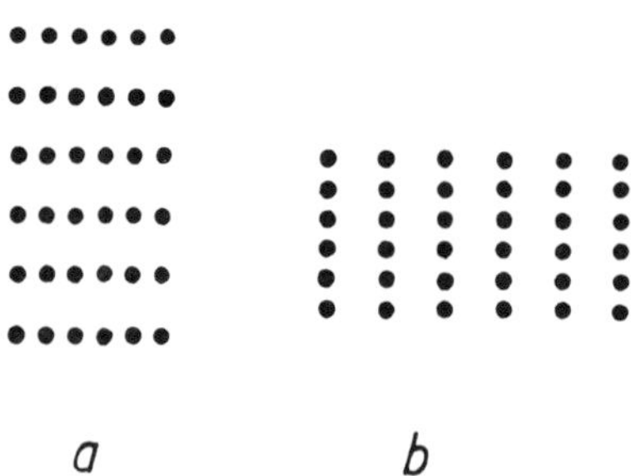

Figure 6.44
A rectangular array of 6 by 6 dots grouped (a) into rows, and (b) into columns.

the horizontal direction, which we will denote by $\mathbf{Numbers}_H$, and the second is **Numbers** applied in the vertical direction, which we will denote by $\mathbf{Numbers}_V$. Thus Fig 6.44a is given by the following nested hierarchy:

$$\mathbf{Point}.\mathbf{Numbers}_H.\mathbf{Numbers}_V$$

Now consider Fig 6.44b. Given the same starting point, the top left-corner dot, the Inertial Strategy requires the dot to be moved to the position that involves the least amount of change. In Fig 6.44b, this means that the dot must be moved vertically. When this strategy is reapplied repeatedly, it generates a column. Notice that the column has been produced by a single *process*, i.e., the repetition of a single generator. Having created a column, one need provide only one further process to create the remainder of the stimuli: the process generated by applying, to the column, the smallest horizontal movement consistent with the stimulus. Therefore, the column is moved successively rightward in the horizontal direction.

There are thus two processes involved in generating Fig 6.44b: the first is **Numbers** applied in the vertical direction, that is, $\mathbf{Numbers}_V$; and the second is **Numbers** applied in the horizontal direction, that is, $\mathbf{Numbers}_H$. Thus Fig 6.44b is given by the following nested hierarchy:

$$\mathbf{Point}.\mathbf{Numbers}_V.\mathbf{Numbers}_H$$

Comparing the group-sequence given for Fig 6.44a with the group-sequence given for Fig 6.44b, we see that they involve the same three groups, **Point**, $\mathbf{Numbers}_H$, and $\mathbf{Numbers}_V$. However, they each have the whole-number groups in a different order. Observe that this order is determined by the Inertial Strategy, i.e., the requirement that the action at any stage should be minimal.

Having established the group-sequences for the two configurations, the Grouping Principle can then be used to predict automatically, from the sequences, what perception will take to be the groupings. According to the principle, the perceptual groupings will be the cosets of left-subsequences. Thus, consider first the group-sequence $\mathbf{Point}.\mathbf{Numbers}_H.\mathbf{Numbers}_V$ for Fig 6.44a. The cosets of left-subsequences are of the form:

$$\mathbf{Point}.g_1.g_2$$
$$\mathbf{Point}.\mathbf{Numbers}_H.g_2$$
$$\mathbf{Point}.\mathbf{Numbers}_H.\mathbf{Numbers}_V$$

The stimulus sets determined by these three types of cosets are, respectively, the dots, the rows, and the whole. These, indeed, are the perceived groupings in Fig 6.44a.

Again, consider the group-sequence **Point**.**Numbers**$_V$.**Numbers**$_H$ for Fig 6.44b. The cosets of left-subsequences are of the form:

$$\mathbf{Point}.g_1.g_2$$

$$\mathbf{Point}.\mathbf{Numbers}_V.g_2$$

$$\mathbf{Point}.\mathbf{Numbers}_V.\mathbf{Numbers}_H$$

The stimulus sets determined by these three types of cosets are, respectively, the dots, the columns, and the whole. Again, these turn out to be the perceived groupings in Fig 6.44b.

We are therefore led to this conclusion: The Inertial Strategy yields the order of the group-sequence, and the Grouping Principle can then automatically extract the groupings from that sequence.

Notice that the Inertial Strategy explains, in exactly the same way, the order of the group-sequence defining a square. The sequence is **Point**.**Translations**.**D**$_4$ rather than **Point**.**D**$_4$.**Translations** (reversing the last two factors), because, a point undergoes less change in moving to the "next point" under **Translations** than under **D**$_4$. Therefore, **Translations** must precede **D**$_4$ in the nested hierarchy. The Grouping Principle then automatically extracts the groupings.

Now let us return to the Gestalt criteria. Fig 6.44 is standardly used to illustrate the criterion known as *proximity*, i.e., that stimuli closer together should be grouped. Clearly, the proximity criterion is implied by the Inertial Strategy; i.e., since histories form groupings, spatial inertia in a history (i.e., minimal spatial change) will result in grouping based on proximity. However, our discussion above showed that, because the Inertial Strategy is a *historically* based principle, it leads, with our other historically based principles, to a much stronger structural analysis than is available from a Gestalt approach. For example, we were able above to extract the groupings as cosets of left-subsequences in the control hierarchy, etc. Furthermore, we shall now see that the Inertial Component implies the other Gestalt criteria listed above.

Let us first consider the criterion of *singularity*. Gestalt psychologists (e.g., Goldmeier, 1936/1972) observed that, if one continuously varies a stimulus along some property dimension, there can appear, at certain isolated points on the dimension, stimuli that are highly distinctive. For example, when one varies a parallelogram along the shear dimension, one obtains a family of parallelograms, i.e., shapes in which opposite pairs of sides are parallel. However, in this family, one of the parallelograms that is created is actually a rectangle. This member has a visual distinctiveness, or conspicuousness, not possessed by the other parallelograms. In Gestalt terms, it is *singular*. Goldmeier (1936/1972) showed that, when a variation of a shape arrives at a singular point, there seems to be a sudden reduction in the number of relevant dimensions to the stimulus. For example, in the case of the varying parallelograms, the feature of *angle* is salient for all parallelograms except for the rectangle.

The phenomenon of singularity is easily explained within our theoretical framework. We simply propose that *a singular stimulus is the history that remains after one of the processes has been removed*; i.e., after one layer of time has been removed. For example, in the family of parallelograms, the rectangle is seen as singular because one less process is involved in its construction. This, of course, explains the dimension-lowering phenomenon described by Goldmeier. It is not that there is a smaller *amount* of process in a history, but that one of the processes is actually missing; i.e., there is one less process-coordinate.

The dimension-lowering property exists, for another reason, which, by the Asymmetry Principle, is an equivalent reason. By that principle, the removal of one of the processes causes a reduction in the atemporal asymmetry of the stimulus. Thus, the dimension is lowered because the number of distinguishabilities in the stimulus is lowered. Notice that, correspondingly, the symmetry group increases. For example, the symmetry group of the parallelogram is of size two, whereas the symmetry group of the rectangle is of size four, and the symmetry group of the square is of size eight.

Singularity, as an evaluative Gestalt criterion, is the requirement that a stimulus be described as being *as near as possible to a singular point*. Thus the criterion follows from the Inertial Component which states that an assigned action should involve the least amount of change. Since, by the Asymmetry Principle, change must have progressed from symmetry to asymmetry, minimization of change must mean nearness to symmetry.

The remaining Gestalt criteria, similarity and stability, also follow from the Inertial Component. *Similarity* means proximity in a non-spatial dimension, and can therefore be shown to follow from the Inertial Component in the same way that proximity was shown to do so, above. *Stability* follows from the Inertial Component because, by the Asymmetry Principle, the component implies nearness to symmetry, and by the Stability Principle (section 1.6), this atemporal symmetry corresponds with *temporal symmetry*, which is, by definition, stability.

6.19 Why the Ordering Stretches.Shears.Rotations

In concentrating on grouping for a while, we have been examining the inferred internal history of a stimulus set. We now need to look more closely at how the internal history is transferred by external processes. In order to do this, we must concentrate for some time on external processes.

We return once again to the fully externalized history of the rotated parallelogram. As we said earlier, this history will be examined in considerable detail because it reveals principles that we will find later to be entirely general, for simple shape as well as complex shape and motion. We therefore beg the reader to be patient while our discussion of this example continues.

Recall that the relevant space of actions in this history is given by the group **Special–Linear**.[15] We must try to understand more fully why subjects removed the processes comprising **Special–Linear** in the order **Rotations** followed by **Shears** followed by **Stretches**, rather than in an alternative order.

Recall that a linear transformation is one that preserves straightness and position of the origin (section 5.12). In particular, when a linear transformation acts, it does not bend straight lines, straight planes, etc. Stretch, shear and rotation, are each examples of linear transformations.

Although all linear transformations preserve the straightness of lines, planes, etc., some have an additional preserving action. They preserve the *orientation* of some lines, planes, etc., that go through the origin. Any such preserved line, plane, etc., is called an *eigenspace*. To illustrate, consider our three transformations, stretch, shear, and rotation.

Stretch. When a stretch acts, there are exactly two straight lines through the origin that do not change orientation. In Fig 6.45a, these are the two bold lines shown; that is, the x- and y-axes. (Note that any other line, through the origin, changes orientation.) Thus, in Fig 6.45a, the x- and y-axes each constitute eigenspaces of the transformation shown. Note

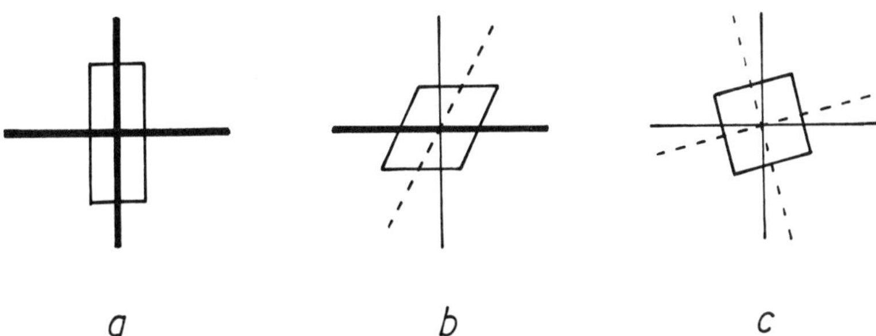

Figure 6.45
(a) A stretch, (b) a shear, and (c) a rotation have respectively two, one, and zero invariant lines through the origin.

that while the axes remain in the same orientation, they undergo change in size; e.g., the y-axis undergoes stretch. This is not a problem. The two axes constitute eigenspaces because they remain in the same orientation, irrespective of whether they change size.

Shear. When a shear acts, there is one and only one line through the origin that stays in the same orientation. For example, in Fig 6.45b, this line is the bold line, the x-axis. Thus in that figure, the x-axis constitutes the eigenspace of the shear transformation shown.

Rotation. Under any rotation, the entire space shifts around the origin as shown in Fig 6.45c. Therefore, every line changes orientation. Thus there is no eigenspace.

The usefulness of the eigenspace concept is that it is related to the amount of change induced by a process. An eigenspace is an invariant. An invariant is a property that remains unchanged. Thus the *amount* of eigenspace (i.e., the dimension of the total eigenspace) is one measure of the lack of change associated with a transformation. That is, a transformation with more eigenspace involves less change.

The crucial thing now to observe is that the order chosen by the subjects, for the processes, stretch, shear, and rotation, is actually the order of the amounts of eigenspace associated with each process. That is, as can be seen from the Fig 6.45, *the successive processes—stretch, shear, and rotation—have respectively two, one, and zero eigenspace lines*. This means that the successive processes have *increasing* amount of change using the eigenspace criterion. Therefore, it seems that the order of the processes conforms to the Inertial Strategy which states that, at each point in a history, one chooses, from the available actions, that action which involves the least amount of change. We will continue to explore this idea in the next section.

6.20 The Interaction Principles

We have noted that subjects appear to order the external processes so that a process is hypothesized as earlier if it involves less change—in accord with the Inertial Strategy. We now have to understand what "less change" means in terms of the configuration that is acted on.

By the Asymmetry Principle, a process is understood as introducing asymmetry. Thus the *change* that a process causes is asymmetry. This means that a process introduces less change if it preserves more symmetries of the configuration.

We can consider a transformation that induces less change in a configuration to be more *allowable* by the configuration because it satisfies the Inertial Component of the History Minimization Principle. Therefore, since transformations that preserve symmetries induce less change, they are more *allowed* by the configuration. We shall say that such transformations are more *structurally allowable* by the configuration. The question now remains as to how a transformation can preserve a symmetry.

To preserve a symmetry of a configuration, one must preserve the axis associated with the symmetry. One way a symmetry axis can often be preserved is to make it an eigenspace, i.e., a preserved subspace, of the applied transformation. If the symmetry axis is a line, then the axis becomes an eigenspace-line, a preserved line. If the symmetry axis is a plane, as it is in a three-dimensional object such as a horse, it becomes an eigenspace-plane, a preserved plane. This leads us to a principle that was initially proposed in Leyton (1984):

FIRST INTERACTION PRINCIPLE. *Symmetry axes of a configuration become the eigenspaces of the most structurally allowable processes applicable to the configuration.*

The principle is called an *interaction* principle because it concerns how the structure of a configuration interacts with the processes that can be applied to the configuration.

The principle will be extensively corroborated for simple and complex shape, and even for motion perception. However, for the moment, observe that, in the history from the square to the rotated parallelogram, the process that was *actually* applied to the square was stretch (i.e., because the stretch was applied first). In this process, the *eigenspaces* of stretch (the bold lines in Fig 6.45a) were the vertical and horizontal *symmetry axes* of the square, in accord with the First Interaction Principle.

The above principle equates two phenomena: *symmetry axes* and *eigenspaces*. The two phenomena have one important property in common: they are both lines that are *invariant* under transformations. However, besides this, they have entirely different properties and roles. In order to understand how non-trivial the act is of forcing an equation between these two concepts, we consider them first separately.

Symmetry Axes. A symmetry axis is a line (plane, etc.) in which every point is sent to itself under the associated transformation (e.g., a rotation or reflection). This is illustrated in Fig 6.46a, with a triangle. Under reflection about the central axis, all the points on either side of the axis move. However, a point such as *P*, on the axis, remains unchanged. Observe that the reflection can be a member of the *internal structure* of the figure because it is capable of generating the whole figure from only half of it.

Eigenspaces. An eigenspace is a line (plane, etc.) of points that usually *move* along that line (plane, etc.) under a linear transformation. For example, consider Fig 6.46b, where an upwards stretch acts on the triangle.

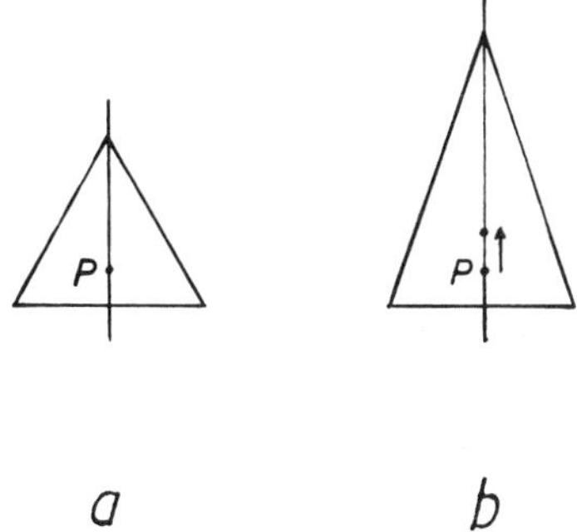

Figure 6.46
The central axis is an invariant line in two senses: (a) as a symmetry axis, and (b) as a line of flexibility.

The central axis is an eigenspace because it is a line that does not change orientation in the transformation. However, points such as P *move* along that line itself. The stretch is a member of the external structure of the triangle; that is, it describes what happens to the triangle.

To summarize: A symmetry axis comprises an array of *static* points; an eigenspace typically comprises an array of *moving* points. Conceptually, therefore, one should consider the First Interaction Principle as stating:

> **Invariant lines that are rigid, under transformations belonging to the internal structure of a configuration, become lines of flexibility, under transformations that can be applied to the configuration.**

The First Interaction Principle is now combined with another idea to yield a theoretical result that will prove to be important to us. Let us return to the square. We saw that the symmetry axes that bisect the sides become the eigenspaces of the stretch group, as in Fig 6.45a. However, these axes are not the only symmetry axes that a square has. The diagonals are also symmetry axes. Nevertheless, as the experiments in section 6.28 corroborate, the side-bisectors are much more salient than the diagonal axes. This means that the most *salient* symmetry axes are those axes that were chosen as eigenspaces.

Quite simply, one can argue the following: Some symmetry axes are more perceptually salient than others. Now because a process is more structurally allowable if it converts a perceived symmetry axis into an eigenspace, one can conclude that the more perceptible, i.e., salient, a symmetry axis is, the more likely it is to be chosen as an eigenspace. That is:

SECOND INTERACTION PRINCIPLE. *The structural allowability of a process increases with the salience of those symmetry axes that are aligned with its eigenspaces.*

Let us now consider the First and Second Interaction Principles in terms of the inferred history of a rotated parallelogram. Recall, as shown in Fig 6.45, that the three processes involved—stretch, shear, rotation—have respectively two, one, and zero eigenspace lines; i.e., lines that do not change orientation in the associated transformation. However, consider now the source of these lines. The salient symmetry axes of the original square are the vertical and horizontal reflection axes; and, as we saw,

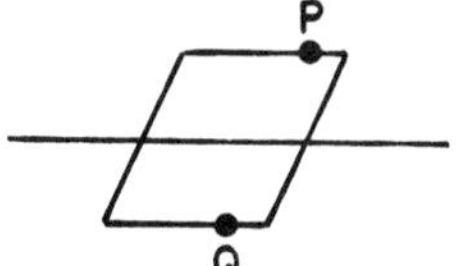

Figure 6.47
A shear preserves bisection.

in accord with the Interaction Principles, these axes become the eigenspaces of stretch, the process that is actually applied to the square. The crucial point to observe now is that, after applying stretch, these lines *remain* symmetry axes in the resultant rectangle. This means that as little damage as possible has been done to the square.

Now consider the application of shear. Shear, as shown in Fig 6.47, has only one eigenspace line; i.e., one line that does not change orientation. However, observe that this line was also a symmetry axis of the square; i.e., it was the horizontal symmetry axis that remained in the rectangle. Thus, in applying the shear, a symmetry axis of the square becomes an eigenspace. Observe also that this served the purpose of preserving an aspect of the symmetry of a square, as follows: Although, after applying shear, the horizontal line is no longer an axis of *reflection*, it is nevertheless an axis of *bisection*; that is, each point P in the top half of the sheared figure (see Fig 6.47) can be matched with a point Q in the bottom half of the figure so that P and Q are *an equal distance* from the horizontal line along a transverse line that is parallel to the left and right sides of the figure. Symmetries of this type have been called skewed symmetries by Kanade (1981), who has shown how to infer from them the full reflection axes that they were prior to the transformation. We discussed in detail the relation between these bisection axes and the prior reflection axes, in section 3.10.

The crucial point is this: In applying the shear such that the eigenspace of shear is a reflection axis of the figure prior to the transformation, an important aspect of the reflectional symmetry is preserved: the bisecting aspect.

Finally, in considering rotation, we find that rotation has no eigenspaces and therefore has none that can be aligned with the symmetry axes of the square.

We can see, therefore, in examining the three transformations, stretch, shear, and rotation, that wherever there is an eigenspace in one of the transformations, it is matched to a salient symmetry axis of the square. That is, the transformations are applied so that the Interaction Principles hold.

Let us now introduce some notation: Suppose one has a sequence G_1 of internal groups, and a sequence G_2 of groups applied to G_1—that is, one has the nested sequence $G_1 . G_2$. Then if G_1 has salient symmetry axes, and G_2 has eigenspaces, we shall indicate that the salient symmetry axes in G_1 are aligned with the eigenspaces in G_2 by inserting the symbol $\wedge$ between the two sequences thus: $G_1 \wedge G_2$. For example, the full process-structure of the square becomes:

$$[[\mathbf{Point.Translations.D_4}]_I \wedge \mathbf{Stretches.Shears.Rotations}]_E$$

Finally, something remarkable emerges. It is that the Interaction Principles imply a temporal *ordering* on the processes applied to an object. Suppose that a number of processes have been inferred as having been applied to an object. Consider two of these processes A and B. If process A has more eigenspace than process B, then A has to precede B because more eigenspace can be aligned with the salient symmetry axes of the configuration. Thus, to maximally apply the Interaction Principles, the processes should be applied in the order of decreasing eigenspace dimension. The effect is to preserve the symmetry axes for the maximal time.

The stretch-shear-rotation example illustrates the argument which we have just presented. The stretch precedes shear because stretch, having two eigenspaces, can preserve both the vertical and horizontal symmetries, whereas the shear, having only one eigenspace, can preserve only one of these symmetries. Observe also that shear is aligned so that its eigenspace is one of the eigenspaces of stretch, and thus it preserves, through a further stage, a symmetry that was preserved under stretch.

6.21 Preliminary Corroboration

Although more complicated corroborations of the Interaction Principles will be presented in later sections, e.g., on motion phenomena, we briefly review, in this section, corroborating experiments presented in Leyton (1986c). We first describe experiments on simple shape, and then experiments on complex shapes such as the outlines of animals, birds, and plants, shown in Fig 6.48.

Figure 6.48
The complex shapes on which the Interaction Principles were tested locally.

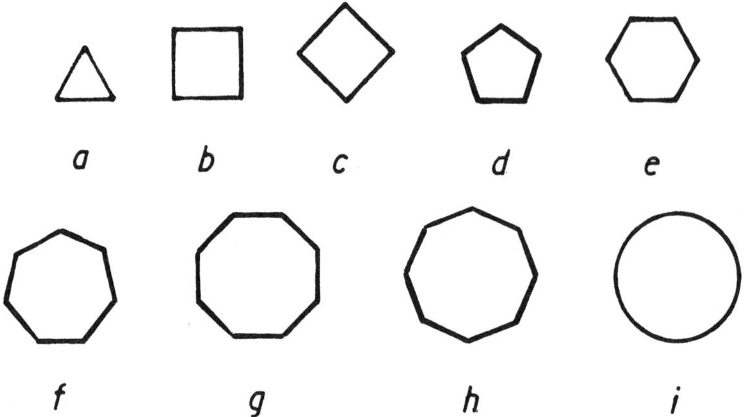

Figure 6.49
The polygons on which the Interaction Principles were tested globally.

In each of the experiments on simple shape, a different means was employed of creating axis-salience within a figure. Given a particular means, the subject had to decide which of the available axes in the figure most allowed an elongation. Choosing a salient axis would corroborate the Interaction Principles. All the test figures used were regular polygons because, with these figures, one can exploit the fact that each presents a "pool" of geometrically equivalent axes from which the subject can make a selection, biased only by the experimentally chosen salience variable. Thus, in one experiment, the vertical gravitational axes of the nine equal-sided polygons in Fig 6.49 were made salient. Note that diamonds were included in the set of polygons as well as squares. Furthermore, shapes we call "crystals" (Fig 6.49h) were included as well as octagons. These shapes are octagons placed on a vertex. Their relation to octagons is the same as that of diamonds to squares, and they are perceptually distinct from octagons. Now, for each test polygon, (e.g., Fig 6.50a) the subject was presented with a set of comparison figures (e.g., Fig 6.50b) where each figure was an elongation of the associated test polygon along one of its axes. Every subject was given 15 sets of test and comparison figures as follows: (1) triangles, (2 & 3) two sets of gravitationally oriented squares; (4 & 5) two sets of gravitationally oriented diamonds; (6) pentagons; (7 & 8) two sets of hexagons; (9) septagons; (10 & 11) two sets of octagons; (12 & 13) two sets of crystals; (14 & 15) two sets of circles. The subject had to select, from each of these sets, that figure which best matched the impression of

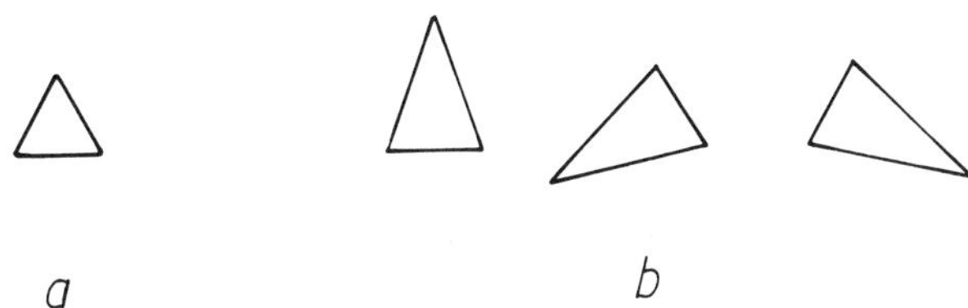

Figure 6.50
The form of the test used for each of the polygons in Fig 6.49.

the corresponding test polygon. The result was that subjects, with statistical significance, chose the elongation that was aligned with the salient axis (the gravitational one in this case), thus corroborating the Interaction Principles [n = 6, 15 comparisons per subject; expected mean = 7.5 sets in which gravitational alignment was chosen; actual mean = 12.5; $t(5) = 9.444$, $p < .0005$, one tailed]. A corresponding set of results was obtained for test polygons where axis-salience was produced by splitting the figure along an axis.

Now let us turn to a corroborating experiment on complex shape (Leyton, 1986c, Experiment 13). Subjects were presented with the 22 drawings of animals, birds, plants, and abstract shapes, shown in Fig 6.48. They were asked to assign directions of flexibility to each figure, at four points in each figure, in this way: At each of the four points, there was a star consisting of four lines. One perpendicular pair of lines (in the star) was a pair of local symmetry axes. The other two lines were diagonal to these and were not symmetry axes themselves. The subjects were asked to choose the line describing the direction of maximal flexibility of the figure in that region. With considerable statistical significance, subjects chose one of the two local symmetry axes as a line of flexibility [$n = 12$; 88 choices per subject; expected mean = 44; actual mean = 77.58; $t(11) = 27.228$, $p < .0005$, one tailed]. This implies that the subjects were converting symmetry axes into eigenspaces.

6.22 The Backwards and Forwards Interaction Principles

We had, in fact, introduced an interaction principle near the beginning of the book (section 1.12), when we were examining the history of complex shapes such as embryos, tumors and clouds. The interaction principle given there was actually a temporally *reversed* version of the First Interac-

Figure 6.51
An example of the use of the backwards version of the First Interaction Principle.

tion Principle, introduced in the last section. The initial version is re-stated here, and called the *backwards* version:

FIRST INTERACTION PRINCIPLE (BACKWARDS VERSION). *Symmetry axes of the configuration are directions along which hypothesized processes are most likely to have acted.*

Thus, the idea is that, given a complex shape in the present, e.g., the shape shown in Fig 6.51, the symmetry axes *in the present* are the directions along which the preceding processes acted. The justification for this principle is the Symmetry Principle, which states that a symmetry cannot be destroyed backwards in time. In particular, for the complex shape shown, this means that, going backwards in time moves the boundary back along the symmetry axis.

In contrast, the First Interaction Principle introduced in the previous section, concerns the *forwards* time direction. The problem that we were considering was this: Given a shape, which processes are more structurally allowable by the shape? This problem arises in the situation where the shape is an inferred past state and we want to decide which of a set of inferred processes is most likely to have been applied first, in the *forwards* time direction.

The forwards version of the First Interaction Principle is justified by the Inertial Component of the History Minimization Principle. The backwards version of the First Interaction Principle is justified by the Symmetry Principle which states that a symmetry in the present is preserved backwards in time.

What is common to the backwards and forwards versions of the First Interaction Principle is the conversion of the symmetry axes of a configuration into eigenspaces of a process. In the rest of the book, this common aspect will be referred to as the First Interaction Principle, and we shall state the principle to be the forwards or backwards version only when necessary. Note that there is, analogously, a forwards and backwards version of the Second Interaction Principle.

6.23 Defining the Cartesian Reference Frame

In order to investigate more deeply the importance of the Interaction Principles, we must examine the concept of a Cartesian reference frame.

It is virtually impossible doing any research on perceptual organization without soon requiring the notion of a Cartesian reference frame. Such frames have been invoked as crucial in diverse perceptual situations. For example, according to Marr & Nishihara (1978), an important stage in the perception of biological shape is the assignment of Cartesian frames to regions of the two-dimensional image; these define what will be taken to be the limbs of an animal. Again, in simple shape, it has been shown that radically different percepts arise from the same figure when a Cartesian frame is placed over the figure in different orientations—a phenomenon known as the orientation-and-form phenomenon (Goldmeier, 1936/1972; Rock, 1973).

It is currently assumed that, despite their strong perceptual effects, Cartesian frames are themselves trivial. The consequence of this assumption is that their effects on a stimulus seem to be magical. That is, the stimulus magically changes when the frame is placed over it in different positions and orientations. No one has attempted to explain why the frames have such effects on stimulus sets. For instance, the shape examples just given—the effects of assigning frames to biological limbs, and the orientation-and-form phenomenon—are completely unexplained. In this chapter, an attempt will be made to develop an understanding of why Cartesian frames have the consequences that they do. In the process of developing this understanding, we shall see that Cartesian frames are not the trivial structures that they have been assumed so far to be.

If one gives someone a blank sheet of paper, and asks them to draw a Cartesian frame on the paper, they will almost certainly draw the frame (1) in the horizontal-vertical orientation [which will be called the *gravitational* orientation to simplify the discussion], (2) such that the horizontal and vertical axes are perpendicular to each other, and (3) such that a unit distance along the horizontal equals a unit distance along the vertical. A frame with these three properties will be called the *canonical Cartesian frame*. That is, the canonical Cartesian frame is (1) in the gravitational orientation, (2) perpendicular, and (3) square.

In terms of representing data (e.g., graphically) in a Cartesian frame, none of these three conditions need actually hold for the chosen frame. One has an infinite number of frames of varying orientation, shearness, and non-squareness that can adequately structure data. However, only one is usually chosen: the canonical one.

We now argue that Cartesian frames are *process* structures. These process structures have an external and internal component as follows.

6.24 The External History of the Cartesian Reference Frame

We argue that the external structure contains three groups as follows:

(1) Given a frame (e.g., printed on a page), in which the x-axis is not horizontal, one sees the frame as a rotated version of the gravitationally aligned frame. This means that one infers the gravitational frame to be a *past state* of the presented frame. Note that this inference is an external one. We conclude therefore that **Rotations** is part of the external history of the frame.

(2) Given a frame (e.g., printed on a page), in which the x-axis and the y-axis are not perpendicular, one sees the frame as a sheared version of a perpendicular frame. This means that one infers the perpendicular frame to be a *past state* of the presented frame. Note, again, that this inference is an external one. We conclude therefore that **Shears** is part of the external history of the frame.

(3) Given a frame (e.g., printed on a page), in which the units on the x-axis are not the same size as the units on the y-axis, one sees the frame as a stretched version of one in which the units on the two axes are the same size (i.e., a square frame). This means that one infers the square frame to be a *past state* of the presented frame. Note, again, that this inference is an external one. We conclude therefore that **Stretches** is part of the external history of the frame.

The above discussion therefore implies that, given an arbitrary Cartesian reference frame, the frame is assigned an external history, that travels through the groups **Stretches.Shears.Rotations**, back to some inferred starting state. This inferred state is the canonical Cartesian frame.

6.25 The Internal History of the Cartesian Reference Frame

We shall now see that the canonical frame, which is obtained as the past state of an arbitrary frame, is like any prototype: It is defined internally by Euclidean processes. This internal structure consists of two components.

(1) Observe that a canonical frame is a regular grid, in the sense that *all* units in the x-direction have the same size; and, furthermore, *all* units in the y-direction have the same size. This means that the horizontal axis of the frame can be generated by the process **Translations** acting horizontally on the origin-point; and, furthermore, the vertical axis of the frame can be generated by the process **Translations** acting vertically on the origin-point; as shown in Fig 6.52a. Thus the entire Cartesian plane can be generated by applying **Translations** × **Translations** to the single origin point. Therefore the plane can be understood as the *trace* of **Translations** × **Translations**. That is, **Translations** × **Translations** is part of the internal structure of the canonical frame. Notice that **Translations** × **Translations** is a Euclidean group, in accord with our claim that all prototypes consist only of Euclidean processes.

(2) Observe that reflection about the vertical axis maps the canonical grid to itself and reflection about the horizontal axis maps the grid to itself;

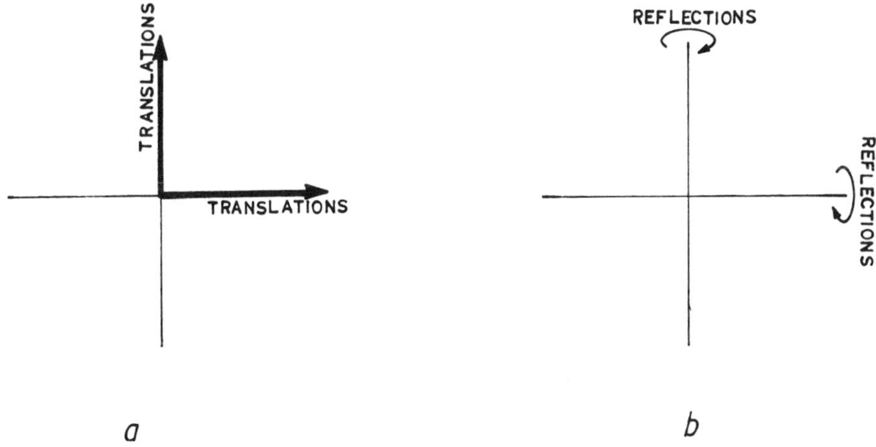

Figure 6.52
The plane can be generated by (a) **Translations** × **Translations**, or (b) **Reflections** × **Reflections**.

i.e., the grid is symmetrical under these two reflections. Internalizing these transformations, we see that reflection about the vertical axis maps one half of the grid to the other, and reflection about the horizontal axis also maps one half of the grid to the other, as shown in Fig 6.52b. Notice that reflection about the vertical axis is given by the group **Reflections**, which consists of two elements: *e*, the identity element, and *v*, the reflection itself. Similarly, reflection about the horizontal axis is given by the group **Reflections**, which correspondingly consists of two elements: *e*, the identity element, and *h*, the reflection about the horizontal axis. Thus, the reflection structure of the canonical reference frame is **Reflections** × **Reflections**; i.e., the vertical **Reflections** group times the horizontal one. Note that, because each of the **Reflections** groups has two elements, their product consists of four elements $\{e, v, h, vh\}$.

The above discussion shows that the internal structure of the canonical Cartesian frame consists of two components **Reflections** × **Reflections** and **Translations** × **Translations**. Observe now that there is a strong relationship between these components. We have said that **Reflections** × **Reflections** identifies reflectional symmetries in the grid. However, the grid is actually **Translations** × **Translations**; i.e., each grid line is generated by a copy of **Translations**. Thus **Reflections** × **Reflections** identifies symmetries in **Translations** × **Translations**. In identifying these symmetries, **Reflections** × **Reflections** is showing that any reflected copy of **Translations** × **Translations** will generate the grid. For example, suppose that, in using **Translations** × **Translations**, one applied horizontal translations in the rightward direction and vertical translations in the the upward direction, as shown in Fig 6.53a. Then one could use the reflection *v* from **Reflections** × **Reflections** to make a copy of this translation structure in which the horizontal direction for translation is reversed, as shown in Fig 6.53b. This copy would equally generate the grid. Because there are four reflections in **Reflections** × **Reflections**, there are four copies of **Translations** × **Translations** that could generate the grid.

Two consequences follow from this phenomenon: First, the **Translations** × **Translations** component is *nested internally* within the **Reflections** × **Reflections** component. (Recall, from section 6.14, that a group $\mathbf{G}_1$ is nested internally within a group $\mathbf{G}_2$ if there is one copy of $\mathbf{G}_1$ for each individual member of $\mathbf{G}_2$.) This means that **Translations** × **Translations** precedes **Reflections** × **Reflections** in the nested hierarchy describing the internal structure of the frame, thus:

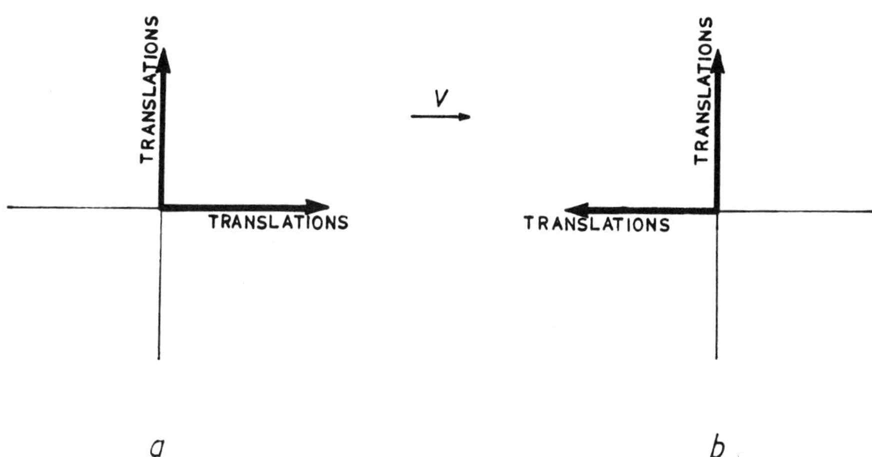

Figure 6.53
The reflection v sends (a) one copy of **Translations** × **Translations** onto (b) another copy of **Translations** × **Translations**.

$$[(\textbf{Translations} \times \textbf{Translations}).(\textbf{Reflections} \times \textbf{Reflections})]_I$$

Second, since the product between the translation component and the reflection component is one in which the reflection component makes *equivalent* (automorphic) copies of the translation component, the product is one that is given a special name in group theory: it is called a *semi-direct product*. That is, a semi-direct product between two groups is a product in which one group makes equivalent copies of the other. The symbol $\times_{sd}$ is used in this book to denote the product. In conclusion, we see that the internal process-structure of the canonical Cartesian frame is given by

$$[\textbf{Point}.(\textbf{Translations} \times \textbf{Translations}) \times_{sd} (\textbf{Reflections} \times \textbf{Reflections})]_I$$

The factor **Point** has been added to the left because the next factor, **Translations** × **Translations**, acts on a point to generate the grid structure. Footnote 16 presents technical aspects of interpreting the various product signs in group sequences.

The reader should understand that we have not dwelt on the semi-direct product relationship between **Translations** × **Translations** and **Reflections** × **Reflections** because of a concern with *mathematics*, but instead because of a concern with *perception*. The semi-direct product captures a highly salient perceptual feature of the Cartesian frame: that the translation structure is perceived to have certain reflectional symmetries, e.g., as illustrated in Fig 6.53. The semi-direct product simply encodes this perceptual phenomenon. As we shall shortly see, the salience of this phenomenon is one of the strongest organizing factors in perception.

6.26 The Psychological Nature of Cartesian Reference Frames

In section 6.24, we argued that an arbitrary Cartesian reference frame is assigned an external history consisting of **Stretches.Shears.Rotations**, such that removal of this history leads backwards in time to the canonical reference frame. We have found that, in accord with the Externalization Principle, the canonical frame is a nested hierarchy of Euclidean processes; in fact, $[\textbf{Point.(Translations} \times \textbf{Translations)} \times_{sd} \textbf{(Reflections} \times \textbf{Reflections)}]_I$. Since this internal group sequence is acted on by the external sequence **Stretches.Shears.Rotations**, the internal sequence is multiplied by the external one in this way:

$$[[\textbf{Point.(Translations} \times \textbf{Translations)} \times_{sd} \textbf{(Reflections} \times \textbf{Reflections)}]_I \\ \textbf{.Stretches.Shears.Rotations}]_E$$

Observe that this is a nested product. That is, the external operations *transfer* the internal structure from the canonical frame onto the non-canonical frames. It is this that allows the non-canonical frames to carry the *memory* of the canonical frame on them.

We need to add one factor to the above sequence: a specification of how the external component is aligned with the internal component. For example, as the sequence is currently given, the **Stretches** could be acting in an arbitrary direction. However, observe that the most structurally allowable stretches would be ones that acted along the x- and y-axes. This is because the frames that resulted from such stretches would actually preserve the reflectional structure of the starting frame; i.e., canonical frame. That is, such stretches would preserve the x- and y-axes as symmetry axes of the **Reflections** × **Reflections** structure. The stretches would do so by converting the symmetry axes into lines of flexibility; i.e., eigenspaces. We conclude therefore that, in accord with the Interaction Principles, the symmetry axes of the **Reflections** × **Reflections** structure are aligned with the eigenspaces of the external processes, and we indicate this by inserting the interaction product symbol, $\wedge$, between the internal and external structures, thus:

$$[[\textbf{Point.(Translations} \times \textbf{Translations)} \times_{sd} \textbf{(Reflections} \times \textbf{Reflections)}]_I \wedge \textbf{Stretches.Shears.Rotations}]_E$$

Our discussion of Cartesian reference frames therefore leads to this proposal:

CARTESIAN FRAME PROPOSAL. *Psychologically, a Cartesian reference frame is the following structure of processes:*

$$[[\textbf{Point.(Translations} \times \textbf{Translations)} \times_{sd} \textbf{(Reflections} \times \textbf{Reflections)}]_I \wedge \textbf{Stretches.Shears.Rotations}]_E$$

It is important to emphasize that this claim is a *psychological* one. We are interested in Cartesian reference frames not as they are defined in mathematics books, but as they are defined by human perceptual systems. Our claim is that they are defined as a structure of *processes*; in fact, as the particular process-structure given above. This claim is not the end-point of an argument but the beginning of one. We shall now go on to show that this claim solves long-standing problems in perception and integrates many diverse perceptual phenomena within a single framework that is is strongly explanatory.

6.27 Imposing a Cartesian Frame

Recall now that we have been developing two structures in our analysis of perceptual organization: (1) The first is $\mathbf{G}_1.\mathbf{G}_2 \ldots \mathbf{G}_n$, the sequence of groups defining the process-structure of the stimulus. There are of course brackets at various points along this sequence, indicating the internal and external subsequences. (2) The other kind of structure is the Cartesian reference frame, which, we have argued, is defined by $[[\textbf{Point.(Translations} \times \textbf{Translations)} \times_{sd} \textbf{(Reflections} \times \textbf{Reflections)}]_I \wedge \textbf{Stretches.Shears.Rotations}]_E$.

The question that will now be addressed is this: What can it mean to *impose* a Cartesian frame on a stimulus set? Conventionally, the imposition of such a frame is understood to be an act rather like this: A pair of axes is drawn on some transparent sheet and the sheet is lain over the stimulus set. Magically, as a consequence of this act, the stimulus set becomes structured in a certain way that leads to perceptual results, such as the identification of limbs on a biological shape, and the orientation-and-form phenomenon. A new view of the imposition of Cartesian frames will now be given, such that the results no longer appear magical.

We have two group sequences, one defining the process-structure of the stimulus and the other defining the frame. The crucial claim to be made is this:

SUBHISTORY PRINCIPLE. *The basis for the psychological impression that a Cartesian reference frame has been imposed on a stimulus set is this: The full process-structure*

$$\mathbf{G}_1.\mathbf{G}_2\ldots\mathbf{G}_n$$

defining the stimulus set, contains the process structure

$$[[\mathbf{Point}.(\mathbf{Translations} \times \mathbf{Translations}) \times_{sd} (\mathbf{Reflections} \times \mathbf{Reflections})]_I \wedge \mathbf{Stretches}.\mathbf{Shears}.\mathbf{Rotation}]_E$$

defining the Cartesian frame.

In other words, some of the processes that are contained in the history of the stimulus set are known to be those that define the Cartesian frame. A helpful metaphor, to understand this principle, is that of "genetic splicing": The imposition of a Cartesian frame on a stimulus set means that the group sequence of the frame is *spliced into* the full group sequence of the stimulus. This ensures that the history defining the stimulus contains the frame processes as part of that history.

6.28 A Solution to the Square-Diamond Problem

We shall, over this and the following sections, apply the principles of the preceding sections to solving several phenomena in perception. We begin with very simple shape and work our way carefully up to complex shape.

The problem with which we begin is the square-diamond problem. This was discovered by Mach (1897), who observed that a square placed with its sides parallel to the gravitational frame, as shown in Fig 6.54a, seems to be phenomenologically unlike the same object placed in the orientation shown in Fig 6.54b. The latter seems more like a "diamond". In the almost

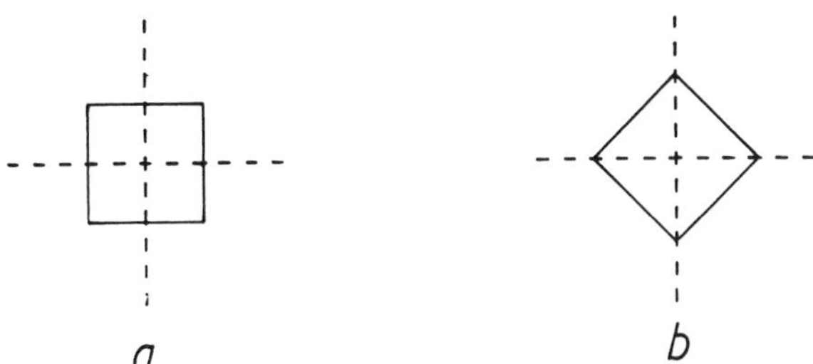

Figure 6.54
The square-diamond effect.

90 years since Mach's discovery, the phenomenon has remained a mystery. Let us now use the theory developed in the previous sections to propose a solution to the problem. It will be argued (a) that the square and diamond are distinguished by the assignments of their internal structures, (b) that they are distinguished by the assignments of the external structures, and (c) that the latter distinction is that predicted from the former by the Interaction Principles. These results are supported in two experiments, as follows.

(1) The difference between the internal structures

The Subhistory Principle (section 6.27) implies that, in assigning the Cartesian frame to a stimulus set, the internal structure of the frame becomes part of the internal structure of the stimulus. The internal structure of the frame is **[Point.(Translations** $\times$ **Translations)** $\times_{sd}$ **(Reflections** $\times$ **Reflections)]**$_I$. It is sufficient to show here that the **Reflections** $\times$ **Reflections** component in this structure is assigned differently to the square, from the way it is to a diamond. The difference can be illustrated as follows. Both figures have the same underlying stimulus set which we will call a *tetragon*. Consider Fig 6.55, a tetragon on which two pairs of symmetry axes have been drawn: (m_1, m_2) the side bisectors, and (n_1, n_2) the angle bisectors. The following experiment shows that **Reflections** $\times$ **Reflections** is assigned to a square so that (m_1, m_2) are its symmetry axes, and **Reflections** $\times$ **Reflections** is assigned to a diamond so that (n_1, n_2) are its symmetry axes.

In Leyton (1986c, Experiment 11), we presented subjects with 32 pairs of tetragons in the form shown in Fig 6.56. Each member of a pair was in the same orientation. Each tetragon had a pair of dots as shown in Fig 6.56.

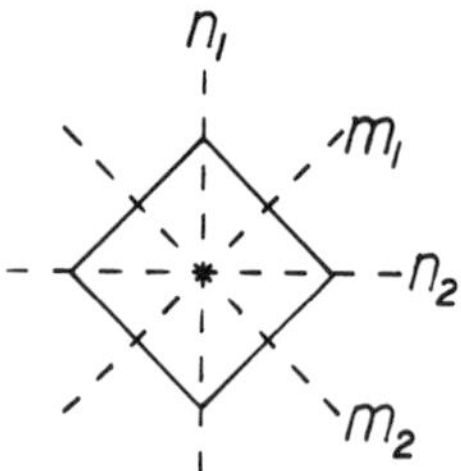

Figure 6.55
The two sets of axes involved in the square-diamond effect.

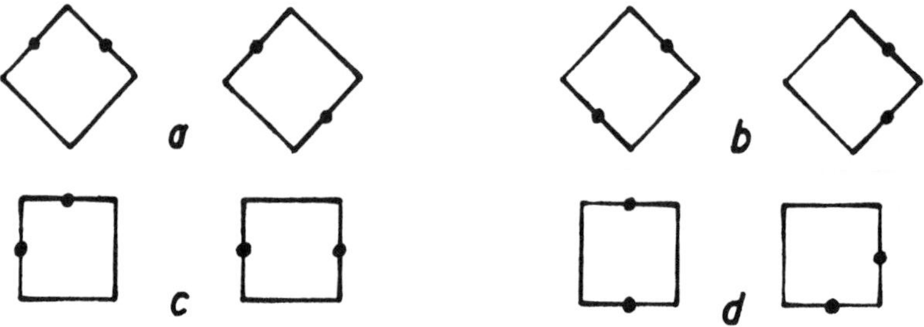

Figure 6.56
Pairs of tetragons in which one member of the pair is seen as a square and the other as a diamond.

The dot pairs were created as follows. The two tetragons always had one dot in a common position. The other dot was in a reflectionally symmetric position on the figure. The axis of symmetry used in one figure was a side-bisector, and the axis used in the other figure was an angle-bisector. For half of the pairs the subjects were asked which member of the pair was more square-like. For the other half of the pairs, the subjects were asked which member of the pair was more diamond-like.

Although the subjects were shown the figures in the conventional square or diamond orientation, half the subjects were asked the "wrong" question: When presented with a pair of conventional squares (e.g., Fig 6.56c), they were asked which member of the pair was more diamond-like. Similarly they were asked the "wrong" question for a pair of conventional diamonds.

The results showed that, with considerable statistical significance, the figures that were symmetric about the side-bisectors were regarded as square-like, and the figures that were symmetric about the angle-bisectors were regarded as diamond-like [$n = 12$; 32 trials per subject; expected mean = 16 trials in accord with experimental hypothesis; actual mean = 31.67; $t(11) = 47.00$, $p < .0005$, one tailed]. This was the case even for subjects given the *wrong question* condition.

Thus the hypothesis that the square and the diamond are distinguished by the assignment of their **Reflections** × **Reflections** structures appears to be corroborated.

(2) The difference between the external group assignments

In section 6.26, it was argued that the Cartesian reference frame is a process-structure in which the external sequence contains **Stretches. Shears.Rotations**. Furthermore, our Subhistory Principle states that this sequence of group factors must occur along the external group sequence of a figure if the figure yields the impression that a frame has "been imposed" on it. We now claim that this sequence is assigned differently for a square than for a diamond. It is sufficient to show a difference in the assignment of the first group factor, **Stretches**.

The claim is that stretches are perceived as more *allowable* along the side-bisectors in a square and along the angle-bisectors in a diamond. Observe that this accords with the Interaction Principles: These directions are the salient axes of the respective internal **Reflections** × **Reflections** structures. In particular, the Second Interaction Principle is relevant here. It states that axis of *salience*, in the internal structure, is translated into *structural allowability* of the external transformations determined by the First Interaction Principle (i.e., the external transformations that have eigenspaces along the symmetry axes).

In Leyton (1986c, Experiment 12), we presented subjects with pairs of figures, of the type shown in Fig 6.57. Each figure was derived from a tetragon in the same orientation, but one figure was a stretch along a side-bisector of the tetragon, and the other was a stretch along an angle-bisector. The results showed that, with considerable statistical significance, the former type was regarded as more square-like and the latter as more diamond-like. [Scores had greater statistical significance than the previous experiment.]

This supports the hypothesis that the square and the diamond are distinguished by the assignments of the structurally allowable external groups. Furthermore the external distinction made is that which is predicted by the

Figure 6.57
Each member of a pair has come from the same tetragon but with a stretch along a different axis.

Interaction Principles from the internal distinction provided by the preceding experiment.

6.29 Alternative Subhistories Conforming to the Subhistory Principle

The Subhistory Principle (section 6.27) states that the imposition of a Cartesian frame means that the process-structure defining the frame is part of that defining the stimulus. It is necessary for us to examine more closely what this implies in terms of the square-diamond phenomenon.

Both the square and the diamond are tetragons, and we can assume that they are given by the same process-hierarchy

$$[[\mathbf{Point.Translations.D_4}]_I \wedge \mathbf{Stretches.Shears.Rotations}]_E$$

We should now carefully look at the Cartesian frame as a subhistory in this sequence.

Internal Structure
According to the Subhistory Principle, the internal sequence defining the frame must be part of the internal sequence defining the tetragon. One of the components of the former sequence is **Reflections** × **Reflections**. Thus, the Subhistory Principle predicts that **Reflections** × **Reflections** must be part of the sequence defining the tetragon. In fact it is. It is contained in the factor $\mathbf{D_4}$.

We shall now see that **Reflections** × **Reflections** corresponds to two different parts of $\mathbf{D_4}$, for the square and the diamond respectively; i.e., two different subhistories.

Reflections × **Reflections** consists of four elements: (1) the identity element, (2) reflection about the vertical axis; (3) reflection about the horizontal axis; (4) the multiple of the previous two reflections—which is

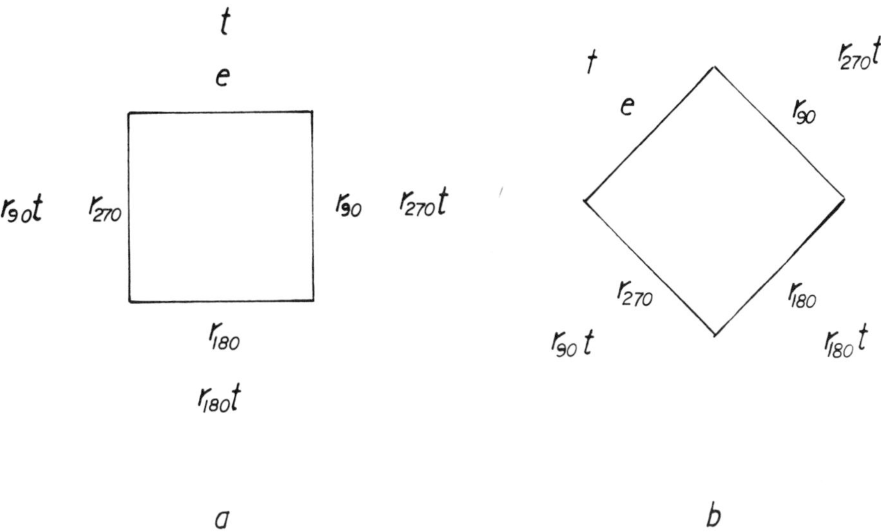

Figure 6.58
$\mathbf{D}_4$ mapped to (a) a square, and (b) a diamond.

actually a rotation. We shall now find these four elements in $\mathbf{D}_4$, when the latter defines a square, and when it defines a diamond.

Consider first a square. Fig 6.58a shows $\mathbf{D}_4$ mapped to the square. We can go through the four elements of **Reflections** × **Reflections** in the order listed in the previous paragraph and find their locations on this figure, thus: (1) The identity element e is located on the top side. (2) Reflection of this side about the vertical axis yields the top side again, and this reflection is created by t. (3) Reflection of the top side about the horizontal axis yields the bottom side, and the operation that obtains this from e is $r_{180}t$, shown on the bottom side. (4) The final operation is the multiple of the previous two reflections, t and $r_{180}t$, and this is r_{180}, which also obtains the bottom side.

Now consider a diamond. Fig 6.58b shows $\mathbf{D}_4$ mapped to the diamond. The four elements of **Reflections** × **Reflections** are located, on this figure, as follows: (1) The identity element e is located on the upper left side. (2) Reflection of this side about the vertical axis yields the upper right side, and this reflection is created by $r_{270}t$. (3) Reflection of the upper left side, e, about the horizontal axis yields the lower left side, and the operation that obtains this is $r_{90}t$, shown on the lower left side. (4) The final operation is the multiple of the previous two reflections, $r_{270}t$ and $r_{90}t$, and this is r_{180}, which obtains the lower right side.

Thus, recalling that $\mathbf{D}_4$ consists of the eight elements,

$$\textit{Rotations}:\quad e,\ r_{90},\ r_{180},\ r_{270}$$

$$\textit{Reflections}:\quad t,\ r_{90}t,\ r_{180}t,\ r_{270}t$$

we can see that, for a square, **Reflections** × **Reflections** selects the elements

$$e,\ r_{180},\ t,\ r_{180}t$$

whereas, for a diamond, **Reflections** × **Reflections** selects the elements,

$$e,\ r_{180},\ r_{90}t,\ r_{270}t$$

Thus, while the Subhistory Principle requires the factor **Reflections** × **Reflections** of the Cartesian frame to be part of the factor $\mathbf{D}_4$ defining the tetragon, **Reflections** × **Reflections** corresponds to two different parts of $\mathbf{D}_4$ for the square and the diamond respectively.

External Structure

The group sequence associated with a square or a diamond is

$$[[\mathbf{Point.Translations.D_4}]_I \wedge \mathbf{Stretches.Shears.Rotations}]_E$$

in which the external sequence **Stretches.Shears.Rotations** acts on the internal sequence **Point.Translations.D$_4$**. However, what we can now understand is that the external sequence acts on the internal sequence in two different ways in the square and diamond respectively.

The Interaction Principle demands that the salient symmetry axes of the internal structure become eigenspaces in the most structurally allowable external processes. The salient symmetry axes are the gravitational axes and these are given by the **Reflections** × **Reflections** component of the Cartesian frame. Thus, the external processes are aligned such that their eigenspaces are the axes of the **Reflections** × **Reflections** component contained within the $\mathbf{D}_4$ structure. However, the **Reflections** × **Reflections** component corresponds to two different parts of the $\mathbf{D}_4$ structure of the square and the diamond respectively. Thus, the **Stretches.Shears.Rotations** component must be aligned with two different substructures of $\mathbf{D}_4$ in the square and the diamond respectively. More specifically, for a square, the eigenspace lines of **Stretches.Shears.Rotations** must be aligned with the symmetry axes of the t and $r_{180}t$ operations in $\mathbf{D}_4$; whereas, for a diamond, the eigenspace lines must be aligned with the symmetry axes of the $r_{90}t$ and $r_{270}t$ operations in $\mathbf{D}_4$.

Thus we see that, while, in the sequence

$$[[\mathbf{Point.Translations.D_4}]_I \wedge \mathbf{Stretches.Shears.Rotations}]_E$$

the operation $\wedge$ connects axes in the internal structure to the eigenspaces in the external structure, the operation $\wedge$ nevertheless can, in different situations, select different symmetry axes in the internal structure to be connected to the eigenspaces.

What we need therefore is a general principled account of how the operation $\wedge$ selects those symmetry axes that are to be aligned with the eigenspaces. In order to develop such an account, we need to solve one of the major problems of perceptual organization: the orientation-and-form problem.

6.30 The Solution to the Orientation-and-Form Problem

The orientation-and-form phenomenon is this: Several stimuli have the following property: If the stimulus is placed in two different orientations, it is seen as two different objects. Subjects might not even notice that the same stimulus underlies the two objects.

One can regard the extensive research on the orientation-and-form problem as amongst the most important in perceptual psychology, for it shows that perception is not a mere *re*-presentation of the external world, but a *description* (Goldmeier, 1936/1972; Rock, 1973). A fundamental principle of Gestalt psychology is that the same stimulus can have several descriptions, each of which constitutes a completely different percept. The orientation-and-form phenomenon is a particular instantiation of that principle. However, while the literature has demonstrated that this phenomenon exists, no theory has emerged that *explains* why it occurs. Since its original discovery by Mach in 1897, the phenomenon has remained essentially a mystery. In fact, it will be argued here that the orientation-and-form problem has not been fully described. Of course, it is usually the case in science that while a problem remains a mystery it is inadequately described. Only its solution provides a structure that clarifies the original problem. Let us therefore bring together parts of the perceptual theory that has been developed here in an attempt to clarify, generalize, and solve the orientation-and-form problem.

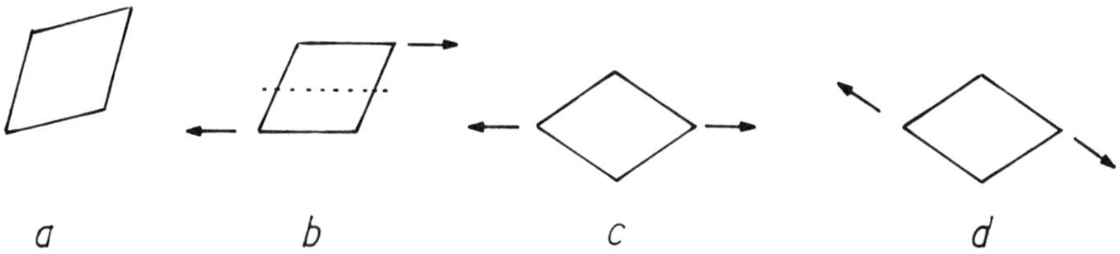

Figure 6.59
The different interpretations of an ambiguous figure depend on the different histories assigned.

The first thing we argue is that the orientation-and-form phenomenon is based on the inference of *process-history*. In fact, we argue that the phenomenon arises from the problem of choosing between alternative process-histories. To illustrate: Consider Fig 6.59a. It has two different interpretations, depending on its orientation. In the orientation shown in Fig 6.59b, it is usually seen as a sheared square—the arrows showing the direction of shear. In the orientation shown in Fig 6.59c, it is usually seen as a stretched diamond—the arrows showing the direction of stretch. Notice that the arrows act in a different direction *relative* to the object in Fig 6.59b, from the direction relative to the object in Fig 6.59c. That is, in Fig 6.59b, the arrows are parallel to the sides of the shape. If, in Fig 6.59c, they were to act parallel to the sides, the resulting percept would be Fig 6.59d. But Fig 6.59d is a difficult interpretation to make. Instead, people choose the interpretation in Fig 6.59c, where the arrows no longer act parallel to the sides but across the corners. Thus the examples in Fig 6.59 appear to support our proposal that the orientation-and-form phenomenon concerns the choice between alternative *process-histories*. How can we explain this?

Let us begin by looking carefully at the label *orientation-and-form*. It brings together two terms *orientation* and *form*. Since we have begun to see that the orientation-and-form phenomenon could be a process-based phenomenon, let us translate these two terms into our process-based framework.

(1) The term "orientation". The orientation of a stimulus is a single value of the stimulus. Therefore, it is a point in the external structure of the stimulus, since the external structure arises from the single state assumption. Because the process associated with orientation is rotation, orientation corresponds with the existence of rotation in the external structure.

That is, the orientation of a stimulus is the coordinate that it has in the process, rotation, in the external structure. However, we argue that orientation is not the only external factor that is involved in the orientation-and-form phenomenon. For example, as Fig 6.59 shows, stretch and shear are also involved; i.e., Fig 6.59b is a sheared square and Fig 6.59c is a stretched diamond. Therefore, we will argue that, instead of the term *orientation*, one should use the more general process-based term, *external structure*.

(2) The term "form". Since the term *form* usually refers to the internal relationships between the parts of a stimulus, we shall argue that, by the term *form*, one actually means the *internal structure*; or more rigorously, the *assignment of internal structure*.

Having translated (and generalized) the terms *orientation* and *form*, respectively, as *external* and *internal* structure, in our process-based system, we can also translate the conventional description of the orientation-and-form phenomenon into our system. The conventional description is this: *Different orientations of the same stimulus result in the stimulus having different forms*. The translation into our system is this:

> *The assignment of different external structures to the same stimulus results in the stimulus having different assignments of internal structure.*

We can now solve the orientation-and-form problem. We must explain how the assignments of different external structures can determine different assignments of internal structure. The solution is in fact provided by our Interaction Principles, as follows: Observe first that, if one assigns different external structures, one could be assigning different eigenspaces, i.e., directions of action. However, the Interaction Principles require that the eigenspaces be aligned with the salient symmetry axes. Therefore, different eigenspaces imply different salient symmetry axes. But this means that the internal structures must be assigned differently, because symmetry is embodied in the internal structure. To restate the argument succinctly: Different external structures can mean different eigenspaces, which must imply different salient symmetry axes, which must imply different internal structures. In this way, different external structures can determine different internal structures. Thus, we argue, it is the connection between the external and internal structure, provided by the Interaction Principles, that solves the orientation-and-form problem.

We shall see that most examples of the orientation-and-form problem involve a *single* internal structure that is assigned in *different ways* to the underlying stimulus. As we said, these different internal assignments are determined by different external assignments. However, what determines the different external assignments? Because the external assignments involve only single-coordinate values in the external processes (by the single-state assumption) the external coordinates are determined purely by the Inertial Component of the History Minimization Principle; i.e., one chooses the external coordinates that involve the least amount of change. In fact, because the internal structure of the alternative histories is the same, the choice between the alternative *full* histories is determined only by the Inertial Component applied to the external coordinates.

Therefore, the orientation-and-form phenomenon, in most examples, involves a procedure that consists of two parts:

PERCEPTUAL PROCEDURE.

(1) *The Interaction Principles determine, from each of the alternative external structures, the assignment of the respective internal structures.*

(2) *In the cases where the internal structures are the same, the full history is chosen by applying the Inertial Component of the History Minimization Principle to the alternative external structures.*

We shall now illustrate the above claims by a number of examples.

We begin with the simplest possible example, by returning to complete our solution to the square-diamond problem. Consider the tetragon in Fig 6.60a. We want to predict whether it will be seen as a square or a diamond. Recall first that both interpretations have the same internal structures. In both cases, this internal structure includes the internal structure of the Cartesian Frame, that is, [**Point.(Translations** × **Translations)** $\times_{sd}$ **(Reflections** × **Reflections)**]$_I$. It is this internal structure that is assigned differently in a square as opposed to a diamond. The assignment of the two reflection axes are shown for a square in Fig 6.60b, and for a diamond in Fig 6.60c.

We now use our solution to the orientation-and-form problem to predict which assignment will be chosen for Fig 6.60a.

Since the same internal structures are assigned in the two interpretations, we need look only at the external coordinates. In the interpretation shown in Fig 6.60b, all external coordinates have zero value; that is, stretch, shear and rotation are all zero. This is because the **Reflections** ×

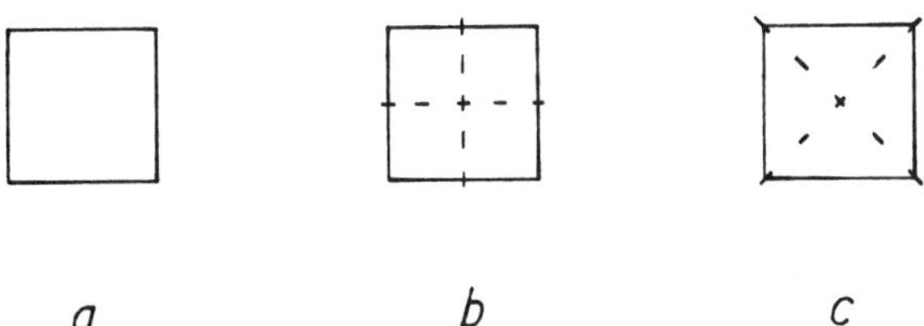

Figure 6.60
Given the tetragon in (a), less rotation is involved for the square interpretation (b) than the diamond interpretation (c).

Reflections structure that has been imposed belongs to the *canonical* Cartesian frame; i.e., it is gravitationally aligned, perpendicular, and equal-sided. Now consider the interpretation in Fig 6.60c. The **Reflections** × **Reflections** structure that has been imposed has zero stretch and shear. However, the frame is not in the gravitational orientation. This means that the frame has a non-zero external coordinate; i.e., in the rotation process. Thus, to compare: The square interpretation (Fig 6.60b) has only zero external coordinates, and the diamond interpretation has a single, non-zero external coordinate.

Now, the Inertial Component of the History Minimization Principle requires one to choose the interpretation that involves the least amount of change. In the present example, this must be the interpretation with the smallest external coordinates, because the internal structures are the same. Since the square has only zero external coordinates and the diamond has a non-zero rotation coordinate, perception will therefore choose the square as the interpretation.

Now consider the tetragon in Fig 6.61a. Again, it can be seen either as a square, as shown in Fig 6.61b, or as a diamond, as shown in Fig 6.61c. We can use exactly the same procedure to decide which interpretation will be chosen. We observe first that the square interpretation (Fig 6.61b) involves a non-zero rotation coordinate and the diamond interpretation (Fig 6.61c) involves only zero external coordinates. Thus we conclude that, in accord with Inertial Component of the History Minimization Principle, the percept will be the diamond.

This completes our solution to the square-diamond problem. That is, we have explained why a tetragon placed in the orientation shown in Fig 6.60a is seen as a square, and why a tetragon placed in the orientation shown in Fig 6.61a is seen as a diamond.

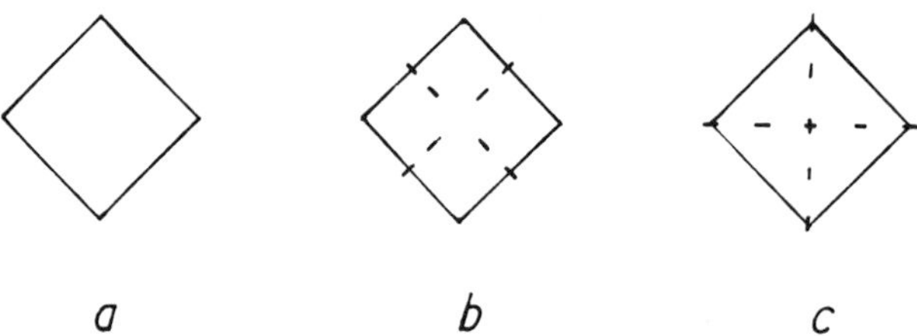

Figure 6.61
Given the tetragon in (a), more rotation is involved for the square interpretation (b) than the diamond interpretation (c).

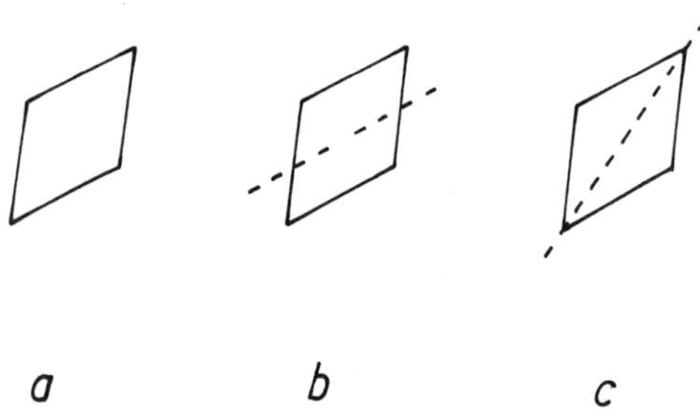

Figure 6.62
The stimulus (a) can be viewed either as a sheared square (b) or a stretched diamond (c).

In the example just described, a single external factor was involved: rotation. Hence the name *orientation* and form. However, we have argued that other external factors determine the outcome, if the external history contains them. Let us now consider the example in Fig 6.62a. The stimulus has perceptual ambiguity: It can be viewed either as a sheared square, as shown in Fig 6.62b, or as a stretched diamond, as shown in Fig 6.62c. That is, processes such as shear and stretch are involved. Let us now use the two-part Perceptual Procedure, defined above, to analyze this situation carefully and to predict which interpretation will be chosen by subjects.

Notice first that both interpretations have the Cartesian frame imposed on them, one of whose axes we have shown in each interpretation. According to our Subhistory Principle, this means that the conjectured history of each stimulus is perceived to contain the Cartesian processes as a subhistory. It will be sufficient in our discussion to concentrate on this subhistory.

Consider first the sheared square interpretation (Fig 6.62b). Observe that if one removes the shear asymmetry, i.e., the inequality between angles, one obtains an equal-angled tetragon. However, why does one postulate that this tetragon was a square, rather than a diamond? The answer is given once again by the first part of the above Perceptual Procedure, i.e., the Interaction Principles. That is, since the shape in Fig 6.62b is sheared, there is a single eigenspace line, the dotted line shown. The Interaction Principles determine that a salient symmetry axis of the tetragon must have been aligned with the eigenspace. Since the eigenspace bisects sides of the presented figure, the salient symmetry axis must have bisected sides of the original figure. Therefore, the double reflection axes of the Cartesian frame were side-bisectors in the original figure. Thus the original figure was a square.

Consider now the stretched diamond interpretation (Fig 6.62c). Observe that if one removes the stretch asymmetry, i.e., the inequality between angles, one obtains an equal-angled tetragon. However, why does one postulate that this tetragon was a diamond, rather than a square? The answer is given by the first part of the above Perceptual Procedure, i.e., the Interaction Principles. That is, since the shape in Fig 6.62c is stretched, the eigenspace lines bisect the angles. The Interaction Principles determine that the salient symmetry axes of the original tetragon must have been aligned with the eigenspaces. Since the eigenspaces bisect angles of the presented figure, the salient symmetry axes must have bisected the angles of the original figure. Therefore, the double reflection axes of the Cartesian frame were angle-bisectors in the original figure. Thus the original figure was a diamond.

Let us now specify the complete external structure of Fig 6.62b. It is sufficient to locate the coordinates of the figure within the **Stretches.Shears.Rotations** structure of the Cartesian frame. Observe first that the interpretation involves no stretch, and thus the figure is at the zero-point (identity element) of the **Stretches** group. Next, observe that the interpretation involves non-zero shear, and thus the figure has some non-zero value y in the **Shears** group. Finally, observe that the interpretation involves rotation of the dotted line from the horizontal. Thus the figure has some coordinate r_1 in the **Rotations** group. This means that the figure has the coordinates

$$(0, y, r_1)$$

in the external structure

Stretches.Shears.Rotations

Now let us specify the complete external structure of Fig 6.62c. Observe first that the interpretation involves non-zero stretch, and thus the figure has a non-zero value x in the **Stretches** group. Next, observe that the interpretation involves zero shear, and thus the figure is at the identity element of the **Shears** group. Finally, observe that the interpretation involves rotation of the dotted line from the horizontal. Thus the figure has some coordinate r_2 in the **Rotations** group. This means that the figure has the coordinates

$$(x, 0, r_2)$$

in the external structure

Stretches.Shears.Rotations

The reason why we have established the external coordinates of the two interpretations, Fig 6.62b and 6.62c, is that part (2) of our Perceptual Procedure states that, if the internal structures are the same, then the choice of interpretation is determined by applying the Inertial Component of the History Minimization Principle to the external structure. The internal structures are indeed the same because both of them are instances of the internal structure of the Cartesian Frame imposed on the respective tetragons.

The Inertial Component requires that the interpretation chosen for the presented figure will be that with the least amount of change. Since change means induced asymmetry, the Inertial Component requires that the history with the greatest structural allowability (i.e., preserved symmetry) is chosen.

Recall now that the left-to-right order in the sequence **Stretches.Shears. Rotations** corresponds to the order of decreasing structural allowability, because the groups successively destroy the gravitational symmetry axes.

With this in mind, let us compare the external coordinates of Fig 6.62b with those of Fig 6.62c. They are respectively,

Sheared square: $(0, y, r_1)$

Stretched diamond: $(x, 0, r_2)$

We observe first that the sheared square has zero value in the most structurally allowable process **Stretches**, and a non-zero value in the less structurally allowable process, **Shears**. This contrasts with the stretched diamond, which has a non-zero coordinate in the most structurally allowable process, **Stretches**, and a zero coordinate in the less structurally allowable process **Shears**. This means that, based on these first two coordinates, the stretched diamond is more structurally allowable than the sheared square.

However, now compare the third coordinate in each case. Both are non-zero rotations. But the stretched diamond involves more rotation than the sheared square, as can be seen by looking at the angle between the dotted line and the horizontal in Fig 6.62b, and comparing it with the angle between the dotted line and the horizontal in Fig 6.62c. The greater angle in Fig 6.62c means greater rotation which means less structural allowability. This means that the stretched diamond loses some of the structural allowability it had in comparison with the sheared square. The structural allowability of the two interpretations therefore turns out to be roughly equal. Thus, by the second part of the Perceptual Procedure, i.e., the use of the Inertial Component of the History Minimization Principle, the two interpretations will be chosen approximately equally.

Let us now manipulate the ambiguous Fig 6.62a in two ways to obtain further results. First we will put the stimulus in the orientation shown in Fig 6.63a. Again this figure has two viable interpretations, the sheared square in Fig 6.63b, or the stretched diamond in Fig 6.63c. Again, we carry out the procedure of establishing and evaluating the external histories. Interpretation b now has only one non-zero coordinate, shear; having zero stretch and rotation. In contrast, interpretation c has two non-zero coordinates: non-zero stretch and non-zero rotation. Therefore, by the Inertial Component, the history in interpretation b is smaller than in interpretation c. Thus we would expect interpretation b to be chosen rather than c.

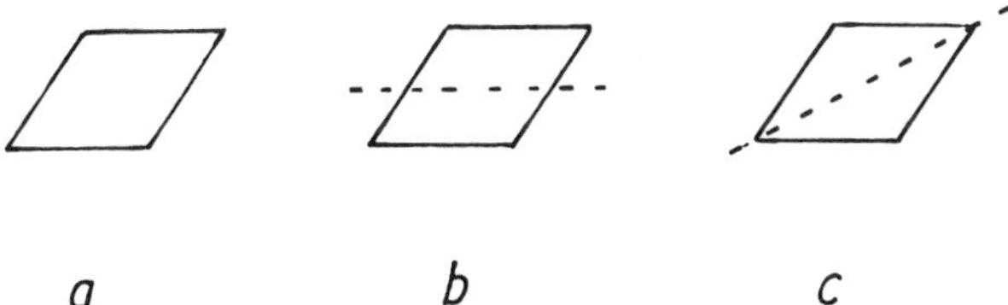

Figure 6.63
The stimulus (a) can be viewed either as a sheared square (b) or a stretched diamond (c).

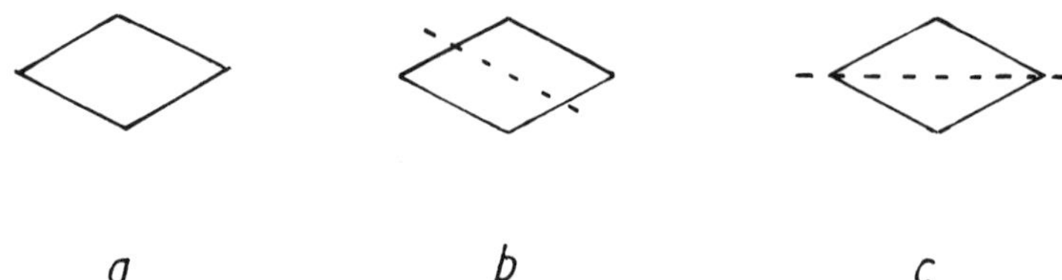

Figure 6.64
The stimulus (a) can be viewed either as a sheared square (b) or a stretched diamond (c).

Alternatively, let us place the same figure in the orientation shown in Fig 6.64a. This again has perceptual ambiguity; i.e., it can be seen as the sheared square, as in Fig 6.64b, or the stretched diamond as in Fig 6.64c. Again, we carry out the procedure of establishing and evaluating the external histories. Interpretation b now has two non-zero coordinates: non-zero shear and non-zero rotation. In contrast, interpretation c now has only non-zero stretch: having zero shear and rotation. Therefore, by the Inertial Component, the history in interpretation c is smaller than in interpretation b. Thus we would expect interpretation c to be chosen rather than b.

We can now return to the example of the orientation-and-form problem presented in Fig 6.59. We had asked the question why Fig 6.59a, when placed in the orientation shown in Fig 6.59b is seen as a sheared square, whereas, when it is placed in the orientation shown in Fig 6.59c, it is seen as a stretched diamond. Using the above argument, the answer can now be given, thus: (1) Consider first Fig 6.59b. Our discussion of Fig 6.63 showed that the sheared square interpretation, in this case, has only one non-zero coordinate, in contrast to the stretched diamond interpretation that could be assigned to this same stimulus. Thus the sheared square interpretation is chosen for Fig 6.59b. (2) Consider now Fig 6.59c. Our discussion of Fig 6.64 showed that the stretched diamond interpretation, in this case, has only one non-zero coordinate, in contrast to the sheared square interpretation that could be assigned to this same stimulus. Thus the stretched diamond interpretation is chosen for Fig 6.59c.

In conclusion, we have seen that the orientation-and-form problem seems to be solved in the following way. (1) Close examination reveals that the problem is one of the assignment of *history*, and thus belongs to our view of perception as the inference of process-history. (2) We saw that the orientation-and-form problem was in fact this: Different external histories determine different assignments of internal history. (3) We found that this determination occurs via the Interaction Principles. (4) In most cases, the salient internal structure is the same in the various interpretations because it is the internal structure of the Cartesian frame. However, the internal structure is *assigned* differently in the different interpretations. (5) This allows the selection of the full histories to be made by the application of the Inertial Component of the History Minimization Principle to the different sets of external coordinates in the alternative interpretations.

6.31 The Cartesian Frame Bundle

We have seen that the process-structure, defining the Cartesian reference frame, is extremely salient. In fact, we now propose that the salience indicates that this process-structure is actually hard-wired into the part of the nervous system responsible for low-level vision. We claim that the nervous system uses these specific hard-wired units to impose the structure on anything it possibly can, and at every level of scale. For example, a square receives the structure not only globally, as described above, but along its edges, as shown in Fig 6.65, where each disc indicates the placement of a Cartesian frame structure. In particular, each edge locally possesses the **Reflections** × **Reflections** structure; i.e., the edge possesses at each point two local axes of reflection, one along the edge and one perpendicular to the edge.

We argue that, not only are Cartesian frames imposed on everything that they can be imposed upon, they are nested within themselves. This is

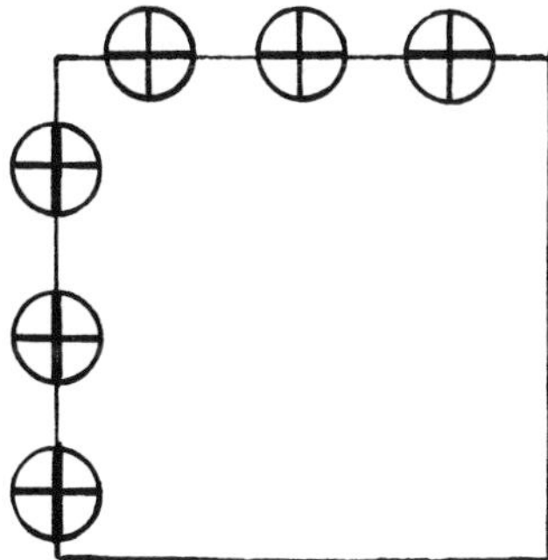

Figure 6.65
The Cartesian frame is assigned not only globally but locally.

achieved in the following way: Given a frame,

$$[[\textbf{Point.(Translations} \times \textbf{Translations)} \times_{sd} \textbf{(Reflections} \times \textbf{Reflections)}]_I \wedge \textbf{Stretches.Shears.Rotations}]_E$$

one substitutes for **Point**, on the far left, another such frame thus:

$$[[\textbf{FRAME.(Translations} \times \textbf{Translations)} \times_{sd} \textbf{(Reflections} \times \textbf{Reflections)}]_I \wedge \textbf{Stretches.Shears.Rotations}]_E$$

where the **FRAME** component stands for the entire group sequence of a Cartesian frame. This substitution can be made again for **Point** on the far left of **FRAME** itself, and so on, recursively, to any level. Any such recursive substitution of Cartesian frames will be called a *Cartesian frame bundle*. We argue that these bundles are hard-wired into the nervous system and are among the most psychologically salient organizational structures.

Let us consider the organizing effect of such bundles. First observe that, in any such bundle, a Cartesian frame is *nested* within another Cartesian frame. Since we have equated *nesting* with *transfer*, this means that a higher-level frame transfers the lower-level frame to different stimulus sets. In fact, the higher-level frame creates a field of lower-level frames. This can be seen by observing, that, in the two-level bundle,

$$[[\textbf{FRAME.(Translations} \times \textbf{Translations)} \times_{sd} \textbf{(Reflections} \times \textbf{Reflections)}]_I \wedge \textbf{Stretches.Shears.Rotations}]_E$$

the **Translations** × **Translations** structure, in the higher frame, transfers the entire lower frame to each point on the plane.

We shall see plenty of evidence for this transfer in the following sections. However, as a preliminary compelling illustration, consider Fig 6.66, which shows a phenomenon discovered by Attneave (1968, 1971). Fig 6.66 is a field of equilateral triangles. Each triangle can be seen as pointing in one of three directions. Nevertheless, the triangles together point in a single direction. Furthermore, if one looks at Fig 6.66 for sufficient time, the direction of pointing will change. However, all the triangles will change at exactly the same time, and into the same direction.

Our explanation for this phenomenon is as follows. We argue that the pointing effect in an individual triangle occurs because the triangle receives a Cartesian frame in a particular way: One of the reflection axes of

Figure 6.66
The field-of-pointing-triangles phenomenon discovered by Attneave.

the frame is aligned with one of the symmetry axes of the triangle, and thus bisects one of the vertices of the triangle. We shall call this Cartesian axis the *bisecting* axis. It is the bisected vertex that becomes the pointing vertex. The pointing effect occurs because the vertex is asymmetrically related to the *other* Cartesian axis; i.e., the latter axis is a reflection axis but has the bisected vertex on one side and no corresponding vertex on the other.[17]

Now let us deal with the fact that, in the field of pointing triangles, all triangles point in the same direction. We argue that the basis for this is that the Cartesian frame of the triangle is translated from one triangle to another by the translation structure of a second-level Cartesian frame. However, observe that this translation could be realized in many alternative ways. Which way is chosen? The answer is given by the Interaction Principles: the salient symmetry axes of a configuration become the eigenspaces of the structurally allowable processes applicable to the configuration. Here the most salient symmetry axis is the bisecting axis defined by the lower Cartesian frame, imposed on a single triangle. The **Translations** × **Translations** factor of the next level must therefore translate a frame along such an axis because the axis thereby becomes an eigenspace (invariant line) of the translation. In this way, the reflection axes become transferred onto the other triangles *in the same orientation.* In particular, the pointing direction is transferred onto the other triangles in the same direction.

The recursive aspect of a frame bundle can result in a structure, e.g., reflection, acting on a copy of itself at a lower level. To illustrate this, we constructed the following example. We combined the pointing triangle with the square-diamond effect as shown in Fig 6.67a and 6.67b. The two figures show exactly the same stimulus oriented differently. In other words, the relationship between the triangles and the tetragon is the same in both

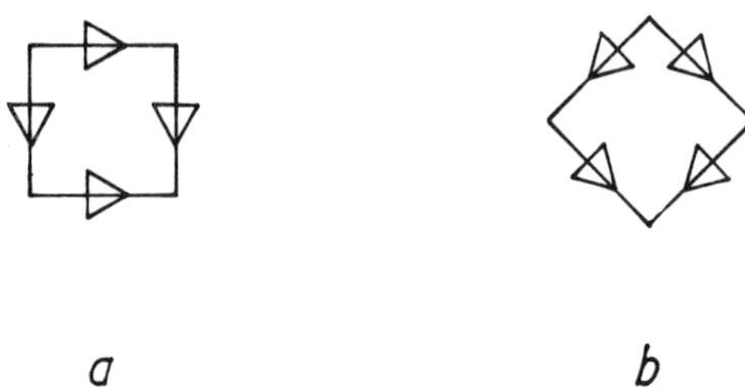

Figure 6.67
Combining the pointing triangles phenomenon with the square-diamond phenomenon.

figures. Nevertheless, the figures are perceived differently. In Fig 6.67a, the triangles on opposite sides are grouped together and are seen as parallel; two pointing vertically and two pointing horizontally. In contrast, in Fig 6.67b, the triangles are not seen as parallel. The two triangles on the top half are grouped together and are seen as pointing away from each other, and the two triangles on the bottom half are grouped together and are seen as pointing towards each other.

The two figures give strong evidence for Cartesian frame bundles in the nervous system. What we have, in these examples, is a two-level bundle. On the lower level, a local frame identifies the salient bisecting symmetry axis of each pointing triangle. On the next level, another Cartesian frame, the gravitational one, transfers the reflection structure of one triangle onto another by *reflecting* that reflection. Thus, in Fig 6.67a, the higher level reflects the triangle structure from one side to its opposite side. In contrast, in Fig 6.67b, the higher level reflects the triangle structure from one side to the adjacent side. The consequence is that the triangles point in parallel directions in Fig 6.67a and in converging and diverging directions in Fig 6.67b.

To restate: What the two figures show is that there are two levels of Cartesian frame, one nested within the other. The lower frame gives the **Reflections** × **Reflections** structure of each triangle, and the upper frame gives a **Reflections** × **Reflections** structure that transfers the lower **Reflections** × **Reflections** structure between triangles.

Of course, there is no limit to such nested hierarchies. Note, for example, that the individual sides of the triangle have a still lower **Reflections** × **Reflections** structure which itself can be exploited by creating examples with further nested levels.

Neurophysiology has uncovered much concerning the low-level workings of the visual nervous system. However, what has been uncovered has been structurally degenerate; e.g., bar-detectors. Although discoveries of this type are important, they will remain structurally degenerate unless neurophysiologists turn their efforts to locating the neurophysiological correlates of structurally more developed theories such as that advanced in the present chapter. The compelling nature of examples such as Fig 6.66 and 6.67, lead one to assume that such effects result from hard-wired units. The structural theory we offer presents a strict characterization of such units and also a strict characterization of their inter-relationships. With these types of characterizations, the neuronal units should not be difficult to locate.

6.32 Motion Perception

The principles we have been developing in the present chapter apply equally to motion perception. We consider a number of examples, as follows:

6.32.1 Induced Motion

Let us begin by examining the following simple phenomenon. Duncker (1929) discovered that, if a rectangular frame like that shown in Fig 6.68, is moved, and a dot within the frame is fixed, then the viewer perceives the motion to be the opposite: That is, the frame is seen as fixed and the dot is seen as moving.

This phenomenon can be understood quite simply as follows. If one were to see the frame as moving and the dot as fixed, one would be seeing a large number of points as moving; i.e., all the points of the frame. In contrast, if one were to see the dot as moving and the frame as fixed, one

Figure 6.68
A rectangle moving and a dot fixed, relative to an observer, are seen as the opposite.

would be seeing a small number of points as moving, i.e., the points comprising the dot. Thus, by the Inertial Component (of the History Minimization Principle), which states that change must be minimized, one must choose the latter interpretation.

Before going onto more complex examples, let us look more closely at this phenomenon. It actually involves a type of symmetry we have not used so far: scaling symmetry. Recall that any symmetry is indistinguishability under transformation. In scaling symmetry the relevant transformation is expansion, i.e., scaling. A large object has greater scaling symmetry than a small object. This is because, given a large object as shown in Fig 6.69a, one is more likely to be able to take an area such as the disc shown within the object, and expand the disc, by some amount, as shown in Fig 6.69b, and still remain within the object. With a smaller object, the disc is more likely to have gone outside the object by this expansion. Thus, larger objects have greater *scaling symmetry*.

We now propose that this fact, i.e., that more object means more scaling symmetry, is actually incorporated within the visual system in the following way:

MASS PRINCIPLE. *Scaling symmetry corresponds perceptually to mass.*

That is, the visual system identifies mass by size, which is scaling symmetry. Notice that, under this view, the *background*, behind a mass, is itself mass.

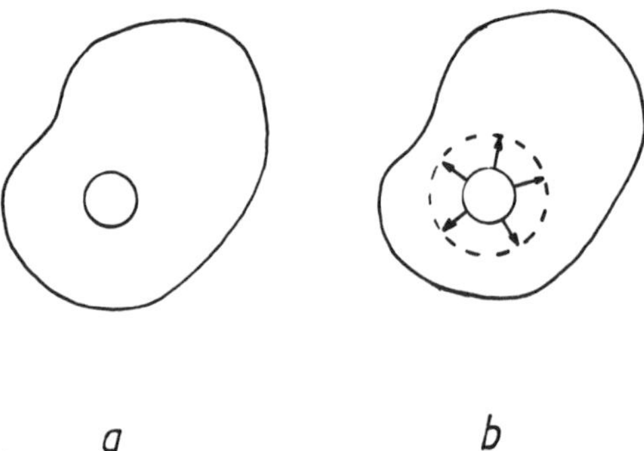

Figure 6.69
A larger object has more scaling symmetry.

What the Mass Principle implies is that, in the particular phenomenon shown in Fig 6.68, the use of the Inertial Component to explain the phenomenon becomes coincident with the explanatory role of inertia in Newtonian mechanics. Newton proposed that an object has an intrinsic resistance to change, i.e., an inertia. Furthermore, this inertia increases with mass, i.e., amount. Under our system, amount is given by scaling symmetry. Thus (all else being equal, e.g., density) scaling symmetry measures Newtonian inertia. Furthermore, the primary function that Newton required of the concept of inertia, i.e., that it describe resistance to change, is the function that it fulfills in the Inertial Component of the History Minimization Principle. That is, resistance to change minimizes history.

Finally, observe that our Inertial Component is much more general than its corresponding statement in Newtonian mechanics. The Newtonian inertia concept corresponds to only one type of symmetry, scaling symmetry. Our Inertial Component applies to any type of symmetry; i.e., it claims that all types of symmetry have inertia—not just mass.

6.32.2 Successive Embeddings of Induced Motion

Duncker (1929) observed that the dot-frame illusion (Fig 6.68) is a phenomenon with which people are quite familiar in their daily lives: When one looks up at the sky on a stormy night, one sees the moon moving quickly through a large cloud. Of course, the moon is actually still and the cloud is actually moving. Nevertheless, one perceives the movement to be the opposite way round. The cloud, having larger scaling symmetry is, under the Inertial Component, required to be at rest relative to the moon which has less scaling symmetry.

Duncker also observed that several such systems can be embedded within each other. For example, if one sees the moon and cloud between tall buildings, as shown in Fig 6.70, then the moon is seen as moving relative to the cloud, but the cloud is seen as moving relative to the buildings.

This of course is simply an embedding of the dot-frame pair within itself. That is, on one level, the moon corresponds to the dot and the cloud corresponds to the frame; and on the next level, the cloud corresponds to the dot and the buildings correspond to the frame.

Examples of this type show that motion is decomposed into nested systems of control, in the same way that we have found for process-structures generally. That is, in the motion case, each level is a group—the group of

Figure 6.70
Successive embeddings of the dot-frame illusion.

motion vectors,[18] from which the vector describing the least change, is chosen. The motion structure is therefore a hierarchy of groups. Furthermore, the structure is truly nested; that is, each level is defined as acting on the previous level, i.e., each level is defined as a set of actions relative to the previous static level.

6.32.3 Johansson (1950) Motion

We have so far considered the motion of two objects with unequal sizes. Now let us remove the unequal size factor and see what we obtain.

In particular, consider Fig 6.71a. It represents two equal dots moving perpendicularly, along the arrows $\mathbf{M}_1\mathbf{M}_2$, backwards and forwards simultaneously. This simple configuration was devised by Johansson (1950) and presented to subjects. The results were surprising. The subjects did not perceive the dots as moving perpendicularly to each other. Instead, they saw the dots as moving to and from each other along the *diagonal line* marked $\mathbf{R}_1\mathbf{R}_2$ in Fig 6.71b. The impression, sometimes voiced, was that the dots were the ends of a diagonal rod that was stretching and contracting. Furthermore, the subjects saw this diagonal line or rod as moving, as a whole, along the opposite diagonal—that given by the arrow marked **C** in Fig 6.71b.

Johansson's subtle analysis was that the "actual" motion, shown in Fig 6.71a, was decomposed into two factors: (1) the motion that the two dots have *relative* to each other, i.e., to and from each other, given by the arrows $\mathbf{R}_1$ and $\mathbf{R}_2$ in Fig 6.71b; and (2) the motion that the two dots have *in common*, given by the arrow **C** in Fig 6.71b. The two components were called, respectively, the *relative motion*, and the *common motion*. The decomposition is shown in Fig 6.71c.

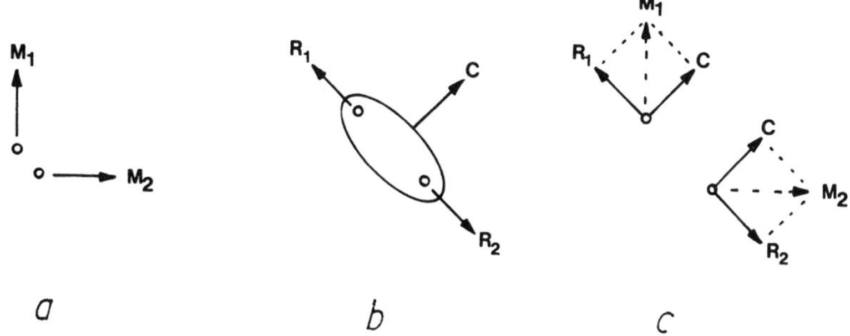

Figure 6.71
Two dots moving as shown in (a) are seen as shown in (b). Johansson's explanation, using vector decomposition, is shown in (c).

Figure 6.72
The two dots in (a) have a symmetry structure shown in (b).

We shall now see that the Johansson motion phenomenon is explained by the principles developed in this chapter.

We begin by examining the two dots themselves, without the motion, as shown in Fig 6.72a. Observe that the pair of dots is reflectionally symmetric about two axes: (1) the line that joins the dots, shown as line *A* in Fig 6.72b; and (2) the line that perpendicularly bisects them, labeled *B* in Fig 6.72b. That is, reflecting the dots about either of these lines sends the dots to themselves. This means that the internal structure of the dots is this:

Point.(Reflections × Reflections)

Now examine the motion of the dot-pair. As we said, the pair is perceived as stretching and contracting while also moving off in the perpendicular direction. That is, the group **Stretches** acts on the dot-pair, and so does the group **Translations**. Thus the full process-structure, involved in the phenomenon, is this:

Point.(Reflections × Reflections).Stretches.Translations

Observe that the **Stretches** and **Translations** processes act *simultaneously* in the phenomenon, i.e., the dot-pair stretches *while* moving. We represent this fact by bracketing together the two processes thus

Point.(Reflections × Reflections).[Stretches.Translations]$_P$

and indexing the brackets with the letter *P*, indicating that the processes act in *parallel*. Thus we have three types of (square) brackets in our notational system: (1) those indicating internal nesting, indexed by *I*; (2) those indicating external nesting, indexed by *E*; (3) those indicating parallelism, indexed by *P*.

We can now show that the Johansson phenomenon obeys the central principles in this chapter. First, observe that, in this phenomenon, the perceived organization conforms to the Interaction Principles, as follows: As noted above, the pair of dots have two symmetry axes shown as *A* and *B* in Fig 6.72b. One of them, axis *A*, becomes the eigenspace-line (invariant line) defining the direction of the relative motion, $\mathbf{R}_1$ and $\mathbf{R}_2$ in Fig 6.71b. The other symmetry axis, *B*, in Fig 6.72b, becomes the eigenspace-line defining the direction of the common motion, **C**, in Fig 6.71b. Note that, as we said in section 6.20, the equation between symmetry axes and eigenspace-lines is non-trivial: Symmetry axes are lines that are rigid under internal structure; whereas eigenspace-lines are lines of flexibility or directions of motion of the configuration.

To indicate the conversion of symmetry axes into eigenspaces in the above group sequence, we insert the interaction product $\wedge$ in the sequence, thus:

Point.(Reflections × Reflections) $\wedge$ **[Stretches.Translations]**$_P$

Observe that this indicates that the axis of one of the **Reflections** groups on the left of the product becomes the eigenspace of the **Stretches** group on the right of the product; and the axis of the other **Reflections** group on the left becomes the eigenspace of the **Translations** group on the right.

Besides obeying the Interaction Principles, we can see that the process-structure obeys the Subhistory Principle, which states that the psychological impression that a Cartesian frame has been imposed on a configuration is due to the occurrence of the frame process-groups within the full process-structure of the configuration. In fact, observe that it is possible to impose the standard Cartesian axes along the motion vectors in Fig 6.71a. Nevertheless, one imposes the axes along the motion vectors in Fig 6.71b. This is because these axes are the **Reflections** × **Reflections** axes and the eigenspace of **Stretches**, in accord with the Subhistory Principle, which states that the Cartesian frame implies these processes. In fact, in accord with the Subhistory Principle, these processes appear along the group sequence given above for the motion organization. Furthermore, it would be a simple matter to show that all the factors of the Cartesian frame occur along the full group sequence of the organization. For example, in a more arbitrary motion configuration, **Shears** could be introduced to describe the extent of shear between the two motion axes.

Notice that the entire group sequence, given above for the motion, is internally nested or parallel. In fact, the component

$$\textbf{Point.(Reflections} \times \textbf{Reflections)}$$

is internally nested within the parallel component. This contrasts with our usual description of the Cartesian frame as having an internal component and an external component, thus:

$$[[\textbf{Point.(Translations} \times \textbf{Translations)} \times_{sd} \textbf{(Reflections} \times \textbf{Reflections)}]_I \wedge \textbf{Stretches.Shears.Rotations}]_E$$

In particular, the **Point.(Reflections** × **Reflections)** component is *internally* nested within the **Stretches** component in the motion structure, and *externally* nested within the **Stretches** component in the Cartesian structure. This does not violate the Subhistory Principle. The principle requires only that the processes (or group factors) of the Cartesian frame are contained within the processes (or group factors) of the organization—not that external and internal nestings are preserved. The alteration of a nesting status from an external to an internal one means only that the observer has had a chance to witness the process through time rather than just at its endpoint. In particular, this change of status of the **Stretches** component in the above case is due simply to the fact that the **Stretches** process is now seen as motion. As we argued in section 3.13, motion is an example of *internal* structure. Thus, the external process becomes an internal one.

Note also that, while the external/internal status need not be preserved from the Cartesian frame, the use of the Interaction Principles does need to be preserved. The principles describe part of the structure of the frame—i.e., how the groups are to be aligned so that they obey the Inertial Strategy. Indeed, we do see that, in the motion phenomenon, the interaction product $\wedge$ is preserved from the Cartesian frame. Furthermore, it occurs in the same place in the motion sequence as it does in the frame.

Now let us consider another aspect of the above group sequence defining the motion. Observe that there is a **Translations** component at the end of the group sequence. Closer examination reveals that this indicates that the perceptual system is actually imposing a Cartesian frame *bundle*, as follows: Note first that the **Translations** factor occurs at the beginning of a Cartesian frame sequence:

$$[[\mathbf{Point}.(\mathbf{Translations} \times \mathbf{Translations}) \times_{sd} (\mathbf{Reflections} \times \mathbf{Reflections})]_I \wedge \mathbf{Stretches}.\mathbf{Shears}.\mathbf{Rotations}]_E$$

Thus one has to suppose that the **Translations** factor in the motion sequence does not come from the same Cartesian frame as the sequence to its left, but comes from the left-hand end of a second-level Cartesian frame, in which the frame, discussed previously, is nested. In fact, this is exactly the perceptual impression made by the stimulus: That is, the plane defined by the axis pair is itself being moved to different points of a plane; i.e., there are two planes that are involved, one that moves, and the other that is fixed relative to the observer and defines the fact that the former plane is moving. *Perceptually*, one therefore has the group sequence defining a Cartesian frame substituted for the **Point** factor in the group sequence of another Cartesian frame. This is exactly the type of structure we described in our discussion of frame bundles.

Finally we should observe that the Johansson motion phenomenon accords with the Grouping Principle, which states that the groupings of an organization are the cosets of left-subsequences in the group sequence. In particular, the cosets of the left-subsequence **Point.(Reflections × Reflections)**, in the above motion sequence, are the separate diagonal occurrences of the rod along the common motion direction.

Complex Shape

6.33 Generalized Cylinders

We now begin our examination of complex shape, which continues for the remainder of the chapter.

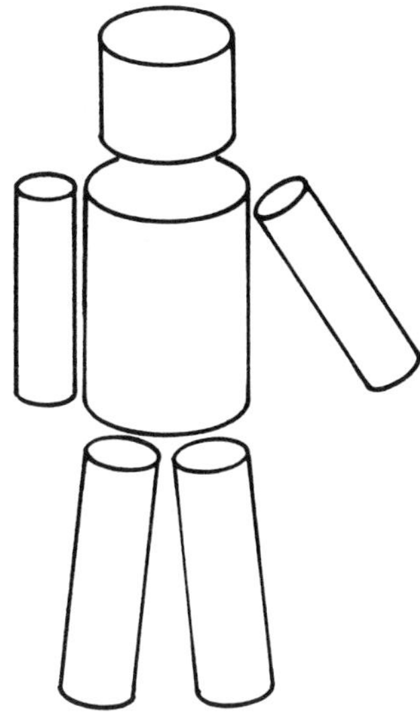

Figure 6.73
A man represented by cylinders.

We start by looking at the most frequently cited representation of the *parts* of a complex shape, and show that this representation conforms to the principles of this chapter. The representation is based on what are called *generalized cylinders*, and was introduced by Binford (1971), and advocated again by Marr & Nishihara (1978). The idea is that each part of a complex shape can be represented by a generalized cylinder. A famous example, from Marr & Nishihara (1978) is shown in Fig 6.73. Each of the man's limbs is given by a generalized cylinder.

The definition of a generalized cylinder is illustrated in Fig 6.74a. One starts (e.g., at the bottom of Fig 6.74a) with a cross-section, which we will always assume to be a circle. This is widened and narrowed as it is moved through space (e.g., upward in Fig 6.74a) along a curve. The widening and narrowing of the cross-section is called the *sweeping rule*. The curve along which the cross-section moves is called the *spine*.

It is easy to see that such a structure is a nested hierarchy of control, as follows: Observe first that the cross-section is circularly symmetric. By the Symmetry-to-Trace Conversion Principle, it can therefore be generated by taking a single point and applying **Rotations** to that point, as shown in Fig 6.74c. However, observe also that the cross-section is reflectionally

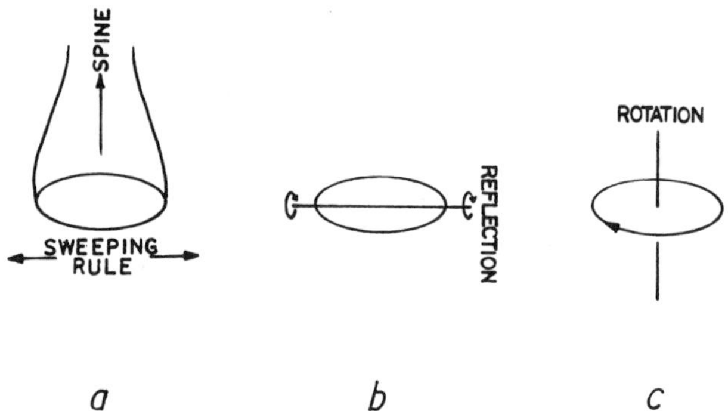

Figure 6.74
(a) The definition of a generalized cylinder. (b & c) The reflectional and rotational structure of the cross section.

symmetric about itself, as shown in Fig 6.74b. That is, the group **Reflections** makes equivalent copies of the rotational structure. We saw, in section 6.25, that a situation, where two groups are combined and one makes equivalent copies of the other, is called a *semi-direct product* and is represented by the symbol $\times_{sd}$. We see therefore that the internal structure of the cross-section is given by

$$\mathbf{Point.(Rotations} \times_{sd} \mathbf{Reflections)}$$

Now let us consider the actions that are applied to the cross-section. There are two such actions: (1) **Stretches**, which defines the expansion and contraction of the cross-section, i.e., the sweeping rule as shown in Fig 6.74a; and (2) **Translations**, which describes the movement of the cross-section, i.e., the spine as shown in Fig 6.74a. The two processes, **Stretches** and **Translations** act in parallel. Thus the process structure defining the cylinder is

$$\mathbf{Point.(Rotations} \times_{sd} \mathbf{Reflections).[Stretches.Translations]}_P$$

We shall now see that a generalized cylinder conforms to the principles of this chapter. First, closer examination reveals that a generalized cylinder conforms to the Interaction Principles; i.e., symmetry axes are converted to eigenspaces, as follows: One symmetry axis is the reflection axis shown in Fig 6.74b. This axis becomes an eigenspace-line, i.e., invariant line, defining the direction of the cross-section expansion (sweeping rule) shown in Fig 6.74a. In fact, each diameter of the cross-section is a reflection axis, and each becomes an eigenspace-line of the sweeping rule. The other type of symmetry axis of the cross-section is the axis of **Rotations**, as shown in Fig 6.74c. This too becomes an eigenspace-line: It is the line along which the cross-section is translated, i.e., the spine, as shown in Fig 6.74a.

To represent this alignment of symmetry axes with eigenspaces, we again insert the interaction product, $\wedge$, in the above group sequence, thus:

$$\mathbf{Point}.(\mathbf{Rotations} \times_{sd} \mathbf{Reflections}) \wedge [\mathbf{Stretches}.\mathbf{Translations}]_P$$

Note that the product $\wedge$ matches the axis of **Rotations**, on the left of the product, with the eigenspace of **Translations**, on the right of the product; and it matches the axis of **Reflections**, on the left of the product, with the eigenspace of **Stretches** on the right of the product.

Let us now consider a remarkable phenomenon. Let us consider both the sequence that was proposed earlier as the structure of the Johansson motion situation, and the sequence just given for generalized cylinders:

Johansson motion: $\mathbf{Point}.(\mathbf{Reflections} \times \mathbf{Reflections}) \wedge [\mathbf{Stretches}.\mathbf{Translations}]_P$

Generalized cylinder: $\mathbf{Point}.(\mathbf{Rotations} \times_{sd} \mathbf{Reflections}) \wedge [\mathbf{Stretches}.\mathbf{Translations}]_P$

What is remarkable is that closer examination reveals that the two sequences are identical—except that the former is a two-dimensional version of the latter. Both begin at the far left with a point. Then, progressing to the next factor, the motion case has **Reflections**, and the cylinder case has **Rotations**. However, if a generalized cylinder were used to describe a two-dimensional situation, the rotational symmetry would be replaced by a reflectional one, thus giving the same component, at this position in the sequence, as in the motion situation. Then, progressing on to the next factor to the right, both sequences have a further **Reflections** component. Finally, the remainders of both sequences consist of the **Stretches** group and the **Translations** group acting in parallel.

Our analysis therefore shows that the two situations, which are currently understood as quite different, are in fact structurally the same. Furthermore, using our concepts, it is instructive to see how they correspond, as follows: Let us line up eigenspaces. In the generalized cylinder, the eigenspaces are the sweeping rule and the spine. By matching these to the eigenspaces of the motion situation, we find that the cylinder's *sweeping*

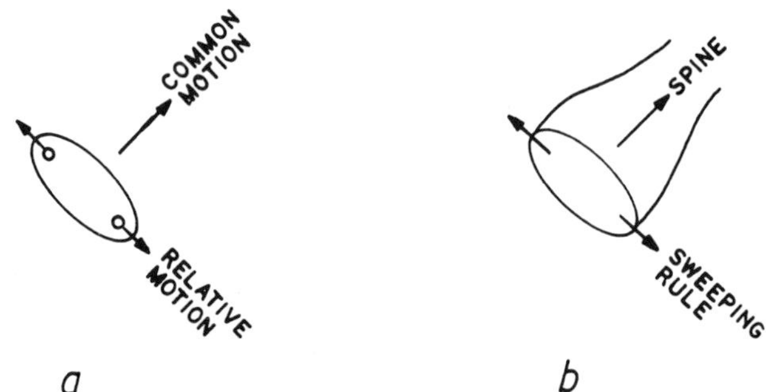

Figure 6.75
A correspondence exists between (a) the Johansson motion structure, and (b) the generalized cylinder structure.

rule corresponds to the *relative motion vector* and the cylinder's *spine* corresponds to the *common motion vector*—as shown by comparing Fig 6.75a, the motion structure, with Fig 6.75b, the cylinder structure.

Let us now consider the following phenomenon. Our Subhistory Principle states that the imposition of a Cartesian frame is the inclusion of the process hierarchy $[[\mathbf{Point.(Translations \times Translations)} \times_{sd} \mathbf{(Reflections \times Reflections)}]_I \wedge \mathbf{Stretches.Shears.Rotations}]_E$ along the full sequence $\mathbf{G}_1.\mathbf{G}_2 \ldots \mathbf{G}_n$ defining the perceptual organization. Considering now the motion and the cylinder situations, we see that certain factors of the Cartesian frame do indeed occur along the sequences given for those percepts. In fact, we shall find in the next section that, as predicted, *all* the Cartesian group factors occur along the latter sequences—when the latter are fully specified. An understanding of this will be crucial to the analysis of complex shape.

However, at this stage, consider the following: In the analysis by Marr & Nishihara (1978), the Cartesian frames that are assigned to various parts of the two-dimensional shape in the retinal image are themselves *two-dimensional.* By the above analysis, we see that this means that the (two-dimensional) shape-outline of each limb is structured in exactly the same way as the Johansson motion phenomenon. In other words, what results is a generalized "ribbon" rather than a generalized cylinder. Now, to form the cylinder, i.e., a limb, one rotates the ribbon about its axis. However, in this rotation, the axis remains over itself; i.e., it becomes an *eigenspace.* In other words, a reflectional symmetry axis of the two-dimensional

Cartesian frame becomes the eigenspace in the rotational movement that creates the limb. Thus we see that the *Interaction Principles* are crucially involved in converting the assigned two-dimensional Cartesian frame into a three-dimensional limb.

Finally consider the following: Observe that in the most famous example of the use of generalized cylinders, Fig 6.73, taken from Marr & Nishihara's (1978) paper, the figure consists not of *generalized* cylinders, but of *perfect* cylinders. We shall now see that this indicates that Marr and Nishihara are using a different type of history from that described above.

To see this, observe that, in the generalized cylinder construction, the entire structure is internal, and, in particular, the **Stretches** process acts within this internal structure; that is, the shape is a trace of the stretching movement over time. Let us however, make one alteration: Let us simply remove **Stretches** from this internal structure and make it act externally. The internal structure then becomes:

$$[\textbf{Point.(Rotations} \times_{sd} \textbf{Reflections)} \wedge \textbf{Translations}]_I$$

Observe that this internal structure generates a perfect cylinder, as shown in Fig 6.76. That is, in the above sequence, the **Point.(Rotations** $\times_{sd}$ **Reflections)** component generates the cross-section shown in Fig 6.76a, and the **Translations** component moves the cross-section through space, as shown in Fig 6.76b, where there is no alteration in the width of the cross-section, because the **Stretches** component is absent. **Stretches** is then added *externally* to produce the deformed cylinder, e.g., as in Fig 6.76c.

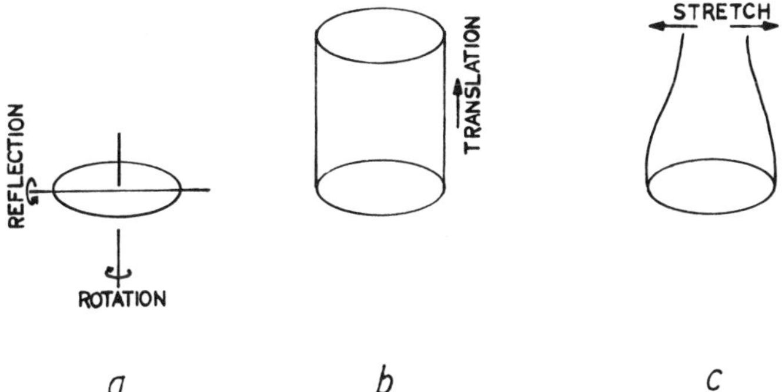

Figure 6.76
The successive stages in the generalized cylinder, except that stretch is externalized.

It is important to observe that, although the history just described produces the same final shape as the generalized cylinder construction of Fig 6.74, this history is an *externalized* one, whereas the generalized cylinder construction is a fully internalized one. More specifically, the history just described involves first the generation of a prototype shape and then a deformation of that shape; whereas the generalized cylinder construction omits a prototype stage, and simply traces out the final figure. The latter scenario is like the production of a pottery jar on a lathe: the jar increases in height as it is deformed.

Correspondingly, in being presented only with the final shape, the inferred *backward* histories would be different under the two scenarios. Under the externalized scenario, the backward history would be the removal first of deformation, thus obtaining a perfect cylinder, and then the undoing of the trace history. In contrast, under the generalized cylinder construction, the backward history would be the undoing of a pure trace history; e.g., as in the reversal of the history of a pottery jar on a lathe.

We can see therefore that, by using perfect cylinders in Fig 6.73, Marr and Nishihara are not using the generalized cylinder construction but rather the externalized history. That is, the man's limbs have undergone a stage of prototypification that is avoided in the generalized cylinder construction.

Furthermore, it is important to observe not only that Marr and Nishihara are using the externalized history defined above, but that this history conforms to our Externalization Principle which states that the purpose of externalization is to reduce the internal structure to a nested hierarchy of Euclidean processes. Perfect cylinders are nested hierarchies of Euclidean processes. Thus, in choosing perfect cylinders, Marr and Nishihara are, without being aware of it, conforming to the strong restriction of the Externalization Principle.

6.34 A Full Process Structure

We shall now argue that there is a major flaw in the use of generalized cylinders in the research literature: Generalized cylinders are viewed as *models*—that the mind possesses independently of any image, and that the mind attempts to impose on an image. We shall argue that this view is incorrect. In contrast, we shall argue that the mind simply applies the Asymmetry and Symmetry Principles to the image, as it does to any image.

When these principles are applied to certain images, they result in structures of the form of generalized cylinders. However, these resulting structures are a product of using the Asymmetry and Symmetry Principles, not of attempting to identify generalized cylinders.

The contrast between our view and the current view is therefore this: In our view, there is a single fundamental set of rules that the mind uses in all situations. In the current view, there is a highly specific model that the visual system finds applicable in certain situations.

In order to argue our point, we therefore need to return to first principles.

The Asymmetry Principle states that an asymmetry in the present arose from a symmetry in the past; and the Symmetry Principle states that a symmetry in the present always existed as such. We now apply these principles to complex shapes such as the shapes of animals, birds, plants, and abstract shapes—like those shown in Fig 6.77.

We had pointed out in Chapter 1 that most complex shapes, e.g., those in Fig 6.77, are not reflectionally symmetric in the standard sense: That is, one cannot take a mirror and place it across the shape such that one half of the shape is reflected onto the other. Nevertheless, a complex shape can be symmetric in a *local* sense which we now examine closely.

Consider two segments of a shape's boundary. For example, Fig 6.78a shows two segments labeled c_1 and c_2 which are part of some complex shape. Notice that one cannot place a mirror across this figure such that it will reflect one curve onto the other. However, observe that the curves have the following reflectional property: A straight mirror shown as **M** in Fig 6.78b will reflect the *tangent* at point A onto the *tangent* at point B. The two tangents are represented by the straight lines $\mathbf{t}_A$ and $\mathbf{t}_B$ respectively.

Why would one be interested in tangents? The reason is that tangents represent the orientation of the curve precisely at the point they touch. For example, tangent $\mathbf{t}_A$ represents the orientation of the curve precisely at A. This means that tangents represent a curve *locally*, i.e., at the point of touching. Now the existence of the mirror **M,** in Fig 6.78b, shows that, *precisely* at point A, the curve is reflectionally symmetric to the curve *precisely* at point B.

With the two sections of curve shown in Fig 6.78a, it happens that, given any point in curve c_1, there is always a unique point on the curve c_2 that is reflectionally symmetric to it in the sense just defined. For example, in Fig

Figure 6.77
Complex shapes.

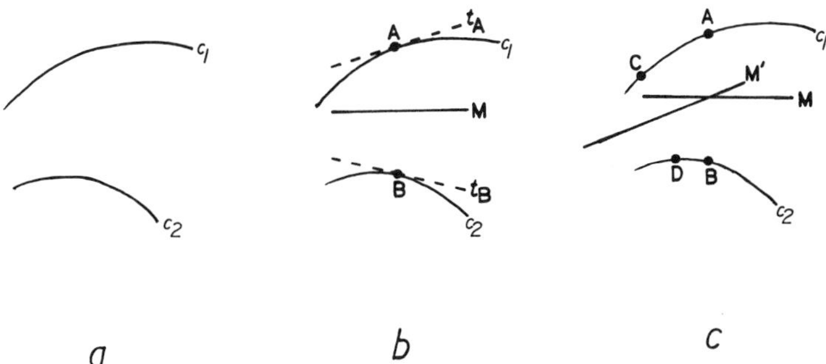

Figure 6.78
(a) Two curve segments c_1 and c_2. (b) There is a local mirror **M** between the tangents at *A* and *B*. (c) There is also a local mirror **M**′ between the tangents at *C* and *D*.

6.78c, the tangent at point *C* is reflectionally symmetric about the mirror **M**′ to the tangent at *D*. Observe that, as shown in Fig 6.78c, the mirror **M**′ has a different orientation from the mirror **M** used for points *A* and *B*.

Note also that, if one chooses any point between *A* and *C* on curve c_1, then its reflectionally symmetric point on curve c_2 will be between points *B* and *D*. Furthermore, its mirror will lie between **M** and **M**′.

This means that, taking all points in order along c_1, the reflectionally symmetric points on curve c_2 are in the same order along c_2. Furthermore, the mirrors change orientation smoothly from one to another. This set of mirrors will be called the *local mirror set* of the two curves.

Now let us apply the Asymmetry and Symmetry Principles. First, the Symmetry Principle states that the symmetries in the present must have existed throughout the past. This means that the points that are locally symmetric in the present—e.g., points *A* and *B* in Fig 6.78b—must have been locally symmetric in the past. Thus we can see that the local mirror set selects the content of the shape that must have existed in the past.

Now let us use the Asymmetry Principle which states that the asymmetries in the present are removed backward in time. In the local symmetry structure defined above, there are two kinds of asymmetry, which we discuss in turn.

ASYMMETRY TYPE 1. As we said in our discussion of Fig 6.78c, the successive mirrors, in the mirror set, smoothly change orientation. That is, the mirrors have distinguishability in orientation. According to the Asymmetry Principle, this distinguishability must be removed backwards in

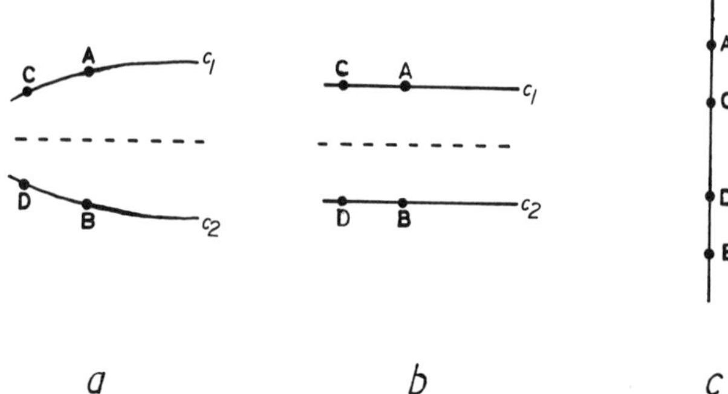

Figure 6.79
(a) Removal of distinguishability between mirrors in Fig 6.78c produces global symmetry. (b & c) are two alternative methods of removing distinguishability between tangents.

time. This means that, at some point in the past, the mirrors did not change orientation, i.e., they were all coincident. That is, there was effectively, only one mirror. This means that the past structure must have been *globally* symmetric. For example, Fig 6.79a shows what results from Fig 6.78c, when the mirrors become indistinguishable; the dotted line along the middle is the resulting single mirror, and the figure is globally symmetric about that mirror. Observe that the backwards transition from Fig 6.78c to Fig 6.79a has removed *bending* from Fig 6.78c.

ASYMMETRY TYPE 2. Fig 6.79a emphasizes that a second type of distinguishability still exists: the distinguishability in orientation between opposite tangents; e.g., between the tangent at A and that at B. There are two ways that this distinguishability can be removed:

Method 1: Make the opposite tangents parallel.

Method 2: Make the opposite tangents collinear.

We now discuss these in turn:

Method 1 results in two parallel straight sides, as shown in Fig 6.79b. This is because there is no other way of making all tangent pairs parallel. Observe that, in going forwards in time from Fig 6.79b to Fig 6.79a, *stretch* has been introduced, i.e., widening and narrowing. Notice that the earlier structure, Fig 6.79b, is a cylinder. We shall therefore say that Method 1 leads to the *cylinder scenario.*

Method 2 chooses instead to make the opposite tangents *collinear*. This results in a structure like Fig 6.79c. That is, curves c_1 and c_2 in Fig 6.79a are assumed to be part of the same curve, and the tangents become aligned on this overall curve. Notice that what has happened in going forwards in time from Fig 6.79c to Fig 6.79a is that the boundary has been pushed out. We shall say therefore that Method 2 leads to the *boundary-pushing scenario*.

Thus, we propose the following:

PROPOSAL. *Removal of distinguishability in orientation between locally symmetric tangents, by making them parallel (Method 1), leads to the cylinder scenario; and, by making them collinear (Method 2), leads to the boundary-pushing scenario.*

We shall see that Methods 1 and 2 are two fundamentally different scenarios for the history of a complex shape. They will be discussed in detail in the remaining sections of this chapter.

The reader should observe that cylinder structures, e.g., those of Marr & Nishihara in Fig 6.73, are structures that result automatically from Method 1. The mind does not have to possess cylinder models that it imposes on the stimulus. The mind simply applies the Asymmetry Principle, and to do so, has to make the virtually trivial choice between two methods for removing distinguishability between opposite tangents: making the tangents parallel or making them collinear. The former method produces cylinders.

6.35 The Cylinder Scenario

Figs 6.78c, 6.79a and 6.79b, show the removal of ASYMMETRY TYPE 1 and ASYMMETRY TYPE 2, successively, in the cylinder scenario. For ease of viewing, we give this history again as Fig 6.80a, b and c.

Observe first that the transition backwards in time from Fig 6.80a to 6.80b is the same in the cylinder scenario as in the boundary-pushing scenario. The difference occurs in the next transition, i.e., from Fig 6.80b to 6.80c, which occurs because of a use of Method 1 as opposed to Method 2.

Despite the fact that the difference between the cylinder scenario and the boundary-pushing scenario occurs only at the second transition, this

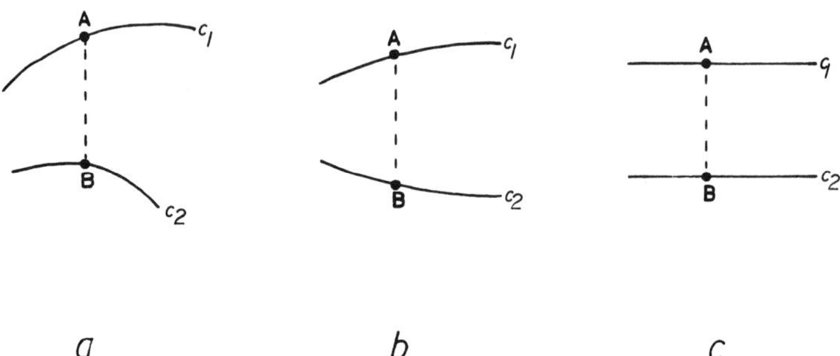

Figure 6.80
Successive removal of asymmetries in the cylinder scenario.

difference has a crucial organizational effect on all stages of the history. The reason is that Method 1 leads to a different *internal structure* from Method 2. Since the internal structure must underlie all stages, Method 1 must determine a different organization to the shape, at all stages.

The internal structure of the cylinder scenario can be understood from Fig 6.80c. This figure represents a perfect cylinder in three dimensions. A perfect cylinder, like any prototype, is a trace structure consisting of a nested hierarchy of Euclidean processes. The hierarchy involves the translation of a cross-section. This internal structure must therefore be transferred to Fig 6.80a. That is, Fig 6.80a must be the translation of a cross-section. Let us therefore attempt to understand how the cross-section structure of Fig 6.80c is transferred to Fig 6.80a. To do so, it suffices to consider Fig 6.80a as a two-dimensional structure.

Observe first that, in the perfect cylinder Fig 6.80c, any pair of points that are reflectionally symmetric locally—for example, *A* and *B*—must lie on the same cross-section (the dotted vertical line in Fig 6.80c). Now recall that this local reflectional symmetry, i.e., the reflectional symmetry between the tangent at *A* and that at *B*, is the *past* of a local reflectional symmetry in Fig 6.80a, that is, across the points labelled *A* and *B* in Fig 6.80a. This means that the cross-sections in Fig 6.80a must be between points that are locally reflectionally symmetric. This is a crucial conclusion, for the following reason: One is presented in the image with only Fig 6.80a, not with Fig 6.80c. Nevertheless, by simply using the Asymmetry and Symmetry Principles in the form of Method 1, one concludes that the *cross-sections* of the *internal structure* of Fig 6.80a must be between points that are locally reflectionally symmetric.

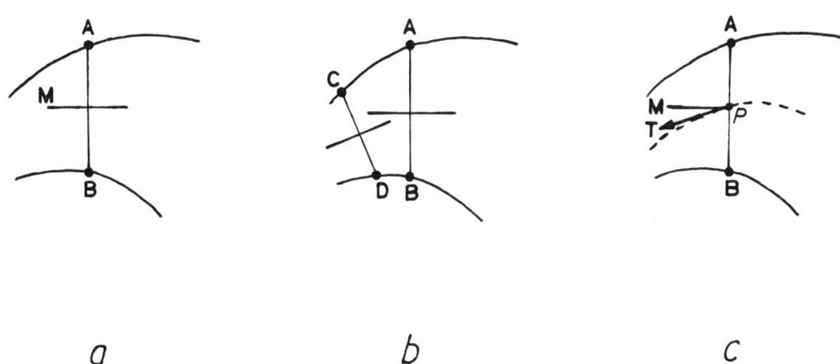

Figure 6.81
The symmetry for (a) one pair of points, (b) for two pairs, and (c) along the entire pair of curves.

Now, the existence of local reflectional symmetry between *A* and *B*, in Fig 6.80a, means that there is a mirror between these points, reflecting their tangents onto each other. This mirror is shown as **M** in Fig 6.81a. But recall also that, as a three-dimensional structure, the cross-section itself contains an axis of reflection, as shown in Fig 6.74b. This means that, in the two-dimensional structure shown in Fig 6.81a, each pair of symmetry points, *A* and *B*, defines a perpendicular pair of symmetry axes: one along **M**, and the other along *AB*. For example, as can be seen in Fig 6.81b, there is not only a pair of such axes for the points *AB*, but also for the points *CD*, and indeed all along the figure.

Locating the symmetry axes, for each point-pair, is crucial, because we can now use the Interaction Principles to fully define the process-structure of the shape. That is, each symmetry axis becomes an eigenspace, i.e., invariant line, of a process; as we shall now see.

The perpendicular axes, for each point-pair, constitute a **Reflections** × **Reflections** coordinate system. We shall now elaborate the processes that this coordinate system undergoes in defining the shape.

Observe first that the cross-section changes in width along the shape in Fig 6.81a. This means that the coordinate system undergoes **Stretches**. Since it is the cross-section that is perceived as widening or narrowing, one of the lines of flexibility, or eigenspaces, of the stretch must be the cross-section itself. This means that the cross-section, as a symmetry axis, is converted into the cross-section, as an eigenspace. That is, the perceived structure accords with the Interaction Principles.

Next, observe that, in the *perfect* cylinder, i.e., the past state shown in Fig 6.80c, the coordinate frame moves along **M**, the mirror. This means that, in the perfect cylinder, the symmetry axis **M** is converted into the eigenspace (invariant line) of the movement; i.e., in accord with the Interaction Principles. Let us now look at what this direction of movement (eigenspace) becomes in the presented shape, e.g., Fig 6.81b. Observe, from Fig 6.81b, that the center of the coordinate system makes some trajectory—shown by the dashed line in Fig 6.81c. This means that, at any point along the trajectory, e.g., the point shown as P in Fig 6.81c, the direction of movement (eigenspace) is given by the arrow tangential to the trajectory, i.e., the arrow **T** in Fig 6.81c. We can see therefore that the vector **T** is a memory of **M**, the direction of movement in the perfect cylinder. Thus, the coordinate system defined by the perpendicular axes, **M** and AB, in Fig 6.81c, is *sheared*, to become the coordinate system defined by the axes **T** and AB.

Finally, observe that the pair of perpendicular axes, **M** and AB, undergo **Rotations**, from one position to the next, as can be seen from Fig 6.81b.

In conclusion, we see that the pair of perpendicular symmetry axes, i.e., the **Reflections** × **Reflections** structure, undergoes the processes

Stretches.Shears.Rotations

Thus we see, remarkably, that the shape is completely characterized by the process-structure defining the Cartesian reference frame, i.e., **Stretches.Shears.Rotations** acting on a **Reflections** × **Reflections** structure.

When we remove the **Stretches.Shears.Rotations** processes from Fig 6.81a, we obtain the perfect cylinder as the past state. This is because the removal of **Stretches** results in a shape with no changing width; the removal of **Shears** results in a shape where the mirror **M** and the axis tangent **T** are coincident; and the removal of **Rotations** results in a shape where the axis is straight. The perfect cylinder is the only resulting shape that fulfills these three criteria.

Let us now elaborate all possible intermediate shapes that can occur in the history back to that past state. This is done by removing the individual processes from **Stretches.Shears.Rotations** in all possible combinations.

Fig 6.82 represents all possible histories between the presented shape and the perfect cylinder in the past. Any such history is obtained by taking a trajectory downward through Fig 6.82. At the top node in Fig 6.82, we have **Stretches.Shears.Rotations**, which gives the presented shape. Then, at

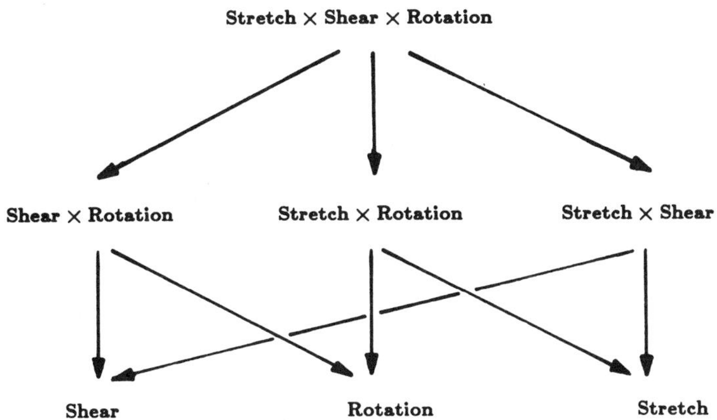

Figure 6.82
All possible intermediate histories in the cylinder scenario.

the middle level, each node represents a shape in which a single process has been removed: either **Stretches**, **Shears** or **Rotations**. The bottom level of nodes represent the shapes in which a further process has been removed, thus leaving only one process in each shape. When this final process is itself removed, one obtains the perfect cylinder. That is, in the final stage of the history, all the nodes at the bottom of Fig 6.82 connect together to a single final node below these nodes, and this final node represents the perfect cylinder.

We have encountered the removal of these same processes previously, in the case of simple shape; e.g., the rotated parallelogram going through its successive stages back to a square. However, the individual transformations were defined *globally*. For example, a single shear acted on the entire rectangle converting it into a parallelogram. In the case of complex shape, the transformations act locally, that is, each local coordinate system within the shape in Fig 6.81a has its own particular amount of transformation.

Let us consider, for a moment, the case where the transformations represent *global* actions. It is important to observe that, in this case, all the nodes in Fig 6.82 represent mathematically realizable shapes. For example, if the top node is a rotated parallelogram, the middle nodes represent, from left to right, respectively: a rotated rhombus, a rotated rectangle, and a parallelogram. Furthermore, the bottom nodes represent, from left to right, respectively: a rhombus, a rotated square, and a rectangle. In fact, although the experimental results established that subjects move downward along the *right-hand* sequence of nodes (i.e., they go to a non-rotated parallelogram to a rectangle, on the way to a square), we see that a shape nevertheless exists at each of the other nodes.

Now let us consider complex shape, i.e., the case where the transformations are used *locally*, as in Fig 6.81. We want to determine the shapes that exist at the nodes in this case. To solve this problem, we proved the following theorem:

LIMITATION THEOREM (Leyton, 1987c). *Let a shape be locally characterized by the decomposition* **Stretches.Shears.Rotations**, *with the eigenspaces aligned along the local reflectional axes, in accord with the Interaction Principles. Then, the removal of one of the processes necessarily involves the removal of one of the other processes.*

What the theorem states is that, if one tries to remove one of the processes, one will find that it has acted in parallel with one of the other processes and has to be removed with that other process. In terms of possible shapes, this means that no shapes can exist at any of the middle-level nodes in Fig 6.82. That is, the backward history has to jump down two levels in Fig 6.82. Furthermore, in Leyton (1987c), we show that no shape can exist at the bottom left node. This leaves only two possible nodes: the rotation-only and stretch-only nodes on the bottom level. The rotation-only node corresponds to shapes that are like worms: They are perfect cylinders except that their axes are bent as shown at the middle node at the bottom of Fig 6.83. The stretch-only node corresponds to shapes like goblets, i.e., straight cylinders in which there is only a widening and narrowing of the cross-section as shown at the bottom-right of Fig 6.83.

Fig 6.83 summarizes all the shapes that can be obtained. The top node represents the initial arbitrary shape. However, of the six nodes below it, only two nodes can have possible shapes.

The Limitation Theorem above therefore gives complete validation to the argument we gave in the previous section, when we first discussed the use of the Asymmetry Principle locally on complex shape. That is, we said that Fig 6.81a contains only two types of asymmetry: ASYMMETRY TYPE 1, the distinguishability in orientation of the mirrors; ASYMMETRY TYPE 2, the distinguishability in orientation of the opposite tangents. We had therefore concluded that the Asymmetry Principle requires the removal of only these two asymmetries. Now observe the following: If

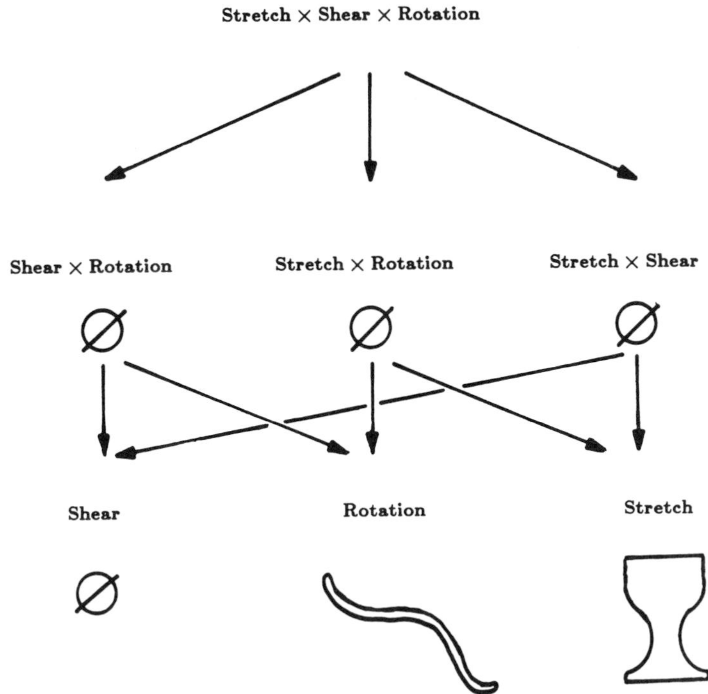

Figure 6.83
Given a shape at the top node, there are only two possible nodes (they are in the bottom row) that can have shapes.

one removes, from the presented shapes, ASYMMETRY TYPE 1, one obtains globally symmetric objects; i.e., the shapes at the bottom right node in Fig 6.83. Alternatively, if one removes, from the presented shape, ASYMMETRY TYPE 2, one obtains the bent cylinder; i.e., the shapes at the bottom middle node of Fig 6.83. The Limitation Theorem therefore validates the proposal that indeed we have to consider only these two types of asymmetry.

6.36 The Boundary-Pushing Scenario

The boundary-pushing scenario is illustrated again in Fig 6.84. The left-most figure has the two forms of asymmetry: ASYMMETRY TYPE 1 (orientational distinguishability between the mirrors) and ASYMME-

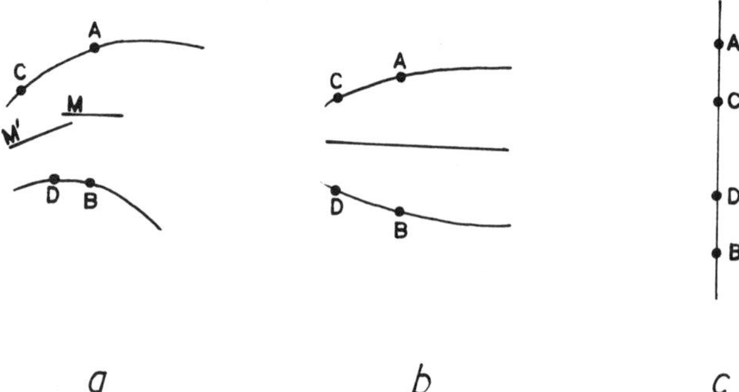

Figure 6.84
The successive removal of asymmetries in the boundary-pushing scenario.

TRY TYPE 2 (orientational distinguishability between opposite tangents). Removing the first type of asymmetry results in Fig 6.84b, which is also obtained under the cylinder scenario. In this stage, one has removed the rotation of the successive mirrors; i.e., the process of *bending* has been removed.

The second type of asymmetry, the orientational distinguishability between opposite tangents, is removed by Method 2, which makes the tangents collinear, rather than parallel as in the cylinder scenario. Making the tangents collinear results in a boundary shape like Fig 6.84c. Now recall that, by "the aligning of tangents", we mean the removal of orientational variation in the tangents. This means that curvature variation is removed. For example, the progression to Fig 6.84c is the removal of a protrusion or indentation, etc.

The removal of curvature variation can be divided into two cases. *Case 1*: If the entire boundary is smooth, then we have the type of situation discussed in detail in Chapter 1, where we examined the shapes of embryos, tumors, clouds, etc., and analyzed them to be the result of pushing the boundary along the axes. Each axis corresponds to a fluctuation in curvature, as illustrated in Fig 6.85. Thus Fig 6.84a could come from any of the fluctuations in Fig 6.85. Notice that, for Case 1, the removal of fluctuations from the overall shape results in a circle; i.e., the tangents are *rotationally* aligned, in contrast to the particular illustration given in Fig 6.84c, where the tangents are *translationally* aligned. *Case 2*: Alternatively, if the boundary is not smooth, i.e., contains corners, the sections between

Figure 6.85
Each axis, in the boundary-pushing scenario, corresponds to a fluctuation of curvature.

the corners become straight, backward in time, as shown in Fig 6.84c. This means, as argued in section 1.19, that one obtains a straight-sided polygon. However, the techniques for the reduction in curvature variation in Case 2 are the same as in Case 1. That is, fluctuation in a straight side is again analyzed as the result of pushing the boundary along the symmetry axis. For ease of exposition, our discussion will proceed as if we are considering only Case 1, when in fact the concepts are equally applicable to Case 2.

The boundary-pushing scenario therefore corresponds to the scenario developed in Chapter 1. Recall that the inference of this scenario used two rules: (1) the Symmetry-Curvature Duality Theorem which states, that, to each curvature extremum, one can assign a unique axis leading to the extremum; and (2) the Interaction Principle (in fact, the Backward First Interaction Principle) which implies that each symmetry axis is the trace of the process that created the extremum. An axis such as that shown in Fig 6.84b is therefore an example of such an axis.

Chapter 1 gave a detailed discussion of this scenario, and we do not need to repeat that discussion here. However, what we will do over the next few sections is to compare the cylinder scenario with the boundary-pushing scenario, in order to obtain a deeper understanding of the structure of shape. Before we begin this comparison, we need to look at previous methods of defining local symmetry.

6.37 The Classical Local Symmetry Analyses

6.37.1 Blum's SAT

The first attempt to define symmetry axes in complex shape was by made by Blum (1973), and his approach is called the *Symmetric Axis Transform* or SAT. The definition of the SAT can be illustrated as follows: Given two curves c_1 and c_2, as shown in Fig 6.86a, one inserts a disc between the curves and expands it until it touches both curves, i.e., at points *A* and *B* in Fig 6.86a. The disc will be maximal in that one cannot expand it further, and remain between the two curves. Points *A* and *B* are then regarded as *symmetrical*, and the circle-center is regarded as the *symmetry point* between points *A* and *B*. If one then traces out the trajectory of such maximal discs, as shown in Fig 6.86b, the centers of the discs will trace out the dashed curve also shown in Fig 6.86b. This curve is called the *symmetry axis* produced by the SAT.

There are, however, problems with the SAT. Because the discs have to fit between the two curves, and, more generally, fit within the boundary of any shape that is being used, the discs often cannot reach certain points that are obviously symmetrical. For example, points *A* and *B* in Fig 6.86c

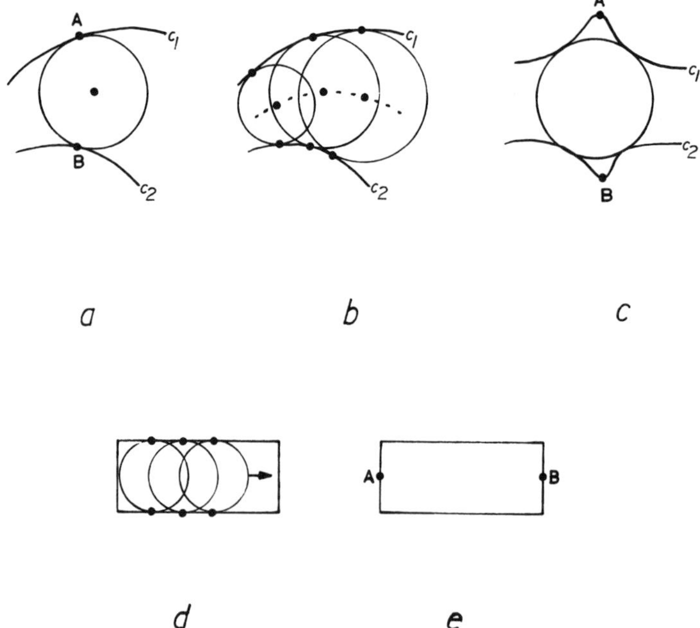

Figure 6.86
(a & b) The construction of the SAT. (c) The inability of the maximal disc to reach boundary points *A* and *B*. (d) The inability of the maximal disc to reach boundary points *A* and *B* simultaneously.

are perceptually symmetrical, and yet a disc cannot be inserted between the curves to touch both *A* and *B* simultaneously. This problem occurs even in simple shapes such as a rectangle: The circles can be rolled along the rectangle as shown in Fig 6.86d. The circle centers can therefore produce the horizontal axis. However, the SAT cannot produce the vertical symmetry axis, for the following reason. The vertical symmetry exists between points such as *A* and *B*, in Fig 6.86e, at the two ends of the rectangle. However, no circle can be squeezed into the rectangle, so that it can touch both *A* and *B* at the same time. Therefore, one does not have a circle center to represent the symmetry point between *A* and *B*.

6.37.2 Brady's SLS

In order to overcome the difficulties inherent in the SAT, Brady (1983) devised a different system, which he called the *Smoothed Local Symmetry* or SLS. The SLS can be illustrated again using the two curves c_1 and c_2 which we have shown in Fig 6.87a. In the SLS, one chooses pairs of opposite points *A* and *B* that have the following property: The angle shown as α in Fig 6.87a is the same size as the angle shown as β. As can be seen, α is the angle between the cross-section and the tangent at *A*; and β is the angle between the cross-section and the tangent at *B*. Any points *A* and *B* that have this equal-angle property are defined as *symmetrical*. The *symmetry point* between *A* and *B* is then designated to be the midpoint *P*, on the cross-section. Correspondingly, the *symmetry axis* is defined to be the

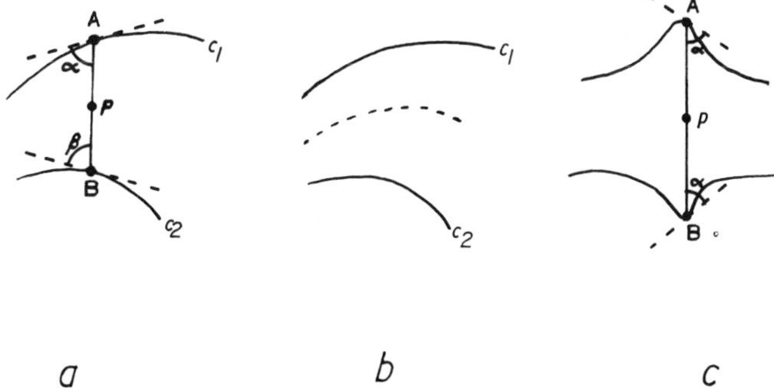

Figure 6.87
(a & b) The construction of the SLS. (c) The ability to reach *A* and *B*.

trajectory of such midpoints, as one goes along curves c_1 and c_2, identifying the successive pairs of symmetric points *A* and *B*. For the two curves c_1 and c_2, the symmetry axis produced by the SLS is shown as the dashed line in Fig 6.87b.

The SLS avoids the problems of the SAT (previous section), as can be seen in Fig 6.87c: Although points *A* and *B* cannot be reached simultaneously by one of the SAT discs, the SLS system will pair them because they have the equal-angle property; i.e., the angles that their tangents make to the cross-section are equal. Thus the SLS will identify the symmetry point *P* between them.

6.37.3 What is the Real Difference Between the SAT and SLS?

After the SAT and SLS were discovered, it was thought that they represented basically different constructions. A controversy arose as to which construction was the better one. However, closer examination reveals that they have a fundamental similarity.

To see this, let us return first to the SAT, and its illustration in Fig 6.86a. It can easily be shown that, if one of the SAT circles exists between points *A* and *B*, as in Fig 6.86a, then *A* and *B* will have the SLS equal-angle property; i.e., the tangent at *A* and the tangent at *B* will make the same angle to the cross-section.

Conversely, let us return to the SLS and its illustration in Fig 6.87a. It can be shown that, if two points, such as *A* and *B* in Fig 6.87a, have the equal-angle property, then one will be able to draw a circle that is tangential to the curves at those two points. The circle might not be *contained between* the two curves, but the equal-angle property will ensure the existence of a circle tangential at *A* and *B*. To illustrate, consider points *A* and *B* in Fig 6.87c. They have the equal-angle property. Therefore, one can draw a circle that is tangential at both *A* and *B*, even though the circle cannot stay completely between the two curves.

From now on, let us drop the original SAT restriction that the circles must be contained entirely between the two curves, and regard the circles as conforming only to the other demand made in the SAT: that the circles are tangential to the curves at two points.

With this minor amendment, we therefore find that the SAT and SLS have a basic similarity: If points *A* and *B* can be linked by one of the SAT

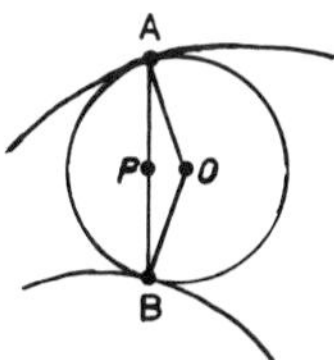

Figure 6.88
The SAT and SLS choose points O and P respectively as the symmetry points.

circles (i.e., a circle that is tangential at A and B), then they will have the equal-angle property of the SLS. Conversely, if points A and B have the equal-angle property of the SLS, then they can be linked by a SAT circle.

This equivalence can be succinctly stated thus: Two points A and B are *symmetric* in the sense defined by the SAT if and only if they are *symmetric* in the sense defined by the SLS.

So much for the major similarity between the SAT and SLS. Now for the major difference between them: Consider Fig 6.88. Points A and B are two points paired by either the SAT or the SLS. The SAT however defines the symmetry point between them to be the circle center O, whereas the SLS defines the symmetry point to be the midpoint P along the chord AB. This difference is important because, in either system, it is the trajectory of symmetry points that defines the symmetry axis. A different symmetry point will mean a different symmetry axis.

Given that the SAT and SLS have a basic similarity and a basic difference, let us now try to address the controversy as to which of the two symmetry analyses is the correct one to apply to a shape. The answer we shall give is that they are both correct. However, in producing this answer, we will find that there are an infinite number of other possible symmetry analyses that are all equally correct.

6.38 The Structure of Local Symmetry Analyses

In order to resolve the debate as to which symmetry analysis correctly describes a shape, we need to develop a more fundamental level of understanding of symmetry analyses than is currently available.

Rather than follow the approaches taken in the literature, the approach we are going to take is to look at how *mirrors* define symmetry. We introduced this approach in section 6.34, and continue it here.

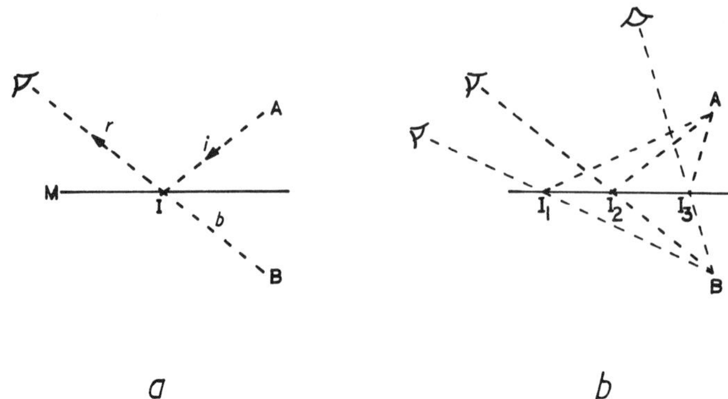

Figure 6.89
(a) The structure of rays defining a reflection. (b) The arbitrariness of the incidence point.

In order to define reflectional symmetry, one needs a mirror. However, not only does one need a mirror but one also needs an *observer*, and a *light-ray*. These three components define symmetry in the way shown in Fig 6.89a. A ray of light, labeled *i*, the *incident ray*, leaves point *A* and hits the mirror at some point *I*, the *incident point*, where it is reflected, becoming the ray labeled *r*, the *reflected ray*, and enters the eye shown at the top left of the figure. The reason why the observer can see *A* and *B* as reflectionally equivalent is that, if one extends the reflected ray, *r*, backward through the mirror—that is, obtaining the ray labeled *b*, the *back ray*—one arrives at *B*. Thus, the ray entering the eye could have started equally at *A* or *B*. Note that the incident ray, *i*, and the back ray, *b*, are reflected copies of each other.

The ability to see the reflection depends on the arrangement of three factors: (1) the observer, (2) the light-ray, and (3) the incidence point. We shall call this arrangement, the *observation arrangement*.

The important thing to notice now is that the observer can stand virtually anywhere and see the reflection. This is because, no matter which position the observer chooses, there is always an observation arrangement that allows him/her to perceive the reflection. For example, Fig 6.89b shows the observer in three different positions. For each of these positions there is some ray that leaves *A* and hits the mirror at some incidence point such that the ray is reflected into the eye. Thus the eye can see the *same* symmetry—that between *A* and *B*—in each of these positions.

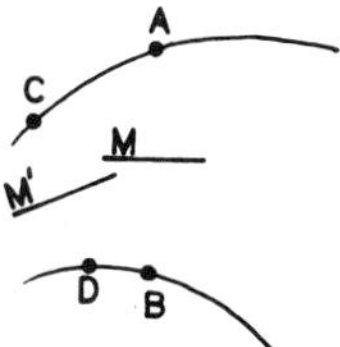

Figure 6.90
M and **M′** are local mirrors; i.e., they satisfy the construction in Fig 6.89a.

Now let us consider reflectional symmetry in a *shape*. If a shape is globally symmetric, a single mirror reflects one half of the shape onto the other half. However, we have seen that, even though a shape might not have global symmetry, it might have local symmetry: that is, the possibility exists, as shown in Fig 6.90, that a mirror **M** reflects the tangent at a point A onto the tangent at a point B.

To define the symmetry between A and B, in Fig 6.90, one requires an observation arrangement as described above; i.e., one requires an observer, a light-ray, and an incidence point.

Now, further along the two curves in Fig 6.90, there is another pair of points, C and D, that have local reflectional symmetry yielded by another mirror. This means that another observation arrangement must be involved in defining that symmetry; i.e., another arrangement of observer, light-ray, and incidence point.

Therefore, to define the local symmetry between the entire curves c_1 and c_2, we need two things: (1) a mirror set, i.e., the collection of mirrors yielded by the two curves; and (2) an observation arrangement for each mirror in the mirror set.

Notice now that, given any mirror in the set, there are an infinite number of observation arrangements that can be used for that particular mirror, as we saw illustrated in Fig 6.89b. We therefore propose the following definition:

DEFINITION OF LOCAL SYMMETRY ANALYSIS. *A* **local symmetry analysis** *is a mirror set together with a rule for selecting an observation arrangement for each mirror in the set. The rule will be called the* **observation selection rule**.

In any situation, the best kind of rule is one that defines some consistency. In the case of the observation selection rule, this could mean that the rule chooses observation arrangements that have some property consistent across all mirrors.

What form would such a consistency take? To answer this, recall first that an observation arrangement consists of three factors: an observer, a light-ray, and an incidence point. Note now that the observation arrangement is determined once one chooses one of these three factors. For example, if one chooses a position for the observer, then this will determine the direction of the light ray, and also the position of the incidence point. Alternatively, if one chooses a position for the incidence point, then this will determine the direction of the light ray, and furthermore, it will determine the observer to be somewhere along the reflected ray; etc.

Thus a *consistent* observation selection rule would be one in which one chooses either a consistent position for the observer, a consistent direction for the ray, or a consistent position for the incidence point.

Using these concepts we can now return to the SAT and SLS, and understand them more fully. What we will argue is that the SAT and SLS are in fact, *observation selection rules*; i.e., rules for the selection of particular observation arrangements. Furthermore, both are rules for the consistent choice of *incidence point*.

Let us return to Fig 6.88. Under the SAT, the circle center O is the point with respect to which A and B are symmetrical. Under the SLS, the chord midpoint P is the point with respect to which A and B are symmetrical. It turns out that both O and P are points on the *mirror* **M** that defines the local symmetry between A and B.

Fig 6.91a shows the point O as it is located on the mirror. The point can be regarded as the incidence point, i.e., the point where the incidence ray, i,

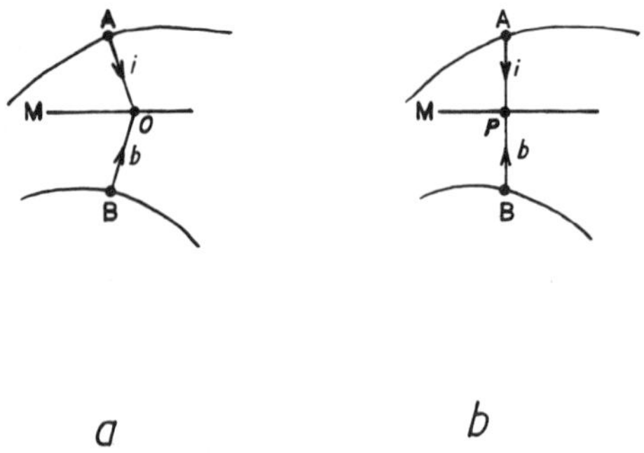

Figure 6.91
The points O and P of the SAT and SLS are two incidence points on the same mirror **M**.

coming from A, and the back ray, b, coming from B, converge, as shown in the figure. Alternatively, Fig 6.91b shows the point P as it is located on the mirror. This point, can be regarded as the incidence point instead, i.e., the point where the incidence ray, i, coming from A, and the back ray, b coming from B, converge, as shown in the figure.

The SAT and SLS can therefore be regarded as consistent rules for the selection of incidence points. That is, the two schemes can be redefined thus:

DEFINITION OF SAT. *Given a curve, the SAT consists of*

(1) the mirror set of the curve;

(2) the following observation selection rule: for each mirror in the mirror set, choose the incidence point to be the center of the associated tangential circle.

DEFINITION OF SLS. *Given a curve, the SLS consists of*

(1) the mirror set of the curve;

(2) the following observation selection rule: for each mirror in the mirror set, choose the incidence point to be the chord midpoint.

We can also understand that what is meant by a *symmetry axis* in the SAT and SLS is the *locus of incidence points*.

With the above concepts, we can now see that there are infinitely many types of symmetry analyses and hence symmetry axes. Each is based on the same mirror set—as are the SAT and SLS. But, like the SAT and SLS, each uses a different observation selection rule. For example, one can construct a symmetry analysis in which the observer has the same relative position with respect to each mirror in the mirror set, or a fixed absolute position, etc. Each type of symmetry analysis produces its own symmetry axis, i.e., trajectory of incidence points. It is reasonable to assume that, in the future, people will explore the structural properties of many different types of symmetry analyses, given that one can now generate systematically the possible types using the above scheme. Conversely, one can assume that people will explore properties that must hold for all symmetry analyses. To my knowledge, the first available such property is the Symmetry-Curvature Duality Theorem, given in Chapter 1, and restated here in the form in which it applies to all symmetry analyses:

SYMMETRY-CURVATURE DUALITY THEOREM. *Given a smooth curve on which there is only a single curvature extremum, the curve's symme-*

try points are paired one on each side of the extremum, and the pairs of symmetry points converge to the extremum.

Alternative statements that will be useful to us are:

COROLLARY 1. *Given a smooth curve on which there is only a single curvature extremum, there is a single trajectory of doubly tangential circles and this trajectory terminates at a circle that is tangential only to the extremum.*

COROLLARY 2. *Given a smooth curve on which there is only a single curvature extremum, the mirror set consists of a single trajectory of mirrors and this trajectory terminates at a mirror that intersects the extremum and is the only mirror that has a self-reflecting point belonging to the curve.*

6.39 Symmetry Analysis for Boundary-Pushing

As we said, it has conventionally been thought that there is a single correct local symmetry analysis. This has led to a controversy as to which of the two existing analyses, the SAT or SLS, is the correct analysis. However, we have seen, in the previous section, that, if one develops a thorough definition of symmetry analysis—a definition that we propose must be carefully grounded in the use of mirror sets and observation selection rules—then one finds that there are infinitely many symmetry analyses. Each is the choice of an observation selection rule on the mirror set yielded by the curve.

We now argue that the current view of a single correct symmetry analysis is wrong for a further reason. We argue that different symmetry analyses serve different *purposes* and are therefore appropriate for different circumstances. In fact, the different symmetry analyses capture different types of information, and thus the type of information required in a situation determines which type of symmetry analysis should be chosen. That is, symmetry analyses should be judged *functionally* relative to situations in which they are to be used.

In this section, we are going to compare symmetry analyses with respect to a particular and crucial function: their ability to capture the boundary-pushing scenario. Recall that, in both the cylinder scenario and the boundary-pushing scenario, one type of asymmetry, the distinguishability between orientations in the mirror set, is removed in the same way—by the removal of bending. However, the other type of asymmetry, the distinguishability between orientations of opposite tangents, is removed by two different methods. In the cylinder scenario, the opposite tangents are made parallel, and, in the boundary-pushing scenario, they are made collinear. In the latter scenario, this means that the boundary is understood as a "fluctuation" and is moved back along the symmetry axis of the fluctuation until the fluctuation is removed.

Thus, to more fully understand the boundary-pushing scenario, we need to understand how to define the appropriate symmetry axis for this scenario.

Recall, from the last section, that a symmetry *axis* is really the trajectory of *incidence points* through the mirror set, as determined by an observation selection rule. Both the SAT and SLS are examples where the observation selection rule is a *consistent* choice of incidence point.

In this section, we are going to argue that the SAT and SLS produce highly inappropriate symmetry axes for the boundary-pushing scenario. We shall argue that a different symmetry analysis, that introduced in Chapter 1, produces much more appropriate symmetry axes. In fact, this new analysis is so appropriate that we will call it *Process-Inferring Symmetry Analysis* or PISA.

The definition of PISA is based on an observation selection rule that chooses a consistent incident point that is different from that chosen in the SAT and SLS. The difference can be seen in Fig 6.92. The SAT chooses the circle center O, and the SLS chooses the chord midpoint P. In contrast, PISA chooses the point shown as Q, that is, the midpoint on the *arc* between A and B. We shall see that this subtle change produces symmetry axes that are fundamentally different from the SAT and SLS axes. For example, the PISA axes can be on different sides of the curve from the other two axes, and can be finite and continuous in situations where the others are infinite and discontinuous.

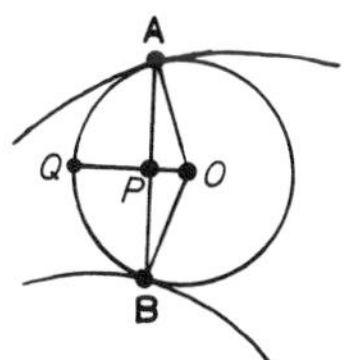

Figure 6.92
O, P, and Q are the incidence points chosen by the SAT, SLS, and PISA respectively.

In this section we will compare the SAT, SLS and PISA with respect to the ability of their axes to represent the boundary-pushing scenario. In this scenario, each curvature extremum, in the shape boundary, is explained by a process acting along the symmetry axis leading to the extremum. Thus, given a curvature extremum, the SAT, SLS, and PISA must each produce a symmetry axis that terminates at the extremum. By the Symmetry-Curvature Duality Theorem (particularly, Corollary 1, section 6.38), this condition is fulfilled by all three analyses.[19] Therefore each analysis produces an axis that can represent some aspect of the process creating the extremum. However, we shall find that the axes produced by the three analyses are radically different.

Our method of comparing the axes will be as follows: Recall, from Chapter 1, that there are four types of curvature extremum that can exist in the boundary, and each is associated with a particular type of process, thus:

M^+	protrusion
m^-	indentation
m^+	squashing
M^-	resistance

These must therefore be the alternative forms that a process can take under the boundary-pushing scenario. We shall now compare the SAT, SLS and PISA with respect to representing these four types of processes. Throughout the comparison we shall use the letters O, P, and Q, to represent the chosen incidence points in the SAT, SLS, and PISA, respectively—as shown in Fig 6.92. Thus, when required, the reader can refer back to the following description of the three symmetry analyses:

SAT:	the axis is the locus of circle centers, O.
SLS:	the axis is the locus of chord midpoints, P.
PISA:	the axis is the locus of arc midpoints, Q.

6.39.1 The Inference of Indentation

We will begin by arguing that PISA infers *indentation* correctly whereas the SAT and SLS do not. Consider the shape shown in Fig 6.93a. It has an indentation at the top. Our purpose is to derive a symmetry axis that correctly represents the trace of the process that created the indentation.

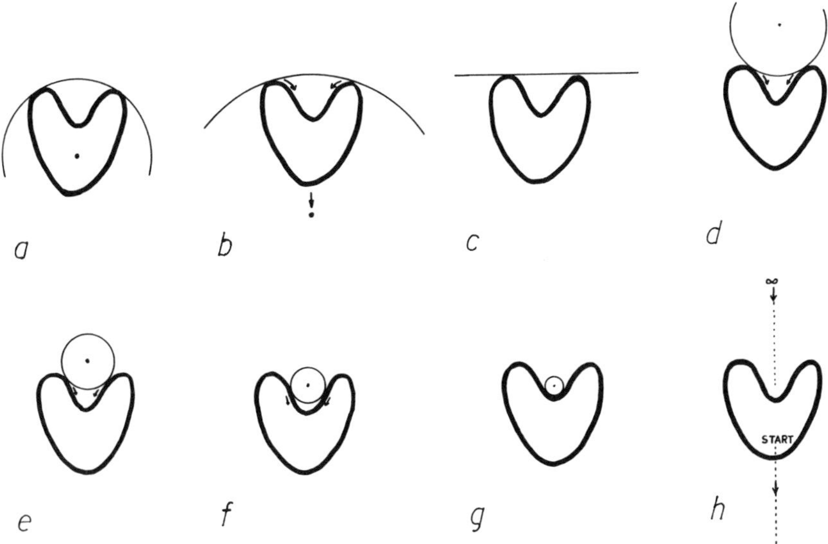

Figure 6.93
The SAT on an indentation produces an infinite, discontinuous axis.

Recall that the trajectory of doubly-tangential circles is the same whichever symmetry analysis is chosen. Observe also that the doubly-tangential circles must work their way down the mouth of the indentation so that each circle is tangential simultaneously to two points; one on either side of the indentation.

We will consider first the SAT: the trajectory of circle centers.

Observe first that, in the present shape, the initial circle has the following strange property: As shown in Fig 6.93a, in order to touch the two end-points of the indentation, the circle must *exscribe* (be outside) the shape. Thus the circle center is below the contour, as represented by the dot. The next thing to observe is that an immediately subsequent circle, in the trajectory, has to look as shown in Fig 6.93b. That is, in order for the circle to work its way into the indentation, i.e., such that the two tangent points are further into the indentation, the circle has to have a wider radius, as shown in Fig 6.93b. However, observe that the circle-center has therefore moved down, as shown by the dot at the bottom of Fig 6.93b.

Now, in order to reach tangent points that are still further in the indentation, the circle has to widen further until it has infinite radius as shown in

Fig 6.93c. Here, the circle-center has gone off the bottom of the page, to infinity.

To continue further into the indentation, the circle then becomes finite again as shown in Fig 6.93d. Thus the circle center has reappeared. However, this time, it has come in from infinity via the top of the page.

Figs 6.93e, 6.93f, and 6.93g show the remainder of the trajectory of circles and their centers, as the circles work their way down into the indentation.

Therefore, the trajectory of circle-centers for the entire sequence is represented in Fig 6.93h. As we have seen, this trajectory, i.e., the SAT-trace, is very badly behaved: It breaks into two *infinite* pieces. Furthermore, the pieces are strongly discontinuous; i.e., they cannot be meaningfully joined because the start-point is below the finish-point.

This type of trace is inappropriate for an indentation, because an indentation is understood to be continuous and finite. We conclude therefore that the SAT is unsuited for the inference of indentation.

Now let us consider the SLS. This symmetry analysis is badly behaved in a different way. We will not illustrate this fully here. However, consider the same sequence of diagrams, i.e., Fig 6.93. Recall that the SLS axis is the trace of cross-section midpoints. Careful examination of Figs 6.93a to 6.93d shows that the cross-section moves upward and then downward on this shape. Thus the SLS actually changes direction. Although the effect is a weak one for this particular shape, it is a very strong one for shapes such as a smooth dumb-bell. More specifically, the effect becomes stronger, the more that the ears of the shape, shown in Fig 6.93a, point away from each other (while still retaining the indentation).

The fact that the SLS axis changes direction makes it an inappropriate representation for the trace of an indenting process. The obvious reason is that indentation is understood as a simple *uni-directional* process *into* the shape.

Now let us consider the suitability of the PISA axis as a trace for indentation. As pointed out earlier, PISA uses the same doubly tangential circles as the SAT and SLS, but selects the circumference midpoint as the symmetry point. Thus in Fig 6.94 we consider the same trajectory of circles, while constructing PISA. However, we now follow the trace of the circumference midpoint. As shown in Fig 6.94a, the point starts just above the mouth of the indentation—which is intuitively appropriate for an indenting process. In Fig 6.94b, it has moved smoothly down, as the circle has

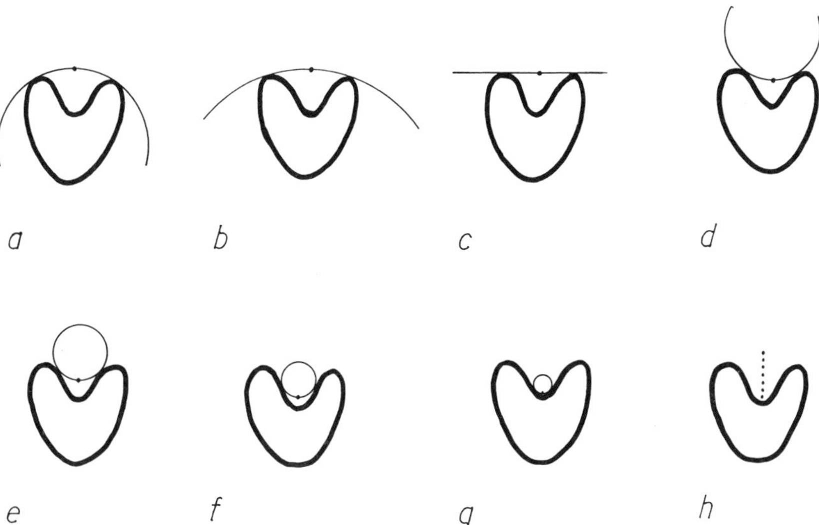

Figure 6.94
The PISA on an indentation produces a finite, continuous, uni-directional axis.

become larger. In Fig 6.94c, as the circle has become infinite, the point has again made a simple movement downward. Again, in Fig 6.94d, even though the circle now faces the other way, the circumference point has nevertheless continued on its simple downward trajectory. In Figs 6.94e, f and g, we see this trajectory completing its course.

Thus, Fig 6.94h shows the entire trace created by PISA. One can see that, unlike the SAT, which involves two disconnected infinite pieces, the trace is finite and connected. Again, unlike the SLS which changes direction, it is uni-directional. In short, the PISA-axis has the required properties of the trace of an indentation process.

6.39.2 The Inference of Squashing

Consider an ellipse. As shown in Fig 6.95a, an ellipse has four extrema E, F, G and H. Since the Asymmetry Principle implies that this shape was deformed from a shape without curvature variation, i.e., a circle, an obvious hypothesis is that it was the result of vertically *squashing* the circle. This means that points *E* and *F*, as placed on the original circle, moved along the arrows shown in Fig 6.95b, arriving at their present position.

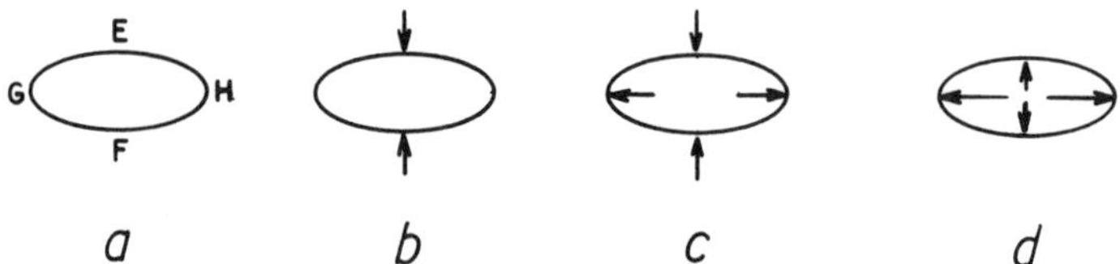

Figure 6.95
The four extrema, on an ellipse, being assigned different processes.

That is, the arrows are the *traces* of those points. Thus the traces are *external* and *inwardly-directed.*[20]

Now consider points *G* and *H* on the ellipse. The most likely interpretation for these extrema is that, as placed on the original undeformed shape, i.e., the circle, they moved outward, creating the traces shown in Fig 6.95c, i.e., *internal* and *outwardly-directed* traces.

We shall now see that, whereas the second pair of traces (at *G* and *H*) could have been created by the SAT or SLS, the first pair of traces (at *E* and *F*) could not. That is, the SAT and SLS cannot infer *squashing* processes.

To see why, observe first that any cross-section of an ellipse must lie inside the ellipse. This means that any SLS-axis (the locus of cross-section midpoints) must lie inside the ellipse. Thus the four SLS-axes, associated with extrema *E, F, G* and *H*, must be as shown in Fig 6.95d. However, observe that *internal* axes at *E* and *F* would not be viable *traces* of processes. A similar type of argument leads to the rejection of the SAT.

It appears therefore that, to infer the most likely traces, i.e., those shown in Fig 6.95c, we need a symmetry analysis that has a rather strange property: It must create internal axes at *G* and *H*, and external axes at *E* and *F*. We shall say that the analysis must be *extremum-sensitive*; i.e., sensitive to the type of extremum to which it is moving—that is, it must *switch sides* depending on the type of extremum.

We shall now see that our new symmetry analysis, PISA, has the property of extremum-sensitivity. Let us return to the ellipse. Fig 6.96a shows the right-half of the ellipse. In Fig 6.96b, the trajectory of doubly-tangential circles, for this segment, has been drawn. Furthermore, a dot on each circle marks the PISA symmetry point, i.e., the circumference point *Q*. Finally, in Fig 6.96c, we show the trace of circumference points *Q*. Thus, under PISA, the axis is as we would want it for extremum *H*; i.e., out-

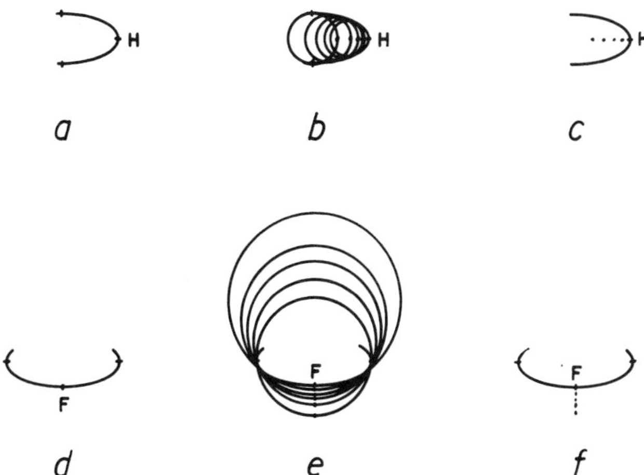

Figure 6.96
(a, b, & c) The right-hand extremum, on an ellipse, being assigned an axis by the PISA. (d, e, & f) The bottom extremum, on an ellipse, being assigned an axis by the PISA.

wardly moving but internal. Note that, for this extremum, the SAT and SLS would each produce an axis with these properties.

In contrast, consider now the bottom half of the ellipse shown in Fig 6.96d. Here the central extremum is *F*. Fig 6.96e shows the doubly tangential circles that arise from this segment. The important thing to observe is that each circle *exscribes* (lies outside) the ellipse. Thus the circumference points *Q* must necessarily lie outside the ellipse. These points are the dots shown on the circles in Fig 6.96e. Finally, Fig 6.96f gives the dots on their own. Thus we see that the axis created by PISA is external and inwardly-directed at *F*. In contrast, the SAT and SLS produce axes above *F*, because the cross-sections lie inside the ellipse.

A mathematical proof that PISA is *generally* extremum-sensitive, i.e., not just on ellipses, can be obtained by putting together a number of proofs in Leyton (1987b). In particular, the latter paper proves that, surrounding any M^+ or m^- extremum, there is a region of curve on which any doubly-tangential circle must inscribe the region. Conversely, the paper also proves that, surrounding any M^- or m^+ extremum, there is a region of curve on which any doubly-tangential circle must exscribe the region. The extremum-sensitivity of the new analysis follows immediately.

6.39.3 The Inference of Protrusion

Indentation and protrusion are figure-ground reversals of each other. However, whereas the SAT and SLS do very badly on indentation, they usually do better on protrusion. The reason is as follows: As can be seen from Fig 6.93, the discontinuity problems with the SAT and SLS, on indentation, arise because, at the entrance to an indentation, the contour switches from being outside to being inside the general body of the shape. In protrusions, this is not usually the case; i.e., the contour remains outside, all along the protrusion.

However, severe problems can arise when protrusions do not have the latter property. For example, let us consider a figure-ground reversal of the shape in Fig 6.93. That is, the shape is now that of a hole with a protrusion extending into it; e.g., a throat with a flap. Because symmetry analyses act on the curve information only, and are therefore impervious to the figure-ground structure, the same topological consequences will result from applying the three symmetry analyses here, as resulted when we considered the structure to be that of an indentation. That is, the SAT and SLS will generate the same problems, and PISA will not. Thus we conclude that PISA is, on the whole, better behaved on protrusions than the SAT and SLS are.

6.39.4 The Inference of Internal Resistance

Internal resistance is the figure-ground reversal of squashing. Furthermore, the topological consequences that resulted from applying the three symmetry analyses to a squashing situation depended on properties of the curve arbitrarily close to the extremum. Thus, these properties will also occur in an internal resistance situation. Therefore, we conclude that the SAT and SLS are incongruous for internal resistance as universally as they are incongruous for squashing. Correspondingly, PISA remains well-behaved.

6.39.5 PISA as Boundary-Pushing

There is another reason why PISA is more appropriate for the description of axes under the boundary-pushing scenario. Note that an axis is supposed to be the trace of a process. By *trace* one means here the trace of some physically significant point. Observe that an extremum on a contour is a physically significant *boundary point*. However, the PISA symmetry point, Q, is also a boundary point: it lies on the circumference of a disc (Fig 6.97). Thus, as illustrated in Fig 6.97, PISA is a trajectory of boundary points Q of which the last is the extremum. Therefore PISA can be regarded as the record of the boundary-movement itself.

We can make these observations more precise. Move the circle along the contour, towards the extremum, and stop the circle at a pair of points A and B, as in Fig 6.97. Because movement towards the extremum represents forward time, the contour up to points A and B can be regarded as representing time up to those points. The question is: At the moment represented by A and B, what was the remainder of the boundary? This information is unavailable because that moment was in the past. Thus, one has to conjecture a remainder that involves the least amount of informational commitment. This remainder is the circle-segment between points A and B, because a circular segment has no curvature-variation and thus has the least information (in the Shannon sense of information, i.e., variety). Now as the circle moves further along the curve, more information is obtained about the contour, while the curve's completion is always a circular arc because, in the absence of further information at that point, the circular completion involves the least informational commitment. Eventually, A and B together reach the extremum, where the least commitment becomes a point.

Thus, PISA can be regarded as representing a recapitulation of boundary movement under a continuous strategy of minimal informational commitment.

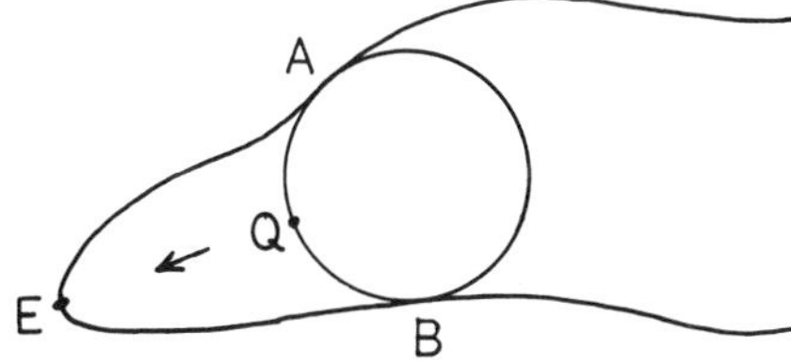

Figure 6.97
The PISA as a boundary point.

6.40 Functional Appropriateness of Symmetry Analyses

If we can regard PISA as tracking a physically significant point, the presumptive curvature extremum, under the process, we can regard the SAT and SLS as each tracking other physically significant points under the process: The SAT tracks the circle center and the SLS tracks the chord midpoint.

However, the strange thing is that, in the boundary-pushing scenario, these latter tracks are meaningless. Consider for example the ellipse in Fig 6.98. The arrows show the axes of the SLS. Although we can regard the SLS as tracking the chord-midpoint, neither of the vertical tracks shown is meaningful under the vertical squashing processes.

The consequence is that, although each symmetry analysis can be regarded as tracking a specific point under a process, we cannot understand each of them as carrying out this function universally across processes. Under some processes the results are meaningless.

This reinforces our claim that the validity of a symmetry analysis depends on its function relative to a particular situation; i.e., in contrast to the current view that there is a single correct symmetry analysis. We argue that, not only does each analysis encode its own specific structural information, but this structural information is meaningful only in particular situations. To understand this more deeply, let us now consider the functional appropriateness of PISA and SLS.

First, we have seen that PISA can be regarded as tracking the boundary under the boundary-pushing scenario. Observe now the type of inference involved in this function: When we are given the shape, we make the assumption that we are presented with the state of the boundary at a single moment. That is, with respect to the boundary, we are making the single-state assumption. What PISA does is to reconstruct the previous states of the boundary. This means that PISA is an *external* inference rule; i.e., it reconstructs the past under the single-state assumption.

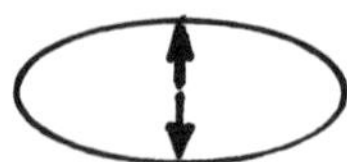

Figure 6.98
SLS axes on an ellipse are meaningless in the boundary-pushing scenario.

Now let us turn to the SLS. As we said, one can regard the SLS as tracking the cross-section over time and under a particular process. Note that this view is different from the view taken by Brady (1983) when he designed the SLS. Brady proposed the SLS as a static description of the shape and a description that is applicable to all shape. The difference between our view and Brady's can be traced ultimately to our dictum "Shape is time". That is, any shape-description is ultimately a process description of the image. Thus, a particular shape-description can be given only where its implied process description is valid.

Therefore, in turning the SLS into a temporal construct, we must try to find the particular type of process-structure to which it could be relevant. The obvious such structure is the cylinder scenario: This is because the SLS is the track of a cross-section midpoint and the cylinder scenario involves the movement of a cross-section.

We first need to check that the *same* cross-section is involved in the SLS as in the cylinder scenario. Recall that the cylinder scenario was constructed as a method of removing ASYMMETRY TYPE 2, the orientational distinguishability between symmetrically opposite tangents; e.g., the tangents at *A* and *B* in Fig 6.99a. The scenario made these tangents parallel, backward in time, i.e., producing a perfect cylinder. But two parallel tangents in the perfect cylinder must have belonged to the two end-points of the same cross-section in that cylinder. Thus, the cross-section between two symmetric points in the present shape can be regarded as *memory* of the cross-section of the perfect cylinder of the past. However, the cross-section in the SLS is also between two symmetric tangents in the present; e.g., those at *A* and *B* in Fig 6.99a. Therefore, the cross-section in the SLS can be regarded as memory of the cross-section in a past cylinder. Therefore, the SLS cross-section can be regarded as the cross-section of the cylinder scenario.

Now let us look at the scenario itself. Recall first that we established it to be the following: The cross-section AB and the mirror **M**, illustrated in Fig 6.99b, form a perpendicular pair of reflection axes; i.e., a **Reflections** × **Reflections** structure. This structure undergoes the processes **Stretches. Shears.Rotations**, as it moves along the axis. The processes are perceived as acting on the reflection pair via the Interaction Principles; i.e., the symmetry axes become eigenspace-lines. This process-structure completely defines the change from the perfect cylinder to the final shape. That is, each

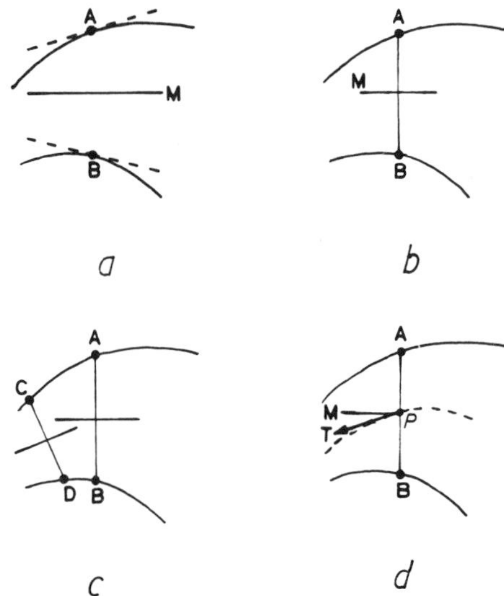

Figure 6.99
(a, b, & c) The cross-sections in the SLS are memory of the reflection structure in the cylinder scenario. (d) The SLS axis is memory of the translation structure in the cylinder scenario.

Reflections × **Reflections** pair in the perfect cylinder undergoes a specific amount of **Stretches, Shears,** and **Rotations** to produce the final shape.

Given this scenario, it is easy to see that the SLS *axis* encodes one aspect of it: the trajectory of the origin of the **Reflections** × **Reflections** structure. For example, the axis goes through the origins of the two **Reflections** × **Reflections** structures shown in Fig 6.99c. This trajectory is the dashed line in Fig 6.99d.

Let us define this structure more precisely. The trajectory of the origin is the trace of the **Translations** process that is part of the *internal* structure of the shape, i.e., the process that actually moves the cross-section along the shape. In short: the SLS axis is the trace of the **Translations** component of the internal structure.

Looking over the discussion in this section, we are therefore led to this conclusion: PISA can be regarded as defining part of the *external* structure of the *boundary-moving* scenario, and the SLS can be regarded as defining part of the *internal* structure of the *cylinder* scenario. Each symmetry analysis is functionally tied to its own particular process-structure, and not only expresses a different type of process, but expresses a different type of inference rule (external vs. internal) within its process-structure.

6.41 Physical Appropriateness of the Two Scenarios

Although this chapter has been concerned with domain-independent rules, we will now, as a short postscript, compare the cylinder scenario to the boundary-pushing scenario, with respect to their physical appropriateness for describing complex shape. Their basic differences lie in the way in which they describe both the process-structure of an individual *part*, and the process-relationships between the parts.

Let us consider complex biological shape, e.g., the shapes of animals. We have seen that the standard approach in Computational Vision, e.g., that of Marr & Nishihara, is to describe the parts of an animal using cylinders. Is this description, however, a viable one?

A cylinder structure is created by the movement of a cross-section along an axis. However, a part of an animal, e.g., an arm or head, is not created in this way. Fig 6.100a illustrates how a head would look while being created by the cylinder scenario: The cross-section would pass the mouth at some stage and later the eyes, like a water-level rising up a tank.

In order to understand the implications of this, we need to observe that the surface of the shape, under the cylinder scenario, is decomposed into two types: (1) surface that will be part of the external surface of the final object; and (2) surface that is a cross-section plane, and will later be internal to the final object. For example, the surface of the head in Fig 6.100b is

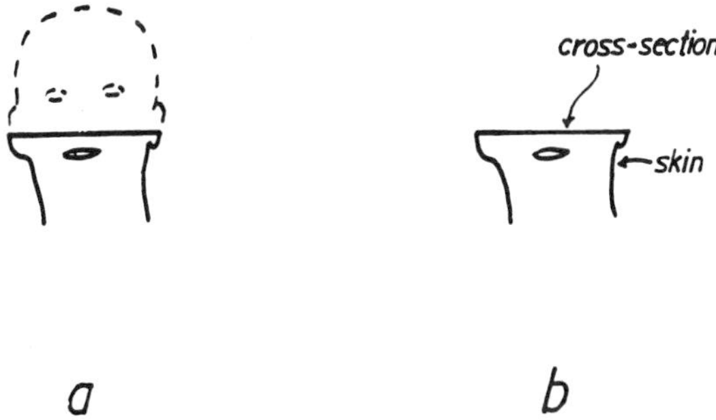

Figure 6.100
The cylinder scenario for a biological part is inappropriate.

decomposed as follows: The sides of the head are external surface, whereas the cross-section, at the top, will later be inside the completed head. We can call the two types of surface: *skin* and *cross-section*, respectively.

Now observe that, under this scenario, as can be seen from the head in Fig 6.100a, the skin appears incrementally, as a ring moving up the side of the part. This, of course, is an inappropriate scenario for biological shape. Furthermore, under this scenario, the *starting* shape, prior to the development of limbs, has the appearance of a torso on which the limbs have been sliced off. More precisely, the initial torso consists of a cylindrical skin surface (the trunk) that contains planar cross-sections (where the limbs will grow). This starting surface must therefore contain (first-order) discontinuities along the rings separating the planes from the cylindrical portions.

In contrast, real biological parts are formed by starting with a smooth homogeneous surface that deforms outward, creating the parts as curvature fluctuations in the smooth surface. At all stages, the surface consists of a single unbroken external skin. These properties are exactly captured in the boundary-pushing scenario. Furthermore, under the latter scenario, the surface is pushed out at points that become the curvature extrema at the tips of limbs. This is exactly what happens in real biological shape, as follows:

French, Bryant, & Bryant (1976) and Bryant, French & Bryant (1981), have shown that, before the formation of a limb, the surface of an embryo, in the region of the future limb, is initially homogeneous, and then becomes re-organized as circularly symmetric around a point in the surface. This point then pushes out becoming the curvature extremum at the tip of a limb. Thus the Symmetry-Curvature Duality Theorem, i.e., the correspondence between single symmetry axes and single curvature extrema, dictates the development of an actual physical limb (i.e., axis) in biology, as it does the boundary-pushing scenario in perception. We conclude that the boundary-pushing scenario is much more appropriate for biological shape than the cylinder scenario.

Cylinder scenarios tend to be correct only in manufactured objects, where growth is not involved. This is why Marr & Nishihara's representation of a man, in Fig 6.73, looks like a manufactured object. It looks crude because we easily infer that its process-structure is not only *incorrect* for a man, but is *degenerate*, in this sense: To *manufacture* a complex object out of generalized cylinders, one usually produces the cylinders separately and then simply *attaches* them to each other. In contrast, the boundary-moving scenario involves the progressive development of the limbs out of a single initial shape. That is, what is simply attachment, in the Marr & Nishihara system, is a complex development over time, in the boundary-movement scenario. The Marr & Nishihara diagram of a man (Fig 6.73) appears crude to us because we do not perceive the diagram as containing the developmental scenario.

6.42 Summary

The history that is assigned to a shape is clearly dependent on the type of materials that are assumed to be involved. Nevertheless, we argue that there are rules that are valid irrespective of the material domain under consideration; i.e., domain-independent rules. For example, our three symmetry based rules—the Asymmetry Principle, the Symmetry Principle, and History Minimization Principle—are domain-independent. So is our Externalization Principle which states that the removal of external history always reduces the internal structure to purely Euclidean processes.

In order to elaborate further domain-independent principles we began by examining in detail the trace-structure of regular polygons. Recall that the importance of these shapes is that they are a major subclass of those that are determined by the Externalization Principle. Because these shapes emerge from the use of the Asymmetry Principle, they have a number of symmetries. In accord with the Symmetry-to-Trace Conversion Principle, these symmetries can be turned into traces. We argued that this conversion is realized in three stages: Stage 1 converts the symmetry transformations, in the external state-transition diagram, into transformations that act internally between the parts of the figure. Stage 2 then takes this internalized state-transition diagram, and *initializes* it; i.e., picks a distinguished starting state. As a result of this choice, each of the other states is labelled by the *actions* that obtained it from the initial state. Finally, Stage 3 ensures that any action is incorporated under another action whenever this is possible, rather than letting the actions be carried out independently. This incorporation structure is fully realized by the subsumption of actions under repetitions of a minimal set of generators, i.e., under the color graph of the group describing the symmetries. In the color graph, each cycle of one color corresponds to a cyclic group, and the latter corresponds to a process. Thus the subsumption stage can be understood as realizing the Subsumption Strategy which states that all actions should be incorporated under processes.

The above theory concerns the *internal* structure of fully externalized histories. Although, the theory applies to any regular polygon, much of the theory was illustrated by considering a square. We also examined the *external* component in the fully externalized history between the square and the rotated parallelogram. Remarkably, we found that, even though the external component is produced by different procedures from the internal component—i.e., removing asymmetries externally rather than internally—the resulting representation of the external history has two important properties in common with the above internal structures: (1) the presented stimulus is placed within an initialized state-transition diagram of a machine; and (2) the actions in the state-transition diagram are subsumed under processes. Thus, the explanation of the presented parallelogram, in terms of rotation, shear, and stretch, is the decomposition of the space of actions into cyclic groups, i.e., processes.

We then briefly surveyed the notion of reference frame in perceptual psychology, and made a claim, the Reference Frame Hypothesis, that was validated in the subsequent sections of the chapter. This hypothesis states that a reference frame is a programmed machine that is imposed on a stimulus set by defining the stimuli as states of a machine, i.e., the fulfillment of the Representational Machine Analogy. In particular, we claimed that a *perceptual* reference frame is an initialized color graph of the input space of a machine. The map that sends the reference frame onto the stimulus set is called the *explanation map*. Those stimuli instantiated in the color graph are called the *instantiated* components of the map. The rest of the frame (color graph) maps to the empty set, but is nevertheless psychologically understood to be present in the sense that the stimuli are seen as embedded in the frame.

We then defined the notion of process-coordinate; i.e., the amount of process that was used in the creation of a stimulus. For example, each point of a regular polygon has two coordinates, one in the translations process tracing a side, and the other in the rotation (or reflection) process that created the orientation of the side. The coordinates are mapped *between* the sides and these mappings are themselves mapped, by external history, to mappings between sides in deformed shapes.

With these concepts one can see that the principles we have extracted concerning the fully externalized histories can apply to alternative histories that are not fully externalized. For example, we saw that the minimal fully internalized history that traces out the parallelogram *directly*, involves the same nested hierarchy as the fully externalized history, but with the external sequence preceding the internal one in the temporal order, i.e., the external one alters the frame in which the internal one will be implemented. We also distinguished between *outcome transfer* and *program transfer*: In the former the *outcome* of a realized history is transferred to another moment and therefore does not have to be carried out again; whereas in the latter, the *program* for a history is transferred to another moment and does have to be carried out to produce an outcome. We saw that a nested hierarchy, i.e., a hierarchy of mappings, yields either interpretation.

We then developed a theory of grouping. We proposed that a grouping is simply a history up to a certain stage, and this means that it must be given by a coset of a left-subsequence in the process-group hierarchy. We then examined how the perceptual system evaluates one grouping as better than its alternatives and thus rejects the alternatives. The Gestalt school offered at least eleven criteria of evaluation. We argued that these criteria fall into two classes, where (1) the members of one class are simply manifestations of the Subsumption Component of the History Minimization Principle, and (2) the members of the other class are simply manifestations of the Inertial Component.

We next turned to an examination of how internal history is transferred by external processes. We first found that the ordering of the **Stretches**, **Shears**, and **Rotations** processes in the external history of a rotated parallelogram corresponds to decreasing eigenspace dimension, and that the eigenspace lines are coincident with the symmetry axes of a square. Generally we proposed two Interaction Principles: (1) The symmetry axes of a configuration become eigenspaces of the most structurally allowable processes applicable to the configuration; and (2) the salience of the symmetry axes corresponds to allowability of the associated eigenspaces. These principles are a method of realizing the Inertial Component of the History Minimization Principle, i.e., the requirement that change be minimal. The above Interaction Principles are forwards versions of a correspondence between symmetry axes and eigenspaces. From Chapter 1, one sees that the backwards versions are also valid. However, rather than following from the Inertial Component, the latter follow from the Symmetry Principle, i.e., the requirement that symmetry be preserved backwards in time.

In order to investigate more deeply the importance of the Interaction Principles we then examined the phenomenon of Cartesian reference frames. Our basic proposal concerning Cartesian reference frames is the Subhistory Principle which states that the basis for the psychological impression that such a frame has been imposed on a stimulus set, is that the full process-structure defining the stimulus set contains the process-structure defining the frame.

We then turned to solving the orientation-and-form problem. We found that the problem is one of the assignment of *history*: Different external histories determine different assignments of internal history. We argued that the assignments are determined via the Interaction Principles. In most cases, the salient internal structure is the same in the various assignments because it is the internal structure of the Cartesian frame. This allows the selection of the histories to be made purely by the application of the Inertial Component of the History Minimization Principle to the different sets of external coordinates in the alternative histories.

We then argued that the process-structure of a Cartesian frame can be perceptually embedded within itself recursively. Furthermore, the succession of embeddings conforms to the Interaction Principles. Such structures were used to explain the phenomenon of fields of pointing triangles.

We then turned to motion perception, and began by looking at Duncker's induced motion phenomenon, in which a moving rectangular frame with a fixed dot is seen as a fixed frame with a moving dot. The phenomenon is easy to explain using the Inertial Component of the History Minimization Principle: Since the frame has more points, to see it move would involve describing more change. We observed also that the frame correspondingly involves greater scaling symmetry, and proposed the Mass Principle which states that scaling symmetry corresponds perceptually to mass. As a result of this, we saw that the Inertial Component of the History Minimization Principle subsumes the explanatory role of inertia in Newton's mechanics, which in effect holds that greater scaling symmetry implies greater resistance to change. The Inertial Component is more general because it implies that any symmetry has resistance to change.

We then considered the Johansson (1950) motion phenomenon, in which the perpendicular motion of two dots is seen as decomposed into the relative motion of the dots along one diagonal, and the common motion of the dots along the opposite diagonal (Fig 6.71a). We saw that the phenomenon obeys the Interaction Principles, i.e., the dot-pair has two reflection axes which become eigenspace-lines of the relative and common motion respectively. Again, the phenomenon obeys the Subhistory Principle which states that the imposition of a Cartesian frame is the inclusion of the process-structure of the frame into the full process-structure of the configuration. We observed that this configuration involves a two-level Cartesian frame bundle. Finally, we observed that the phenomenon obeys the Grouping Principle, because the individual rod groupings are cosets of left-subsequences as predicted by the principle.

Our discussion then turned to complex shape, which occupied us for the rest of the chapter. We began by examining generalized cylinders. Elaborating the process-structure of such cylinders, we found that they obey the Interaction Principles: the reflectional symmetry axes of the cross-section become the eigenspaces of the sweeping rule (i.e., the widening and narrowing of the cylinder), and the rotational axis of the cross-section becomes the eigenspace of the spine (the curve along which the cross-section is translated).

Remarkably, we found that generalized cylinders have the same structure as the Johansson motion phenomenon. In fact, the *sweeping rule* and *spine* of the cylinder correspond, respectively, to the *relative motion* and *common motion* vectors of the Johansson phenomenon.

Generalized cylinders are a purely internal structure. However, one can convert the **Stretches** component, in the cylinder, from an internal process to an external one, thus producing a deformed-cylinder scenario. Unlike the generalized-cylinder scenario, this alternative one has an intervening prototype, i.e., the perfect cylinder. In fact, Marr and Nishihara use this scenario rather than the generalized-cylinder one, which they claim they use. Observe that the deformation scenario conforms to the Externalization Principle, which states that the purpose of externalization is to reduce the internal structure to a nested hierarchy of Euclidean processes.

The current view of the cognitive role of generalized cylinders is that they are models that cognition possesses. In contrast, we argued that generalized cylinders do not exist as models, but are the *result* of applying the Asymmetry and Symmetry Principles. Our argument was as follows. There are two types of asymmetry in any part of the boundary associated with a single axis: Type 1, the distinguishability in orientation of the local mirrors defining the axis; and Type 2, the distinguishability in orientation between opposite tangents of the boundary. After the removal of Type 1, the removal of Type 2 can be realized by one of two methods: Method 1, in which the tangents are made parallel, leading to the cylinder scenario; and Method 2, in which the tangents are made collinear, leading to the boundary-pushing scenario.

Method 1 implies the following internal organization. The double-reflection structure consisting of the cross-section and its perpendicular undergoes (1) **Stretches** as the cross-section is widened and narrowed, (2) **Shears** as the local symmetry axis of the perfect cylinder becomes the tangent to the curved axis of the spine, (3) **Rotations** as the spine curves. Thus we saw that the shape is characterized by the factors defining the Cartesian frame. We then considered what happens to a shape when one tries to remove each of the factors of **Stretches.Shears.Rotations** individually and successively in all possible ways. In the use of these processes globally, a shape results at each possible stage. However, according to our Limitation Theorem, when one uses these processes locally, only two possible shapes can result: globally symmetric shapes like goblets and locally parallel shapes like worms.

We then considered the boundary-pushing scenario, i.e., the removal of Type 2 asymmetry, the orientational variation in the boundary. This turns out to be the scenario that was investigated in detail in Chapter 1. Thus it was unnecessary to consider this scenario further on its own. However, we then began a comparison of this scenario with the cylinder scenario.

It was first necessary to develop a theory of the structure of local-symmetry analyses. We argued that a local symmetry analysis should be regarded a mirror set together with a rule for selecting an observation arrangement for each mirror in the set. The rule is called the observation selection rule. Given this view of local symmetry, it was then possible to understand that the SAT and SLS are local symmetry analyses each defined by a particular consistent observation selection rule.

We then attempted to determine the appropriate symmetry analysis for the boundary-pushing scenario. Since there are four types of extrema, there are four types of processes in this scenario: protrusion, indentation, squashing and internal resistance. We compared the SAT, SLS, and PISA, for each of the processes and found that, where the SAT and SLS have serious topological problems, PISA does not. PISA is also appropriate for the scenario because it can be regarded as the trace of the boundary under a continuous strategy of minimal informational commitment.

With respect to symmetry analyses generally, we argued that, not only does each symmetry analysis encode a different structural factor but that factor can be either meaningful or meaningless in different situations. In particular, we saw that PISA can be regarded as defining part of the external structure of the boundary-moving scenario, and the SLS can be regarded as defining part of the internal structure of the cylinder scenario. Finally, we saw that, while the cylinder scenario might be appropriate for some manufactured objects, the boundary-pushing scenario is appropriate for biological shape. In fact, although cylinder representations have been widely advocated for biological shape, they actually look crude because, as historical representations, they are incorrect and degenerate, in contrast to the boundary-pushing scenario.

7 Linguistics

7.1 Introduction

In this chapter, we are going to argue that language is process-history, and that understanding language is the unpacking of that history. In other words, language understanding is an example of the *process-inference problem*—the central problem of this book. To put it another way, language conforms to our Representation-is-Explanation Principle (section 4.1); i.e., human beings represent linguistic utterances as states in a process-history that causally explains those utterances. We proposed in section 4.1 that the Representation-is-Explanation Principle *characterizes cognition* in this sense: A system is a cognitive system if it can be described as having a level in which the medium that it processes consists of causal explanations. Thus, in accord with this principle, we argue that linguistic representations are causal explanations.

Our view of language dictates a structural approach to syntax that is, in important ways, different from that of the Chomskian school. Furthermore, we shall argue that certain central results produced by that school, results that are not fully motivated or are described in a contradictory fashion, gain a clarity and proper foundation when understood or redescribed within the view that we shall try to elaborate.

Let us consider an individual sentence. Our view is that a sentence is an *embodiment of time*. It is process-memory. Think of a sentence as a geological entity—for example, a rock compounded of lava fused with other types of deposits, layered, scratched, scarred, deformed, twisted, etc. Our proposal is that, when one hears a sentence, one unpacks time—in much the same way as one does when one examines a rock formation and peels away the successive deposits, scars, deformations, etc.

Much of contemporary syntax is concerned with what the *constituents* of a sentence are and how they are organized. In this chapter we shall argue that constituents are the layers of time, and that constituent structure is process-history.

This equation between constituent structure and process-history is actually obscured in current formal theories of syntax. One might object by saying that current theories are *generative* and thus *process-based*. However, we shall argue that the current view of generativity is an inappropriate one and actually obscures the process-history comprising a sentence, rather than revealing it. It is when one sees the sentence *literally* as compounded of layers of time, like a geological or archeological relic, that one can gain a clearer understanding of how to approach and analyze syntactic structure. The generative rules that then emerge are different from the conventional ones, both structurally and functionally.

7.2 Current Concepts of Constituent Structure

The study of syntax is founded on the notion of *constituent structure*. In this section, we briefly recapitulate some basic concepts concerning such structure, for the reader unfamiliar with syntax. The concepts we review are: (1) the notion that a syntactic category is a distributional unit; (2) the use of Phrase-Structure Rules to define constituent structure; (3) some basic ideas of X-Bar Syntax. Those readers familiar with these concepts can proceed to section 7.3, where we return to the notion of process-history.

(1) Syntactic Categories

Syntax is the branch of linguistics that is concerned with the way in which words and phrases are distributed through a sentence. To illustrate this, consider the following sentence:

The decision on the divorce could not be made by the lawyer

In this sentence, the word *the* occurs three times in three particular positions. The word could not have occurred in other positions. For example, it could not have occurred in any of the positions marked by * as follows:

The decision * on the divorce * could * not * be * made *
by the lawyer *

Thus the word *the* is highly restricted with respect to its *distribution*.

On the other hand, in the very positions in which the word *the* is allowed to occur, one can substitute another word for it, the word *a*, thus:

A decision on *a* divorce could not be made by *a* lawyer

Thus the words *the* and *a* have the same distributional properties; i.e., they can occur at the same positions in sentences. Entities that share the same distributional properties are said to belong to the same *syntactic category*. Thus the words *the* and *a* belong to the same syntactic category because they can appear, interchangeably, in the same set of positions in sentences.

Words are not the only entities that can belong to syntactic categories. Phrases also have distributional properties, i.e., they occur at certain positions and not in others. For example, consider the phrase *the angry actor* in this sentence:

THE ANGRY ACTOR turned to look at the director

It occurs at the beginning of the sentence, but it cannot be placed at most of the other positions in the sentence. Nevertheless, it can be placed in the position held by *the director* at the end of the sentence. Furthermore, *the director* can be placed in the position that is held by *the angry actor* at the beginning of the sentence. Thus, we can simply interchange *the director* and *the angry actor* thus:

THE DIRECTOR turned to look at THE ANGRY ACTOR

This means that *the director* and *the angry actor* have the same distributional properties; they can occur in the same set of positions as each other. Thus, we can again say that they belong to the same *syntactic category*.

Generally, therefore, one can define a syntactic category thus:

> A *syntactic category* is a set of entities that have the same distribution; i.e., the entities can occur interchangeably in a number of positions within a sentence and within different sentences.

(2) Phrase-Structure Rules

Standard arguments in linguistics show that the entities or constituents that form a sentence are hierarchically organized. The hierarchical organization of a sentence is given by what are called *Phrase-Structure Rules*. Such a rule specifies how a constituent is broken down into its sub-constituents. The Phrase-Structure Rules are applied successively starting with the largest entity, the whole sentence, breaking this down into constituents, then breaking these constituents into their constituents, then these constituents into their constituents, and so on, until one has broken everything down into the smallest allowable constituents, the single words.

Thus for example, let us consider the following sentence:

The restless teacher would wait in the cold classroom

The entire sentence is denoted by the letter S. The first Phrase-Structure Rule specifies that S is broken down like this:

$$S \rightarrow NP\ AUX\ VP$$

This rule reads, from left to right, as follows: "S is broken down into the Noun Phrase, NP; followed by the Auxiliary, AUX; followed by the Verb Phrase, VP." In the above sentence, these three constituents are:

NP = The restless teacher
AUX = would
VP = wait in the cold classroom

Now, these constituents can, in turn be broken down into their sub-constituents, by further rules. Thus the Noun Phrase, NP, can be broken down using this rule:

$$NP \rightarrow DET\ A\ N$$

From left to right, this sequence of symbols reads "the Noun Phrase, NP, is broken down into the Determiner, DET, followed by the Adjective, A, followed by the Noun, N." In our example, where the NP is *the restless teacher*, these constituents are

DET = the
A = restless
N = teacher

Now recall that, in our sentence, *The restless teacher would wait in the cold classroom*, there is a constituent called a Verb Phrase, VP; that is, the phrase, *wait in the cold classroom.* This constituent is itself broken down by the following rule:

$$VP \rightarrow V\ PP$$

From left to right, this sequence of symbols reads "the Verb Phrase, VP, is broken down into the Verb, V, followed by the Prepositional Phrase, PP." In our example, where the VP is *wait in the cold classroom*, these constituents are

V = wait,
PP = in the cold classroom.

There is now only one constituent left that has not been broken down into single words—the final constituent mentioned, the Prepositional Phrase, PP; that is, *in the cold classroom*. This constituent is itself broken down by the following rule:

$$PP \rightarrow P\ NP$$

From left to right, this sequence of symbols reads "the Prepositional Phrase, PP, is broken down into the Preposition, P, followed by the Noun Phrase, NP." In our example, where the PP is *in the cold classroom* these constituents are

P = in

NP = the cold classroom.

Now observe that although the last mentioned constituent, the NP, *the cold classroom*, is not itself broken into words, we already have a rule that will break it into words. The rule is

$$NP \rightarrow DET\ A\ N$$

given earlier, where DET is the determiner, A is the Adjective, and N is the noun. Applying this rule to our final Noun Phrase, *the cold classroom*, these constituents are

DET = the,

A = cold

N = classroom.

Using the rules given above, the Phrase-Structure Rules, we have given a complete specification of how the overall sentence S is broken down successively into constituents. The four rules we used are

$$S \rightarrow NP\ AUX\ VP$$

$$NP \rightarrow DET\ A\ N$$

$$VP \rightarrow V\ PP$$

$$PP \rightarrow P\ NP$$

Let us finally emphasize the close relation between syntactic categories and Phrase-Structure Rules. As we said, a syntactic category is a set of entities with the same distributional properties. A Phrase-Structure Rule specifies how a constituent is broken down into its sub-constituents. The relationship between these two ideas is that the constituents correspond to syntactic categories. That is, the constituents are defined as *distributional* items, and these distributional items are structured hierarchically by the Phrase-Structure Rules. For example, at the end of the previous paragraph, we listed a number of Phrase-Structure Rules. Each rule specifies how a certain type of constituent is broken down into a set of constituents. What we should observe is that, in any of the rules, the constituents mentioned, for example, NP, VP, PP, DET, N, AUX V, P, etc., are all syntactic categories; i.e., items that are defined distributionally.

(3) X-Bar Syntax
Like Phrase-Structure Grammar, X-Bar Syntax is a means of describing the hierarchical constituent structure of a sentence by defining how constituents are broken down into sub-constituents. X-Bar Syntax is now widely favored as a replacement for Phrase-Structure Grammar. The reasons are as follows:

Intermediate Level Constituents

In Phrase-Structure Grammar, constituents are either word-level categories such as Nouns (N), Verbs (V), Prepositions (P), Adjectives (A), Adverbs (ADV), etc., or phrase-level categories such as Noun Phrases (NP), Verb Phrases (VP), Prepositional Phrases (PP), Adjectival Phrases (AP), Adverbial Phrases (ADVP), etc. In X-Bar Syntax, there are categories intermediate between these two levels.

For example, consider the phrase,

a wise ruler.

It is a Noun Phrase; and a Noun Phrase is considered to be a syntactic category because it can occur as a structural unit at a number of points in a sentence and in different sentence types. However, this property holds also for the "sub-phrase"

wise ruler,

i.e., without the initial word *a*. For example, one has sentences like:

WISE RULER though he is, he loves practical jokes.

and

He is a patient man and WISE RULER.

Several distributional arguments like this persuade linguists that there are intermediate levels of constituents between Noun Phrases like *a wise ruler* and Nouns like *ruler*. In fact, any phrase is believed to contain a particular constituent called a *Head* and intermediate levels that are represented by "bars" on the Head, as can be illustrated thus: In a Noun Phrase, the Head is the Noun and is represented by N; the intermediate levels are represented by $\overline{\text{N}}$, $\overline{\overline{\text{N}}}$, . . . , etc., called respectively, N-bar, N-double-bar, . . . , etc. Thus, for example, in the phrase, *the wise ruler*, the Head is *ruler*, which is a Noun, N. The intermediate constituent *wise ruler* is a $\overline{\text{N}}$; and the whole phrase *the wise ruler* is a $\overline{\overline{\text{N}}}$.

Separation of Constituency Rules from Linear-Order Rules

In Phrase-Structure Grammar, the rules specify not only how a constituent is broken down into its sub-constituents, but how the sub-constituents are ordered. For example, the rule

$$\text{NP} \rightarrow \text{DET A N}$$

specifies not only that a Noun Phrase such as *the strong horse* breaks down into a Determiner *the*, an Adjective *strong*, and a Noun *horse*, but that the Determiner precedes the Adjective, and the Adjective precedes the Noun. However, in different languages, the constituents occur in different orders. Thus it has been argued, starting with Staal (1967) and Sanders (1969, 1970a, 1970b, 1971, 1972a, 1972b), that any constituency rule (such as that above) should not contain information as to the order of constituents, and there should be a separate set of rules that determine the linear order of constituents in a language. Thus the above rule should specify only an unordered list of sub-constituents:

$$\text{NP} \rightarrow \text{DET, A, N}$$

i.e., commas are inserted to indicate an unordered list in contrast to an ordered one.

Category Neutrality in X-Bar Syntax

In X-Bar Syntax, the number of intermediate constituents is not arbitrary. For example, Noun Phrases are generally assumed to have a Head, N, and only two subsequent levels of constituent, $\overline{\text{N}}$ and $\overline{\overline{\text{N}}}$, where the $\overline{\overline{\text{N}}}$ is the final level, i.e., the Noun Phrase itself. Thus the phrase

the student of physics at the party

previously called a Noun Phrase is now called a $\bar{\bar{N}}$, and is broken down by three successive rules. First, $\bar{\bar{N}}$, the whole phrase, is broken down thus:

$$\bar{\bar{N}} \rightarrow DET,\ \bar{N}$$

where DET is *the*, and $\bar{N}$ is *student of physics at the party*. The DET is called a *Specifier*. The second constituent, $\bar{N}$, is then broken down thus:

$$\bar{N} \rightarrow \bar{N},\ PP$$

where the $\bar{N}$ after the arrow is *student of physics* and the PP is *at the party*. The second constituent is called an *Adjunct*. Finally, the $\bar{N}$, *student of physics*, is broken down by the rule

$$\bar{N} \rightarrow N,\ PP$$

where N is *student* and PP is *of physics*. The second constituent is called a *complement*.

To recapitulate: (1) $\bar{\bar{N}}$ is broken down into $\bar{N}$ and a *Specifier*; (2) $\bar{N}$ is broken down into a second $\bar{N}$ and an *Adjunct*; and (3) the second $\bar{N}$ is broken down into N, the Head, and a *Complement*.

We can now state one of the central principles of most versions of X-Bar Syntax (e.g., Stolwell, 1981; Chomsky, 1986a, b). The principle is this: The constituent structure just described is true not just for Noun Phrases but for all other phrase categories; i.e., for Verb Phrases, Prepositional Phrases, Adjectival Phrases, Adverbial Phrases, etc. That is, any phrase is of the form $\bar{\bar{X}}$, where X is the *Head* of the phrase, and $\bar{\bar{X}}$ is broken down by three successive rules. First, $\bar{\bar{X}}$ is broken down into $\bar{X}$ and a Specifier, thus:

$$\bar{\bar{X}} \rightarrow \bar{X},\ Sp$$

Then, $\bar{X}$ is broken down into a second $\bar{X}$ and an Adjunct, Ad, thus:

$$\bar{X} \rightarrow \bar{X},\ Ad$$

Finally, the second $\bar{X}$ is broken down into the Head X and a Complement, Co, thus:

$$\bar{X} \rightarrow X,\ Co$$

(Strictly speaking: (1) In the first rule, Sp is optional, (2) the second rule is optional as a whole; and (3) in the third rule, Co can consist of several complements.)

7.3 Replacing Current Constituency Rules

In this section, we are going to argue that Phrase-Structure Grammar and X-Bar Syntax are *inappropriate* to describe constituent structure. Both types of rules are most closely associated with the approach to syntax invented by Noam Chomsky—and we will thus refer to both types of rules as Chomskian Constituency Rules. We will argue that, although the Chomskian Constituency Rules form generative systems, they are insufficiently generative, and when they are generative, they are generative in the wrong way. We will propose a generative system that has entirely different properties, and is more fully generative.

To begin: It is clear that a sentence consists of a number of distinguishabilities. For example, consider the phrase *the angry student*. The words, *the*, *angry*, and *student* are distinguishable from each other, in many ways. In particular, they are syntactically distinguishable in that *the* is a Determiner, *angry* is an Adjective, and *student* is a Noun.

Observe now that Phrase-Structure Grammar accounts for the phrase *the angry man* by using the rule

$$NP \rightarrow DET, A, N$$

That is, the grammar dictates that the above rule was involved in generating the phrase, *the angry man*.

Observe that the rule starts with a single item NP, and produces a number of distinguishable items, DET, A, N. That is, the rule *creates distinguishabilities*. This means that the Phrase-Structure Rule conforms to our Asymmetry Principle: distinguishabilities are seen as arising from prior indistinguishabilities. This is true of any Phrase-Structure Rule, or any of the rules of X-Bar Syntax; i.e., they generate distinguishabilities from indistinguishabilities. Thus we can argue that the Asymmetry Principle is at work here on a more fundamental level than the rules themselves. That is, despite there being several different rules, and indeed two alternative rule systems—the Phrase Structure Rules and X-Bar Syntax—all the rules conform to the Asymmetry Principle. Although this point might at first seem trivial, its importance will become evident as the discussion proceeds.

The rules represent the intuitions that linguists have about syntactic structure. We see that instinctively, the linguists use the Asymmetry Principle—the progression from indistinguishability to distinguishability—as a basis for all the rules. However, are the linguists correct in the particular rules that they choose? Could they be choosing more appropriate rules that equally conform to the Asymmetry Principle?

To examine this question, consider again the rule

NP → DET, A, N

as part of the generation of the phrase *the angry student*. Observe that the assignment of this rule is an example of *external* inference. In going backwards, through the generation, the set of three items, DET, A, N, are replaced by an item NP not contained among the three. That is, in inferring what generatively preceded DET, A, N, one goes outside the set DET, A, N; i.e., one makes an external inference. Note that the strict definition of external inference is that it is based on the *single-state assumption*. In the above case, because the state—DET, A, N—does not contain the immediately prior state, NP, or any state prior to that, the state contains only itself; i.e., it conforms to the single-state assumption.

However, observe now that, when one examines the meaning of the phrase *the angry student*, one sees that the relationships within the phrase are directionally *asymmetric*, in this sense: The word *angry* describes a property of *student*; not vice versa, i.e., *student* does not describe a property of *angry*. In standard terminology, one says that *angry* **modifies** *student*. Similarly, we observe that the word *the* describes a property of *angry student*; i.e., it indicates that a particular angry student is being referred to, rather than any angry student. Conversely, *angry student* does not describe a property of the word *the*. Thus, we can say that the word *the* **modifies** *angry student*; but not vice versa.

This means that the phrase is built-up by starting with the constituent

student

then applying the modification *angry* to it, thus obtaining

angry student

and finally applying the modification *the*, thus obtaining

the angry student.

In terms of the corresponding syntactic symbols, one starts with a Noun,

N

then applies the Adjective, A, obtaining

$$\mathrm{A(N)}$$

and finally applies the Determiner, DET, obtaining,

$$\mathrm{DET(A(N))}.$$

The sequence of steps in the generation is

$$\mathrm{N} \rightarrow \mathrm{A(N)} \rightarrow \mathrm{DET(A(N))}$$

or, in terms of the actual words,

student → angry student → the angry student

Observe now that the final state, *the angry student*, contains the previous state, *angry student*, which contains the still prior state, *student*. Equivalently, the final state, DET(A(N)), contains the previous state A(N), which contains the still prior state, N. Because the final state DET(A(N)) contains each of the successive preceding states, this means that the final state is seen as a *trace* of the previous states. Thus one makes the multi-state assumption about the final state. But this means that the backwards progression, from the final state to the previous ones, is carried out by *internal* inference.

We are therefore led to this crucial conclusion: The historical structure defined by the above modification sequence is fundamentally different from the historical structure defined by Phrase-Structure Rules. The modification structure produces *internal* history, and Phrase-Structure Rules produce *external* history. That is, the progression backwards, through the generation sequence, is done by making *internal* inference in the case of modification structure, and *external* inference in the case of Phrase-Structure Rules.

Let us return to modification structure. As we said, according to the modification approach, the phrase *the angry student*, is generated thus:

$$\mathrm{N} \rightarrow \mathrm{A(N)} \rightarrow \mathrm{DET(A(N))}$$

Observe that the right-most component, DET(A(N)), is actually the full phrase, NP. That is, we have

$$\mathrm{N} \rightarrow \mathrm{A(N)} \rightarrow \mathrm{DET(A(N))} = \mathrm{NP}$$

This sequence embodies the following rule:

> **If you start with N, then apply A, and then apply DET, you will obtain NP.**

We shall call such a rule, a *Modification Rule.*

Two properties characterize a Modification Rule: (1) Such a rule specifies a constituent (NP, in the above case) as the outcome of the successive application of other constituents. This, as we shall see in greater detail, makes a Modification Rule profoundly different from a Phrase-Structure Rule. (2) A Modification Rule specifies that the successively applied constituents are all present in the outcome constituent; i.e., the outcome is a *trace* of the applied constituents.

Let us, for a moment, consider a notational issue. The particular Modification Rule, given above, is notated thus:

$$\mathrm{N} \rightarrow \mathrm{A(N)} \rightarrow \mathrm{DET(A(N))} = \mathrm{NP}$$

Observe that, in this notation, the segment after the final arrow, that is

$$\mathrm{DET(A(N))} = \mathrm{NP}$$

contains all the information contained in the entire line. This is because it records the fact that, if one starts with N, and then applies A, and then applies DET, one obtains NP. We shall call the sequence

$$\mathrm{N} \rightarrow \mathrm{A(N)} \rightarrow \mathrm{DET(A(N))} = \mathrm{NP}$$

the *long form* notation, and

$$\mathrm{DET(A(N))} = \mathrm{NP}$$

the *short form* notation. Both contain all the information specified by the Modification Rule.

Let us now compare, in detail, the Phrase-Structure Rule

$$\mathrm{NP} \rightarrow \mathrm{DET,\ A,\ N}$$

to the Modification Rule

$$\mathrm{N} \rightarrow \mathrm{A(N)} \rightarrow \mathrm{DET(A(N))} = \mathrm{NP}$$

so that we can see how fundamentally different Modification Rules are from Phrase-Structure Rules.

The first thing to observe is that both, the Phrase-Structure Rule and the Modification Rule, specify a relationship between a *constituent*, NP, and a set of *sub-constituents*, DET, A, N. Thus, both specify what linguists call *constituent structure*, and they do so between the same elements. However, they do so very differently, as follows:

First, under the Phrase-Structure Rule, the sub-constituents come into existence all at the same time, whereas, under the Modification Rule, they are introduced successively. Another way of saying this is that, under the Modification Rule, the sub-constituents are generatively related to each other; whereas under the Phrase-Structure Rule, they are not.

This is what we meant in section 7.1 by saying that Chomskian syntax misses important aspects of the generativity in constituent structure. That is, Chomskian syntax is not generative enough. A Phrase-Structure Rule fails to distinguish between the different generative levels that exist between sub-constituents, by virtue of the modifying relationships between sub-constituents. A Modification Rule encodes the sub-constituents as belonging to different generative levels.

In terms of *constituent* structure, therefore, the sub-constituents in a Phrase-Structure Rule belong to the same level of *constituency*, whereas, under a Modification Rule, the sub-constituents belong to different *constituent* levels. In fact, in a Phrase-Structure Rule, the sub-constituents are not only within the same level of the constituent structure but the structure within the level is trivial: It is merely a list of the sub-constituents; i.e., the sub-constituents are structurally dissociated from each other. In contrast, under a Modification Rule, the sub-constituents are operators acting on each other, and they are therefore structurally related to each other. In fact, they are nested within each other.

Finally, consider the directionality of the arrows in the two types of rules, the Phrase-Structure Rule,

$$\text{NP} \rightarrow \text{DET, A, N}$$

and the Modification Rule,

$$\text{N} \rightarrow \text{A(N)} \rightarrow \text{DET(A(N))} = \text{NP}$$

In the former, the arrow goes from the constituent, NP, to the sub-constituents, DET, A, N. In the latter, the arrows go in the reverse direction, from the sub-constituents to the constituent. This reversal of arrows reflects several of the differences listed above, but it also reflects the fact that a Phrase-Structure Rule specifies how a constituent is *broken down* into sub-constituents, whereas the Modification Rule specifies how sub-constituents are *built up* into a constituent.

In conclusion therefore, we can see that there are six major differences between a Phrase-Structure Rule and a Modification Rule:

(1) **Phrase-Structure Rule:** The rule creates *external history.*
Modification Rule: The rule creates *internal history.*

(2) **Phrase-Structure Rule:** The sub-constituents are not generatively related to each other, and come into existence at the same time.
Modification Rule: The sub-constituents are generatively related to each other, and come into existence successively.

(3) **Phrase-Structure Rule:** The sub-constituents are assigned to the same *constituent* level.
Modification Rule: The sub-constituents are assigned to different *constituent* levels.

(4) **Phrase-Structure Rule:** The sub-constituents are structurally dissociated from each other, being simply members of a set.
Modification Rule: The sub-constituents are structurally related to each other, being operators on each other. In fact, they are *nested* operators.

(5) **Phrase-Structure Rule:** The rule maps constituents to sub-constituents.
Modification Rule: The rule maps sub-constituents to constituents.

(6) **Phrase-Structure Rule:** Constituent structure is one of *breaking down.*
Modification Rule: Constituent structure is one of *building up.*

Now let us turn to X-Bar Syntax. Recall that there are three successive generative rules in X-Bar Syntax:

$$\bar{\bar{\bar{X}}} \rightarrow \bar{\bar{X}},\ \mathrm{Sp}$$
$$\bar{\bar{X}} \rightarrow \bar{X},\ \mathrm{Ad}$$
$$\bar{X} \rightarrow X,\ \mathrm{Co}$$

One might at first think that these rules express a structure of modification. After all, the very notation, X, $\bar{X}$, $\bar{\bar{X}}$, seems to indicate that a constituent, the Head, X, is being modified. However, this notation gives merely the illusion of modification, because no actual modification is being expressed by the rules. To see this, let us replace the notation, X, $\bar{X}$, $\bar{\bar{X}}$, by a notation that does not have the aura of modification. This will enable us to see more clearly whether the rules actually do express modification. Replacing the symbols X, $\bar{X}$, $\bar{\bar{X}}$, in the above three rules, we can re-state the rules as:

$$Q \rightarrow R,\ S$$
$$R \rightarrow R,\ T$$
$$R \rightarrow U,\ V$$

This version of the three rules carries the same syntactic, i.e., *distributional*, information as the original version. However, we can now see that one cannot examine the symbols on the right-hand side of the last rule (given here as U and V) and decide which is the Head. This means that the rules do not tell us whether the higher constituent, Q, is an expansion of U, or an expansion of V (i.e., whether Q is a "projection" of U, or a "projection" of V). In our terms, the rules do not allow us to see whether Q is a modification of U or a modification of V.

When we introduced the notion of modification, earlier, we justified it on semantic grounds; e.g., in the phrase *angry student*, the word *angry* adds a property to *student* and not vice versa. However, we can now see that modification can also be understood as a syntactic property, that of expanding a Head. This expansion is not properly defined within the three generative rules of X-Bar Syntax.

We have seen that the problem arises in X-Bar Syntax because, when we replace the prejudicial symbols X, $\bar{X}$, $\bar{\bar{X}}$, the final rule (R → U, V) does not inform us which is the Head constituent. The ultimate source of this problem is the structure of this type of rule: a rule that re-writes a constituent as a list of sub-constituents. All rules of Phrase-Structure Grammar and X-Bar Syntax are of this form. When discussing Phrase-Structure Rules we listed six main properties of these rules that are drawbacks in comparison with Modification Rules. We can now see that these six properties apply also to the rules of X-Bar Syntax. In particular, we will now list the properties again, and the reader will be able see that the failure to distinguish the Head, is a consequence of the six properties. The properties are: (1) The rule creates external history. (2) The sub-constituents are not generatively related to each other, and come into existence at the same time. (3) The sub-constituents are assigned to the same constituent level. (4) The sub-constituents are structurally dissociated from each other, being simply members of a set. (5) The rule maps constituents to sub-constituents. (6) Constituent structure is one of breaking down.

Finally, observe that, corresponding to the three rules of X-Bar Syntax, there are three rules in Modification Grammar. We state these rules here in reverse order to their corresponding ones in X-Bar Syntax, because the Modification Rules must begin with the Head X.

$$X \rightarrow Co(X) = \bar{X}$$
$$\bar{X} \rightarrow Ad(\bar{X}) = \bar{X}$$
$$\bar{X} \rightarrow Sp(\bar{X}) = \bar{\bar{X}}$$

where the symbols Co, Ad, and Sp are the constituents mentioned in the three rules of X-Bar Syntax, but are here used as *operators*. As an example of the use of all three rules, one generates the constituent $\bar{\bar{X}}$ by the following sequence of successive modifications:

$$X \rightarrow Co(X) \rightarrow Ad(Co(X)) \rightarrow Sp(Ad(Co(X)))$$

Although the above three rules are the Modification Rules that *correspond* to the three rules of X-Bar Syntax, they are very different from the X-Bar rules. For example, they have the six crucial properties we listed earlier for Modification Rules. Thus they avoid the six drawbacks we listed for X-Bar Syntax. In particular, the Modification Rules have an unambiguous Head, since the initial element to which the modifications are applied is the Head.

We conclude, from this section, that the Chomskian Constituency Rules (i.e., the Phrase-Structure Rules and the rules of X-Bar Syntax) are incapable of expressing the full generativity of syntax, and that Modification Rules express this complete generativity.

7.4 Reviewing Further Concepts

In the next section, we are going to continue with our analysis of sentences as process-structures, when we start to look at *transformations*. In order to proceed however, we need to review the following concepts, for readers unfamiliar with them: (1) subcategorization, (2) linear order rules, and

(3) case-roles. The reader who is familiar with these concepts can pass directly to section 7.5.

(1) Subcategorization

Linguists argue that a grammar contains a dictionary, called a *Lexicon*, in which all the words used by the speaker are listed. As in a conventional dictionary, one finds in the Lexicon that, next to any word, there is a description of the properties of that word. The properties are *idiosyncratic* to the word because, if they were not—e.g., if they were general syntactic rules—then there would be no reason to list them together with the particular word.

One of the idiosyncratic properties of a word is the syntactic category to which it belongs. For example, the word *put* belongs to the category, Verb. We cannot infer this fact from general syntactic rules. Therefore, one has to list this fact in the entry for *put* in the Lexicon.

In order to *use* a word it is necessary to know which category it belongs to; for example, to use the word *put*, we must know that it is a Verb, not a Noun. However, to use a word, we must know still more particular information. For example, to use the verb *put* we need to know that we cannot use it without a *Direct Object*. For example, the following is not a grammatical sentence:

*I put.

(In Linguistics, the symbol *, preceding an example, means that the example is ungrammatical.) We have to add a Direct Object that specifies what is "put". But even this is not enough, because the following is not a grammatical sentence:

*I put the cup.

We have to add a Prepositional Phrase, for example,

I put the cup on the table.

This sentence is, at last, grammatical.

Thus, to use the word *put*, we need to know not only that it is a Verb, but that it is a particular kind of Verb—one that must take a Direct Object and a Prepositional Phrase. Another way of saying this is as follows: The *category*, Verb, contains a *subcategory*, those verbs that take a Direct Object and a Prepositional Phrase. To use the word *put*, we need to know that it belongs to this particular *subcategory*. Therefore, next to the word *put* in the Lexicon, we need to specify not only the *category* to which *put* belongs but its *subcategory*. The latter information is called the *subcategorization information*.

How should the subcategorization information be specified? To answer this, consider other verbs. Observe that there are verbs like *go* and *come* that do not have to take a Direct Object or Prepositional Phrase. For example, the following are grammatical:

I go.

I come.

Observe again that there are words like *hit* and *invite* that take a Direct Object, but do not have to take a Prepositional Phrase, thus:

I hit Bill.

I invited Bill.

Finally, as we saw, there are verbs, such as *put*, that require both a Direct Object and a Prepositional Phrase.

It seems therefore that specifying the range of constituents that the verb takes is a generally useful method of subcategorizing a verb. This is the method chosen for all types of words; i.e., any Noun, Adjective, Adverb, Preposition, etc., is subcategorized by the range of constituents it takes. In fact, it is generally argued that the only constituents one has to consider, in subcategorizing a word, are its *Complement* constituents. To understand what this means, recall that, in X-Bar Syntax, there are three rules:

$$\bar{\bar{X}} \rightarrow \bar{X},\ Sp$$

$$\bar{X} \rightarrow \bar{X},\ Ad$$

$$\bar{X} \rightarrow X,\ Co$$

The third rule states that a word X, together with its Complements, Co, produces $\bar{X}$. Linguists tend to believe that, in subcategorizing a word X, one needs to consider only the Complements, Co, that it takes in producing $\bar{X}$. That is,

> SUBCATEGORIZATION PRINCIPLE. A word X is subcategorized by the range of idiosyncratic Complement (Co) constituents that it takes under the rule
>
> $$\bar{X} \rightarrow X,\ Co$$

(2) Linear Order Rules

We noted, in section 7.2, that it is currently argued that rules defining the linear order of elements in a sentence should be separate from rules defining constituent structure; e.g., the rules of Phrase-Structure Grammar or X-Bar Syntax.

One linear order rule that has been suggested (e.g., by Stolwell, 1981) can be stated informally thus:

> PERIPHERY PRINCIPLE: The Head, X, appears as the left-most or right-most constituent of X-Bar.

This principle is supposed to be universal across languages. In addition, it is generally believed that each particular language selects whether X is to be the left-most or right-most constituent in that language. For example, for English, one has

> HEAD-FIRST PRINCIPLE: In English, the Head precedes its Complements.

We can demonstrate how these rules work together with constituent structure rules (e.g., X-Bar Syntax) as follows: Recall that the third rule in X-Bar Syntax is

$$\bar{X} \rightarrow X, \mathrm{Co}$$

This rule states that the constituent $\bar{X}$ is made up of constituents X and Co. But the rule does not specify the order in which X and Co should occur. In English, the order is given by the above Head-First Principle, which specifies that X must precede Co in the linear order of the sentence.

The same applies to our Modification Grammar. For example, corresponding to the above rule in X-Bar Syntax, there is the Modification Rule

$$X \rightarrow \mathrm{Co}(X) = \bar{X}$$

The notation for the constructed component Co(X) should not be interpreted as meaning that Co precedes X. It means that Co *acts* on X as a modifier. Ordering rules then determine the linear position of Co and X.

Finally, we should observe that, because it is believed that linear order principles form an independent set of rules, it is also believed that order information is not required in the subcategorization information that is listed for each word in the Lexicon. Recall that, given a word, its subcategorization information is the list of (idiosyncratic) complements that the word takes. The current view is that the order of these complements need not be specified in the Lexicon because the linear order rules specify order for all constituents generally.

(3) Case-Roles

In a classic paper, Fillmore (1968) argued that a (syntactic) constituent analysis can often miss important structural regularities that are evident within sentences, and across sentences.

For example, consider the two sentences,

The child rolled the ball along the ground.

The ball rolled along the ground.

The constituent, *the ball*, occurs in both sentences. However, according to Phrase-Structure Grammar or X-Bar Syntax, *the ball* has a different constituent status in the two sentences. In the first sentence, it is the Noun Phrase Complement of the verb *rolled*, whereas, in the second sentence it is the Noun Phrase Subject of the sentence. However, observe that, when one considers the *meaning* of the two sentences, one finds that, in both sentences, *the ball* has the same relationship to the verb *roll*: it is the physical object that undergoes the action of rolling. Linguistically, one says that it is the PATIENT of the action.

The concept of PATIENT is an example of what is called a *case-role*: a role with respect to the action described by the verb. According to Fillmore, the Noun Phrases associated with a verb are each assigned a case-role. For example, in the sentence

John gave Mary the book in the library.

the four constituents, *John*, *Mary*, *the book*, *the library*, are each assigned a different case-role, i.e., a role with respect to the action *gave*.

John is the AGENT, the instigator of the action;

Mary is the DATIVE, the recipient of the action;

the book is the PATIENT, the object undergoing the action;

the library is the LOCATIVE, the location of the action.

In contemporary syntactic theory, case-roles are often called θ-roles. We shall however retain the term "case-role" to avoid obscuring the meaning of the term, for the uninitiated reader.

In the assignment of case-roles, the following principle is fundamental:

> CASE-ROLE CRITERION: Each NP associated with the verb has exactly one and only one case-role, and each case-role is assigned to one and only one NP associated with the verb.

Because the case-roles assigned to NPs in a sentence are prescribed by the particular verb used in the sentence, it is generally assumed that the case-roles are listed in the Lexicon under the entry for the particular verb. For example, in the Lexicon, the entry for the verb *put* specifies that it has associated with it an AGENT (someone that puts) a PATIENT (something that is put), and a LOCATIVE (a place where the something is put). Thus, the assumption is that, in the Lexicon, the entry for a verb carries information on both subcategorization and the associated case-roles.[1]

7.5 Transformations Are Based on the Asymmetry and Symmetry Principles

In section 7.3, we began to consider sentences as process-structures. We examined the Chomskian Constituency Rules, i.e., the rules of Phrase-Structure Grammar and X-Bar Syntax, and argued that they are insufficiently generative to capture constituent structure, and when they are generative, they are generative in the wrong way. We proposed instead that constituent structure be characterized by Modification Rules. Such rules have several properties that enable them to capture the full generativity of constituent structure.

The three systems—Phrase-Structure Grammar, X-Bar Syntax, and Modification Grammar—are each attempts to characterize the hierarchical constituent structure of a sentence; i.e., to express the relationship between a constituent and its sub-constituents. However, there is evidence to suggest that hierarchical constituent structure is not the only generativity that exists in a sentence. Other operations are required to characterize syntactic structure. These operations, called *transformations*, have the effect of *moving* constituents around, rather than expressing a constituent in terms of its sub-constituents.

In this section, we are going to argue that transformations are manifestations of the general principles proposed in this book. We will do so by examining a standard argument used for justifying the existence of transformations, and showing that our general rules for process-inference are in fact the basis for this argument.

Consider the sentence

The lamp was put on the desk.

The standard proposal is that the NP, *the lamp*, originated from a position after the verb *put* and was moved to the front of the sentence.

One argument that is standardly used to justify this proposal is based on subcategorization facts. The subcategorization specifications for the verb *put* demand that it must take two Complements: a NP and a PP, i.e., the NP describing *what* was put, and the PP describing *where* it was put. Since the NP and PP are Complements of the Head *put*, the Head-First Principle requires that the NP and PP must follow *put* in the linear order of the sentence. However, when we look at the words after *put* in the sentence, we find only the phrase *on the desk*. This is clearly the PP of *put*. Therefore, we can manage to find the PP Complement of *put*. However, we cannot find the NP Complement of *put*. The obvious solution to this problem is that the phrase, *the lamp*, at the beginning of the sentence, was originally positioned after the verb *put*, where it was the NP Complement of *put*.

Case-role considerations reinforce this conclusion. Observe that, even positioned at the beginning of the sentence, *the lamp* has the PATIENT case-role; i.e., it is the object to which the action of putting was applied. Generally, the NP Complement of *put* is assigned the PATIENT case-role. Therefore *the lamp* has the appropriate case-role for being considered to be the NP Complement of *put*. This completes one of the standard arguments used to support the idea that *the lamp* was moved to the front position from a position after the verb.

One component of this argument should be made explicit. In *both* the above sentence and the configuration from which it was derived, it was assumed that *put* conforms to the subcategorization properties and case-role requirements listed for *put* in the Lexicon; i.e., that *put* must take a NP and PP Complement and that the NP must be in the PATIENT case-role. The assumption was crucial. If we had not made this assumption, there would have been no reason to hunt for the "missing" NP, and postulate a position from which it was initially moved.

This assumption is in fact stated generally in linguistics as:

PROJECTION PRINCIPLE: Any syntactic structure must project from the Lexicon, in that any syntactic structure must observe the subcategorization and case-role requirements given in the Lexicon.

We can therefore state the standard argument, described above, as involving three steps:

(1) The Projection Principle: Lexical properties hold at all syntactic levels.

(2) Ordering rules require the PATIENT NP, which is the Complement of *put*, to be placed after *put*.

(3) Therefore, the PATIENT NP, at the beginning of the sentence must have originated from a position after *put*.

We shall now take this argument and show that it is based on our general principles of process-inference.

First we take stage (1) the Projection Principle which states that the subcategorization and case-role properties in the Lexicon hold for all syntactic levels. This principle is, in fact, the requirement that the subcategorization and case-role properties are *indistinguishable* across syntactic levels. That is, the principle is describing a *symmetry* across syntactic levels. In particular, the presence of the PATIENT NP of the verb *put*, is invariant across all levels, and must be instantiated by the same phrase (the lamp) across all levels.

Now, in stage (2) above, the linguistic ordering principles require the PATIENT NP to be positioned after the verb *put*. Thus, while the presence of the PATIENT NP, and its instantiation as *the lamp*, is a symmetry (indistinguishability) across all syntactic levels, its *position* is not. Its position constitutes an *asymmetry* or *distinguishability* across levels; i.e., there is a disparity between the position of the PATIENT NP in the presented sentence and its position in the configuration prescribed by the ordering rules.

At this point, one moves on to stage (3) in the above argument. However, given what we have just been saying, we can now see that stage (3) is actually based on our *Asymmetry Principle*, which states that any disparity is explained by a process that created the disparity and therefore is removed by going backwards in time. That is, in the present situation, the disparity between the two positions for the NP is removed backwards in time, and using internal inference, the NP, in the past, must have held one of the two positions. Furthermore, in accord with the Asymmetry Principle, a process is hypothesized as having been responsible for the disparity in the positions of the NP. Since the disparity is positional, the hypothesized process must have been *movement*.

Thus we see that the argument identifies two sorts of content in a sentence:

SYMMETRY CONTENT:	The items required by the subcategorization and case-role conditions.
ASYMMETRY CONTENT:	The positional disparity of those items.

The conjecturing of transformations (i.e., movements) deals with the second content, and can now be seen to be an instance of our Asymmetry Principle. The first content is actually dealt with by the Projection Principle, which states that the items cannot be destroyed across time. Given this, we can now understand that the Projection Principle is an example of our Symmetry Principle; i.e., it states that the symmetry content cannot be destroyed across time.

In conclusion, we can see that the standard argument, given earlier, for the existence of transformations, is based on the use of our Asymmetry and Symmetry Principles.

7.6 Stratification of Movement

We have seen throughout this book that the mind stratifies histories into layers. For example, the history assigned to a rotated parallelogram proceeds backwards to a non-rotated parallelogram, then to a rectangle, and then to a square. That is, the history is not a *direct* transition from the rotated parallelogram to a square, but is decomposed into stages.

The movement of constituents in a sentence is similarly stratified, rather than being direct. For example, consider the sentence

What did you say that you would do?

Using the standard argument discussed in the previous section, for the existence of transformations, the first word *what* seems to have been moved to the beginning of the sentence from the end of the sentence.

That is, subcategorization and case-role specifications in the Lexicon indicate that *what* is the PATIENT NP that is taken by the verb *do* (now situated at the end of the sentence).

Thus the word *what* has moved over the entire sentence, from the last to the first position. Has this movement taken place in a single step? The standard answer is that it has not: The movement was stratified into two stages and the stratification occurred at the clause boundaries. These boundaries are indicated by the brackets in the following:

What did [you say t_2 that [you would do t_1]]

Also indicated here are the successive positions of *what*: That is, the symbol t_1 at the end, i.e., in the inner-most clause, indicates the first position held by *what*, and the symbol t_2 in the middle clause indicates the second position held by *what* on its way to the front of the sentence.

The general rule that seems to determine stratification of movement was proposed by Chomsky (e.g., 1981) and is stated here informally as:

> SUBJACENCY CONDITION: A single movement cannot be made across more than one clause boundary.[2]

The main argument in favor of this principle is that, when one examines sentences in which a constituent has moved over two clause boundaries between which there is no position at which the constituent could have temporarily landed, one finds that the sentences are ungrammatical.

In conclusion then we see that the movement history of a constituent is stratified, and this stratification occurs across the clause boundaries.

7.7 Two Types of Modification Rules

The important thing we should now observe is that a transformation is a *modification operation*: it modifies the *position* of a constituent; it moves it from one position in a sentence to another.

Thus, surveying sections 7.3 and 7.5, we can now see that we have two types of Modification Rules, as follows: In section 7.3, we introduced Modification Rules to describe the hierarchical structure of constituents. These rules take the form of constituents modifying other constituents. Because, such rules have the effect of building up layers of constituents, we shall now say that they describe processes of *accumulation*.

In contrast, transformations rearrange constituents; i.e., they describe processes of *movement*. In short: our two types of Modification Rules describe two different types of processes:

(1) Accumulation

(2) Movement

Thus, we can think of a sentence as analogous to a geological formation, in this way: The Accumulation Rules describe the successive depositing of the geological layers; the Movement Rules describe shifts in the geological layers, as layers deep within the formation are extracted and brought to a higher level—sometimes to the surface—by volcanos, by mining, etc.

We should now observe that Accumulation Rules and Movement Rules have three important properties in common:

(1) Both types of Modification Rules are based on the Asymmetry Principle. They each are an attempt to explain present distinguishability as having arisen from past indistinguishability, by processes that created that distinguishability. In the case of Accumulation Rules, the distinguishability is between constituents in a sentence; and this distinguishability is explained as having arisen from successively fewer constituents backwards in time, by accumulation. In the case of Movement Rules, the distinguishability is between the constituent order prescribed by linear order rules, and the constituent order in the presented sentence. This distinguishability between orders is explained as having arisen from a single order via movement.

(2) Both types of Modification Rules are internal. In the case of Accumulation Rules, the distinguishability between two constituents in the present is removed by choosing one of them as the past state. That is, the constituents are given the multi-state assumption, and inference is therefore internal. In the case of Movement Rules, the distinguishability between two linear orders is removed by assuming that one of them is the past state. Note that this requires that the present contain both orders even though only one order is given in the presented sentence. The existence of the other order within the present is due to the linear order rules. We know that this latter order is present, in the mind of the person, because, according to linguistic theory, it is only by comparing the presented order with the latter order that a disparity can be found and movement can be hypothesized.

(3) Both types of Modification Rules are stratified, i.e., nested. In the case of Accumulation Rules, the constituents are successively applied; for example, as in

$$N \rightarrow A(N) \rightarrow DET(A(N))$$

In the case of Movement Rules, the movement operations are stratified by the Subjacency Principle, i.e., stratified across clause boundaries, as in the example:

What did [you say t_2 that [you would do t_1]]

Finally, the Accumulation Rules are nested within the Movement Rules. That is, the accumulation structure is built up as successive layers of time, in the form of successive constituents, and the movement structure is built on top of this, as additional layers of time, in the form of successive movements.

In conclusion, we can see that the complete grammar is a nested hierarchy of control. All constituents are consigned to nested layers of time, and the system is therefore completely generative, in contrast to the Chomskian system.

7.8 Atemporal or Temporal Interpretations of Generativity

We ended the previous section by referring to the generative structure of sentences as embodying *layers of time*, i.e., processes. However, at this point in the chapter, the assumption that generativity embodies *temporal* structure is premature, as follows:

Chomsky has repeatedly asserted that generativity describes the structural relationships in syntax without necessarily describing temporal relationships. This assertion can ultimately be traced back to his distinction between *competence* and *performance* (Chomsky, 1965). A speaker's *competence* is the speaker's knowledge of language. In contrast, *performance* is the speaker's use of that knowledge—in speaking, listening, etc. In performance, generativity can gain a temporal aspect. For example, in speaking, generativity might correspond to the successive temporal stages in producing a sentence. Alternatively, in listening, generativity (in its reverse direction) might correspond to the temporal retracing of transformations through a sentence. However, both of these temporal embodiments of generativity are introduced by performance rather than by competence.

Competence, i.e., the speaker's or listener's knowledge of language, can be regarded as providing a sentence with a system of syntactic relationships somewhat like a scaffold. The structure of a physical scaffold, e.g., for a building, can be characterized generatively; but this generativity does not imply anything temporal in the scaffold. The scaffold itself is a static frozen object. Similarly, generativity, in Chomskian syntax, describes the way the linguistic constituents are held together in a cognitive network of relationships, not actual temporal progression.

In contrast to this atemporal view, we are going to argue that generative relations in grammar are cognitively represented as temporal ones. The sentence is understood literally as a relic of *time*, a piece of archeology. That is, *time* is locked into a sentence, and the mind tries to unlock time from it. Furthermore, this unlocking of time gives access to the causal structure of the environment that produced the sentence. That is, the mind understands the generativity not only as representing time but as being propelled by causal interactions between systems in the environment.

We shall call the two interpretations of generativity described above, the *atemporal* and the *temporal* interpretations respectively. They can be recapitulated briefly as follows:

> **Atemporal interpretation.** *The generative relations characterize static structural relationships; i.e., they are simply part of an interconnecting network of cognitive relationships.*

> **Temporal interpretation.** *The generative relations are temporal relationships; i.e., they are part of a history that is assigned to the sentence. This history is understood as generated by causal interactions in the environment.*

Our purpose, in the rest of this chapter, is to argue that the latter is the correct interpretation.

7.9 Functionalist Considerations

Our aim is to show that the generative structure of sentences is a product of the mind's attempt to assign a *causal* explanation to a sentence. Indeed, we shall argue that, because Chomsky does not introduce notions of *causality*, he is unable to fulfill even his own aim of giving a satisfactory structural characterization of syntax; i.e., he is unable to give a complete structural definition of generativity.

In order to show both, that the generative structure belongs to a causal system, and that a complete structural characterization of generativity requires causal notions, we need to introduce concepts from *pragmatics*. Let us recall first that, following the three-fold distinction of Morris (1938), the study of grammar can be regarded as divided into the analysis of three phenomena:

(1) Syntax—configural properties of constituents

(2) Semantics—the relationship of constituents to the entities to which they refer

(3) Pragmatics—contextual properties

We have seen that, in its analysis of syntax, contemporary transformational grammar not only involves an examination of syntax itself, but is forced to include an examination of semantics—because transformational grammar cannot account for syntactic regularities by the use of purely syntactic arguments. For example, the Case-Role Criterion and Projection Principle are used centrally in the understanding of syntactic phenomena, and yet both principles are based on the *semantic* notion of case-role. Case-roles are semantic because they concern how constituents refer to entities in the world, i.e., entities such as the AGENT and PATIENT of an action. Thus transformational grammar has come to recognize the controlling role that semantics has in syntax.

However, what has not been recognized by transformational grammarians is the controlling role that *pragmatics* (the third component in the above triple) has in syntax. Nevertheless, independently of the transformational grammarians, certain schools of linguistics have been developing an analysis of wide-ranging grammatical phenomena that seem to show that a study of pragmatics is crucial to an understanding of syntax.

An important concept, in this analysis, is the distinction between the *topic* of a sentence, and the *comment*. Essentially, the *topic* is the entity that is being talked about, and the *comment* is the information that the sentence adds about this entity. For example, in the sentence,

John walked to the store

it is likely that *John* was the entity that was being discussed in the surrounding discourse, and *walked to the store* is the comment that the sentence added about *John*.

A more general way of understanding the distinction between topic and comment is this: *Topic* is information that is already shared by the speaker and listener, and *comment* is information that is added by the speaker. For example, consider the following two sentences which are supposed to be two successive sentences spoken by the same speaker:

That man's name is John

John is a reasonable individual

In the first sentence *John* is a comment; it is information that is added by the sentence about the topic *that man* or *that man's name*. Having been introduced in the first sentence, the new information, *John*, becomes old information; i.e., shared information, and can thus become the basis upon which new information can be added in the second sentence. That is, generally, new information, once introduced, becomes shared or old information, about which new information can be added. The terms *topic* and *comment* are alternative terms for *old information* and *new information*, respectively. Thus, in the above example, *John* is introduced in the first sentence and is therefore *comment*, at that stage. However, once it has been introduced, it becomes old information and can thus be *topic* in the second sentence, where it can itself receive the comment *is a reasonable individual*. Therefore, the *topic-comment structure* of a sentence is often called the *information structure* of the sentence.[3]

The basic distinction between topic and comment was made by Mathesius (1915), a leading linguist of the Prague school,[4] and was later taken up by the American functionalists (see Li, 1976) and also the British functionalists led by Halliday (1967, 1968). We shall concentrate on the approach taken by the American functionalists.

The study of topic and comment belongs to pragmatics because it concerns *context*, in fact, the assumed mutual knowledge; i.e., presuppositions. The idea is that the *syntax* of a sentence informs the listener as to which part of the sentence the speaker regards as topic, i.e., assumed information, and which part of the sentence the speaker regards as comment on that topic, i.e., information added to that topic. This approach to the analysis of syntax is called *functionalist* because it examines the *functions* of syntax, e.g., the encoding of topic and comment. The basic claim of functionalists is that one requires an examination of the *functions* of syntax in order to gain insight into the *structure* of syntax.

7.10 The Coding of Topic

As we have said, the *topic* of a sentence is the constituent that is presupposed and therefore is the constituent *under discussion*. In fact, in any sentence, there is a graded ranking of the topicality of constituents. The constituent that is most topical—that which is the basis on which the entire sentence adds information— is called the *primary topic*.

Bolinger (1952) argued that, in English, the word-order (i.e., left to right) tends to be the order of topicality, starting with the left-most constituent which is the primary topic. In fact, as one works one's way along an English sentence, each successive constituent tends to be a comment on the preceding part of the sentence. As Bollinger states: "Elements, as they are added, one by one to form a sentence, progressively limit the semantic range of all that preceded" (Bollinger, 1952, p279). Thus, because each constituent adds a comment on the preceding material, the preceding material can be regarded as the topic relative to the added constituent. The consequence is that the order of a sentence represents the topic-to-comment hierarchy.

Bollinger's claim was inverted by Givon (1979) to make the following profound point: A listener can *deduce* what the topic is of a sentence (in English) by looking at the word order. Generally, Givon proposed that each of the world's languages has a code in the sentence that informs the listener what the topic is. In English, the code is given by the word order.

However, word order is not necessarily the means used by other languages to encode topicality. A standard example given is the Philippine languages. Here, a prefix *?ang* is attached to the word that is the primary topic. This word can occur in any position in a sentence.

It is worth examining, in detail, an example discussed by Givon (1979). This example gives the translation of three English sentences into a Philippine language called Bikol (closely related to Tagalog). The English sentences are

(Ex 7.1) (a) The man gave the book to the woman

(b) The book was given by the man to the woman

(c) The woman was given the book by the man

What is important to recognize is that each of the three sentences describes exactly the same action in the world—a man giving a book to a woman. This means that the assignments of *case-roles* (participants in the action) must be the same in each sentence. The case-roles are assigned in this way:

man = AGENT (instigator of the action)
book = PATIENT (object of the action)
woman = DATIVE (recipient of the action)

What is different in the three English sentences is which of the nouns is made the primary topic; i.e., the constituent about which the entire sentence adds information. Because primary topic tends to be coded, in English, by word-order, we can easily infer that the primary topic in Ex 7.1a is *man*; in 7.1b it is *book*; and in 7.1c it is *woman*.

To restate: Each of the sentences describes the same event and therefore has the same case-role assignments. However, the speaker has chosen to make a different case-role the topic of each sentence. The topic chosen is indicated by the word-order in English versions.

Now let us turn to the Philippine language, Bikol. In contrast to English, Bikol keeps the same word-order for each sentence. The word-order in Bikol is

give man book woman

Translated into Bikol, this becomes

-ta?ó -laláke -libro -babáye
-give -man -book -woman

Now, because the word order is the same in Bikol, irrespective of which noun is chosen as topic, the speakers must use some means other than word order to indicate topic. They do so by placing a prefix *?ang* on the noun they wish to make topic. Thus, the three English sentences given in Ex 7.1 above become:

(a) -ta?ó **?ang**-laláke -libro -babáye
-give TOP-man -book -woman

(b) -ta?ó -laláke **?ang**-libro -babáye
-give -man TOP-book -woman

(c) -ta?ó -laláke -libro **?ang**-babáye
-give -man -book TOP-woman

That is, because *man* is the topic in the first English sentence, the Bikol word for man, *-laláke*, receives the prefix *?ang* in the first sentence. Again, because *book* is the topic in the second English sentence, the Bikol word for book, *-libro*, receives the prefix *?ang* in the second sentence. Finally, because *woman* is the topic in the third English sentence, the Bikol word for woman, *-babáye*, receives the prefix *?ang* in the third sentence.

Bikol then indicates the case-roles of the other nouns by adding the following prefixes

PATIENT = ning-
DATIVE = sa-
AGENT = kang-

That is, inserting these prefixes, we obtain:

(a) -ta?ó ?ang-laláke ning-libro sa-babáye
-give TOP-man PAT-book DAT-woman

(b) -ta?ó kang-laláke ?ang-libro sa-babáye
-give AGT-man TOP-book DAT-woman

(c) -ta?ó-(h)an kang-laláke ning-libro ?ang-babáye
-give AGT-man PAT-book TOP-woman

where AGT means AGENT; PAT means PATIENT; and DAT means DATIVE.

Observe finally that because the *topic* noun carries the topic prefix *?ang*, and not a case-role prefix, the listener cannot determine the case-role of that noun. Bikol solves this problem by placing a prefix on the verb—a prefix that informs the listener which case-role should be assigned to the topic noun. We can now present the complete translations of the above three English sentences (as given by Givon, 1979, p153):

(Ex 7.2) (a) AGENT as topic
nag-ta?ó ?ang-laláke ning-libro sa-babáye
AGT-give TOP-man PAT-book DAT-woman
'The man gave a book to the woman.'

(b) PATIENT as topic

na-ta?ó	kang-laláke	?ang-libro	sa-babáye
PAT-give	AGT-man	TOP-book	DAT-woman

'The book was given to the woman by the man.'

(c) DATIVE as topic

na-ta?ó-(h)an	kang-laláke	ning-libro	?ang-babáye
DAT-give	AGT-man	PAT-book	TOP-woman

'The woman was given a book by the man.'

To restate the main points illustrated by these translations: In Bikol, (1) the order of the words is the same in all three examples; (2) the topic of the sentence is marked by the prefix *?ang*; (3) the case-roles of the other nouns are marked by case-role prefixes attached to the nouns; and (4) the case-role of the topic noun is indicated by the prefix on the verb.

Finally, let us consider how one would change topic in English and in Bikol. In English, topic is given by word order. Thus, changing topic is given by changing the word order. This is what happens in the transitions between the three English sentences in Ex 7.1. However, in Bikol, word order remains the same and therefore changing topic cannot be given by changing the word order. Instead, as we can see in the transitions between the three Bikol sentences in Ex 7.2, changing topic in Bikol is given as follows: If one wishes to change topic from a noun X to a noun Y, one goes through the following three steps:

PROCEDURE P

(1) Move the topic prefix *?ang* from X to Y, supplanting the case-role prefix on Y.

(2) Encode the case-role of X by a case-role prefix on X.

(3) Change the prefix on the verb from indicating the case-role of X to indicating the case-role of Y

To summarize this section: We have examined aspects of the notion of *topic*. The topic of a sentence is the constituent that is shared or old information and is therefore used by the sentence to add new information; i.e., comment. We have seen that different grammatical devices are used in different languages to encode the topic of a sentence. English uses word order, making the left-most constituent, the primary topic, and Bikol uses a scheme based on prefixes. Finally, we have seen that if one wants to change topic in English, one changes word order, and if one wants to change topic in Bikol one applies Procedure P.

7.11 Topic and Transformations

A long-standing problem arising from transformational grammar has been that of generalizing Chomsky's system to many of the world's languages. For example, Chomsky views transformations as changing word order. However, it is clear that there are languages (such as Bikol) that do not have corresponding changes in word order. However, functionalist linguists have solved the problem of generalization in a rather elegant way. Recall that word order in English denotes topicality ranking. Thus, because a transformation in English alters word order, it must alter topicality. Now, if one regards transformations, *generally*, as altering topicality ranking, rather than word order, one obtains a view of transformations that is valid for both English and Philippine languages. For example, we have seen, in Bikol, that Procedure P changes the topicality ranking of the words. However, the order of the words has not been changed. Nevertheless, the *effect* is the same as a word-order transformation in English (i.e., the encoding of linguistically significant regularities across sentences).

The functionalist view therefore is that one should understand a transformation as altering topicality ranking, and that this alteration can be achieved *in different ways*: e.g., by changing word order in English and by changing prefixes in Bikol. Let us now illustrate these notions in greater detail, by looking at specific transformations.

(1) Passivization
According to the functionalist view, starting with the initial proposals by the Prague school, passivization has a topicalizing function (Mathesius, 1915; Givon, 1979). Givon puts it thus:

> Passivization is the process by which a **non-agent** is promoted to the role of **main topic** of the sentence (Givon, 1979, p186).

(We will assume in this proposal that the promoted constituent remains within the case-system of the verb.)

One way to test this view is to introduce a topic via a question and to examine the possible replies to the question. Thus consider the following examples, taken from Williams (1977) and Creider (1979):

(Ex 7.3) (a) What did John do?
(b) #The artwork was done by John.
(c) John did the artwork.

(Ex 7.4) (a) Who was the artwork done by?
(b) The artwork was done by John.
(c) #John did the artwork.

When one is given the question Ex 7.3a, the passive reply in Ex 7.3b seems inappropriate, and the active reply in Ex 7.3c seems appropriate, even though both describe the same action—John doing the artwork. The explanation is this: In the question Ex 7.3a, the word *John* is introduced into the discussion. Therefore, *John* can be made a topic of conversation, as is done in the active Ex 7.3c, by putting *John* at the front of the sentence. In the passive, Ex 7.3b, *John* is mentioned, but the topic is *artwork* because the latter is placed at the front of that sentence. Making *artwork* the topic, however, is inappropriate because it has not been introduced into the conversation by the preceding question.

In contrast, consider the question given in Ex 7.4a. Here, the passive reply in Ex 7.4b now seems appropriate, and the active reply in Ex 7.4c seems inappropriate, even though both describe the same action. The explanation is this: In the question Ex 7.4a, the word *artwork* is introduced into the discussion. Therefore, *artwork* can be made a topic of conversation, as is done in the passive Ex 7.4b, by putting *artwork* at the front of the sentence. In the active, Ex 7.4c, *artwork* is mentioned, but the topic is *John* because the latter is placed at the front of that sentence. Making *John* the topic, however, is inappropriate because *John* has not been introduced into the conversation by the preceding question.

Observe now that all four replies in Ex 7.3 and 7.4 describe the same action in the world—John doing the artwork. This means that, in all four replies, the case-roles are assigned in the same way; i.e., *John* is the AGENT and *artwork* is the PATIENT. However, we saw that the fronting of *John* is appropriate only where it has been previously introduced for discussion and can thus be made a topic (i.e., Ex 7.3c). Now the fronting of *John* means that the AGENT has been fronted, and this in turn means that one has an *active* sentence. Thus, the active sentence is appropriate where the AGENT is topic.

Now consider *artwork*, which is the PATIENT. We saw that the fronting of *artwork* is appropriate only where it has been previously introduced for discussion and can thus be made a topic (i.e., Ex 7.4b). The fronting of *artwork* means that the PATIENT has been fronted, and this in turn means that one has a *passive* sentence. Thus, the passive sentence is appropriate where the PATIENT is topic.

The above discussion therefore appears to corroborate Givon's proposal that passivization means the promotion of a non-AGENT case-role to primary topic. Observe that, in contrast to Chomsky's definition of passivization in terms of *word-order*, Givon's definition of passivization is in terms of communicative *function*; i.e., the encoding of *topic* on a non-AGENT case-role.

Recall also that, unlike Chomsky's view of transformations, generally, as altering word-order, the functionalist view, in terms of altering topic, is applicable also to languages such as Philippine, where word-order remains fixed despite other changes. In fact, when we now consider our previous examples from Bikol, we discover that Procedure P, in Bikol, is in fact *passivization*, by Givon's definition, because it allows the promotion of non-AGENT to primary topic. Thus one discovers the process in Philippine that corresponds to passivization in English. Chomsky's approach would not make this discovery possible because the approach excludes consideration of pragmatic function.

Extensive support for the functionalist view of the passive has been put forward by Givon (1979, Chapter 4), who examines several types of passive construction in the world's languages. For example, one broad class of languages he discusses are those that he characterizes as having a "topic-shift" typology. Included under this type are Bahasa Indonesian and two subgroups of the African Bantu languages: (i) Lunda-Ndembu, Lovale, and Kimbundu; and (ii) Dzamba, Lingala, and Likila. In Bahasa Indonesian, for example, there are two passive rules that historically arose from "topic-shift" constructions—constructions where the fronted topic has left the case-role system of the verb. Again, examining the language Kimbundu, Givon (1979, p. 188) observes the following: In the topic-shift constructions of Kimbundu the verb takes a double prefix in which the first prefix agrees with the subject and the second prefix with the fronted topic. In the passive constructions, however, the first prefix neutralizes to third-person plural, and the second prefix agrees with the subject. But this means that the second prefix position in the passive is occupied by the topic prefix in the topic-shift constructions. This demonstrates the topicalizing role of passivization in Kimbundu. Again, besides considering languages of the topic-shift typology, and also the Philippine typology, Givon also considers languages having what he calls the "agent-deletion" typology. The deletion of AGENT can occur in Latin, Turkish, Dutch, and Mojave, where in most cases the verb neutralizes to third-person singular (recall the above neutralization to third-person plural in Kimbundu). Givon argues that, when the AGENT is deleted, the topic must pass to a non-AGENT case-role. For further discussion see also Foley (1976) and Kirsner (1973, 1976).

(2) Dative Shift

We now turn from passivization, to examine another transformation called *dative-shift* (Ross, 1967). An example is the following:

(Ex 7.5) (a) John gave the book to Mary.

(b) John gave Mary the book.

As can be seen, the positions of the PATIENT, *the book*, and the DATIVE, *Mary*, have been reversed, in going from the first sentence to the second.

Again, under the functionalist position, this means a change in the topicality of the moved constituents, as we shall see (Creider, 1979; Givon, 1979). Since, according to Bolinger (1952), the left-to-right word order in English represents topic-to-comment order, *book* is more topical than *Mary* in Ex 7.5a, but *Mary* is more topical than *book* in Ex 7.5b. Givon gives a particular means of demonstrating this. He places an Adverbial clause in front of each of the two sentences, and lets the clause introduce one of the constituents *book* or *Mary*, hence establishing the prior topicality of the particular constituent. Thus consider this example:

(Ex 7.6) (a) When he found it, John gave the book to Mary.

(b) #When he found it, John gave Mary the book.

Ex 7.6a seems more appropriate than Ex 7.6b. The explanation is this: Because the PATIENT *book* has been introduced by *it* in the prior Adverbial clause, it is more appropriate to have *book* earlier in the main clause, as in Ex 7.6a, because an earlier position means greater topicality.

In contrast, consider this example:

(Ex 7.7) (a) When he found her, John gave Mary the book.
(b) #When he found her, John gave the book to Mary.

Here, Ex 7.7a seems more appropriate, the explanation being this: Because the DATIVE, *Mary*, has been introduced by *her* in the prior Adverbial clause, it is more appropriate to have *Mary* earlier in the main clause, as in Ex 7.7a, because an earlier position in the sentence recognizes this greater topicality.

(3) Other Transformations
Creider (1979) examines many transformations, such as about-movement, though-movement, particle-movement, Adverb-fronting, Subject-Subject raising, etc., and shows that each of them induces a re-ranking of the constituents with respect to topicality value.

Let us now summarize the content of this section: A long-standing problem with transformational grammar has been that, as it is expressed at the present time, it does not generalize to several of the world's languages. For example, although Chomsky defines transformations as changing word order, several languages have means of structural alteration without changing word order. However, we have seen that, according to the functionalist view, transformations can be understood as acting on the topic-to-comment hierarchy. Under this view, transformations identify a significant process that has a much more general applicability than the conventional transformational approach of Chomsky.

7.12 Functional Stratification of Transformations

We have seen repeatedly throughout this book that process-histories are stratified into layers of time. We saw this not only in perception but in transformational grammar where movement is stratified across clause boundaries. With the functionalist view that transformations are more general than just word-order changes, it is worth seeing that stratification occurs in a more general way than that allowed in Chomskian grammar.

To do this, we consider a procedure known as *relativization*. This can be illustrated in English by the following example:

(Ex 7.8) (a) John gave the car to the man.

(b) The man to whom John gave the car.

The DATIVE *the man* has been shifted from its position at the end of Ex 7.8a to the first position of Ex 7.8b by *relativization*. Generally, relativization is a process by which a particular constituent is chosen such that the remainder of the clause is organized as a restriction on the domain defined by the constituent. Here Ex 7.8b restricts *the man* to have been the one to whom John gave the car.

Now, languages have constraints on the types of constituents that can be the basis of relativization (Keenan & Comrie, 1977). For example, several Malayo-Polynesian languages such as Malagasy, Javanese, Iban, Minang-Kabau, and Toba Batak have relativization strategies applicable only to the constituent that is primary topic (in fact, the subject). In particular, the Philippine language Bikol, which was discussed in section 7.10, is one such language. Let us look at the consequences of this constraint in Bikol.

Because Bikol has a relativization strategy that is applicable only to the constituent that is primary topic, then, in order to relativize with respect to a constituent that is *not* a primary topic, one has to first *promote* the constituent to primary topic; and this will allow one to use the relativization strategy. The first phase, the promotion phase, is the passivization rule, Procedure P, discussed in section 7.10; i.e., placing the topic marker on the previously non-topicalized constituent, etc.

Thus, consider again the three examples shown earlier in Ex 7.2. In these examples, the topic marker was placed, respectively, on the AGENT, PATIENT, and DATIVE constituents, thus converting them each into primary topics. Each of these three examples is therefore an initial stage that is required before one can relativize the AGENT, PATIENT, DATIVE, respectively. In the second stage, one relativizes with respect to the constituent that holds the topic prefix. Thus, applying this strategy to each of the three examples in Ex 7.2, one obtains, respectively, the following three examples from Givon (1979, p. 155):

(Ex 7.9)

(a) AGENT

marái ?ang-laláke na nag-ta?ó ning-libro sa-babáye

good TOP-man that AGT-give PAT-book DAT-woman

'The man who gave a book to the woman is good.'

(b) PATIENT

marái ?ang-libro na na-ta?ó kang-laláke sa-babáye

good TOP-book that PAT-give AGT-man DAT-woman

'The book that the man gave to the woman is good.'

(c) DATIVE

marái ?ang-babáye na na-ta?o-hán kang-laláke ning-libro

good TOP-woman that DAT-give AGT-man PAT-book

'The woman to whom the man gave a book is good.'

'The woman who was given a book by the man is good.'

We see therefore that relativization of a non-topic constituent is *stratified* into two successive stages: (1) the promotion of the constituent to primary topic (passivization), and (2) relativization on the new primary topic. Note, as can be seen from the above examples, the actual relativization stage involves word-order change, i.e., the movement of the primary topic to the front of the relativizing clause. However, the decomposition of the overall procedure involves two stages, the first of which, passivization, does not involve change in word order. This means that one cannot really characterize the overall procedure as a word-order change. However, one *can* regard the overall procedure as successive changes in the topic-comment structure, as follows: We have seen that the first stage, passivization, can be regarded as such, because it changes the topicality of a constituent, using the topicalizing prefix. Furthermore, the second stage, relativization, can be regarded as such because, by definition, relativization is a means of reorganizing a clause so that the clause becomes a restriction on the domain of a constituent.

Thus, whereas the stratification is not a decomposition into two word-order changes, and thus cannot be classified using Chomsky's view of transformations, it is completely definable under the functionalist perspective in terms of a stratification into successive topic-comment manipulations.

7.13 Differences Between Topic-Comment Structure and Chomskian Constituency Structure

We have argued, following the functionalist linguists, that, in order to obtain grammatical rules that are universal, i.e., valid across the world's languages, one has to invoke functional considerations, in particular, the *topic-comment* structure of a sentence.

In this section, we will see that topic-comment syntax is structurally different from syntax as understood in Chomskian theory. In particular, we shall compare the hierarchical constituent structure as prescribed by the topic-comment structure with the hierarchical constituent structure as prescribed by the Chomskian Constituency Rules; i.e., the rules of Phrase-Structure Grammar and X-Bar Syntax.

The comparison will follow the same argument as that which we used when comparing Chomskian Constituency Rules with Modification Rules. However, additional clarity is obtained when this comparison is made explicitly in terms of the notion of *comment*.

Thus to begin: Consider the phrase, *the angry student*. In Phrase-Structure Grammar, the phrase would be generated by the rule:

$$NP \rightarrow DET, A, N$$

We observed, in section 7.3, that the assignment of this rule to the given phrase is an example of *external* inference. That is, going backwards, through the generation, the three items DET, A, N, are replaced by an item NP not contained among the three. Note also that, because the second state—DET, A, N—does not contain the immediately prior state, NP, or any state prior to that, the state contains only itself; that is, it conforms to the single-state assumption; i.e., the history is external.

However, now consider the way in which the *comments* are structured in the same phrase, *the angry student*. The comments, i.e., additions of information, are actually applied sequentially thus:

student → angry student → the angry student

That is, *angry* is a comment on *student*; and *the* is a comment on *angry student*.

Now observe that the final structure produced by these comments, i.e., *the angry student*, contains the previous structure *angry student*, which contains the still prior structure, *student*. Because the final state contains each of the successive preceding states, this means that the final state is seen as a *trace* of the previous states. Thus, when one is presented with the structure, *the angry student*, one makes the multi-state assumption about its structure of comments. But this means that the backwards progression, through the succession of comments, is carried out by *internal* inference.

This leads us to the following crucial conclusion: The generativity embodied in a phrase by virtue of its structure of comments, i.e., its structure of *information-adding*, is fundamentally different from the generativity defined by Phrase-Structure Rules. The structure of comments, in the phrase, embodies *internal* history, and Phrase-Structure Rules embody *external* history.

In relation to this, observe also that, under the Phrase-Structure Rule,

$$NP \rightarrow DET, \ A, \ N$$

the sub-constituents come into existence all at the same time, whereas the sub-constituents, understood as comments, are introduced successively, thus:

student → angry student → the angry student

That is, in the configuration of comments, the sub-constituents are *generatively* related to each other; whereas under the Phrase-Structure Rule, they are not.

Thus, we can re-emphasize here a point made in section 7.3, that Chomskian syntax does not contain important aspects of the generativity in constituent structure: A Phrase-Structure Rule does not distinguish between the different generative levels that exist between sub-constituents, by virtue of the *comment* or *informational* relationships between sub-constituents. A comment structure, i.e., structure of information-adding, necessarily assigns the sub-constituents to different generative levels. That is, Chomskian syntax is less generative than comment structure.

Therefore, the sub-constituents in a Phrase-Structure Rule belong to the same level of *constituency*, whereas, in the comment structure, consisting of the same sub-constituents, these sub-constituents belong to different *constituent* levels. In fact, in a Phrase-Structure Rule, the sub-constituents are not only assigned to the same level of the constituent structure, but the structure within that level is trivial: It is merely a list of the sub-constituents; i.e., the sub-constituents are structurally dissociated from each other. In contrast, considering the same phrase as a structure of comments, i.e., information-adding, the sub-constituents are comments acting on each other, and they are therefore structurally related to each other. In fact, they are nested within each other.

Let us finally, consider the directionality of the arrows in the two types of structure: the Phrase-Structure Rule,

$$NP \rightarrow DET, A, N$$

and the comment structure,

student → angry student → the angry student

Notice that, in the former, the arrow goes from the constituent, NP, to the sub-constituents, DET, A, N. However, in the latter, the arrows go in the reverse direction, from the sub-constituents to the constituent. This reversal of arrows reflects the fact that a Phrase-Structure Rule specifies how a constituent is *broken down* into sub-constituents, whereas, in the configuration of comments, the sub-constituents are *built up* into a constituent.

The comparisons we have made above between Phrase-Structure Rules and comment structure, apply also between X-Bar Syntax and comment structure. We can therefore see that these comparisons exist in fact between the Chomskian Constituency Rules (Phrase-Structure or X-Bar) and comment structure. These comparisons can be summarized in the following six points:

(1) **Chomskian Constituency Rule:** The rule creates *external* history.
Comment Structure: The structure embodies *internal* history.

(2) **Chomskian Constituency Rule:** The sub-constituents are not generatively related to each other, and come into existence at the same time.
Comment Structure: The sub-constituents are generatively related to each other, and come into existence as successive comments.

(3) **Chomskian Constituency Rule:** The sub-constituents are assigned to the same *constituent* level.
Comment Structure: The sub-constituents are assigned to different *constituent* levels.

(4) **Chomskian Constituency Rule:** The sub-constituents are structurally dissociated from each other, being simply members of a set.
Comment Structure: The sub-constituents are structurally related to each other, being *comments* on each other. In fact, they are *nested* comments.

(5) **Chomskian Constituency Rule:** The rule maps constituents to sub-constituents.
Comment Structure: The structure maps sub-constituents to constituents.

(6) **Chomskian Constituency Rule:** Constituent structure is one of *breaking down.*
Comment Structure: Constituent structure is one of *building up.*

7.14 Comment Structure and Syntactic Structure

We saw in the previous section that the generativity embodied in a phrase by virtue of its structure of comments, i.e., information-adding, is fundamentally different from the generativity defined by Chomskian constituency rules. The question we consider in this section is whether this difference should exist.

The Chomskian constituency rules are *syntactic* rules; i.e., they describe the *distributional* properties of constituents. In contrast, the comment structure is a *pragmatic* organization of constituents; i.e., an organization in terms of certain communicative functions. Should there be a correspondence between constituent structure viewed from a syntactic viewpoint, and constituent structure viewed from a pragmatic viewpoint?

In one sense, it would not be surprising to find a correspondence: the syntax ultimately exists for a single purpose—to encode communicative functions. But which communicative functions? Is there evidence that the communicative function of comment in embodied in syntactic phenomena? We would argue that there is, as follows.

Recall that, in section 7.3, we noted that the generative rules of X-Bar Syntax do not express actual modification despite giving the illusion that they do by using the notation, X, $\overline{X}$, $\overline{\overline{X}}$. We saw that, if one replaces the notation X, $\overline{X}$, $\overline{\overline{X}}$, by non-prejudicial notation, one cannot infer, from the rules, which constituent is the Head, i.e., which constituent is being modified.

Linguists deduce the Head by using a rule outside the system of generative rules. This additional rule states that the Head X, must have the same *syntactic features* as its modifications $\overline{X}$ and $\overline{\overline{X}}$. The *syntactic features* of a syntactic category give a description of the category in terms of a restricted set of supercategory labels. For example, Chomsky (1974) has suggested that the four major word-level categories, Verb, Adjective, Noun, Preposition, can be analyzed as configurations of just two binary syntactic labels: [$\pm$V], which denotes *verbal* or *non-verbal*; and [$\pm$N], which denotes *nominal* or *non-nominal*. Thus, a Verb is [+V, −N]; an Adjective is [+V, +N]; a Noun is [−V, +N]; and a Preposition is [−V, −N].

Now the additional rule in X-Bar theory states that the syntactic features of a Head X are shared by $\bar{X}$ and $\bar{\bar{X}}$. In this way, one can identify the Head produced by the generative rules. That is, although the generative rules do not determine which constituent is to be regarded as the Head, and which constituents are to be regarded as the modifiers, when one applies the additional rule, one can then establish the Head and modification status of the various constituents. This means that, although the generative rules do not express modification, the syntactic theory, *as a whole*, does.

What we argue therefore is that the modification expressed by the syntactic theory, as a whole, is the syntactic manifestation of the comment structure. This claim is not difficult to corroborate: X-Bar Syntax always chooses the Head of a Noun Phrase, Verb Phrase, Adjectival Phrase, Adverbial Phrase, etc., to be the element to which modifications are applied, and these modifications are always functionally in the form of comments on the Head.

We can now see that X-Bar Theory has a considerable weakness. Whereas the theory, *as a whole*, expresses the modification structure, the latter structure is not expressed in the generative rules themselves. The major weakness is that modification is a *generative* phenomenon and is therefore what the generative rules *should* express.

We thus need to replace the generative rules of X-Bar Syntax with alternative generative rules that fully express the generativity in the modification structure. We propose that the alternative rules should be the rules we introduced in section 7.3, i.e., the Modification Rules, or more precisely, one of the two types of Modification Rules introduced in this chapter: the Accumulation Rules. In short, we argue:

> *The generativity inherent to the comment structure of a sentence is expressed* **syntactically** *in the Accumulation Rules. Furthermore, these rules define the hierarchical constituent structure of the syntax of a sentence.*

Observe now that, given a Phrase-Structure Rule such as

$$NP \rightarrow DET,\ A,\ N$$

the associated Accumulation Rule

$$N \rightarrow A(N) \rightarrow DET(A(N)) = NP$$

is also expressed in terms of syntactic categories, for example: N, A, DET, NP. However, in the Accumulation Rule, the syntactic categories are *functions*, in the mathematical sense, i.e., entities like f in the expression $f(x)$. But recall now, from section 7.2, that a syntactic category is a collection of elements with the same *distributional* properties. This means that, under our view, a syntactic category is a set of functions that have the same distributional properties. Thus, syntax, as the study of distributional properties, is the study of the distribution of these functions. It is a conclusion of this section that the hierarchical distribution of these functions is determined by Accumulation Rules.

7.15 Fitting Accumulation Rules Together with Movement Rules

In the preceding section, we examined Accumulation Rules and proposed that they simultaneously have two roles: (1) They define the hierarchical constituent structure of a sentence; and (2) they are the syntactic expression of the comment structure of a sentence. We can can put these two roles together by saying that the Accumulation Rules define *comment syntax*, i.e., the *syntax by which information is added.*

In sections 7.4 to 7.6, however, we saw that, besides Accumulation Rules, there are Modification Rules of another type involved in defining the generative structure of a sentence: Movement Rules. Whereas Accumulation Rules build up hierarchies of constituents, Movement Rules re-arrange the constituents.

Now, as we saw in section 7.4, Movement Rules, or *transformations*, are defined in Chomskian syntax as changing word order. However, we saw that this view does not generalize to languages such as those of the Philippine typology. Nevertheless, we also saw that there is a view of transformations that does generalize across languages: Transformations, or Movement Rules, re-arrange the topic-comment structure. For example, passivization moves the non-AGENT to the primary topic position of the sentence. This definition of passive applies to English, where passivization is realized by a word order change, and to Philippine where passivization is realized by attaching the *?ang* prefix, and does not involve change in word order. What is common to both languages is that the non-AGENT has been moved to a different topic-comment status.

The above view, that Movement Rules re-arrange the topic-comment structure, complements exactly the view we have developed of Accumulation Rules—the rules that build up the hierarchical constituent structure—as follows: We have argued that the hierarchical constituent structure should be understood as reflecting the comment structure, the structure of information-adding. That is, constituents correspond to comments, and are applied to each other using the Accumulation Rules in a way that they syntactically correspond to the action of commenting on each other.

We can now see how Movement Rules, as re-arrangements of constituents, can be understood as re-arranging comment structure: These rules re-arrange the constituent structure as determined by the Accumulation Rules, and the latter encode the comment structure. Thus, the re-arrangements created by Movement Rules, are re-arrangements of constituents with respect to position in the comment hierarchy.

Re-arrangement of position in the comment hierarchy can include the following: A constituent might be removed from being a comment on a constituent A, to becoming a comment on constituent B. Alternatively, the comment relationship between two constituents might be reversed. This latter situation can be illustrated with passivization. In an active sentence, such as

Bill kicked the door

the PATIENT, *the door*, is a comment on the Verb, *kicked*, i.e., specifying *what* was kicked (that is, syntactically *the door* is the complement of *kicked*). In the passive version,

The door was kicked by Bill

the PATIENT, which is still *the door*, is now the primary topic of the sentence and the verb *kicked* is now a comment on *the door* (i.e., structurally *kicked* is now part of the predicate of *the door*). Thus, whereas, in the active sentence, the PATIENT was a comment on the Verb, in the passive sentence, the verb is a comment on the PATIENT. That is, the comment relationship has been *reversed.*

Now in English, the comment hierarchy is encoded (in part) by the linear order rules. Therefore, re-arrangement of constituents with respect to their comment structure can be encoded by re-arrangement in linear order. However, this particular means of encoding is not a necessary one. We saw that, in Philippine, comment hierarchy is encoded (in part) by the prefixes. Thus re-arrangement of constituents with respect to their comment structure is encoded by re-arrangement of prefixes.

We must therefore distinguish between constituent structure as determined by Accumulation Rules and constituent structure as determined by linear order rules. The latter rules are merely a possible means of encoding the former. Movement Rules act on the constituent structure determined by the Accumulation Rules, not on the linear order rules, even though it may be possible to encode their action in terms of the linear order rules.

We can therefore see that Movement Rules are locked into the Accumulation Rules. They are defined in terms of structure as determined by those rules. We can express this by summarizing the position we have been elaborating:

> **Accumulation Rules**. These rules (1) define the hierarchical constituent structure of a sentence, and (2) express the comment structure. In fact, we put these two roles together by saying that the rules define the *comment syntax*, i.e., the syntax by which information is added.
>
> **Movement Rules**. These rules (1) define re-arrangements in the hierarchical structure produced by the Accumulation Rules, and (2) express re-arrangements of the comment structure. In fact, we can put these two roles together by saying that the rules re-arrange the *comment syntax*.

7.16 Language as Memory

We can now return to considering the nature of generativity in syntax. Recall that, in section 7.8, we had distinguished two interpretations of syntactic generativity. On the one hand, generativity can be given an *atemporal* interpretation, in which it is simply a means of characterizing the static structure of relationships in a sentence, i.e., as analogous to a scaffold. Alternatively, generativity can be given a temporal interpretation in which the generative relationships are understood as expressing actual temporal history. Furthermore, this history can be understood as generated by causal interactions.

We now have constructed the machinery to show that the temporal interpretation is the cognitively appropriate one. In this section, we shall argue that the generative rules are a means used to represent actual *temporal* history rather than static structure; and, in the next section, we shall argue that this history is understood as fundamentally *causal*.

In the machinery we have put together, two crucial factors have been (1) understanding that syntax has a more extensive generative structure than is identified in the Chomskian system; (2) understanding that this fuller generative structure is the embodiment of discourse function. Note that, without discovering generativity to be more extensive, we could not have understood the generativity to be functional; i.e., the less generative Phrase-Structure Rules do not embody the functionality of Accumulation Rules.

Realizing that the generativity embodies function, we are now able to see that the generativity embodies time, as follows: We have seen that the generativity has two syntactic forms, (1) the Accumulation Rules, and (2) the Movement Rules. As syntactic constructs, these rules do not necessarily embody time, and are equally amenable to the atemporal or temporal interpretations. However, understood *functionally*, these two constructs are strongly temporal. Let us consider them in turn.

The Accumulation Rules encode the comment structure of a sentence. A comment is the addition of new information to old, i.e., shared, information. Thus, *Accumulation Rules actually represent the means by which information is accumulated.* The accumulation is literally temporal. At some moment in time, only certain information is shared. At a subsequent moment, new information is added, and this new information, once uttered, becomes shared information. Thus the total shared information increases temporally. This, of course, is not surprising: a central purpose of communicating is to increase the shared information.

Each Accumulation Rule encodes the increase in *information*: A rule prescribes a succession of comments, and each comment is added information about the previously accumulated comments. Thus each rule structurally expresses the increase of shared information. Note that each rule creates internal history. Thus the use of the rule backwards in time leads back to the shared information on which the new information is built. The temporal nature of this structure is emphasized by the fact that linguists refer to the shared information as *old* information, and the comment as *new* information—the terms *old* and *new* being temporal terms referring literally to the successive states of the information over time.

We are therefore led to this important conclusion:

> *The generativity inherent in the hierarchical constituent structure of a sentence is temporal. The generativity is given by the Accumulation Rules each of which prescribes a means by which information can be successively increased over time. The generativity of hierarchical constituent structure is therefore the temporal process of accumulating information.*

Recall earlier that we described sentences using a geological analogy. We can now understand this analogy more clearly: We had said that a sentence is like a geological formation in that it represents successive layers of deposits. We can now understand that the deposits in a sentence are *deposits of information.* Information is placed upon some already existing layer of information. This newly added layer then provides the bed on which a new layer of information is placed. When one takes a cross-section through the layers, one has literally a record of time. Any layer, no matter how deeply embedded, was the outer-most layer at some earlier point in time. Just as the layered structure of a geological formation is a record of how the material content of the formation was built, the constituent structure of a sentence is a record of how the information content of the sentence was built.

Let us now turn to the generativity created by Movement Rules. These rules express re-arrangements of constituents with respect to position in the comment hierarchy. We shall now argue that this generativity is actually temporal. The argument will be given first in a simple form, for non-linguistic readers, and then, in a more technical form, for readers more used to dealing with pragmatic concepts.

First, the simple form: Recall that discourse is structured to introduce new information and to encode its status as it shifts from being new information to becoming shared information. In section 7.11, we saw that Movement Rules express exactly this shift in status. For example, passivization shifts a PATIENT from comment to topic—a change in status that is encoded in English by a movement of the PATIENT left-ward in a sentence—and is encoded in Bikol by attaching the topic prefix *ʔang* to the PATIENT. Thus, Movement Rules correspond to the promotion of a constituent from some comment level to a level of greater topicality; i.e., from new information to old. Thus, Movement Rules encode actual *temporal* change; i.e., change *over time*. This means that the generativity created by Movement Rules corresponds to actual temporal processes.

Let us now give the more technical version of the argument. Movement Rules correspond to changes in *presupposed comment relationships*. Such relationships might be instantiated in the prior discourse, but need not be. For example, they can be comment relationships that are assumed to normally hold between particular types of constituents. For instance, it is generally presupposed that a PATIENT, when it occurs, will be a comment on the verb. Passivization expresses a change with respect to this presupposition. It shifts the PATIENT to the topic position, thus defining the PATIENT as old information—but the changed status of PATIENT is with respect to its expected status as new information rather than its instantiated status as such in the prior discourse.[5] Now, since presupposed relationships are, by definition, *temporally prior* to new relationships, the Movement Rules necessarily correspond to actual temporal changes. Thus, the generativity expressed by Movement Rules corresponds to actual temporal processes.

We have seen, in this section, that the generative rules of syntax are not simply the expression of the static scaffold of relationships between constituents in a sentence. The Accumulation Rules express the *temporal* process of accumulating information. The Modification Rules express the *temporal* process of altering informational status. The important conclusion we are led to in this section is that sentences are pieces of process-memory, archeological relics, in which time is locked. When the listener assigns a grammatical structure to the sentence, he or she is performing an archaeological analysis on the sentence, i.e., unlocking time from the sentence. The sentence is this process-history.

7.17 Language as Causal History

In the preceding sections, we have seen that syntactic structure conforms to our Asymmetry Principle, which states that distinguishability in the present is understood as arising from indistinguishability in the past. The Accumulation Rules and Movement Rules each instantiate this principle in a different way: the Accumulation Rules are hypothesized as building up, over time, the distinguishabilities represented by the constituents; the

Movement Rules create, over time, disparities between comment levels held by a single constituent.

While we have used the Asymmetry Principle in this chapter, we have not made use of the other principle, in Chapter 1, based on the notion of asymmetry: the Second System Principle. This principle states that a system that has increased its asymmetry over time must have interacted with a second system. Let us now use the Second System Principle in relation to the linguistic issues of this chapter.

We have seen that syntax involves successive increases in asymmetry over time. This means, by the Second System Principle, that an additional system is understood as having been involved in producing the asymmetry increase. What is the second system?

The answer is that the second system consists of the intentions of the speaker. These intentions are *causal actions* on the syntactic structure. The causal actions introduce the asymmetries. The asymmetries are therefore the *effects* of the actions; i.e., effects of the intentions.

For example, consider those asymmetries created by shifts in topic; i.e., transformations. These asymmetries express different speech acts; e.g., ordering, inquiring, declaring, etc. For instance, consider the change in going from the statement

(Ex 7.10a) Mahler should conduct the symphony.

to the question:

(Ex 7.10b) Should Mahler conduct the symphony?

What has happened in the transition is that the modal, *should*, has been moved to the front of the sentence, making it the topic of discussion (Halliday, 1985). This shift in topic expresses the *intention* of inquiry. The *intention* is the second system that caused the syntactic alteration; i.e., the shift of the modal.

Thus, under the above causal framework, the syntactic asymmetries can be used by the listener to *infer* the intentions of the speaker. Each syntactic asymmetry is an *effect* of a particular intention acting on the syntax, in the same way that a dent in a door is the effect of an impact of a certain force in a certain direction. The *type* of syntactic asymmetry can be used to infer the *particular* intention that caused the asymmetry in the same way as a dent can be used to infer the size and direction of the impacting force that produced the dent. Thus, for example, the disparity between the two positions of the modal *should* in Ex 7.10a and Ex 7.10b indicates that the *type* of intention that produced Ex 7.10b was *inquiry*.

Thus, we can now fully extend our analogy of sentences as pieces of archaeology. We had said previously that, when a listener assigns a grammatical structure to a sentence, he or she is performing an archaeological analysis on the sentence, in the sense that he or she is retracing past states of that sentence. However, we can now see that, by unlocking time from the sentence, the listener is able to trace the sentence back not just to its past states, but to its *causal* origins. These causal origins are the intentions that produced the sentence. The listener's role is therefore analogous to the archaeologist's in a more profound sense: The archaeologist examines objects in order to retrieve not just the past states, but the causal factors, e.g., the social and economic conditions which were responsible for the objects. The archaeological objects are *relics* of those social and economic conditions in the same way that sentences are relics of a speaker's intentions. And, in the same way that an archaeologist unlocks time from those relics in order to retrieve the causal conditions, the listener unpacks time from a sentence in order to retrieve the intentions of speakers.

7.18 Summary

In this chapter, we have argued that understanding a sentence is an example of the *process-inference problem*—the central concern of this book. That is, we have seen that language conforms to our Representation-is-Explanation Principle (section 4.1); i.e., human beings represent linguistic utterances as states in a process-history that causally explains those utterances. In relation to this proposal, recall that we regarded the Representation-is-Explanation Principle as *characterizing* cognition in this way: We proposed that a system is a *cognitive* system if it can be described as having a level in which the medium that it processes consists of causal explanations. Thus, in accord with this principle, we have seen that linguistic representations are causal explanations.

Our argument proceeded in this way. We showed that Chomskian Constituency Rules do not capture the full syntactic generativity of hierarchical constituent structure. For example, they cannot identify the Head of a phrase, and thus establish the modifying relationships between constituents. We therefore suggested an alternative set of constituency rules—the Modification Rules, which explicitly encode the modifying relationships and are thus more fully generative than the Chomskian Rules. Modification Rules differ from the Chomskian Rules in six major ways that allow them not only to be more thoroughly generative, but to be generative in a very different way.

Our constituency rules are in fact one of two types of Modification Rules. They are Accumulation Rules. The other type are the Movement Rules. We saw that a standard argument used by linguists to justify the existence of movement operations is, in fact, based on our Asymmetry and Symmetry Principles, where the Asymmetry Principle acts on distinguishabilities in positional information and the Symmetry Principle is instantiated by the Projection Principle. We also saw that Accumulation Rules and Movement Rules share three fundamental properties: they are conjectured in order to satisfy the Asymmetry Principle; they are both internal; and their use is stratified.

In Chomsky's view of syntax, generativity is understood as *atemporal*—temporal aspects being consigned to performance issues. However, we argued that syntactic generativity does in fact correspond to actual temporal process. In order to argue this, we required functional considerations: We found that Accumulation Rules correspond to the temporal process of accumulating shared information; and Movement Rules correspond to the temporal process by which new information is promoted to shared information.

Finally, we argued that, in accord with the Second System Principle, an additional system does exist that causes the successive increases in syntactic asymmetry over time. The additional system consists of the intentions of the speaker. Therefore, in assigning a grammatical structure to a sentence, the listener is not only retrieving the history of the sentence itself but the causal interactions that created the history, i.e., the interactions between the syntax and the speaker's intentions.

In short, we argued that the relationship of a listener to a sentence is that of an archeologist to a relic: the sentence is a piece of history, and the listener attempts to unpack this history not only by retracing the temporal development of the sentence, i.e., its syntax, but by inferring the the causal interactions that produced that development.

8 Art

8.1 Introduction

Art is perhaps the most inexplicable phenomenon of the human species. One can explain all other higher mental activity, such as science or mathematics, by simply pointing to the obvious benefits they have conferred. But why do people pursue art so passionately—whether as an audience for art, or as creators of art? Why do people travel great distances to attend concerts and see art exhibitions? Why do people cover the walls of their apartments and their offices with paintings? Why do people listen to music almost endlessly—in their homes, their cars, in restaurants, in shops, etc? Why do people devote considerable effort to making all their environments aesthetically pleasing? Why are paintings the most expensive objects in human history?

We are going to argue that art has a profoundly important role for survival. Our main thesis will be that art is intimately related to the phenomenon of *causal explanation*. In fact, we shall claim that *the aesthetic response is the mind's evaluation of causal explanation.*

This brings aesthetic phenomena into the very center of cognition. By our Representation-is-Explanation Principle, any cognitive representation is a causal explanation. Thus we shall argue that aesthetics is the mind's evaluation of cognitive representation. Again, since we have argued that the *content* of cognition is causal explanation, we will therefore have to conclude that aesthetics is the evaluation of cognitive content.

We shall find that aesthetics is central to the main topic of this book: the history-recovery problem, i.e., the problem of assigning causal explanation to a stimulus set. Since we shall argue that aesthetics is the evaluation of causal explanation, we must conclude that aesthetics is the evaluation of success in solving the history-recovery problem.

On a broad level, our argument will proceed as follows. We shall first examine a number of very different paintings, and analyze them in detail in order to show that what people regard as the artistic content of an art-work is the process-history recovered from the work. More precisely, the aesthetic response to an art-work is a response to the causal explanation assigned to the work's shape structure, i.e., to the asymmetries in the work.

We shall then argue that art-works contain the maximal amount of asymmetry that an individual is capable of causally explaining. In fact, we shall see that the desire for art-works appears to be part of a general desire that human beings have for obtaining asymmetry. This has been demonstrated by a large body of psychological data that shows (1) that, in all individuals, there is an *asymmetry-seeking drive*, a drive to continually find or increase asymmetry; and (2) that aesthetic response evaluates the success of this drive, i.e., aesthetic pleasure increases with the greater amount of asymmetry that is found.

In fact, we shall see that the asymmetry-seeking drive attempts to push the amount of asymmetry, in a situation, up to the maximal level that the individual can *causally explain*. The conclusion we shall draw from this is that, rather than understanding that the aesthetic response evaluates the amount of *asymmetry* which has been obtained in a situation, one should more correctly understand that the aesthetic response evaluates the amount of *causal explanation* which has been assigned to that situation. That is, we shall argue that the aesthetic *content* of a situation is the *causal explanation* that has been assigned to it. However, our Representation-is-Explanation Principle implies that, in the representation of any situation, the *cognitive* content is the *causal explanation*. Thus, we will be forced to conclude that *aesthetic content and cognitive content are the same.*

8.2 Art-Works

As was stated above, the main proposition that will be offered and substantiated, in this chapter, is that aesthetic response is a response to causal explanation. In order to substantiate this proposition, it will be valuable first to show that aesthetic response, in the particular case of *art-works*, is a response to causal explanation; i.e., the causal explanations assigned to the works. Since art-works are valuable to us only because of the aesthetic response they induce in us, this will imply that our response to an art-work is confined *purely* to the causal explanation we have assigned to it. In other words, an art-work is, for us, purely its causal explanation. Since this causal explanation is the process-history we have inferred from its shape-structure—the following conclusion will result:

FIRST ART-WORK PRINCIPLE. *An art-work is the process-history inferred from it.*

Accordingly, we shall see that, when people talk of the *meaning* of an art-work, they are actually referring to the process-history they have inferred from it. Furthermore, their experience of the work, while they are actually perceiving it, is the experience of deriving process-history from its shape-structure. That is:

COROLLARY OF THE FIRST ART-WORK PRINCIPLE. *The* **meaning** *of an art-work is the process-history inferred from it. The* **experience** *of an art-work is the experience of inferring that history, i.e., the experience of solving its history-recovery problem.*

We regard the First Art-Work Principle and its Corollary as so important to the argument we are presenting concerning cognition that we will devote considerable amount of effort to corroborating them. Our corroboration will consist of taking four very different paintings and analyzing their structure systematically and in great detail. The analyses will reveal that the paintings are exactly defined by their inferred process-history.

8.3 Using the Extrema-Based Rules

The different chapters of this book have developed rules by which the different features of the stimulus are used to extract process-history. In particular, in Chapters 1 and 2, we elaborated a system of rules to show how curvature extrema provide a rich source of process-history. We shall now take the extrema-based rules of Chapters 1 and 2, and apply them, in detail, to paintings. In fact, we shall use the rules mainly to analyze the *external* structure derivable from the curvature extrema in the paintings. A more complete analysis would look also at the internal structure, and would use also the rules developed in the other chapters of this book. This more complete form of analysis will be carried out in a book we are preparing that is devoted entirely to the structure of paintings. However, we believe that the analyses we carry out in the present chapter—analyses purely of the external structure derivable from curvature extrema—will persuade the reader that the process-structure of a painting defines the perceived composition and meaning of the painting. This conclusion will substantially corroborate (1) the First Art-Work Principle, which states that an art-work is the process-history inferred from it, and (2) the Corollary of the First Art-Work Principle which states that the meaning of the painting is the inferred history.

Before we begin, let us recall that the rules we developed, for the extraction of history from curvature extrema, begin with two basic rules. The first rule is illustrated in Fig 8.1. The rules states:

Figure 8.1
To each curvature extremum there is a symmetry axis leading to the extremum.

SYMMETRY-CURVATURE DUALITY THEOREM. *Any section of curve, that has one and only one curvature extremum, has one and only one symmetry axis. This axis is forced to terminate at the extremum itself.*

Thus, to each extremum in Fig 8.1, one has been able to assign a unique symmetry axis leading to the extremum.

The second rule is the following:

INTERACTION PRINCIPLE (backwards version). *Symmetry axes are the directions along which processes are hypothesized as most likely to have acted.*

Thus, according to this principle, a process acted along each axis shown in Fig 8.1. This rule is based on the Symmetry Principle, which states that symmetry must be preserved backwards in time.

Now, according to the Asymmetry Principle, the variation in curvature around the curve in Fig 8.1 must disappear backwards in time. Therefore, forwards in time, the curve is understood as having been deformed along the axes shown and in the direction of the resulting extrema.

Now let us take a number of paintings and apply these rules to the paintings.

8.4 Picasso's *Woman Ironing*

The first painting we shall take is Picasso's *Woman Ironing*, shown in Fig 8.2. The meaning of the painting is obvious. It shows a woman, bruised and pummelled by the outside world, bent over by exhaustion, and using what little strength she has left to hold herself upright and to carry out a simple task that has become too burdensome.

We shall now argue that the meaning of the painting is conveyed in its process-structure, and this structure is established by our rules for the recovery of history from curvature extrema.

Figure 8.2
Pablo Picasso, *Woman Ironing* (1904). Oil on canvas, 45¾ × 28¾″ (116.2 × 73 cm). Gift, Justin K. Thannhauser, 1978. Solomon R. Guggenheim Museum, New York. 78.2514 T41. Photographer: David Heald. Photograph © The Solomon R. Guggenheim Foundation.

8.4.1 The Main Curvature Extrema

Let us begin by asking the basic question that anyone writing about a painting asks: What is the composition? Everything in a painting is conveyed through the composition. When one establishes the compositional structure, one finds the key to the painting.

What we find, remarkably, is that the compositional structure of the painting is given by our rules for the extraction of history from curvature extrema. In fact, the compositional structure, as we will see with any painting, is its process-structure.

To begin: We simply apply our rules. The first rule, the Symmetry-Curvature Duality Theorem tells us that, to each curvature extremum, there is a unique symmetry axis terminating at the extremum. Thus, we shall simply go through the painting and locate all the curvature extrema, and draw in the associated axes. Then the second rule, the backwards Interaction Principle, tells us to assign a process along each axis. Therefore we will draw an arrow coincident with each axis.

Thus, to complete this section, we will now go through the main extrema of the painting. In the following sections, we will use our list of extrema to discuss the composition of the work.

(1) The Shoulder. The most pronounced curvature extremum of the entire painting is the top of the shoulder. In Fig 8.3, we have, in accord with our rules, placed a process along the symmetry axis terminating at this extremum; i.e., the upward arrow in the shoulder.

(2) The Arm-Pit. Another significant curvature extremum is the arm-pit under the chin. Again, in accord with our rules, we have placed an arrow along the symmetry axis terminating at this extremum.

(3) The Upper Boundary of the Head. The curve defining the upper boundary of the head is an arch. This curve, in fact, starts at the chin, moves along the jaw-line, then the ear, then along the upper line of the head, and down the left edge of the hair. This arch has a *curvature extremum* at the top. In Fig 8.3, we have shown the resulting process—that which terminates at the extremum. Observe that this vertical process is reinforced by the vertical line defining the left side of the arm.

(4 & 5) The Chin and Crown of the Head. The chin and the crown of the head are each curvature extrema, and each therefore has a process terminating at it, as shown in Fig 8.3. One can indicate these two processes

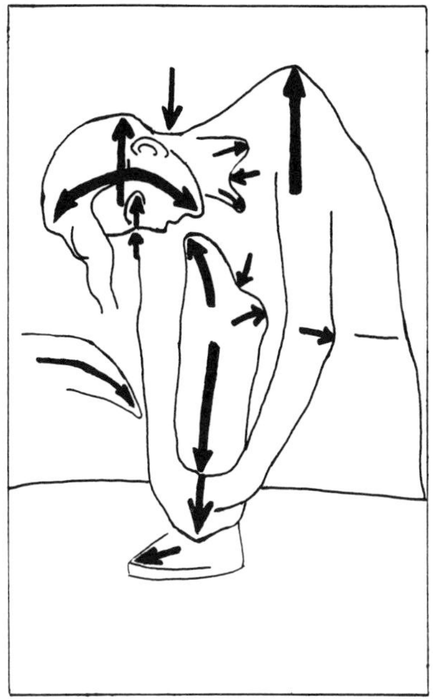

Figure 8.3
The main curvature extrema, with the associated processes, on Picasso's *Woman Ironing*.

by a single double-headed arrow—the arched arrow shown—because a single symmetry axis goes through the head from the chin to the crown. This axis defines the symmetry between two curves: the upper arch boundary of the head, and the lower boundary giving the face.

(6, 7, & 8) Eye-Socket, Bridge of Nose, Lateral Head Axis. Returning to the photograph of the painting, Fig 8.2, we see that the eye-socket is given by a shadow that has a number of extrema that echo various other structures in the composition. For example, the right side of the socket is given by a curved line that echoes the hairlines across the forehead. The left side of the socket echoes the left side of the head, etc. For ease of drawing, in Fig 8.3, we have indicated only one of the extrema on the socket: the upper most point. According to our rules, it has an associated upward process terminating at the extremum, and given by the arrow drawn in the eye-socket. Now observe that the bridge of the nose is a curvature extremum, and therefore also has an upward arrow terminating at it, as shown. The arrow drawn in the eye-socket continues the arrow in the bridge of the nose. Lastly, observe that the lateral arched arrow—from chin to crown—has a curvature extremum in the middle of it. This extremum must also have an upward process terminating at it—which we have not drawn in, for ease of reading the diagram. This latter upward process lies along, and continues, the upward processes in the eye-socket and nose-bridge. Note that the processes continue the left-hand line of the arm beneath.

(9) The Ear. The ear has the same structure as the head. This means that it has the same upward process and the same lateral arch-like process that we have defined for the head.

(10) The Hand-Mass. The curves formed by the two arms terminate at the hand-mass, whose most outer outline forms an extremum—one of the most important of the entire painting. In Fig 8.3, we have, in accord with our rules, placed a downward arrow along the symmetry axis terminating at this extremum.

(11) The Upper Line of the Neck. The dip in the upper outline of the neck is an extremum. Thus, in accord with our rules, we have placed a process along the symmetry axis terminating at this extremum—the downward arrow shown.

(12) The Neck-Shadows. In returning to the actual painting, Fig 8.2, we see that the dark shadow on the neck has a number of structurally important extrema. Three of the extrema, with the three associated processes are shown in Fig 8.3. The upper-most process points towards the shoulder; the middle process points to the head and reinforces the leftward direction of the head; and the lowest process points down, reinforcing the downward arrow in the chin, and the curved movement of the back.

(13) The Waist. The left-side of the waist (viewer's left) defines an extremum, and the associated process has been drawn in Fig 8.3.

(14) The Breast. The breast defines an extremum, and we have also drawn the associated process.

(15) The Elbow. The elbow on the right side of the painting defines an extremum and the associated process has been drawn.

(16) The Iron. The main extremum on the iron produces an arrow as shown. The arrow not only defines the direction of the ironing, but in so doing defines a movement transverse to the strong vertical movements.

(17) The Drapery. The drapery on the left-hand side, half-way up the picture, contains a "horn" shape. In accord with our rules, the extremum on this horn defines an arrow leading to that extremum, as shown. This movement reinforces both the bent structure in the body and the downward movement of the arms.

8.4.2 Meaning is History

Recall our statement of the meaning of the painting: "a woman bruised and pummelled by the outside world, bent over by exhaustion, and using what little strength she has left to hold herself upright and to carry out a simple task that has become too burdensome".

We shall now see that this description of the meaning is embodied in the process-structure of the painting.

The most dominant process is the upward one terminating at the tip of the shoulder (Fig 8.3). This describes the push of the woman upwards against immense exhaustion. The push is also reinforced by the upward arrows in the head. The exhaustion itself is given by the downward process leading to the top line of the neck, the elongated downward process leading to the finger mass, and the downward turning processes in the head leading to the chin and crown.

Besides the *upward* struggle against the *downward* exhaustion, another major theme in the painting is the opposition of the *outward* forces of the woman against the *inward* forces of her environment—an opposition that symbolizes her struggle against the environment. This can be illustrated in the following way. Observe that the painting is built around two identical shapes. One of the shapes is the outer solid bold line shown in Fig 8.4; i.e., the outline of the woman—along the outer edge of the shoulder, the neck, the head, and the arms. The other is an almost *mirror-image* of that shape and is given by the inner solid bold line in Fig 8.4; i.e., the inner outline of the woman.

Because the two shapes are almost mirror images of each other, there is a correspondence between the extrema on one shape and the extrema on the other. The correspondence is this:

OUTER SHAPE ↔ INNER SHAPE
shoulder ↔ armpit
neck ↔ breast
head ↔ waist
lower hand-line ↔ upper hand-line

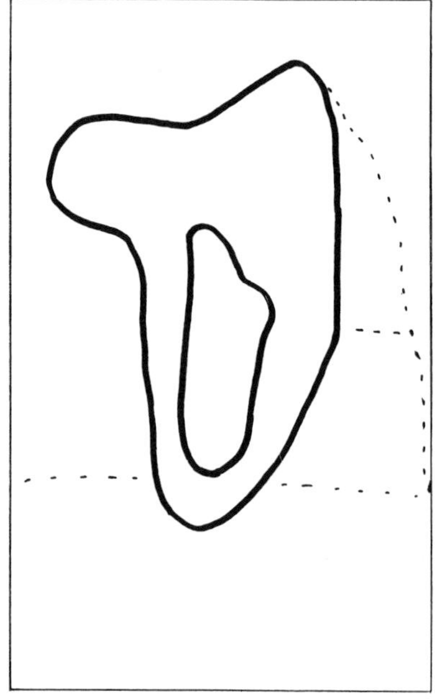

Figure 8.4
Two important related shapes in Picasso's *Woman Ironing*.

It is crucial to observe that the two shapes are figure-ground reversals of each other; i.e., the outer shape is the outline of solid figure and the inner shape is the outline of background or environment.

This means that the corresponding *processes* in the two shapes are figure-ground reversals of each other: In particular, a *protrusion*, in one shape, is an *indentation*, in the other; and vice versa. Thus we have:

OUTER SHAPE ↔ INNER SHAPE

protrusion: shoulder ↔ indentation: armpit

indentation: neck ↔ protrusion: breast

protrusion: head ↔ indentation: waist

protrusion: lower hand-line ↔ indentation: upper hand-line

Now recall our statement above that a major theme of the painting is the struggle of the outward forces of the woman against the inward forces of the environment. In each of the cases listed in the above correspondence, a protrusion represents an outward force of the woman, and an indentation represents the inward force of the environment. Thus the struggle is captured in the fight between protrusions (woman) and indentations (environment). In addition, Picasso creates an irony by using exactly the same shape to represent the two sides of the struggle. That is, by making the outline of the woman a figure-ground reversal of the outline of the environment, he converts any outward force of the woman into an opposing inward force of the environment. The forces with which she fights become the forces against which she fights.

In conclusion, we see that the process-structure of the painting embodies the meaning of the painting: (1) the woman's *upward push* against her own *downward exhaustion*; (2) the woman's *outward fight* against the environment's *inward attack*; and (3) the *ironic mimicking* of her outward fight by the inward attack. Because the meaning of the painting is embodied in the inferred process-structure, we see here a corroboration of (1) our First Art-Work Principle which states that an art-work is the process-history inferred from it, and (2) the Corollary of this principle which states that the meaning of an art-work is the inferred process-history.

8.4.3 Extraction of Further History from Picasso's *Woman Ironing*

In the previous two sections, we have seen how our two basic rules for the extraction of history from curvature extrema—the Symmetry-Curvature Duality Theorem and the backwards Interaction Principle—provide, without any difficulty, major compositional factors in Picasso's painting. Recall now that, in Chapter 2, we established an additional set of rules—to be reviewed shortly—in order to extract still further history from curvature extrema. We shall now see that Picasso also uses these additional rules in structuring the paintings.

The additional rules define the *changing* and *ordering* of processes at curvature extrema. It is first necessary to recall that there are four types of curvature extrema as illustrated in Fig 8.5. The extrema are two extremes of sharpness, M^+ and m^-, which can be seen alternating along the top of Fig 8.5a; and two extremes of flatness, m^+ and M^-, which can be seen respectively at the sides of Fig 8.5a and in the middle of the bay in Fig 8.5b.

Recall now that, despite the fact that the four extrema describe merely degrees of curvature, they each, remarkably, correspond to a *process* term

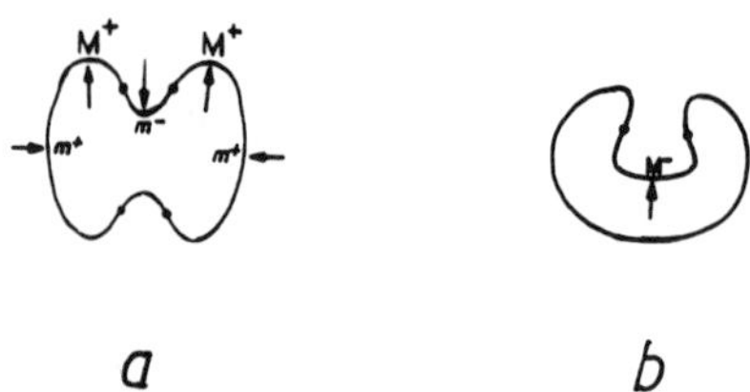

Figure 8.5
Illustration of the four kinds of curvature extrema.

in English—a fact we captured in our Semantic Interpretation Rule:

SEMANTIC INTERPRETATION RULE

M^+ protrusion

m^- indentation

m^+ squashing

M^- internal resistance

In Chapter 2, the additional rules we developed constitute a grammar that describes how these processes can change over time by continuing or bifurcating. The rules are expressed in terms of changes at the corresponding extrema. There are six such rules:

PROCESS GRAMMAR

A squashing continues till it indents:

$Cm^+ : m^+ \rightarrow Om^-O$

An internal resistance continues till it protrudes:

$CM^- : M^- \rightarrow OM^+O$

A protrusion bifurcates:

$BM^+ : M^+ \rightarrow M^+m^+M^+$

An indentation bifurcates:

$Bm^- : m^- \rightarrow m^-M^-m^-$

A protrusion is introduced:

$Bm^+ : m^+ \rightarrow m^+M^+m^+$

An indentation is introduced:

$BM^- : M^- \rightarrow M^-m^-M^-$

Recall also that, in Chapter 2 (sections 2.20-2.22), we proposed a principle that orders the emergence of processes in a history and also orders the grammatical rules that determine the successive emergence. The principle is this:

SIZE-IS-TIME HEURISTIC. *In the absence of information to the contrary, size corresponds to time. That is, the larger the boundary movement, the more likely it is to have (1) started earlier, and (2) taken longer to develop.*

One realizes this principle—i.e., one obtains the ordering of processes and grammatical rules—by blurring or de-blurring. As one blurs a shape, the curvature variation disappears in the order of smaller details before larger. Conversely, given the blurred shape, when one de-blurs it, the curvature variation re-appears, in the order of larger details before smaller. Note that we argued that processes are conjectured to explain the introduction of curvature variation. Thus de-blurring introduces what processes introduce, and corresponds to the forward time direction. Blurring therefore corresponds to the reversal of time.

Now let us return to Picasso's painting. First consider the tip of the shoulder. In Fig 8.3, we drew the tip as a M^+ extremum. The associated process, in accord with the Semantic Interpretation Rule, was shown as a protrusion, as given by the upward arrow. This diagram represents a coarse, i.e., slightly blurred, version of the painting, i.e., omitting some of the finer detail. Fig 8.6a represents the shoulder structure under this coarse view.

Now, when one looks at the painting itself, that is, Fig 8.2, and examines the shoulder in greater detail, one finds that the shoulder has a flattening at the top. This means that its structure is that shown in Fig 8.6b. First observe that the flattening is given by a m^+ extremum at the center. Two

Figure 8.6
De-blurring at the shoulder in Picasso's *Woman Ironing*.

M^+ extrema flank the central m^+. Observe that, by our Semantic Interpretation Rule, the m^+ extremum must mean a *squashing* process, i.e., as shown by the downward arrow leading to the extremum. Furthermore, the M^+ extrema must each be *protrusions*; i.e., as shown by the two arrows leading up to these two extrema (in Fig 8.6b).

Now, in going from the blurred version, Fig 8.6a, to the de-blurred version Fig 8.6b, the extrema have changed from the single M^+, in Fig 8.6a, to the string of extrema $M^+m^+M^+$, in Fig 8.6b. That is, we have:

$$M^+ \rightarrow M^+m^+M^+$$

But this conforms exactly to the third rule in the above Process Grammar. That is:

A protrusion bifurcates:

$$BM^+ : M^+ \rightarrow M^+m^+M^+$$

That is, in de-blurring, i.e., going from Fig 8.6a to Fig 8.6b, the protrusion arrow has split, one branch going to the left and the other to the right, while a squashing arrow has been introduced in the center from above.

Now let us use our Size-is-Time Heuristic. This principle implies that de-blurring recapitulates time—in the forward time direction. Thus, the transition from Fig 8.6a to Fig 8.6b represents forward time.

Some consideration reveals that this transition actually embodies a major theme of the painting. Recall that we said that a major theme is the fight of the outward forces of the woman against the inward counter-forces of the environment. Fig 8.6a represents an outward force of the woman. In Fig 8.6b, an inward counter-force of the environment has been introduced; i.e., in flattening the tip of the shoulder, the environment is literally *squashing* her intent. Note that, in accord with our Semantic Interpretation Rule, *squashing* is the process corresponding to a m^+ extremum.

In the *narrative* story of the painting, the transition from Fig 8.6a to Fig 8.6b represents forward time, because the woman's shoulder was smooth and rounded earlier in her life. It is time and its punishing forces that have caused a flattening. Thus, the temporal order from Fig 8.6a to Fig 8.6b, as contained in the story of the painting, is the order predicted by our Size-is-Time Heuristic.

In conclusion, we see that the *process-structure* of the shoulder, as

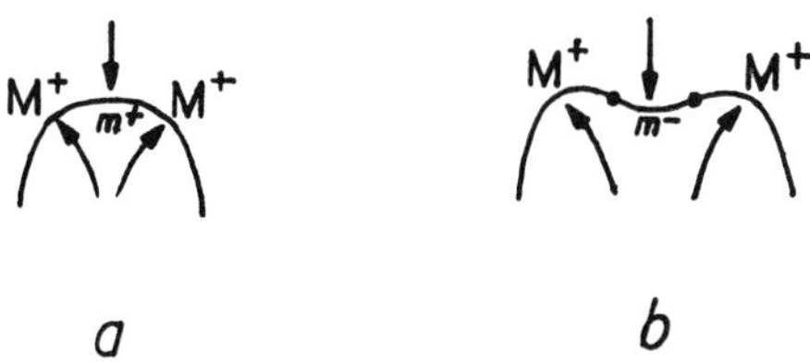

Figure 8.7
Further de-blurring in the shoulder of Picasso's *Woman Ironing*.

captured by our grammatical rule

$$BM^+ : M^+ \rightarrow M^+m^+M^+$$

together with our Size-is-Time Heuristic, embodies a central theme of the painting. That is, we have, once again, a corroboration of our First Art-Work Principle (Corollary), which states that the meaning of an art-work is the history inferred from it, and the experience of an art-work is the experience of inferring that history.

Let us consider the shoulder still further. In Fig 8.6b, we drew the center of the top as a m^+ extremum. The associated process, in accord with the Semantic Interpretation Rule, was shown as a squashing, and was given by the downward arrow. As we said, this diagram represents a de-blurred version of the painting.

While Fig 8.6b shows the work at some level of de-blurring, one can actually de-blur the painting still further. Thus, let us go back to the painting itself, that is, Fig 8.2, and examine the shoulder in still greater detail. What we find is that the shoulder has a dip in the top. This means that its structure is that shown in Fig 8.7b. First observe that the dip is given by a m^- extremum at the center. By our Semantic Interpretation Rule, this extremum must mean the presence of an *indentation* process. Notice also that, in Fig 8.7b, two dots have been placed on either side of the m^- extremum, indicating the introduction of two points of 0 curvature; i.e., two points of complete flatness.

Fig 8.7a shows our previous level of de-blurring, i.e., with the introduction of the downward squashing process. Fig 8.7b shows what results from the additional level of de-blurring. Now, in going from Fig 8.7a to Fig 8.7b, the central m^+ extremum has changed to a m^- flanked on either side by 0 in Fig 8.7b. That is, we have:

$$m^+ \rightarrow Om^-O$$

But this conforms exactly to the first rule in the above Process Grammar. That is:

A squashing continues till it indents:

$$Cm^+ : m^+ \rightarrow Om^-O$$

That is, in de-blurring, i.e., going from Fig 8.7a to Fig 8.7b, the squashing arrow has continued till it created an indentation.

Now recall that our Size-is-Time Heuristic implies that de-blurring recapitulates time—in the forward time direction. Thus, the transition from Fig 8.7a to Fig 8.7b represents forward time. This transition elaborates further the major theme of the painting: the woman against the inward forces of the environment. In Fig 8.7b, the force squashing the woman's intent has continued further until it has actually invaded her.

Note also that in the *narrative* story of the painting, the transition from Fig 8.7a to Fig 8.7b represents forward time, because the woman's shoulder, which was once merely squashed or bruised is now actually penetrated by the environment.

In conclusion, we see that the *process-structure* of the shoulder is captured by a sequence of grammatical rules

$$BM^+ : M^+ \rightarrow M^+m^+M^+$$

followed by

$$Cm^+ : m^+ \rightarrow Om^-O$$

together with our Size-is-Time Heuristic; and that this embodies a central theme of the painting. That is, we have, once again, a corroboration of our First Art-Work Principle, which states that an art-work is the process-history inferred from it.

Now let us turn to another part of the painting, the upper boundary of the neck. In Fig 8.3, we drew this region as a m^- extremum. The associated process, in accord with the Semantic Interpretation Rule, was shown as an indentation, given by the downward arrow. This diagram represents a coarse, i.e., slightly blurred, version of the painting; that is, omitting some of the finer detail. Fig 8.8a shows the neck structure under this coarse view.

Now, when one looks at the painting itself, that is, Fig 8.2, and examines the neck in greater detail, one finds that the neck is actually flattened in the middle. This means that its structure is that shown in Fig 8.8b. First

Figure 8.8
De-blurring the neck in Picasso's *Woman Ironing*.

observe that the flattening is given by a M^- extremum at the center. Two m^- extrema flank the central M^-. Observe that, by our Semantic Interpretation Rule, the M^- extremum must mean a *resistance* process, i.e., that shown as the upward arrow leading to the extremum. Furthermore, the m^- extrema must each be *indentations*; i.e., those shown by the two arrows leading down to these two extrema.

Now, in going from the blurred version, Fig 8.8a, to the de-blurred version Fig 8.8b, the extrema have changed from the single m^-, in Fig 8.8a, to the string of extrema $m^- M^- m^-$, in Fig 8.8b. That is, we have:

$$m^- \rightarrow m^- M^- m^-$$

But this conforms exactly to the fourth rule in the above Process Grammar. That is:

An indentation bifurcates:

$$Bm^- : m^- \rightarrow m^- M^- m^-$$

That is, in de-blurring, i.e., going from Fig 8.8a to Fig 8.8b, the indentation arrow has split, one branch going to the left and the other to the right, while a resistance arrow has been introduced in the center from below.

Now let us use our Size-is-Time Heuristic. This principle implies that de-blurring recapitulates time—in the forward time direction. Thus, the transition from Fig 8.8a to Fig 8.8b represents forward time.

Some consideration reveals that this transition again embodies this major theme: the outward forces of the woman against the inward forces of the environment. Fig 8.8a represents an inward force of the environment. In Fig 8.8b, the outward force of the woman has been introduced; i.e., actually *resisting* the force of the environment. Note that, in accord with our Semantic Interpretation Rule, resisting is the process corresponding to a M^- extremum.

Figure 8.9
Further de-blurring at the neck in Picasso's *Woman Ironing*.

In fact, let us examine the neck still further. In Fig 8.8b, we drew the center of the neck as a M^- extremum. As we said, this diagram represents a de-blurred version of the painting. However, while Fig 8.8b shows the painting at some level of de-blurring, one can actually de-blur the work still further. Thus, let us go back to the painting itself, that is, Fig 8.2, and examine the neck in still greater detail.

What one finds is that the neck has a slight hump in the center. This means that its structure is that shown in Fig 8.9b. First observe that the hump is given by a M^+ extremum at the center. By our Semantic Interpretation Rule, this extremum must mean the presence of a *protruding* process. Notice also that, in Fig 8.9b, two dots have been placed on either side of the M^+ extremum, indicating the introduction of two points of 0 curvature; i.e., two points of complete flatness.

Fig 8.9a shows our previous level of de-blurring, i.e., with the presence of the upward resisting process. Fig 8.9b shows what results from the additional level of de-blurring. Now, in going from Fig 8.9a to Fig 8.9b, the central M^- extremum has changed to the string 0 followed by M^+ followed by 0, in Fig 8.9b. That is, we have:

$$M^- \rightarrow OM^+O$$

But this conforms exactly to the second rule in the above Process Grammar. That is:

An internal resistance continues till it protrudes:

$$CM^- : M^- \rightarrow OM^+O$$

That is, in de-blurring, i.e., going from Fig 8.9a to Fig 8.9b, the resisting arrow has continued till it created a protrusion. Thus once again we have an embodiment of the major theme: the outward forces of the woman against the inward forces of the environment.

In conclusion, we see that the *process-structure* of the neck is captured by a sequence of grammatical rules

$$Bm^- : m^- \rightarrow m^- M^- m^-$$

followed by

$$CM^- : M^- \rightarrow OM^+O$$

together with our Size-is-Time Heuristic; and that this embodies a central theme of the painting. That is, we have, once again, a corroboration of our First Art-Work Principle, which states that an art-work is the process-history inferred from it.

When one examines the painting, one sees these kinds of structures repeatedly. For example, the grammatical operations apply to the upper line of the hands, converting the line from the blurred version shown in Fig 8.3—which has a well-rounded downward extremum—to the flattened version shown in the de-blurred painting, Fig 8.2—which indicates the upward resistance of the iron. The reader should observe that the entire outline of the woman's body is permeated with structures of the type we have just been describing, even down to the smallest perceptible level. With these pervasive subtleties of line, Picasso communicates his theme that the woman is bruised and pummelled by time. The bruising and pummelling are captured by the rules of our process grammar, and again, support our principle that the meaning of an art-work is the history inferred from it.

8.4.4 The Bending Process in *Woman Ironing*

Not all the processes defining Picasso's *Woman Ironing* are extrema-based. A major process is that of bending. It is embodied in the bent line of the woman's back, the bent symmetry axis through her body, the bent upper outline of her head, the bent "horn" of drapery on the left-side, etc.

What is important for us to understand is that any bent line is *asymmetric*; i.e., there is distinguishability between one side which is concave, and the other which is convex. Thus, by the Asymmetry Principle, a bent line implies a past state in which this distinguishability between sides did not exist; i.e., the line was straight. It is this fact that makes one understand the line as having undergone the process of *bending*.

Note that, having identified this bending aspect, we have now found, in the structure of the painting, the final factor in our statement of the meaning of the painting: "a woman bruised and pummelled by the outside world *bent over* by exhaustion . . . etc."

The crucial point is that, once again, we have found that the meaning of the painting is embodied in the process-structure one infers from the shapes in the painting. That is, once again, we have a corroboration of our First Art-Work Principle, which states that an art-work is the history inferred from it.

Close examination reveals that the bent outlines come in pairs: the bent outer and inner outlines of her torso; the bent upper and lower outlines of her head; the bent upper and lower outlines of the drapery horn. Each of these pairs produces a bent symmetry axis. In fact, Picasso often reinforces the symmetry axis with a shadow-line.

The structure of a symmetry axis in relation to its two outlines is given by a moving Cartesian frame, in the manner described towards the end of Chapter 6; i.e., the bent aspect is given by the rotation of the frame as it moves along the axis. Thus, from Chapter 6, we have the machinery to give a full process analysis of these bending structures, should we choose to do so at this point. However, to do so would disrupt the discussion we are carrying out of *extrema-based* processes in painting—on which we will concentrate throughout our present analysis.

8.5 Picasso's *Demoiselles d'Avignon*

We will now examine another painting by Picasso, the *Demoiselles d'Avignon*, shown in Fig 8.10. This painting is so stylistically different from the *Woman Ironing*, that, were one unfamiliar with the two works, one would assume, not only that they were painted by different artists, but that the artists were of two very different genres. In choosing to examine *Demoiselles d'Avignon*, we could not choose a more different work from *Woman Ironing*. Nevertheless, we will find, surprisingly, that the rules we have used above, for *Woman Ironing*, not only apply to *Demoiselles d'Avignon*, but reveal major aspects of the composition and meaning of the work.

Figure 8.10
Pablo Picasso, *Les Demoiselles d'Avignon*. Paris (begun May, reworked July 1907). Oil on canvas, 8′ × 7′ 8″ (243.9 × 233.7 cm). Collection, The Museum of Modern Art, New York. Aquired through the Lillie P. Bliss Bequest. Photograph © The Museum of Modern Art, New York.

8.5.1 The Structure of *Demoiselles d'Avignon*

Demoiselles d'Avignon is often considered to be the most significant of Picasso's life and one of the most significant in the history of art.

In order to understand the composition of this painting, observe that the circle is an important shape in its structure. First, the entire composition is circular—the overall circle cutting the corners of the square canvas, and centering on a point roughly coincident with the pubic area of the central woman.

However, besides this overall circularity of the composition, there are a multitude of circles within the painting. Four such circles are shown in Fig 8.11. The easiest to see initially are the upper two. If one goes back to the photograph of the painting, Fig 8.10, one can see that the upper *right* circle has been almost completely drawn in by Picasso with an explicit circular line. To be aware of the upper *left* circle, again look at the photograph of the painting (Fig 8.10): concentrate on the second woman from

Figure 8.11
Four circles in Picasso's *Les Demoiselles d'Avignon.*

the left and try to see the reflectional symmetry between the circular arc drawn by Picasso across her stomach, and the circular shadow line above her—emerging from the elbow of the woman next to her. The symmetry between these two arcs is emphasized by the horizontal line of her mouth (although the circle center is slightly below).

Besides the four circles shown in Fig 8.11, there are a multitude of other circles. For example, several of them center on, and encompass the seated woman, as shown in Fig 8.12. These circles intersect with circles based in the lower left of the painting, around the legs and hips of the two women on that side. In addition, much larger circular arcs cut across the painting in several directions. The many circles criss-cross each other creating a complex set of movements and counter movements that pull the eye in constantly diverging directions. It is as if Picasso is *trying to pull the eye apart*.

Now, recall that the removal of curvature variation from any complex smooth shape leads to a circle—the shape without curvature variation.

Figure 8.12
Circles centering on and encompassing the seated woman in Picasso's *Les Demoiselles d'Avignon*.

This means that, according to the Asymmetry Principle, the ultimate past state, in the external history of a complex smooth shape, is a circle. Thus a circle represents an object before processes have acted.

Given a circle, as one begins to add processes creating extrema, the first shape that one can produce (without any segments of constant curvature) is an ellipse—as shown at the top left of Fig 1.9 of Chapter 1.

Since one obtains an ellipse from a circle, by stretching, we can say that an ellipse embodies the following process theme:

1st PROCESS THEME: Stretching

Observe that ellipses occur many times in the painting. One ellipse is shown on the lower left in Fig 8.13. By returning to the photograph of the painting, Fig 8.10, one can see that Picasso actually draws this ellipse in, with an explicit line on the canvas. Another ellipse is shown in the upper right of Fig 8.13. To see this ellipse on the actual painting, Fig 8.10, concentrate on the middle woman and note the reflectional symmetry between

Figure 8.13
Two ellipses in Picasso's *Les Demoiselles d'Avignon*.

two curves: (1) the upper line of her upper arm, (2) the curved line that descends rightward from her lower breast.

An ellipse crucially involves stretching along its major axis. As we shall see, stretching represents *liberation* in this painting. However, observe that an example of stretching must be the horizontal stretching of the upper oval in Fig 8.13. This example has a remarkably profound meaning. The *cause* of the stretching appears to be the *sight* of the upper right woman, as she looks in. Thus, the *liberating* effect of stretching is created here, by *seeing*. The identification of *liberation* and *seeing* is crucially important in this painting, as we shall observe later. In accord with this, recall that we have noted previously that the painting seems to "pull the eye apart".

In order to more deeply understand the meaning of the painting, we must study the historical structure of the ovals in greater detail. There are in fact a multitude of ovals in the painting. To study them more fully, we must have a historical analysis of *pointed* ovals; i.e., ovals of the form shown in Fig 8.14.

Let Fig 8.15a represent the top half of an oval. In accord with our rules, there is an upward process that has created the top extremum shown. Note that, in accord with the Second System Principle of Chapter 1, the

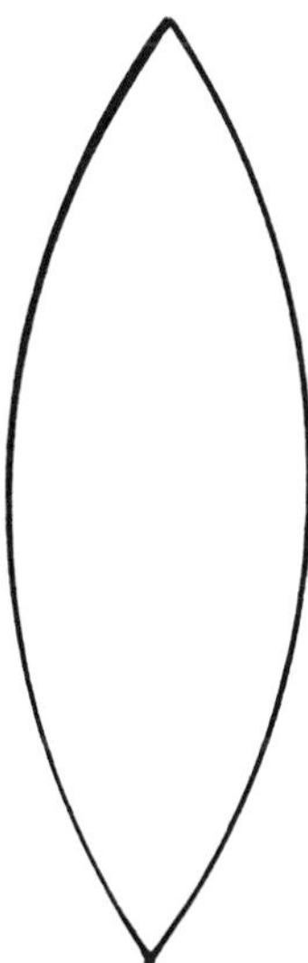

Figure 8.14
A pointed oval.

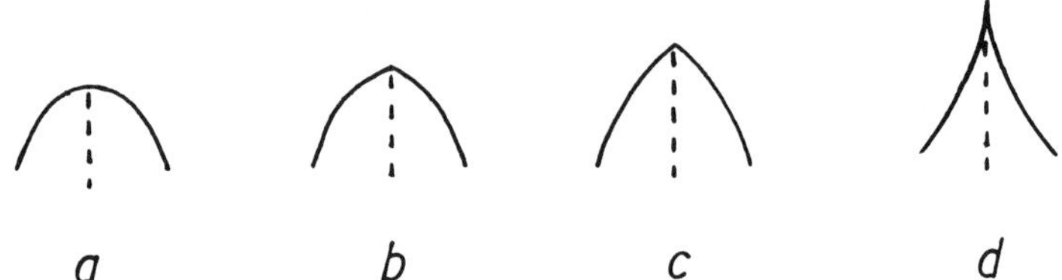

Figure 8.15
The successive changes from a smooth extremum through a kink to a cusp, in accord with the scenario of Hayes & Leyton (1989).

top extremum, being an asymmetry, must have been created by a second object acting upward against the boundary: e.g., if the shape were hollow, then the second system could have been a vertical rod pushing the boundary upward; alternatively, if the shape were the solid top of a finger, the second system would have been the internal flesh and bone, pushing upward against the skin during growth.

We are going to consider the type of scenario that leads to a sharp point at the top of the shape. Such a scenario was analyzed in depth by Hayes & Leyton (1989). The authors considered the following: Suppose that the object pushing upward against the boundary were sharp, e.g., a knife. Initially, because any boundary has some rigidity in it, the extremum formed would be smooth as shown in Fig 8.15a. However, if the object continued to push, the boundary would become pointed, i.e., a kink would be introduced at the top, as shown in Fig 8.15b. Thus, in the formation of a kink, one has the following process theme:

2nd PROCESS THEME: Overcoming Rigidity

As we shall see later, this process-theme embodies a significant aspect of the meaning of *Demoiselles d'Avignon*—in accord with the Corollary of our First Art-Work Principle which states that the meaning of a painting is the process-history inferred from it.

Now let us return to Fig 8.15. One should observe that, at the moment when the kink is introduced, i.e., Fig 8.15b, the curve immediately on either side of the sharp point remains bending outwards. However, suppose the sharp object continued to push upwards against the boundary. The curve immediately on either side would begin to flatten, as shown in Fig 8.15c. Finally, as the sharp object continued to push upwards, the curve on either side would begin to bend inwards and a cusp would be formed as shown in Fig 8.15d. That is, in the transition from the kink in Fig 8.15b to the cusp in Fig 8.15d, the curve on either side of the sharp point would change from bending out to bending in.

In their paper, Hayes and Leyton analyzed how many operations need to be added to the Process Grammar to handle the formation of kinks and cusps. Surprisingly, the authors found that only one operation needs to be added—that which produces the kink.[1] However, subsequent to the formation of the kink, the remaining operations of the Process Grammar suffice to create the cusp.

What is important for us at this stage of our discussion is that Hayes and Leyton observed that the Symmetry-Curvature Duality Theorem applies to sharp points, as it does to curvature extrema (which are smooth). That is, given a sharp point, a symmetry axis must exist that is directed towards, and terminates at, the point. In accord with this, we have drawn a symmetry axis in each of the stages shown in Fig 8.15. Now recall also that our Interaction Principle (backwards version) states that the axis must have been the trace of a process. This means that, if we are presented with any one of the four shapes in Fig 8.15, we can conclude that a process acted along the symmetry axis and created the particular type of tip—smooth extremum, kink, or cusp—which is contained in the shape.

Let us now continue with our analysis of *Demoiselles d'Avignon.* Recall that we began by observing that circles are structurally important in the painting; and we noted that any complex smooth shape has the circle as the ultimate past state in its external history. Recall also that, given a circle, as one begins to add processes creating extrema, the first shape that one can produce is an ellipse. An ellipse is a stretched version of a circle. We found that the ellipse is an important structural theme in the painting.

After our discussion of the Hayes and Leyton paper, we can now continue this history—from the circle to the ellipse—still further forward in time. The stretching processes in the ellipse can now be extended to create sharp points at the ends of the ellipse, as shown in the transition from Fig 8.16a to Fig 8.16b. That is, one creates the pointed oval.

Pointed ovals occur throughout the painting, at many levels. For example, in the overall structure of the painting, the composition is divided into two large pointed ovals as shown in Fig 8.17. In fact, these are not the largest ovals that divide the painting. There is a larger one on the right formed by the two drapery edges that converge off the top of the painting and that extend downward and converge off the bottom of the painting.

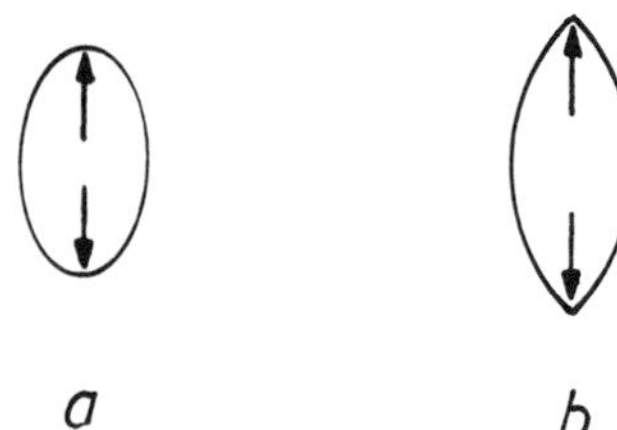

Figure 8.16
An oval changing to a pointed oval.

Figure 8.17
Two large pointed ovals dividing Picasso's *Les Demoiselles d'Avignon*.

Figure 8.18
(a) A pointed oval pulled by a hand, (b) a pointed oval pulled by a knee, and (c) a trace of pointed ovals, in Picasso's *Les Demoiselles d'Avignon*.

On a slightly smaller scale, a pointed oval is pulled up by a hand, as shown in Fig 8.18a, and by a knee, as shown in Fig 8.18b. Note that there is a *trace* of ovals across the bottom of the painting, as shown in Fig 8.18c.

On a smaller scale, pointed ovals are, of course, much in evidence as the limbs of the women; e.g., the upper arms of the central two women, the limbs comprising the more visible leg of the seated woman, etc. Also, at this level of scale, more elongated versions of ovals occur in the drapery, for example, around the protruding knee of the second woman from the left.

On a still smaller scale, all the eyes and all the mouths of the women are pointed ovals.

Figure 8.19
A system of pointed ovals bursting upwards from the lower left, in Picasso's *Les Demoiselles d'Avignon*.

Now observe that the ovals are combined across many levels of scale. For example, Fig 8.19 shows a system of ovals that bursts up from the lower left, fans out towards the middle, and culminates in the upper left. Observe the way in which the upper arm of the second woman is an oval that exactly crosses the curvature extremum of the large oval which, on one side, is formed by the outline of her shoulders down through the skirt of the middle woman, and, on the other side, descends along the edge of the torso of the woman on the left. Since the symmetry axis of the upper arm is aligned with the symmetry axis of the extremum of this larger oval, the upper arm appears to pull the extremum of the larger oval upward, and in fact, seems to be responsible for creating that extremum. This again corroborates our view that curvature extrema are seen as created by processes. In this case, the alignment of two processes serves to make the impression of process-history even stronger.

Before we continue, it is important to note that any great painting structurally unifies the positive shapes—those representing objects—with the

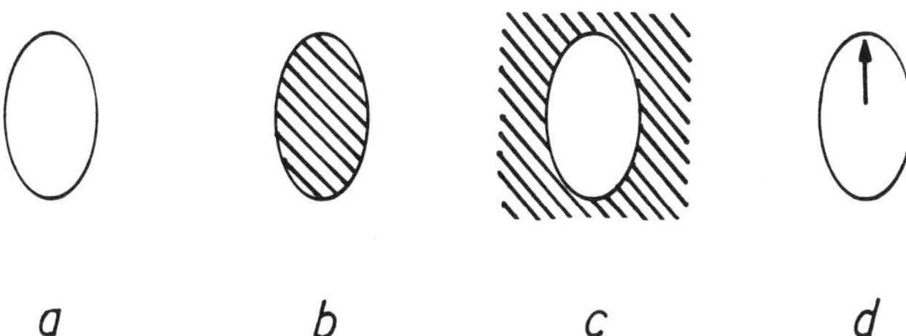

Figure 8.20
(a) An oval can be either (b) solid, or (c) a hole. However, in (d), one sees that the same process will be inferred in the same position in both cases.

negative shapes—those representing the spaces between objects. Furthermore, the shapes formed by the negative areas are as compositionally important as the shapes formed by the positive areas. There is, so-to-speak, a "democracy of shape." Part of Picasso's genius, or indeed the genius of any great painter, is the ability of the artist to ensure that every shape in the painting—whether the shape is positive or negative—has a structural role in the painting. Positive and negative shapes are repetitions and modifications of each other so that they interlock and give the painting a sense of tightness, power and unity. Furthermore, each shape—whether positive or negative—carries, with it, part of the meaning of the painting.

Given the equal significance of negative and positive shapes, it is important for us to observe that our rules, for the inference of process-history from curvature extrema, apply equally to positive shapes as to negative shapes. Thus consider the oval in Fig 8.20a. It can be either solid (positive), as in Fig 8.20b, or a hole (negative), as in Fig 8.20c. The important thing to observe is that a process-trace leading to an extremum, e.g., the top arrow shown in Fig 8.20d, will occur in exactly the same place in the solid object, Fig 8.20b, and the hole, Fig 8.20c. This is because the process-trace is the symmetry axis yielded by the *curve*, and the curve is the same in both cases. Thus, figure-ground reversal, i.e., the change from solid to hole, does not change the process-traces yielded by a shape.

What does change, in figure-ground reversal, are the *labels* one gives to extrema. What is called a "maximum" is subsequently called a "minimum" (and vice versa). Furthermore, what is called "positive" curvature is subsequently called "negative" curvature. That is, one has these changes:

$$M^+ \leftrightarrow m^-$$
$$M^- \leftrightarrow m^+$$

In addition, the process terms "protrusion" and "indentation" are interchanged; and the process terms "squashing" and "internal resistance" are interchanged. That is, one has these changes:

Protrusion ↔ Indentation
Squashing ↔ Internal Resistance

Now, given the equal importance of positive shapes (solids) and negative shapes (holes) in a painting, and given the fact that the process-traces, in our theory, are the same whether the shape is solid or a hole, we can now see that what we have so far said about the structural roles of ovals in Picasso's painting applies equally whether the ovals are positive (solid) or negative (holes). We shall therefore, for a while, consider the ovals as holes.

Let us first briefly recapitulate the historical sequence we have discussed so far in this section. The sequence started with a circle—the shape without extrema. The next stage was the oval which, in particular, has two outward stretching processes each terminating at an extremum, as shown in Fig 8.16a. In the next stage, these processes continue and break the rigidity of the boundary, i.e., the extrema turn into kinks, as shown in Fig 8.16b.

Now, when we return to the sequence defined by Hayes & Leyton (1989), shown in Fig 8.15, we find that there is a final stage: The outward processes have continued further, creating cusps; i.e., the curve on either side of each kink has changed from bending out to bending in—as shown in going from the kink in Fig 8.15b to the cusp in Fig 8.15d.

If both the outward processes on an oval—the processes shown as vertical in Fig 8.21a—have continued in this way, the resulting shape will necessarily be diamond-like, as shown in Fig 8.21d. This is because, in order for the curve to change from bending out to bending in, on either side of each of the cusps, a negative minimum, m^-, needs to be introduced at the two sides of the oval. It is worth examining this in more detail because we shall find later that this structural feature embodies significant aspects of the *meaning* of the painting.

One has started with the oval shown in Fig 8.21a. Note that we are now considering it to be a *hole*. Recall that an oval has four extrema as shown

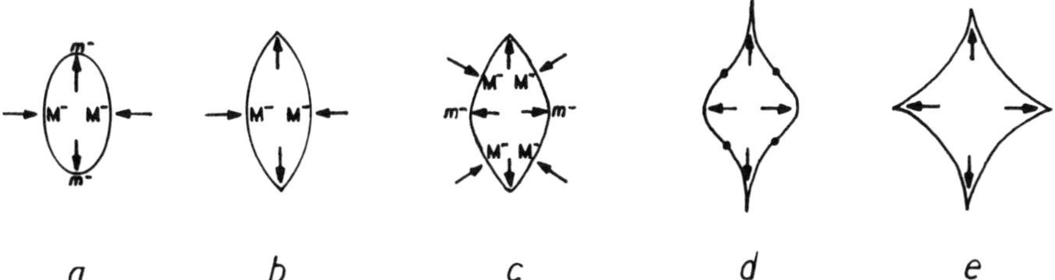

Figure 8.21
The progression of an oval to a diamond.

in Fig 8.21a. Therefore, there are not only the two outward stretching processes at the top and bottom, but the two inward compressing processes terminating at the M^- extrema in the middle of the oval. These processes represent *resistance* pushing inward from the solid surrounding the hole.

Now, in the stage subsequent to the oval, the two outward processes continue and create the kinks, at the top and bottom, as shown in Fig 8.21b. However, in progressing further to the cusp, the curve on either side of each kink has to flatten. In order for this flattening to occur, the two flattest points, the two M^- in the middle of the oval, must move towards the kinks. But since there are two kinks (top and bottom), each M^- must move simultaneously in two directions. This means that each must split into two copies of itself, as shown in Fig 8.21c. In addition, two m^- extrema must necessarily be introduced in the middle. Each of these is created by an outward *indentation*, because the shape is a hole and the process *indents* into the surrounding solid material. Therefore, on each side of the oval, we have an example of the sixth operation of the process-grammar:

An indentation is introduced:

$$BM^- : M^- \rightarrow M^- m^- M^-$$

Subsequently, each M^- continues moving, reaching the top and bottom sharp points and causing the curve to flatten more on either side of the sharp points. Eventually, the curve on either side changes from bending out to bending in, creating a cusp at the top and bottom. Therefore one obtains a diamond as shown in Fig 8.21d. Note that the horizontal outward processes can themselves continue and create two side kinks, as shown in Fig 8.21e. Indeed they can continue further and become cusps.

We have just seen that, central to the formation of the diamond from the oval, is the following process-history: The sideways inward resisting forces, shown in Fig 8.21a, are overcome by the sideways outward forces that pull the shape open laterally. This history involves two important components:

3rd PROCESS THEME: Overcoming Resistance

4th PROCESS THEME: Pulling Open

As we shall see later, both these themes embody significant aspects of the meaning of the painting—in accord with the Corollary of our First Art-Work Principle, which states that the meaning of a painting is the process-history inferred from it. At the present stage, it is important to

Figure 8.22
Some significant diamonds in Picasso's *Les Demoiselles d'Avignon*.

observe that the process themes were produced by the use of the rules for the inference of process-history from curvature extrema.

Diamonds are structurally significant in the painting. The most important is that on the right, given by the large opening curtain, as shown in Fig 8.22. Note that its two horizontal vertices are (1) the pubic region of the central woman, and (2) the right most knee/elbow of the painting. Note how the bottom left side of this diamond is a line that goes along symmetry axes through curvature extrema (e.g., the knee of the seated woman), and the other three sides are explicitly drawn in by Picasso. We have also shown in Fig 8.22, two significant concentric diamonds on the left.

Observe that the diamonds are pulled open still further. For example, the larger of the two diamonds on the left is stretched completely across the canvas to the buttock on the right, as shown in Fig 8.23. Note how its upper vertex is pulled by the hand, and its lower vertex by the foot. By examining this diagram in relation to the painting, in Fig 8.10, the reader will see how clearly Picasso draws in this diamond.

Figure 8.23
A diamond stretched to the right in Picasso's *Les Demoiselles d'Avignon*.

Figure 8.24
A diamond stretched to the left in Picasso's *Les Demoiselles d'Avignon.*

Now, we have seen throughout our discussion that Picasso always balances structures on the left with structures on the right. Thus, the stretching of the left diamond to the right, as shown in Fig 8.23, is balanced by the stretching of the right diamond to the left, as shown in Fig 8.24.

The diamond structures we have been examining arise from the four curvature extrema of an oval. These four extrema were themselves introduced into the original circle. But many more extrema can be introduced into a circle, and each of these can be extended into a kink or cusp. In fact, recall that we observed that this painting has an overall circular composition—the circle being roughly coincident with the square edge of the canvas. This circle has several kinks and cusps introduced into it. For

Figure 8.25
8.25 Kinks and cusps introduced into the overall circular composition of Picasso's *Les Demoiselles d'Avignon*.

example, notice the sequence of three cusps down the left-hand side of the painting, as indicated in Fig 8.25, and also the three cusps down the right-hand side. These *pull open* the circle, in accord with the fourth process theme.

We have looked at the pulling open of both the circle and the oval by the *introduction* of processes. The introduction of processes is given by the fifth and sixth operations of our Process Grammar—the operations that define the introduction of a protrusion and indentation respectively. In particular, the pulling open of the oval hole used the sixth operation.

However, the pulling open theme in the painting can occur as a result of the use of the third and fourth operations of the grammar—the operations that define the *bifurcation* of a protrusion or indentation. In particular, we will consider the use of the third operation in this painting. To do so, we will return to considering the oval as a *solid* shape.

Thus consider the oval, as shown in Fig 8.26a, and again note its vertical stretching processes. When the top process bifurcates, one obtains Fig

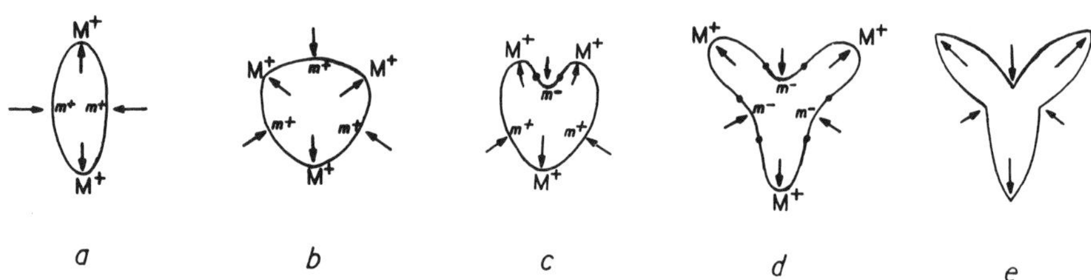

Figure 8.26
The oval undergoing "branching out."

8.26b, i.e., a widening of the top end. In terms of the extrema involved, the single M^+, at the top of Fig 8.26a, has changed into the string of extrema $M^+m^+M^+$, at the top of Fig 8.26b. That is, we have an example of the third rule in our Process Grammar:

A protrusion bifurcates:

$$BM^+ : M^+ \rightarrow M^+m^+M^+$$

Notice also that a new downward squashing process has been introduced at the top of Fig 8.26b. If this process were to continue, it would begin to indent as shown in Fig 8.26c. Note that two dots have been placed on the curve in Fig 8.26c, indicating the introduction of two zeros of curvature, i.e., two points where the curve is completely flat. Thus the transition from Fig 8.26b to Fig 8.26c can be specified by saying that the central m^+ extremum, at the top of Fig 8.26b, has changed to the string: 0 followed by m^- followed by 0, in Fig 8.26c (where each 0 corresponds to a dot). That is, we have an example of the first rule in our Process Grammar:

A squashing continues till it indents:

$$Cm^+ : m^+ \rightarrow Om^-O$$

Observe that this operation can happen also in the side-squashing processes; i.e., they can continue and produce indentations as shown in Fig 8.26d. The consequence will be that the two upper protrusions bend out.

Now, if one looks at the initial oval, Fig 8.26a, one sees that the overall effect has been that the top protrusion in the oval has *branched out*. That is, not only has the protrusion bifurcated into branches, but the branches have moved *outwards*. As we shall see, this overall transition is an impor-

Figure 8.27
A powerful upward movement from the bottom culminating in branching out, in Picasso's *Les Demoiselles d'Avignon.*

tant one in the painting; thus we give it a label:

5th PROCESS THEME: Branching out.

Observe also that, since negative shapes are as important as positive ones, the downward indenting process in the center of Fig 8.26c, must also be important. Its effect is to tear apart the widened upper end of Fig 8.26b. Thus we give it the following label:

6th PROCESS THEME: Tearing Apart.

Now observe that, if all extrema in Fig 8.26d became sharp points, one would obtain Fig 8.26e.

The *branching out* theme is a powerful one in *Demoiselles D'Avignon.* Its most important example occupies the center of the painting as shown in Fig 8.27. The shape in Fig 8.26e occurs as the torso and arms of the central woman in Fig 8.27. The lower protrusion of Fig 8.27 is her pubic area, and the two upper branches are her arms.

This shape in the painting appears as an upward explosion of growth. In fact, the entire upward movement starts with the curvature extremum of the table at the bottom of the painting, given by the lowest arrow in Fig 8.27. Note that the inference of this arrow is provided by our rules: The upward curvature extremum of the table implies, by our Symmetry-Curvature Duality Theorem, a symmetry axis leading to the extremum; and, in accord with our Interaction Principles, the axis is the direction of a process.

This arrow is then continued by an upward arrow in the leg—again defined by our inference rules: The upward curvature extremum of the *knee* implies, by our Symmetry-Curvature Duality Theorem, a symmetry axis leading to that extremum; and again, by our Interaction Principles, this axis is the direction of a process.

Finally, the upward movement is caught by the upward branching from the woman's chest into her arms.

The branching-out theme can be regarded as a different version of the pulling-open theme. Structurally, they differ in that the branching-out theme involves the bifurcation of a process and the pulling-open theme involves the introduction of a process. Nevertheless, one of the effects of a branching-out is to pull something apart.

What is being pulled apart by the branching-out shown in Fig 8.27? We have noted several times that structures on the left of the painting are balanced by structures on the right. For example, in Fig 8.11, we saw that circles on the left are echoed by circles on the right. Again, in Fig 8.17, we saw that the painting is structured by two massive left and right ovals. What the branching-out structure in Fig 8.27 does then is to tear the painting into two halves. Note how this theme is echoed by other examples along the central vertical of the painting. First of all, as shown in Fig 8.28, the upward arrow leading to the knee actually continues and bifurcates into the hips; i.e., in accord with our rules, these upward bifurcating arrows run along symmetry axes in the hips and lead to the two curvature extrema at the top of the hips. This structure echoes exactly the structure in the arms. Second, observe that the downward arrow in the belly, leading to the pubic region, continues and bifurcates into the legs of the women on either side—shown by the two downward arrows in Fig 8.28. Again, these arrows are defined in accord with our inference rules, i.e., they are given by symmetry axes leading to extrema. Notice that the process that *tears* these latter two branches apart is the upward arrow of the knee.

Figure 8.28
Branching out occurring downwards around the knee, and upwards into the hips, in Picasso's *Les Demoiselles d'Avignon.*

8.5.2 The Meaning of Demoiselles d'Avignon

Having examined the history embodied in the composition of the painting, we now turn to the actual meaning of the painting. Recall that the Corollary of our First Art-Work Principle states that the meaning of any painting is its historical structure.

The history, we described in the last section, started with a circle. This shape was first stretched outward into an oval. The outward processes then acted further, causing a breaking in the rigidity of the boundary, turning the two ends of the oval into kinks. The result was the pointed oval. Then, the oval was pulled open, sideways, against its compressive inward forces. The result was the diamond shape; and this diamond was itself pulled further open, e.g., laterally across the canvas. An alternative history for the oval was that in which one of its two outward processes bifurcated, causing a widening of that end of the oval. Then an opposing inward force tore apart the widened end, allowing the shape to branch out.

Thus, as we observed, this history involves six process themes:

1st PROCESS THEME:	**Stretching.**
2nd PROCESS THEME:	**Overcoming Rigidity.**
3rd PROCESS THEME:	**Overcoming Resistance.**
4th PROCESS THEME:	**Pulling open.**
5th PROCESS THEME:	**Branching out.**
6th PROCESS THEME:	**Tearing apart.**

We claim that the painting is *about* these themes. That is, the painting is about *stretching*, *overcoming rigidity*, *overcoming resistance*, *pulling open*, *branching out*, *and tearing apart*.

When one examines these six themes, one realizes that they have a common core: They are the themes of *revolution*. In a revolution, the past—that is, the old order—is stretched; its rigidity is overcome; so is its resistance; an entrance to the future is pulled open; through this entrance, the forces of freedom branch out; the old contract is torn to pieces.

Picasso was of course strongly aware of the revolutionary character of the work in relation to the preceding history of art. The work is both a discussion of revolution and an act of revolution.

But what is undergoing a revolution in this work? To answer this question, one must remember that *Demoiselles d'Avignon* begins the cubist period, the most important single movement in twentieth-century art. Most of the major artistic developments of modern art arise from analytic or synthetic cubism: complex multiple perspectives, the dissolution of three-dimensional form, the consequent abstraction, all-over composition, surface flatness, differing levels of seeing, self-reference in seeing, the inclusion of literary elements such as words, the democracy of materials, the assemblage of found objects, etc. Not only are the discoveries of cubism pervasive throughout subsequent art, but they are pervasive throughout all other media—film, video, newspaper cartoons, advertising, etc.

What all these developments attest to is that cubism was first and foremost a revolution in *sight*. Indeed, so effective and total was this revolution in seeing that the media that use its discoveries—from film to advertising—are completely comprehensible to the general public, and indeed actively sought by the public.

Demoiselles d'Avignon, the first painting of the cubist period, begins the revolution in sight. Each of the six process themes that we elaborated concerns actions on shape—but the shape on which they operate represents sight. That is, going through the six themes: (1) sight is stretched; (2) its rigidity is broken; (3) its classical resistance is overcome; (4) its containing structure pulled apart; (5) the paths of sight branch around objects; (6) sight is torn apart.

Now, as we observed earlier, the ultimate past *shape* on which these actions operate is the circle, a significant shape of the painting. When one looks at the painting and discovers any one of the major circles, one has, at the moment of discovery, a sudden sensation of calmness, as one's eye expands to fill the circle. It is as if one's eye equates its own circular structure with the circle in the painting. This sensation is very strongly felt when looking at the actual painting in the museum, because, in the painting's massive size, the circles can have their full effect.

Thus, in acting on the circle, the various processes in the painting act on the circle of the eye. As we said earlier, when viewing the painting, one has the strong sensation that *one's eye is pulled apart*. This sensation, in the viewer's eye, matches the large-scale historical revolution in sight that was to be instituted by cubism.

The revolutionary action of the painting, while being central to its meaning for the viewer, as well as its meaning for modern art, was an intensely personal experience for Picasso himself. During this period, Picasso seems to have turned inward and become particularly concerned with his own creativity. That is, not only was he engrossed in the products of his creativity, but he was deeply pre-occupied with, and mystified by, the creativity itself.

If an artist becomes centrally concerned with some issue, then, quite naturally, the issue becomes the subject of his or her work. While *Demoiselles D'Avignon* was the revolutionary outpouring of Picasso's creativity, so intense was this creativity that paradoxically, it became the subject of the painting. Creativity is again, the common core that runs through the six process themes that define the history inferred from the painting's shapes. Of course, the way in which these themes embody revolutionary action is identical to the way in which they embody creativity. In life, there is often no distinction between revolutionary action and creativity.

Less obviously, we find that, in this painting, Picasso seems to have equated three concepts: *sight*, *creativity*, and *sex*. Let us begin by considering the first two: In cubism, Picasso is the first artist to discover that sight is a creative act. Sight is an active exploration and manipulation of objects. Sight constructs the world, and in so doing, is a free agent.

Consider now the second two concepts: creativity and sex. It seems that, in this period of his life, while being so intensely focussed on the mystery of his own artistic fertility, Picasso identified it with the sexuality and fertility of women. This type of identification is not uncommon for an artist. For example, we see it in the sketch-books of da Vinci, in the latter's obsessive studies of pregnancy.

The sexual content of *Demoiselles d'Avignon* is obvious. But why? Is it simply that the women are naked?

Recall our claim that the meaning of a painting is embodied in the history derived from its shape structure. Recall that we defined six process themes in Picasso's painting: *stretching*, *overcoming rigidity*, *overcoming resistance*, *pulling open*, *branching out*, *tearing apart*. We have seen how these themes can be regarded as concerning both sight and the revolutionary aspects of creativity. But they can equally well be regarded as concerning the sexuality of women at the moment of intercourse and conception.

As an example, it is worth dwelling on the fourth process theme: the pulling-open theme. Since this theme is so central to the shape-structure of the painting, and since the painting has such a strong sexual content, it is difficult not to conclude that this process-theme embodies the parting of vaginal folds at the moment of sexual entrance into a woman. After all, the woman on the lower right of the painting sits with her legs wide open. Furthermore, her significance is undeniable. She is turning to face us. Her strong stare leads us into the painting.

Now, just as the image of her face has been presented to us only by the action of her turning, there is a way in which the whole painting is the image of her opening vagina which is hidden from us because her pelvis faces away from us, but which has been turned to face us in the entire structure of the painting. After all, the upper figure on the right is pulling open folds that encompass the full length of the painting. The large successively opening diamond-shape formed by these folds was shown in Figs 8.22 and 8.24. Once again, we see that the process-history inferred from the shape structure of the painting—the history of pulling open—embodies the meaning of the painting. That is, we see again that the experience of the painting is the experience of solving the process-recovery problem.

Now recall our noting that, in this period of his life, Picasso became deeply concerned with the mystery of his own creativity and that, furthermore, he identified his creativity with the sexuality of women.

This figure on the upper right, who opens an entrance into the scene, clearly embodies the sexual meaning of the painting. But, in so doing, it seems to be the case that, in this figure—who could be either male or female—Picasso is representing himself at this crucial moment of self-consciousness and artistic discovery. By painting this figure as *staring* into the sexual womb, Picasso identifies the act of sexual penetration with sight. Sight enters and fertilizes. Sight is the source of Picasso's creativity. The figure pulls apart the folds and the path that lies opened is the future of art. Picasso has equated sight, creativity, and sex. This triple identity is contained in each shape of the painting. Each shape carries a role in each of three stories that Picasso makes one. And the story itself is elaborated in the shape structure of the painting. From the shapes, we *recover* the story—the history, that is the meaning of the painting. And the recovered history is Picasso's personal crisis and the crisis in modern art.

8.6 Picasso's *Still Life on a Pedestal Table*

We will now examine a third painting by Picasso, the *Still Life on a Pedestal Table* (1931) shown in Fig 8.29. This painting is again stylistically so different from the *Demoiselles d'Avignon* and *Woman Ironing*, that, again, were one unfamiliar with the three works, one would assume, not only that they were painted by three different artists, but that the artists were of three very different genres. Nevertheless, we will find that the rules we have used for the previous two paintings again reveal major aspects of the composition and meaning of the work.

The entire painting is a structure of curvature extrema. For example, viewing the painting simply as a *mosaic* of shapes, Fig 8.30 shows only a small fraction of the curvature extrema with their inferred processes. However, any painting is more than a *mosaic* of shapes. For example, shapes combine with other shapes to form larger shapes. These larger shapes yield curvature extrema and hence processes. Therefore, Fig 8.30 hints only at some of the small-scale structure of the painting and fails to capture the structure yielded by curvature extrema in the large-scale structure. In Fig 8.31, we have extracted some of the large-scale structure yielded by the curvature extrema. The figure omits many obvious arrows, in order to make the diagram more legible.

To understand the structure of the painting more fully, let us proceed more systematically.

In section 8.5.1, we saw that the structure of *Demoiselles d'Avignon* involved a circle stretching to an oval, then to a pointed oval, which was

Figure 8.29
Pablo Picasso, *Still Life on a Pedestal Table*. Paris (1931). Oil on canvas, $76\frac{3}{8}'' \times 51\frac{1}{4}''$ (194 × 130/cm). Picasso Museum, Paris. © Photo R.M.N.

Figure 8.30
Some of the extrema-based processes arising in Picasso's *Still Life on a Pedestal Table* viewed simply as a mosaic of shapes.

Figure 8.31
Some of the large-scale structure yielded by extrema in Picasso's *Still Life on a Pedestal Table*.

then pulled open to become a diamond, or in which one end bifurcated and branched under the action of an inward tearing process. The *Still Life on a Pedestal Table* involves a very different history from a circle, as follows:

Circles, or approximate circles, are given explicitly throughout the painting, for example, in the fruit and the pedestal legs. As we have noted previously, a circle, when stretched, produces an oval. There are several ovals in the painting, some of which have been shown in Fig 8.32. These are themselves arranged around two ovals shown in Fig 8.33, which occur as two successively larger arches. The arches are formed by applying our inference rules to the ovals in Fig 8.32: That is, the lines comprising the arches are the symmetry axes leading to extrema in the ovals in Fig 8.32. Observe, in fact, from the painting itself, in Fig 8.29, that only half of each oval in Fig 8.32 has been fully drawn by Picasso (although the remainder of each oval is partially drawn in). This gives each oval a strong directionality towards the drawn-in end. The consequence is that the arches they create, those in Fig 8.33, have strong directionality along their lines. In fact, as indicated, the upper arch starts from the lower right—ascending and then descending to the lower left—while the lower arch flows in the opposing direction. The arrow-heads in Fig 8.33 indicate the positions of some of the main extrema generating the two opposite flows.

The major difference between the circle history in *Demoiselles d'Avignon* and the circle history in *Still Life* is what happens to the oval. Unlike the strong pulling-open and branching scenario in *Demoiselles*, the oval in *Still Life* undergoes *bending*. For example, a bent oval is shown in Fig 8.34. The transition from an oval to a bent oval is shown in Fig 8.35. What has happened is that one of the squashing processes, the lower one in Fig 8.35a, has continued and caused an indentation shown at the bottom in Fig 8.35b. The change is described by the first rule in our Process Grammar:

A squashing continues till it indents:

$$Cm^{+} : m^{+} \rightarrow Om^{-}O$$

Now, in the painting *Still Life*, the indenting process continues further, causing a kink to occur, as illustrated in Fig 8.36. This event is, of course, inferred by the use of the kink rule of Hayes & Leyton (1989), discussed in section 8.5.1. Furthermore, in accord with the scenario of Hayes and Leyton, the kinking process continues, producing a cusp. In the case of the

Figure 8.32
Some ovals in Picasso's *Still Life on a Pedestal Table.*

Figure 8.33
Two flows arising from the ovals in Fig 8.32, the arrows being created by extrema. (Picasso's *Still Life on a Pedestal Table.*)

Figure 8.34
A bent oval in Picasso's *Still Life on a Pedestal Table*.

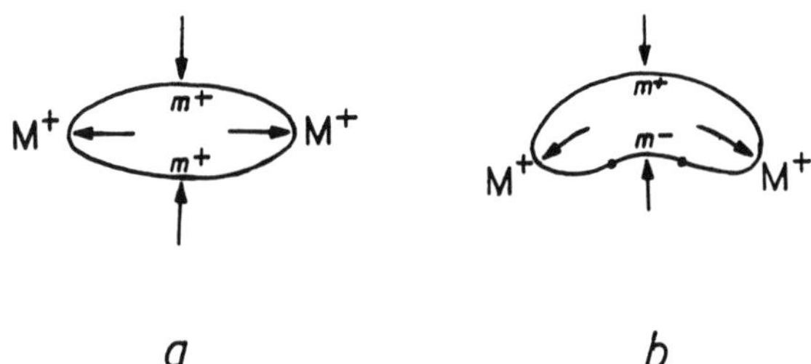

Figure 8.35
The transition from an oval to a bent oval.

Figure 8.36
A kinked oval in Picasso's *Still Life on a Pedestal Table.*

oval, this results in a *creased oval*, which is an important shape in the painting. Three creased ovals, on the right side, are shown in Fig 8.37. Each arrow indicates the cusp defining each crease.

An alternative history of the circle, in this painting, is that in which the circle changes to a pointed circle, as shown in the transition from Fig 8.38a to Fig 8.38b. The sharp point then pushes further out, to become a cusp, as shown in Fig 8.38c. It then bends to become a paisley, as shown in Fig 8.38d. Examples of the successive stages are illustrated in Fig 8.39. The pointed circle occurs in the top of the jug, at the top left. The cusped circle occurs to the left of the jug. The paisley occurs to the right of the jug, in the handle. In fact, this paisley interlocks with a paisley that points upward from the lower spiral of the jug-handle. Observe also that the bending which creates a paisley echoes the bent-creased oval history in the painting. In fact, Picasso combines these histories in the central leg of the table, forming a creased paisley, as shown at the bottom of Fig 8.39.

This completes our discussion of the histories arising from the circle in

Figure 8.37
Some creased ovals in Picasso's *Still Life on a Pedestal Table.*

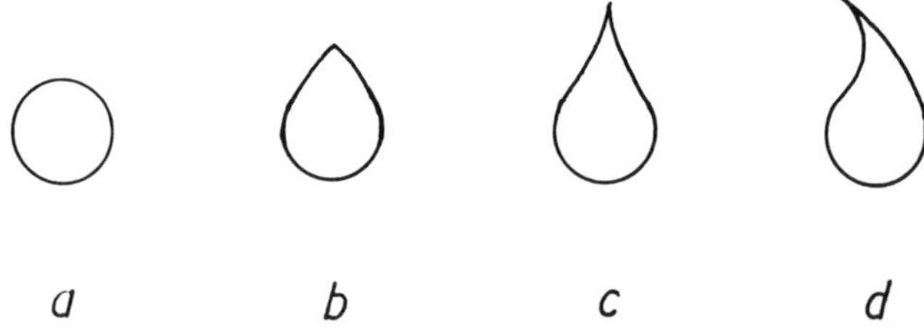

Figure 8.38
(a) A circle changing into (b) a pointing circle, then (c) a cusped circle, and finally (d) a paisley.

Figure 8.39
The shapes of Fig 8.38 occurring in Picasso's *Still Life on a Pedestal Table.*

Still Life. At each stage, the only rules that were used were the rules for the extraction of history from curvature extrema.

In addition to the histories based on the circle, there is a contrasting history based on the straight line. Whereas a circle is the ultimate past state for any smooth closed curve, the straight line is the ultimate past state for any smooth curve that is not closed. This is because the straight line is the shape that results from removing all asymmetries from a non-closed curve—i.e., in accord with the Asymmetry Principle.

Observe that the edges of the canvas are straight lines. Straight lines also occur in the floor and in the striped sections of the table cloth in the middle and left.

Throughout the painting, straight lines undergo bending. For example, the entire painting is divided down the middle by a large wavy line, as shown in Fig 8.40. Furthermore, the canvas is divided horizontally by two bands of wavy lines, which we have shown by the two horizontal wavy lines in Fig 8.40. Notice that, as each line crosses another, it creates a

Figure 8.40
Large wavy lines dividing Picasso's *Still Life on a Pedestal Table.*

curvature extremum in the other. Thus each is the symmetry axis along which the process, creating the extremum, acts, in accord with our rules for the extraction of process-history from curvature extrema.

In addition to these lines with large waves, there are lines with small waves, running like rivulets across the canvas—an example being shown in Fig 8.41. In accord with our rules, the waves, or curvature extrema, are created by the arrows, as shown.

In the painting, a line can also undergo successively increasing bending along itself, producing a spiral, as exemplified by the jug handle. Alternatively, a line can undergo kinking and cusping as exemplified by the three white lines that are drawn across the table cloth (Fig 8.29). The kink and cusp in the line-history echo the pointed and cusped circle in the circle-history, serving to tie these two histories together by common actions. Furthermore, the spirals in the line-history echo the paisley in the circle-history, thus again tying these two histories together by the common history of bending.

Figure 8.41
Lines with more rapid waves running through Picasso's *Still Life on a Pedestal Table.*

Having considered two major types of history in the painting—the history starting with the *circle*, and the history starting with the *line*—we will consider finally the third major type of history in the painting: that starting with the *square*. Observe that it is related to the line-history, because the internal structure of a square consists of lines, in accord with the principles of Chapters 2 and 6. Thus, we can regard the histories from a square as part of the histories from a line.

Generally, several times in the painting, four of the lines in the line-history are arranged so that they create a four-sided polygon. For example, the four lines that are the edges of the canvas create the canvas rectangle.

In accord with the Asymmetry Principle, any four-sided polygon has an external history that starts with a square. That is, a square is the past state that is obtained by equalizing all angles and sides. Squares are present in the painting in the tiled floor.

As we said, histories starting with the square are an important theme of the painting. In these histories, the square deforms to other four-sided

Figure 8.42
8.42 A horizontal band of wavy squares running through Picasso's *Still Life on a Pedestal Table*.

polygons. For example, a stretch is needed to create the canvas rectangle. This vertical stretch helps to define the upward theme of *offering* that is exemplified by the jug.

In the same way that the edges of the canvas yield the canvas rectangle, the large-scale wavy lines, dividing the canvas, as shown in Fig 8.40, produce a mosaic of six large four-sided polygons, all of whose past states must be squares, by the Asymmetry Principle. We shall therefore call them "wavy squares".

The wavy square is the single most common shape in the painting, undergoing a myriad of contortions, and subdividing every level of the work. For example, Fig 8.42 shows a band of wavy squares moving and changing shape across the painting. This band is itself divided into bands of wavy squares by the double molding lines that cut horizontally through it. The number of wavy squares in this region alone is enormous. For example, the central wavy square of the band in Fig 8.42, contains at least ten wavy squares, due to its various subdivisions. Furthermore, this

Figure 8.43
8.43 A vertical band of wavy squares running through Picasso's *Still Life on a Pedestal Table*.

central square is part of a vertical band of wavy squares that descends through the painting, as shown in Fig 8.43.

These bands are trace structures of wavy squares, thus giving a nested structure of control: That is, not only are the bands composed out of traces of wavy squares, but the wavy squares are themselves composed out of traces of lines, in accord with the principles of internal structure given in Chapter 6.

Given a wavy square, we observe that, not only are each of its wavy *sides* defined by processes leading to curvature extrema in the sides—hence giving the temporal sense of waving—but the *corners* follow the kinking→cusping scenario of Hayes & Leyton (1989). For example, observe that several corners, of the wavy squares in Figs 8.42 and 8.43, are cusps. Indeed, as shown in Fig 8.44, the large wavy square that holds the entire painting together has cusps as each of its four corners. In accord with our rules, the processes, leading along the symmetry axes to these cusps, pull the painting outwards in four directions.

Figure 8.44
8.44 A large wavy square used by Picasso to hold *Still Life on a Pedestal Table* together.

These outward cusping movements give the painting the theme of *sprouting*. They echo the cusping that occurs also in the history of the circle, thus tying together the circle-history and the square history with common actions. Finally, observe that the sprouting theme in the circle-history embodies sprouting in nature, since the history is traced back to circles that represent the various fruit in the painting.

8.7 Raphael's *Alba Madonna*

We have, in the previous sections, taken three paintings by Picasso from three very different periods of the artist's life, and shown that, despite the fact that they have very different styles, they all conform to our rules for the extraction of history from curvature extrema—and these rules exhibit fundamental aspects of the composition and meaning of the paintings.

Since we have looked only at paintings by Picasso, the reader might think, however, that our rules apply only to Picasso's work, or to work in a modern vein, with its flat, angular, strongly abstract, character. Furthermore, one might argue that process-history is more naturally central to the violent composition of a modernist painter. This is not the case. Thus to show the relevance of our rules to any period and style of art, we shall now consider a painting by Raphael, whose style cannot be more different from Picasso's. In contrast to Picasso's energy and violence, Raphael's paintings are completely suffused with a melodic gentleness—an infinitely subtle tenderness of line and form.

We will consider Raphael's *Alba Madonna*, shown in Fig 8.45. The canvas is circular in form—a constraint that is difficult for any artist to

Figure 8.45
Raphael, *The Alba Madonna* (c.1510). Oil on wood, transferred to canvas, diameter: .945 ($37\frac{1}{4}''$). National Gallery of Art, Washington. Andrew W. Mellon Collection. Photograph © National Gallery of Art, Washington.

handle, because a circular composition, if not visually anchored in the rectilinear gravitational frame, has a tendency to perceptually roll out of control; but if anchored too strongly, negates the value of having a circular composition at all. Thus the circular form extracts from Raphael all his artistic resources. The result is a painting of immense complexity, that attests to the fact that Raphael is one of the outstanding masters of composition.

8.7.1 Histories from the Straight Line

Like Picasso's *Still Life*, Raphael's *Alba Madonna* consists of histories derived from the straight line, and histories derived from the circle. However, there is a switch. Whereas, in the Picasso, the straight line arises from the edges of the canvas, and the circle is inserted into the painting as an additional contrasting factor—in the Raphael, it is the circle that is presented as the canvas edge, and the straight line that is inserted as the additional factor.

We first consider histories based on the straight line, and then turn to histories based on the circle.

Straight lines occur explicitly throughout the painting. For example, some of the roughly horizontal straight lines are shown in Fig 8.46. But, systems of parallel straight lines occur in other orientations. From straight lines, three types of histories follow in the painting. We elaborate them in turn thus:

1. Straight Line → Curvature Extremum → Bow → Opened Ovals & Convergence

In the painting, a straight line can undergo pushing by a single force creating a single curvature extremum in the line. For example, when the horizontal line shown, near the bottom of Fig 8.47, is pushed upward, it creates the curvature extremum shown. This, of course, gives the large pyramid structure of the painting. An encompassing pyramid occurs often in classical painting, particularly paintings of the Madonna with child. In such paintings, the Madonna appears to rise upward from the horizontal, a symbol of growth and fertility; and also a symbol of upward movement to heaven. The rising movement is given, of course, by our rules for the extraction of history from curvature extrema: That is, the head of the Madonna is the curvature extremum at the apex of the pyramid; and,

Figure 8.46
Some of the roughly horizontal lines in Raphael's *Alba Madonna*.

Figure 8.47
A horizontal line pushed upward in Raphael's *Alba Madonna*.

Figure 8.48
A downward process in Raphael's *Alba Madonna*.

by our rules, there is an upward process terminating at that extremum. Thus, the most frequent organizing principle of classical painting—the pyramid and its upward rising effect—is explained by our rules.

In Raphael's painting, an upward movement is not the only movement created by curvature extrema. There is a downward plunging movement created by a downward-facing extremum, as shown in Fig 8.48. Furthermore, there is a rightward movement across the painting, leading to the curvature extremum that is the Madonna's elbow; and a leftward movement across the painting leading to a curvature extremum at the neck of St. John (the child on the far left)—to be discussed later.

Each of these four movements, just described, is in fact given in stages, by the *successive* movement of extrema. This gives each a *bow* structure—one of the major organizing structures of the painting. For example, as shown in Fig 8.49, the upward and downward movements are given by successive lines explicitly drawn in the painting by the artist—as can be seen by examining the photograph of the painting in Fig 8.45.

Figure 8.49
Downward and upward bows in Raphael's *Alba Madonna*.

These two bows continue further upward and downward, crossing over each other, as shown in Fig 8.50. Observe now that, when a line from the upward bow crosses a line from the downward bow, its area of intersection is a pointed oval. Furthermore, as the bows continue upward and downward, the result is a succession of opening ovals. Now, the inferred starting state of an oval is a circle. Thus in the crossing bows, Raphael ties together the two classes of histories in this painting: histories from a straight line and histories from a circle (to be examined later).

Let us continue examining the bow structure. We have seen that bows occur upward and downward in the painting. In addition, Fig 8.51 shows a rightward bow moving to the Madonna's elbow. Observe that the bow is derived from a vertical straight line that goes between the Madonna's ankle and the upper patch of clouds. In accord with the Asymmetry Principle, this straight line is the ultimate past state of the bow; i.e., the bow before the asymmetries were added. Thus, visually, forward time moves from the straight line rightward to the elbow. Counterbalancing this, as

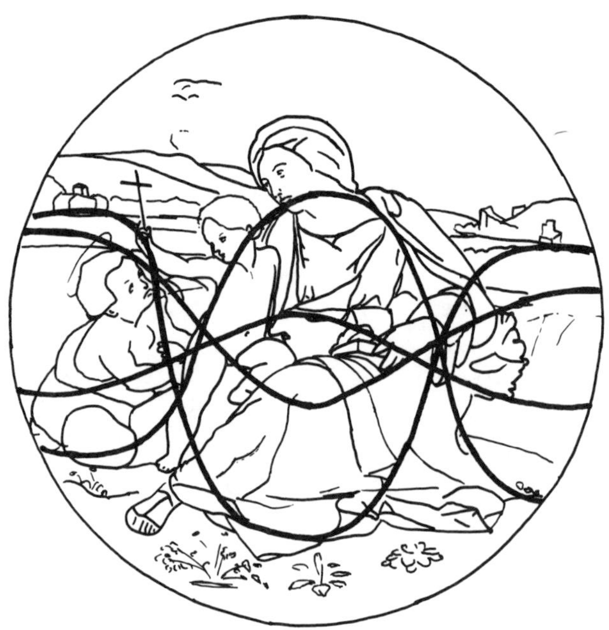

Figure 8.50
The continuation of the bows in Fig 8.49.

Figure 8.51
A rightward bow in Raphael's *Alba Madonna*.

Figure 8.52
A leftward bow in Raphael's *Alba Madonna.*

shown in Fig 8.52, there is a leftward bow terminating in St John's neck giving the sense of leftward thrusting, and thus forward time, from right to left.

We should observe that the sequences of extrema creating bows do not only move along straight lines. For example, Fig 8.53 shows a large sequence of extrema dipping downward to the horizontal base of the pyramid. In addition, Fig 8.54 shows a sequence of extrema that spill beautifully over the knee of the Madonna.

Now observe the following. A bow typically has two "knot" points, where the lines meet. For example, in Fig 8.51, the two knot points are the Madonna's ankle and the upper cloud mass. It is important to study what happens at a knot point.

Consider Fig 8.55a. It shows the knot point at the Madonna's ankle. Since, with this bow, the ultimate starting state is the vertical line (recall Fig 8.51), the successive lines of the bow move *around* the knot point as shown in Fig 8.55a. That is, *forward time* involves movement around the

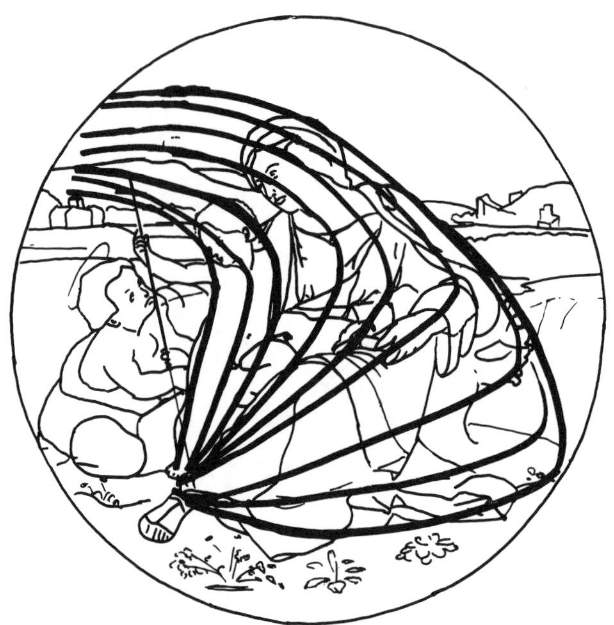

Figure 8.53
A bow dipping to the horizontal base of the pyramid, in Raphael's *Alba Madonna*.

Figure 8.54
A sequence of extrema spilling over the knee of the Madonna, in Raphael's *Alba Madonna*.

Figure 8.55
At a knot-point, there is (a) rotation, and (b) convergence.

knot. This is true of the knot point of any bow. Thus we have this conclusion: *at a knot point of a bow, there is the process of rotation—the successive lines rotate around the knot point.*

However, visually, there is another equally important history involved in a knot point. The succession of lines around the point, produce a succession of curvature extrema as shown in Fig 8.55b. Now, according to our rules for the inference of process-history from curvature extrema, each of the extrema at the knot, has a process leading to and terminating at the extremum. For the knot point in Fig 8.55b, the processes are indicated by the arrows. Since all the extrema, at the knot point, share the same termination point, the result is that the processes converge into the knot point. We shall call such a structure, a *convergence*. Thus, any knot point visually implies two process-structures: a process of *rotation around*, and a process of *convergence into*, the knot point.

The convergence structure is one of the most powerful organizing principles of this painting. It is, in fact, the terminal point of several of the histories, as we shall see. It will be discussed in much greater detail shortly.

2. Straight Line → Curvature Extremum → Kink → Bow of Kinks → Opened Diamonds & Convergence

In history 1, above, the second stage is the formation of curvature extrema. Subsequent to this stage, one can either continue with history 1, or with an alternative history which we shall now give.

In this alternative history, the extremum undergoes kinking, in accord with the rule for kink-formation developed by Hayes & Leyton (1989). Raphael often places a kink near a (rounded) curvature extremum. For example, observe that, while the outer edge of the Madonna's elbow is a rounded extremum, the inner edge forms a kink (and also a cusp). Subsequent to its formation, the kink can undergo the same set of processes given in history 1, as follows: First, the kink undergoes continuation, producing a *bow* of kinks. Then bows from opposite directions eventually overlap. Observe that, whereas the overlapping bows of rounded extrema produced *opening ovals*, overlapping bows of kinks produce *opening diamonds*. These exist on every level of scale in the painting. For example, the pages of the book, at the center of the painting, together with the surrounding drapery folds, define an opening set of diamonds—as can be seen by comparing Fig 8.56 with the photograph of the painting in Fig 8.45.

Finally, we observe that, as we saw previously with bows of rounded extrema, the bows of kinks not only define *rotation around* their knot points, but *convergence into* those points. For example, the pages of the book not only imply the opening, i.e., rotation, of those pages, but the convergence of the pages leftward and rightward to their knot points.

3. Straight Line → Rotated Line → Convergence

Having followed histories that start with straight lines and progress through

Figure 8.56
A set of opening diamonds in Raphael's *Alba Madonna*.

curvature extrema, we now return to straight lines, and follow alternative histories.

Thus, let us return to the straight lines shown in Fig 8.46. It is important to observe that Raphael makes these lines differ slightly in orientation. For example, the upper horizon lines dip on the right below the horizontal. The effect is to start a rotational movement in the picture plane.

For example, the lines shown in Fig 8.57 start at the horizontal and rotate anti-clockwise. Notice that we have not shown the true horizontal line in Fig 8.57. Raphael has not drawn the horizontal line in the painting, because the viewer easily supplies it with the gravitational frame. What Raphael gives is a major line that is almost horizontal—the line in Fig 8.57 that is nearest to the horizontal. This slightly dipping line is a major line in the work, as can be seen by looking at the photograph of the painting in Fig 8.45; that is, Raphael explicitly draws this line as the strongly emphasized waist-line of the Madonna which continues through the neck of St. John (the child on the far left). The non-horizontality of this line starts the rotational movement.

Figure 8.57
Anti-clockwise rotating lines in Raphael's *Alba Madonna.*

Similarly, observe, from the photograph of the painting (Fig 8.45) the almost vertical line descending from the Madonna's chin. This has also been drawn in Fig 8.57. Its role with respect to the true gravitational vertical is the same as the role of the waist-line with respect to the true gravitational horizontal; i.e., it starts an anti-clockwise rotational movement.

Against the anti-clockwise movements, there are clockwise movements. For example, observe, in the photograph Fig 8.45, the straight line that runs leftward along the bottom edge of the Madonna's shirt and through the point where the crucifix meets the cheek of St. John. Like the waist-line, this line is also slightly off the horizontal—but in the opposite direction. It therefore starts a clockwise rotation. In fact, the entire group of figures sets up a gentle sway between clockwise and anti-clockwise movements around the center of the painting.

Important centers of rotating lines occur at other points in the painting. The most important is the point mentioned previously as the knot point at the ankle of the Madonna. More precisely, the point is between the drapery at the ankle and the tip of the big toe of the Christ child. We saw that, because it is a knot point for a bow, rotation is defined around it. However, the bow lines coming from a knot-point are curved. Nevertheless, this point is also a center for powerful *straight* lines that converge there. They are shown in Fig 8.58. These lines are set up by curvature extrema in accord with our rules. Each arrow drawn in Fig 8.58 is a symmetry axis leading to an extremum at the tip of the arrow.

Now, we saw in Fig 8.57, that the center of the painting is the intersection point for several major straight lines. Since this is the true center of the painting, another center such as the Madonna's ankle is seen as a translated version of the true center. Visually, it is as if the center of the painting has been forcibly pulled down to the ankle along a diagonal path. In fact, observe from the photograph of the painting, Fig 8.45, that there is a downward diagonal wedge sandwiched between Christ's straight leg and the Madonna's leg. In accord with our rules, the symmetry axis of the wedge defines a downward *process* leading to the ankle. The wedge therefore creates the movement that pulls the true center down to the ankle.

The true center is also pulled in other directions. For example, it is pulled leftwards to a point just behind the ear of St. John, as shown in Fig 8.59. The line of pulling is the nearly horizontal line of which the Madonna's waist-line is a part—a line we discussed earlier. Observe how

Figure 8.58
The convergence at the ankle of the Madonna, in Raphael's *Alba Madonna*.

Figure 8.59
The convergence at the ear of St. John, in Raphael's *Alba Madonna*.

this left-ward pull is again defined by curvature extrema—the three major ones being given by the three left-ward arrow-heads in Fig 8.59. The other arrow-heads indicate other extrema-based processes creating the convergence point. For readability we have omitted several additional extrema-based arrows terminating at this point, e.g., that created by the fur on the back of St. John.

The true center is pulled also rightwards, along the same, almost-horizontal, line, to become the convergence point shown in Fig 8.60. In fact, major convergence points are arranged at the five vertices of a pentagon defined by the group of figures. The five vertices are (1) the Madonna's ankle on the left; (2) its reflectional opposite on the right; (3) St. John's ear; (4) its reflectional opposite, the point shown in Fig 8.60; (5) the Madonna's temple.

Convergence points pervade the painting at every level of scale. For example, other major convergence points are (1) the point where the crucifix meets the cheek of St. John; (2) the bob at the back of the Madonna's

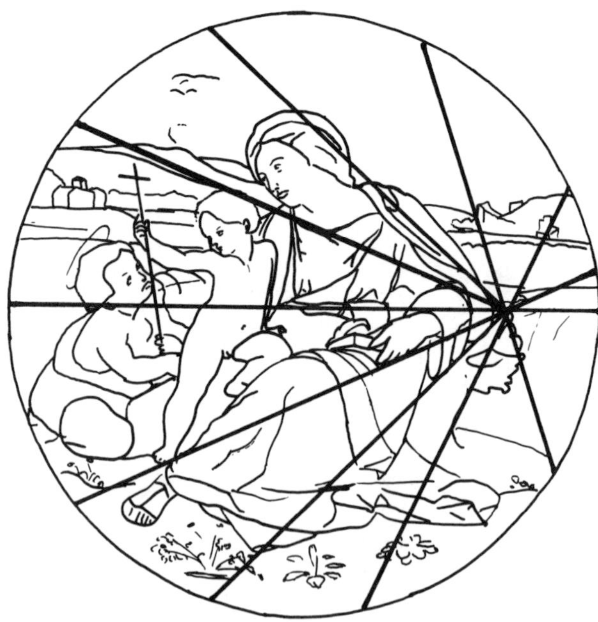

Figure 8.60
The convergence point near the elbow of the Madonna, in Raphael's *Alba Madonna*.

neck; (3) her chin. Notice, from the photograph, Fig 8.45, how many lines are pulled together at her chin: the downward lines of the hills, the line of her shoulder, the vertical and diagonal lines on the left side of her shirt, the shadow line of her jaw, the edge-line of her face.

Between every pair of lines, there is a curvature extremum. By our rules, the extremum implies a process directed towards it. It is the collection of these processes, defined by all the pairs of lines, that gives the intersection point the sense of actual *convergence*; *i.e.*, *inward pulling*.

Quite clearly, convergence has an important meaning in the painting. It represents the *drawing in* of the Madonna, and therefore mother-church. In fact, the head of the Madonna, alone, has at least eight convergence points—besides the chin and the back of the neck, there are the ear, the crown of the head, the four corners of the eyes. Each of these pull us into them, and we cycle from one to another. By being pulled into this gentle face, we experience the warmth of mother-church. By being cycled from one convergence point to another, we experience the church as laying down life's journey as a series of points of convergence and comfort.

Thus, in the drawing-in structure of the curvature extrema, we again see corroboration of the Corollary of our First Art-Work Principle, which states that the meaning of a painting is the history inferred from its shape-structure, and the experience of the painting is the experience of recovering that history.

4. Straight Line → Translated Line → Band → Crossing Bands → Bands of Sheared Squares → Wavy Bands of Sheared Squares

Besides undergoing rotations, the straight lines undergo translations. The translated lines form the internal structure of *bands*, as shown in Fig 8.46. When these bands are of different orientations, they cross each other and form bands of sheared squares. When these latter bands undergo extrema-based processes, they become wavy bands, as shown in Fig 8.61.

We saw that, in his *Still Life*, Picasso also had bands of wavy squares, Fig 8.42. Notice that, in Fig 8.61, as in the Picasso, the bands go from one edge of the canvas to the other, and they also disregard the divisions of foreground and background. This is an important device, in both paintings, for knitting together foreground and background, and for pulling both of these into the plane of the canvas. In addition, the sheared aspect of the squares reinforces the swaying, rotational theme discussed previously.

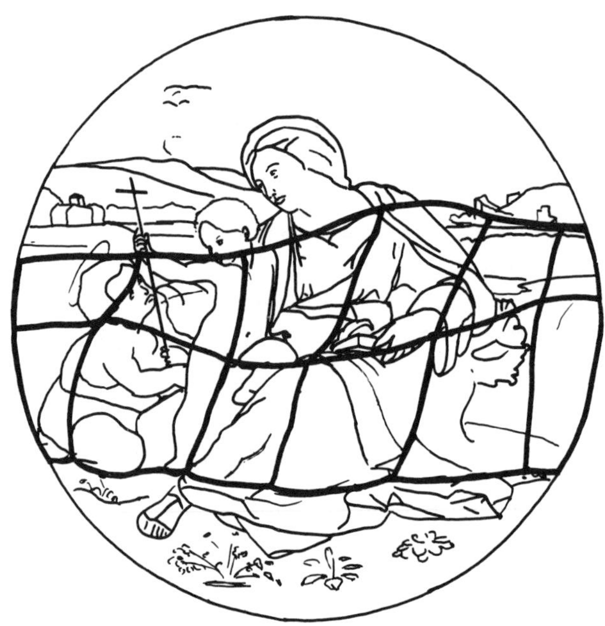

Figure 8.61
Some wavy bands in Raphael's *Alba Madonna*.

8.7.2 Histories from the Circle

As we said above, the *Alba Madonna* consists of histories derived from the straight line and histories derived from the circle. Having elaborated the former, we now elaborate the latter.

5. Circle → Oval → Elongated → Pointed → Cusped → Curled → Rotated → Convergence

The circle is the ultimate starting shape of any smooth closed shape. Its major importance to the painting is explicitly announced by the fact that Raphael chose it as the edge of the canvas. However, it also exists within the canvas: e.g., the two circles drawn in heavier lines on the left and right in Fig 8.62.

As we noted previously, the first shape (with non-constant curvature) that can be derived from a circle, is the oval. Ovals are present throughout the painting. Indeed, the two circles shown in Fig 8.62 stretch horizontally

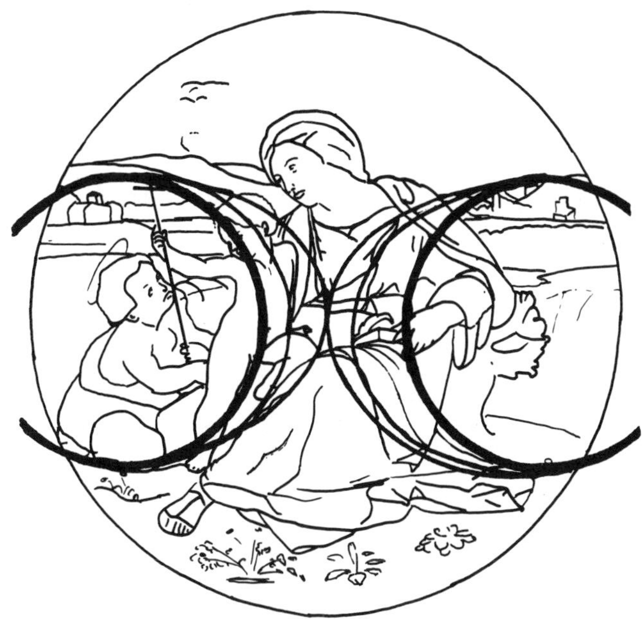

Figure 8.62
Two circles stretching horizontally in Raphael's *Alba Madonna.*

becoming ovals, as shown. Ovals also appear as the face of the Madonna, the bodily limbs of all three people, etc.

The oval then stretches further becoming *elongated*, as shown in Fig 8.63a; then *pointed*, as shown in Fig 8.63b; then *cusped*, as shown in Fig 8.63c; and then *curled*, as shown in Fig 8.63d. These successive transitions are all given by rules we have discussed previously; that is, the successive transition from Fig 8.63a to 8.63b to 8.63c is an example of the successive kinking and cusping scenario of Hayes & Leyton (1989); and the final curled shape is obtained by bending, applied to each end, in accord with the rules developed in section 6.34.

For simplicity of discussion, we shall refer to all four shapes in Fig 8.63 as *elongated ovals*, except when wishing to refer to one of them specifically.

Fig 8.64 shows some of the elongated ovals of the painting. Before we begin to look at them, the reader should note that the arrow, just off the painting on the upper left, indicates the line between land and sky. Now notice how the elongated ovals in the hills, i.e., below the sky-line, are

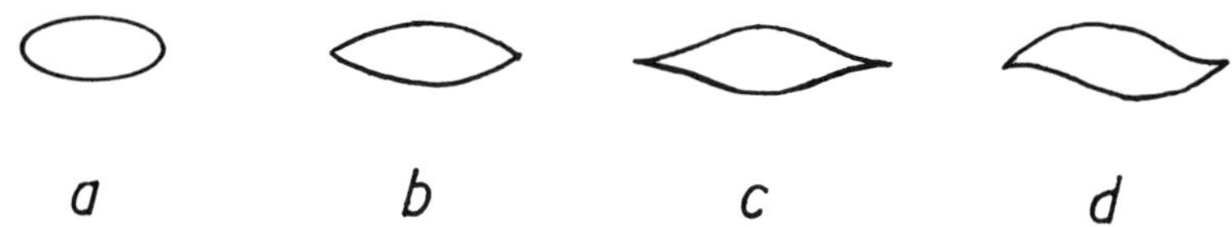

Figure 8.63
(a) An oval, undergoing (b) kinking, (c) cusping, and (d) curling—in accord with the scenario in Hayes & Leyton (1989).

Figure 8.64
Some of the elongated ovals in Raphael's *Alba Madonna*.

pulled into the face of the Madonna, and then flow through ovals in her neck, out through ovals in the landscape on the right. Observe also, in Fig 8.64, that the faint clouds above the skyline on the left (compare with the photograph, Fig 8.45) demarcate elongated ovals in the sky, that are themselves pulled into the Madonna's hair and terminate in the bob at the back of her neck.

The hair alone is a profusion of curled ovals that wrap around each other and are themselves grouped into a single curled oval—the outline of her hair. The same is true of the gown over the upper half of her arm on the right and the transparent sleeve over her forearm on the right. That is, each of these are split into elongated ovals that are themselves grouped into elongated ovals. Note also that St. John, the child on the far left, is a profusion of elongated ovals extending through his limbs, muscles, and clothing.

Fig 8.64 presents elongated ovals on roughly one level of scale. Elongated ovals occur on a much smaller scale as can be seen from the photograph, Fig 8.45; the smaller-scale elongated ovals occurring as the petals of the flowers at the bottom; the minute horizontal gradations in the sky; the anatomical details such the eyes, lips, fingers, and toes; the minute details of the clothes, e.g., on the Madonna's chest; the fine layers of leaves in the band of trees to the right of the Madonna's elbow.

At the other extreme of size, large-scale elongated ovals criss-cross and bind together the entire painting. For example, there is the large elongated oval that starts with the Madonna's foot and extends diagonally up to her elbow. Again, the profusion of elongated ovals through the hills on the left and right (Fig 8.64) are themselves bound together into single elongated ovals each of which extends through the entire width of the painting, as shown in the upper part of Fig 8.65. Notice how the entire sky, as a single shape—the top shape in Fig 8.65, bounded at the top by the edge of the canvas—is an elongated oval, a curled oval. This shape imitates the curled ovals in the Madonna's hair, the curled oval of her forehead, the curled ovals of her eyes, the curled oval of Christ's hair, the ten or more curled ovals within the small upper patch of clouds, etc.

Now consider a curled oval in more detail, e.g., that in Fig 8.66a. As can be seen from Fig 8.66a, the two ends of the shape both imply rotation, and in fact this rotation is in the same direction for both ends, either both clockwise or both anti-clockwise.

Figure 8.65
Large elongated ovals in Raphael's *Alba Madonna*.

Figure 8.66
(a) A curled oval implies rotation. (b) Rotating the curled oval.

Thus the curled ovals, in the painting, all define rotation. For example, the sky, the Madonna's hair, her eyes, etc., all define rotation in the anti-clockwise direction.

Curled ovals therefore connect the histories based on the circle to the histories based on the straight line, by implying a common process: rotation. Thus, although the two sets of histories have ultimately different starting states, Raphael subtly unifies them by introducing commonalities in the actions applied in the two histories.

Finally, we should observe that a curled oval not only implies rotation around its center, but can be rotated around *one end*, producing a convergence, as shown in Fig 8.66b. One can see examples in the hair of the Madonna, and at the bow in the waist of St. John.

6. Circle → Oval → Pointed → Diamond

This alternative history of a circle, via an oval, is one that also occurred in *Demoiselles d'Avignon*. Since it was discussed in detail there, it need not be discussed again here; recall for example Fig 8.21.

7. Circle → Oval → Triangular Oval → True Triangle → Externalized (Equilateral) Triangle → Compressed or Widened (Isosceles) Triangle → Sheared → Indented → Cusped → Rotated → Convergence

An alternative history from the circle, via the oval, is as follows. Recall that an oval involves four processes as shown in Fig 8.67a; that is, two outward protruding processes and two inward squashing processes. When a protrusion is introduced into this shape, as shown in the upper center of Fig 8.67b, the squashing process bifurcates around the new protrusion.

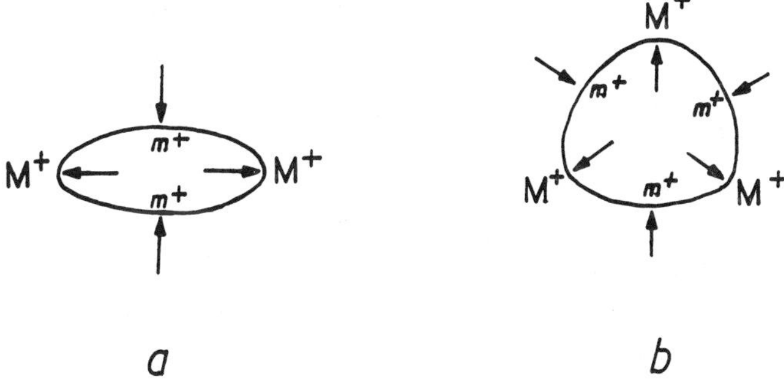

Figure 8.67
The operation of introducing a protrusion, and thus creating a triangular oval.

Figure 8.68
Some triangular ovals in the painting.

The structure is given by the fifth operation of our process-grammar:

A protrusion is introduced:

$$Bm^+ : m^+ \rightarrow m^+M^+m^+$$

The resultant shape is a triangular oval. Fig 8.68 shows some triangular ovals, in the painting, on the middle level of scale. On a smaller scale, triangular ovals include several limbs—e.g., the thigh of St. John, the thighs and calves of Christ's legs—the transparent drapery on the Madonna's forearm, and an inverted triangular oval of shadow on her neck—see the photograph of the painting, Fig 8.45.

The three protruding processes of a triangular oval can continue and create kinking, in accord with the rules of Hayes & Leyton (1989). The result is a true, i.e., *pointed*, triangle—which occurs throughout the painting and which we will now study.

The occurrence of a true triangle, in the history of a circle, creates the following situation: When a triangle is externalized, it produces an equilat-

Figure 8.69
An equilateral triangle in Raphael's *Alba Madonna*.

eral triangle. Note that externalization goes in the direction of backwards time—although, to save introducing a complexity in notation, we have indicated it by a forward arrow in the heading of this subsection. One might have given the equilateral triangle as a new starting node, i.e., with the same starting status as the circle. However, the power of the circular canvas edge makes any triangle in the painting seem derivable from the circle, by kinking. In an ordinary rectangular canvas, this would not be the case.

The equilateral triangle occurs, for example, as shown in Fig 8.69. Note the slightly rotated orientation—adding to the gentle rocking feeling of the composition. Notice also how the left side is the important line descending along the edge of the heads and through the convergence point where St. John's cheek touches the crucifix.

Subsequently, the equilateral triangle is compressed or widened producing isosceles triangles, as in Fig 8.70.

The isosceles triangle then undergoes the three stages shown in Fig 8.71. It is sheared, producing Fig 8.71b. Then the kinking processes that created

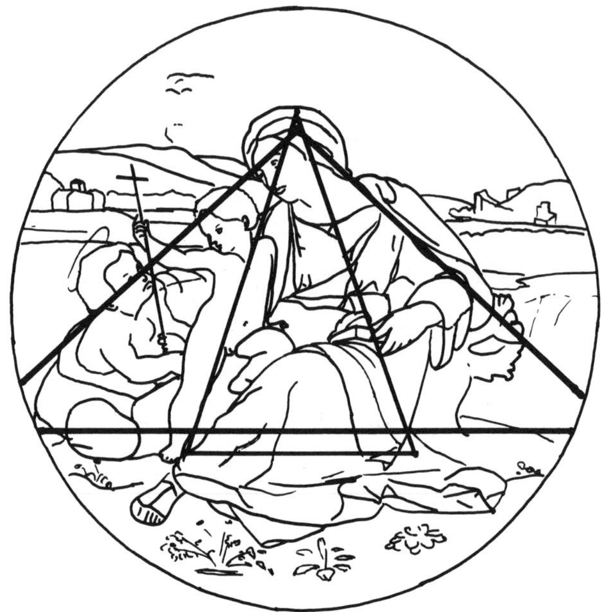

Figure 8.70
An equilateral triangle is compressed and widened, in Raphael's *Alba Madonna*.

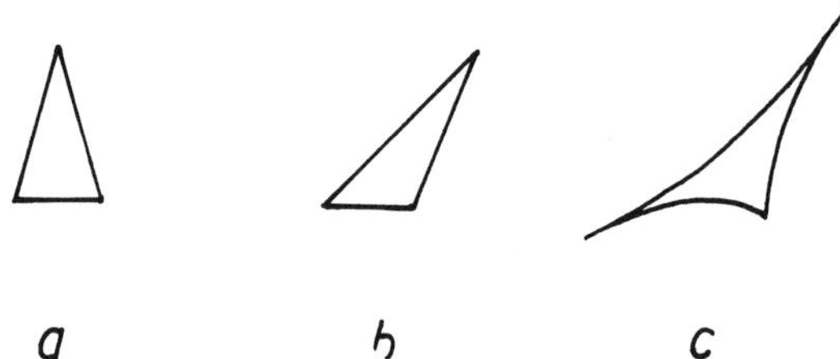

Figure 8.71
(a) An isosceles triangle undergoing (b) shearing and (c) cusping—in accord with the scenario of Hayes & Leyton (1989).

Figure 8.72
Some of the cusped triangles in Raphael's *Alba Madonna*.

the vertices continue, producing cusps, as shown in Fig 8.71c—in accord with the scenario of Hayes & Leyton (1989), discussed in section 8.5.1. Cusping might occur at only one of the vertices; but it could occur at two or all three of them.

Triangles with cusped vertices occur profusely throughout the painting, on all levels of scale. For example, Fig 8.72 shows several cusped triangles on one level of scale.

Finally, throughout the painting, triangles are rotated around a vertex, resulting in a convergence. For example, observe, in Fig 8.72, the cusped triangles that rotate around a point in the Madonna's chin. The vertices at this point each define an extremum, and the associated processes point into the chin, thus creating the convergence. Often, in the painting, a convergence of cusped triangles, at a point, can echo a convergence of larger triangles, at the same point. Thus, the convergence we discussed, of cusped triangles at the chin, is a small-scale version of a set of large triangles that converge on the chin, and whose outer edges are the edges of the canvas, as shown in Fig 8.73.

Figure 8.73
Large cusped triangles dividing Raphael's *Alba Madonna*.

Alternatively, in the history just defined, the cusping stage can be omitted. That is, isosceles triangles can be sheared and rotated, as shown in Fig 8.74. The rotated system is then re-interpreted as a convergence.

8. Triangle → . . . → Divided Triangle → Divergence

Any triangle in the above history can undergo a division, in the following way. Consider the triangle shown in Fig 8.75. The bottom extremum has a symmetry axis, and thus a process is assumed to have created that extremum, as indicated by the arrow. However, the symmetry axis that was used to create the extremum, can subsequently be used to create the upward extremum shown in Fig 8.76. Observe that, in going from Fig 8.75 to 8.76, the single triangle of Fig 8.75 has been divided, producing the pair of triangles in Fig 8.76. Subsequently, these triangles can themselves undergo division along their symmetry axes. For example, the right-hand triangle in Fig 8.76 can itself divide as shown in Fig 8.77.

The consequence of this is two-fold. First, one has the creation of the *convergence* structure: Each dividing triangle produces two inward vertices

Figure 8.74
A rotated-triangle structure in Raphael's *Alba Madonna*.

Figure 8.75
A triangle, with one of its inferred processes, in Raphael's *Alba Madonna*.

Figure 8.76
The division of a triangle in Raphael's *Alba Madonna.*

Figure 8.77
The additional division of a triangle in Raphael's *Alba Madonna.*

where there was previously only one; e.g., compare the successively increasing number of extrema at the ankle, in Figs 8.75, 8.76, 8.77.

However, the other effect is to create a *divergence*; e.g., the extrema at the Madonna's head and shoulder in Fig 8.77 were created by outward arrows. In fact, the appearance of divergence should not be a surprise to us. For example, in the rotated triangle structure in Fig 8.74, one infers not only a convergence due to the inward pointing extrema, but a divergence due to the outward pointing extrema around the edge.

Divergences occur not only throughout the large-scale structure of the painting, but in many small details, e.g., the plants at the bottom. Clearly, the meaning of a divergence is *outgrowth*. The central role of the Madonna, in the Christian mythology, is the birth that she gives. And the birth brings not only physical life but spiritual life. This process of spiritual emergence and flowering is captured in the divergences of the painting. However, since the divergences constitute *process-structure* that is *inferred* from the shapes of the painting, we see once again that it is the inferred process-structure that embodies the *meaning* of an art-work. Furthermore, we find, once again, that the *experience* of an art-work is actually the experience of solving the process-recovery problem.

8.7.3 The Structure of the *Alba Madonna*

In Fig 8.78, we have mapped out the historical structure embodied in the *Alba Madonna*. The figure is given in two parts: the histories descending from the straight line, and the histories descending from the circle.

This painting is the last work we shall analyze in this book. Having completed our study of individual paintings, we now return to the main argument of this chapter.

8.8 The Second and Third Art-Work Principles

In the above sections, we analyzed, in considerable detail, four very different paintings. We found that, in each case, they substantially corroborated our First Art-Work Principle, which states that an art-work is the process-history inferred from it. That is, we repeatedly found that what one regards as the *composition* and *meaning* of a painting turn out to be the process-history inferred from it. Due to limitation of space, we

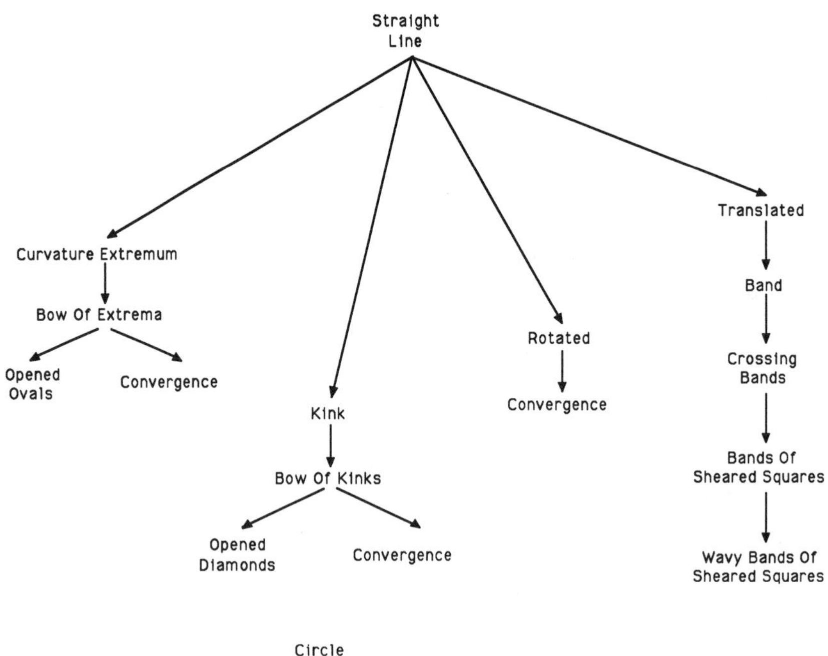

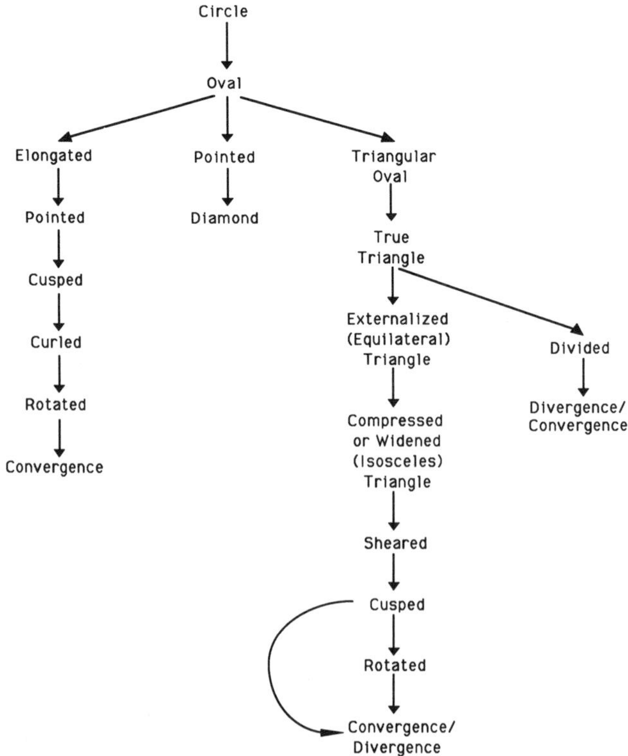

Figure 8.78
Histories from a straight line and histories from a circle, in Raphael's *Alba Madonna*.

concentrated mainly on the process-recovery rules given in the first two chapters of this book. However, in our forthcoming book on the structure of paintings, we will also show in detail how the process-recovery rules in the remaining chapters provide powerful insight into the perception of paintings.

The primary principle substantiated in these analyses is our First Art-Work Principle. However, closer examination reveals that the principle, on its own, does not fully distinguish art-works from non art-works. Two additional principles are needed to establish the distinction.

To see that the First Art-Work Principle does not fully distinguish art-works, consider the statement of the principle: An art-work is the process-history inferred from it. Now recall that, in this book, we have claimed that *all* perception is process-recovery; i.e., the perception of *anything* is the process-history inferred from it. Thus, the First Art-Work Principle claims simply that perceiving an art-work is not different from perceiving anything else; i.e., perceiving an art-work, like perceiving anything else, is recovering process-history.

Thus, the First Art-Work Principle does not create a distinction between art-works and other things; it expresses their commonality. This means that one must seek additional principles that distinguish art-works. We claim that the following two principles provide the required distinguishing criteria. In fact, we argue that they fully define what an art-work is:

SECOND ART-WORK PRINCIPLE. *An art-work allows the* **maximal amount of history** *that a person is capable of inferring from any percept; i.e., the* **maximal amount of causal explanation** *that a person is capable of assigning to any percept.*

This principle has the following corollary:

COROLLARY OF THE SECOND ART-WORK PRINCIPLE. *An art-work contains the maximal amount of* **asymmetry** *from which a person is capable of extracting history; i.e., to which a person is capable of assigning a causal explanation.*

The Second Art-Work Principle and its Corollary will be corroborated in the following sections. The final principle is this:

THIRD ART-WORK PRINCIPLE. *The entire set of histories in an art-work is symmetrically organized.*

This principle appears to lead to a contradiction, but does not. Recall first that the Corollary of the Second Art-Work Principle states that an art-work contains the maximal amount of *asymmetry* from which a person is capable of extracting history. The Third Art-Work Principle implies that the asymmetries must themselves be symmetrically organized. To see that this is not a contradiction, consider the paintings that we analyzed. We, in fact, saw this principle evidenced much in our analyses: For example, we saw that in Raphael's *Alba Madonna*, the bows pulling from left to right (Fig 8.51) were balanced by bows pulling from right to left (Fig 8.52). That is, the two sets of *asymmetries* were arranged symmetrically. Similarly, in Picasso's *Demoiselles d'Avignon*, the diamonds opening from left to right (Fig 8.23) were balanced by diamonds opening from right to left (Fig 8.24).

The symmetrical organization of histories or asymmetries, in a painting, has always been the *primary* organizing factor in an art-work. Using the principles of this book, we can explain this as follows: Recall that our Symmetry Principle states that a symmetry, in the present, is assumed to have always existed. In fact, more strongly, our Second Duality Principle (section 1.6) states that a symmetry in the present is assumed not only to have existed throughout the past but also to exist throughout the future. This principle led to our Stability Principle (section 1.6), which states that, the more symmetric a configuration is, the more stable it is.

Thus by organizing the histories in an art-work, *symmetrically*, the artist makes the histories seem stable, non-accidental, essential, and necessary.

In contemporary art, the stabilizing effect of symmetrizing the histories has been exploited to a still more powerful level than in classical art. In abstract expressionist paintings, such as those of de Kooning, paint is smeared, slopped, scraped, scratched, and scored into the canvas, giving the painting an unfinished, untidy, and deliberately filthy appearance. Nevertheless, de Kooning's powerful sense of compositional symmetry, learned from his extensive studies of classical art, gives the very accidental features of his painting a sense of non-accidentality, of permanence, of inevitability, and enduring strength. For example, by symmetrically organizing "accidental" marks that he has used in the smeared, untidy, portrait of a woman, the portrait assumes a powerful sense of eternity, a monumentality, equal to a woman painted, for example, by da Vinci.

8.9 The Preference for Stimulus Complexity

In the following sections, we shall attempt to corroborate the Second Art-Work Principle, and our discussion will lead us to one of the most important ideas of this book.

Of the three Art-Work Principles, the second one is the only one that we have not yet corroborated. The other two have been strongly corroborated by our detailed analyses of paintings. This is because they can be directly supported by looking purely at paintings. The Second Art-Work Principle requires that paintings are *compared* with other stimulus situations. This is because the Second Art-Work Principle states that art-works contain the maximal amount of asymmetry from which a person can extract history; i.e., to which a person can assign causal explanation. In other words, the principle implies that one will not find other stimulus situations in which the person can extract more history.

Before we begin, observe first that, because we are concerned with showing that art-works contain the maximal amount of asymmetry from which a person can extract history, we are now concerned with the *amount* of asymmetry contained in a stimulus situation. Now recall that the amount of asymmetry is often referred to by a single term, *complexity*. That is, complexity is the *quantification* of asymmetry. Thus, we are concerned with situations that have the maximum complexity from which a person can extract history.

In order to corroborate the Second Art-Work Principle, we will examine one of the most remarkable bodies of research in the history of experimental psychology: the extensive research on stimulus complexity. Whereas any large-scale body of experimental research usually has several contradictory results, and contains several major points on which researchers disagree, the research on stimulus complexity, despite being carried out by scientists from many different areas—animal learning studies, child studies, neurophysiology, psychophysics, sociological studies—all led to a single incontrovertible result; the following:

> **COMPLEXITY-LEVEL HYPOTHESIS.** (1) Each individual has a preferred level of environmental complexity, a level that the individual seeks, and finds extremely appetitive. (2) As environmental complexity moves away from this level, i.e.,

> as complexity is raised or lowered from this level, the environment becomes more and more aversive.

Let us examine the two parts of the Complexity Level Hypothesis in turn.

8.9.1 Complexity Level Hypothesis, Part 1: Optimal Complexity

The first part of the Complexity-Level Hypothesis states that each individual has a preferred level of environmental complexity, a level that the individual seeks, and finds extremely appetitive. In order to understand the considerable significance of this statement let us contrast it with some of the most commonly-held beliefs about perception and behavior. These beliefs, while being widely held by the lay public, were in fact adopted by a branch of psychology known as behaviorism—and, we shall, in the following discussion, describe these beliefs using simple terms taken from behaviorism, terms that we shall explain quite easily. Nevertheless, despite the fact that behaviorism has declined in popularity, the beliefs are still commonly-held by the lay public.

First observe that it is generally believed that an animal has a set of "primary drives", such as hunger, thirst, and sex. These drives are called *primary* because the animal does not satisfy them in order to achieve another goal it has set itself; i.e., satisfying them is not *secondary* to another goal. That is, satisfying each of them is an end in itself. Thus, one can say that a primary drive is associated with a *primary goal*; e.g., the primary drive, hunger, is associated with the primary goal, eating food.

Observe that it is generally believed that, when an animal satisfies a primary drive, there follows a *reduction* in that drive; e.g., eating causes a reduction in hunger—an experience we all have, every day.

Now because primary goals are, by definition, the ultimate goals of an animal, it seems reasonable to conclude that the animal's *perceptual* and *behavioral mechanisms* are used in the service of obtaining those goals. For example, an animal has the capacity to *perceive* and *behave* in order to construct a path that will lead to food, water, sex, etc. This means that perception and behavior are *secondary* to the primary goals; i.e., in the service of the latter. Also, it seems to be the case that, in attempting to obtain a goal, perception and behavior attempt to construct the path that is the most efficient, e.g., that involves the least number of wasteful steps.

Finally, it is generally believed that a path that consistently leads to a goal is *learned*; i.e., this path becomes the path that the animal takes whenever it needs to reach the goal. There is no reason why another path should be taken. Thus, if a path is successful, it is *repeated.*

We can summarize the above commonly-held beliefs about perception and behavior, as follows:

COMMONLY-HELD BELIEFS ABOUT PERCEPTION AND BEHAVIOR:

(1) Satisfying a primary drive—i.e., achieving a primary goal—*reduces* the drive. For example, eating reduces hunger.

(2) Perception and behavior serve the achievement of goals; i.e., they are *secondary* to the achievement of goals.

(3) Perception and behavior attempt to achieve goals by the most efficient paths.

(4) Successful paths, i.e., those that obtain goals and do so most efficiently, are repeated.

So entrenched are these beliefs that, in most behavioral research, the beliefs have been, largely, unchallenged. However, in the 1950s, a body of research began to emerge that tested whether these beliefs were actually valid. This research demonstrated the existence of a powerful phenomenon—that animals search for complexity—and the phenomenon seriously undermines the above beliefs, as we shall see. It is to this research that we will now turn.

Even before the 1950s, an occasional researcher was insightful enough to construct a subtle experiment that challenged the common view.

Consider the following simple, yet important, experiment. Krechevsky (1937) gave healthy rats a choice between two routes to a goal. One of the routes was more direct; the other was longer and presented more variety. The surprising result was that the animals preferred the *less direct* route to the goal. It appeared that an important factor in determining the subjects' behavior, was not only the final goal but also the *complexity level* of the path to the goal; a path with *greater complexity*—greater asymmetry—was preferred.

This simple result is, in fact, profoundly important for the way we view animal behavior. Observe that it violates the third belief listed above—the belief that perception and behavior attempt to achieve goals by the most efficient paths. If animals choose to obtain goals by more complex means than necessary, then they are being inefficient.

We should also observe that the experiment implies also that the second belief in the above list is incorrect—something that will be demonstrated much more strongly later. The second belief is that perception and behavior merely serve the achievment of goals; i.e., they are purely *secondary* to the achievement of goals. However, the experiment implies that the animal has a need to *increase perceiving and behaving*, independently of the goal being sought—because a more complex route requires more *perceiving* and more *behaving*.

Let us examine this more closely. If an animal's only aim were to reach the final goal, then there would be no reason to choose one route over the other. The fact that there is a preference for one type of route over another, means that there is another goal influencing behavior—the goal that determines the choice between routes. Since this goal is an independent one it must, by definition, be classed as a *primary* goal. However, the goal seems to be simply that of increasing perception and behavior; i.e., increasing their complexity. Thus we must conclude that *perceiving* and *behaving* are not merely secondary to other goals. There is a *primary* goal to perceive and behave; that is, a primary goal to increase perceptual and behavioral complexity, i.e., asymmetry. This claim will be demonstrated much more powerfully later.

Let us now consider an experiment that undermines all the commonly-held beliefs given above. It does so by completely eliminating the presence of the standard primary goals—food, water, sex, etc. Dember, Earl, and Paradise (1957) constructed a maze in the shape of a figure 8, where one of the loops provided more visual interest than the other. The rats were allowed to wander freely about the maze. It was found that the majority of the animals preferred being in the loop with greater visual complexity, i.e., asymmetry. Moreover, when the trials were repeated on the next day, the rats that had not previously shown that preference shifted to the loop with greater complexity. In addition, when a similar experiment was performed over a series of five days, no reversal in the shift took place. Thus, the experiments showed (1) that animals prefer environments that are more complex, and (2) that any "change in preference will be from the less to the more complex path" (Dember, Earl, and Paradise, 1957, p514).

How does this experiment violate each of the four beliefs listed above? First, observe that, again, it seems to violate the second belief that perceiving and behaving are secondary to the achievement of other goals. No other goals are present, and yet there is a preference for increased perceptual complexity (asymmetry).

Now observe that, if indeed there were a primary goal to obtain perceptual complexity, then this experiment would violate the first belief in the above list, which states that, when a primary goal is achieved, it reduces the drive to obtain it; e.g., when food is eaten, it reduces hunger. In the experiment just described, no rats showed a shift from the complex loop to the simple loop. Yet, such a shift *should* have occurred if obtaining complexity reduces the need for having complexity, just as obtaining food reduces the need for having food. On the contrary, the experiment shows a shift from simpler to more complex environments.

This, in itself, also violates the fourth belief in the above list: that successful paths are repeated. If an animal, when given a chance, shifts to more asymmetric environments it is not repeating its previous successful paths. Indeed the very phenomenon of asymmetry means a breakdown of repetition, i.e., symmetry. Thus the animal, when given a chance, undermines repetition.

The experiment we have just described has an important property common to a number of experiments that have revealed the preference that exists for complexity. The experiment used one of the standard components of learning experiments: a maze. However, what was different from the usual learning experiments was that it omitted a factor that learning experiments always included: a *standard* primary goal such as food. In learning experiments, the rat learned how to use the maze to obtain the goal. Thus the maze was *secondary* to the goal; i.e., in the service of obtaining the goal. It was only when researchers decided to leave out the goal that the discovery was made that the maze itself had a *primary* value for the rat, its perceptual complexity.

This type of empirical demonstration was tried with a number of the components of standard learning experiments. That is, some researcher had the insight to take a component that had been assumed only to have a secondary role, and to show that it had a primary appetitive value for the animal—its stimulus complexity.

For example, one of the arrangements in standard learning experiments was a rat pressing a bar, when a light came on, in order to obtain food;

i.e., the light *signaled* the presence of food. However, Kish (1955) tried an anomalous variation of this arrangement. He omitted the food and set up an experimental situation in which there was only the bar-pressing and the light. He decided to investigate what would happen if the only consequence for the rats, of pressing the bar, was that the light came on. Conventionally, it would be assumed that the situation would offer nothing appetitive to involve the rat. However, the remarkable result was that the animals found the change in illumination rewarding in itself. They would learn to perform tasks that led only to this reward. The inevitable conclusion, forced upon the psychologist, is that change in illumination—and, more generally, stimulus complexity—is a goal in its own right, i.e., a *primary* goal.

The experiment just described demonstrates something additional to the experiments we had discussed prior to this. Generally, in animal-learning experiments, the most powerful way of showing that some factor, G, is a primary goal, is to show that the animal will *learn* a behavior when G is the only goal; that is, G is a primary *reward.* For example, we know that sugar can act as a primary goal because a circus-animal can be trained to do all manner of tasks if one rewards it with a sugar cube on the successful completion of a task. Thus, a compelling demonstration that perceptual complexity is a primary goal would be if one substituted perceptual complexity for a sugar cube; i.e., if one showed that an animal can *learn* something when the only reward is the presentation of perceptual complexity, i.e., perceptual asymmetry.

The experiment just described was one of the first experiments to show exactly this. In the experiment, animals learned to press a bar when the only reward was a light coming-on. Observe that "a light coming-on" is an *asymmetry*, a distinguishability between the absence of light and the presence of light. The animals did not press the bar to keep the light on, but to see a *change* in illumination, i.e., an *asymmetry* in illumination.

The fact that rats learn a behavior when the only reward is an asymmetry means that asymmetry is a primary reward; i.e., an asymmetry can be used to train animals in the same way that a sugar cube can.

This conclusion has been supported many times. For example, Moon and Lodahl (1956) found that monkeys also learn to press a bar when change of illumination is the only reward. In more elaborate experiments, by Butler (1953) and Butler and Harlow (1954), monkeys learned a task when the only reward was a brief look at a complex stimulus such as a moving train. Indeed, the monkeys adhered to the tasks *for several hours.* Again, in an experiment by Montgomery and Segall (1955), rats learned a task where the only reward was the opportunity to select a more complex maze. All these experiments demonstrate that perceptual complexity—that is, perceptual asymmetry—is a primary reward, just like food and sex; i.e., that asymmetry can be used to train animals without any other inducement.

There are, in fact, additional ways of showing that asymmetry is a primary goal. One important way is to make it *compete* with a standard primary goal such as food. In a lengthy study, Desmond Morris (1962) introduced chimpanzees to the activity of painting; i.e., he gave them paint-brushes, paint, and paper—and showed them how to put paint on the paper. The consequences were remarkable. The animals became obsessed with painting. For example, while painting, they could not be induced away from this occupation, even by the most enticing food; and they would throw a tantrum if an attempt was made to deprive them of their paint-brushes.

It must be observed that what painting produces are asymmetries. As the colors appear off the end of the brush, asymmetries are created between these colors and those already on the paper, between the lines, between the shapes, etc. The chimpanzees were completely absorbed by these perceptual events. In fact, what the study showed was that these asymmetries are as powerfully appetitive as any other known reward such as food.

Let us end this section by attempting to gain a succinct understanding of what has been discussed. We have reviewed data that support the first part of the Complexity-Level Hypothesis, which states that each individual has a preferred level of environmental complexity, a level that he or she seeks, and finds extremely appetitive. We can state this hypothesis more succinctly: *asymmetry is a primary goal.* This means that we can speak of an *asymmetry-seeking drive*, and consider it to be a primary drive.

The Complexity-Level Hypothesis, and its supporting data, violate four common beliefs about perception and behavior. Let us consider how each of these beliefs, in turn, is violated:

(1) The first common belief is that satisfying a primary drive, i.e., achieving a primary goal, reduces the drive; e.g., eating reduces hunger. The asymmetry-seeking drive violates this belief because satisfying the drive does not

reduce it. For example, we saw that rats, when given the chance, shift from simple to more complex environments and that this shift never reverses; i.e., there is never a *reduction* in asymmetry-seeking.

(2) The second common belief is that perception and behavior are *secondary* to (i.e., serve) the achievement of goals. However, the fact that asymmetry-seeking is a primary goal means that perception and behavior are not secondary, because perceiving and behaving, which always increase asymmetry, are goals in their own right. This was demonstrated by all the experiments. In particular, it was demonstrated that asymmetry is a powerful reward for which animals will do considerable work; e.g., monkeys will work for hours to get a glimpse at a toy train. Furthermore, the asymmetry-goal can compete strongly with, and over-ride other primary goals; e.g., chimpanzees while painting cannot be induced by food to stop painting, and they throw tantrums when forcible attempts are made to stop them.

(3) The third common belief is that perception and behavior attempt to achieve goals by the most efficient paths. The asymmetry-seeking drive violates this belief, because, under the drive, the animal takes more complex paths to goals than are necessary.

(4) The final common belief is that successful paths are *repeated*. The asymmetry-seeking drive again violates this belief, because, under this drive, animals are non-repetitive in behaviors, despite the apparent benefits of repeating successful behaviors.[2]

8.9.2 Complexity Level Hypothesis, Part 2: Insufficient Complexity

In the previous section, we reviewed evidence that shows that animals find complexity highly appetitive. In the present section we review data that show that animals find low complexity aversive—and that low complexity can be so destructive that animals can die as a result of receiving too little complexity.

In a sense, the aversiveness of low complexity should not be a surprise to us: We have all experienced boredom—which is an aversive response to slightly lowered complexity. When complexity is diminished further, the aversiveness increases to levels much greater than simply boredom, as we shall see.

First of all we should note that low complexity is so aversive that it is used as a major threat and punishment in prisons—where it is called *solitary confinement*. Such confinement was particularly severe in wartime prison camps, where a prisoner could be put in a completely darkened container that was small enough to restrict bodily movement. In other words, *perceiving* and *behaving* could not take place. Despite the fact that the prisoner might be given a regular supply of food and water, the aversiveness was extreme—thus reinforcing the claim that perceiving and behaving are *primary* drives; i.e., independent of other goals. Furthermore, it is known that solitary confinement is not only aversive, but, when prolonged, it has a destructive effect on one's intellectual and emotional capacities.

Low complexity can be induced only by force or artificial circumstances. Even the promise of rewards is not sufficient to persuade people to remain in a low complexity environment. For example, consider the following remarkable data:

In a study by Bexton, Heron, & Scott (1954), college students were offered financial inducements to be in an experiment that any student would regard as the perfect one in which to participate: They were paid to take a complete rest. All they had to do was lie on a bed, in a small cubicle, in which their eyes were covered and their hands isolated from stimuli by cuffs. The students were considerably motivated to participate—feeling that they were being paid to take a much-needed vacation from studying.

However, to the students' surprise, they found the task highly aversive. After a short while, they became restless. They began to whistle, talk to themselves, and make random movements with their limbs—all attempts to increase complexity. More alarmingly, they later began to hallucinate—presumably an involuntary attempt to increase complexity. After a while, the hallucinations became horrifying. The required task—simply lying there—was so aversive that students could not be induced to persist with it. Afterwards, when given simple intelligence tests, it was found that they had undergone intellectual impairment.

The severely damaging effects of low complexity were evidenced in the study by Spitz (1945, 1946a, 1946b), of children in certain orphan institutions. These institutions had a practice of raising children in cribs enclosed by white sheets, thus eliminating any visual complexity. Furthermore, the children had no toys to play with, and almost no contact with adults.

As a consequence, the children developed severe depression, apathy, and unresponsiveness. So detrimental were the effects that many of them actually died. Again, we should observe that these destructive effects occurred despite the fact that basic needs such as food and water were taken care of.

A massive amount of research has been done on the aversiveness and detrimental effects of low complexity; i.e., sensory deprivation. A book summarizing much of this work is Zubek (1969). This, and the above data reinforce the conclusion that perceiving cannot be regarded merely as secondary to other goals. Perceptual complexity is a primary requirement and goal, the deprivation of which is highly aversive and destructive for the animal.

8.9.3 Complexity Level Hypothesis, Part 2: Excessive Complexity

We have seen that, when animals are given a chance to increase the complexity of their environments, they do so. Is there, however, a limit past which increased complexity is no longer appetitive? If so, what is the factor that determines that limit?

Recall first that the term "complexity" means *amount of asymmetry*. Recall also that, according to our Asymmetry Principle, when an asymmetry is presented, the mind attempts to give it a causal explanation. Clearly, since the brain is finite, there must be a maximum amount of asymmetry that the mind is capable of explaining at any one time. That is, if the amount of asymmetry is increased past this level, it exceeds the mind's capacity to explain it. It would seem reasonable to conjecture that, at this level, the asymmetry ceases to be appetitive.

Indeed, this conjecture seems to be correct. When an environment has more asymmetry than the individual can explain, it becomes, by definition, *chaotic*, and the individual starts to experience *anxiety*. If the chaos is increased or prolonged, the anxiety increases to terror.

For example, in an experiment by Hunt & Brady (1955) two sets of animals were administered electric shocks. The first set were given the shocks when they produced a particular response; and the second set were given the shocks randomly. Thus, in the former case, the causal structure of the environment was comprehensible: the animals could perceive that their responses *caused* the shocks. In the latter case, the environment was not causally comprehensible: shocks sometimes arrived when the animals responded and sometimes arrived when the animals did not respond; i.e., the animals had no means of identifying the cause of the shocks. The results, in this second case, were that the animals gave up attempting to act, and exhibited extreme anxiety: they cowered and went into freezing postures.

Generally, in the literature on stress, anxiety is understood to be the emotional state that results from an environment that is incomprehensible or out of control (Lazarus & Folkman, 1984). Thus, putting together our overall discussion on asymmetry-seeking, we see that animals search for an increase in the amount of asymmetry *up to some level*—a level that they find highly appetitive; and this level is the maximal level to which they can assign *causal explanation*. Beyond this level, the asymmetry ceases to be appetitive and becomes aversive, i.e., it creates anxiety and even terror.

Realizing that it is the factor of *causal explanation* that determines the limit of preferred complexity is important. It shows that it is not asymmetry, per se, that is appetitive. For if asymmetry were appetitive, then continuing to increase it without limit would continue to increase the appetitiveness. Since the limit is set by the ability to give causal explanation, it is not asymmetry that is appetitive, but causal explanation itself. When causal explanation ceases to be present, the appetitiveness ceases to be present.

We must conclude therefore that what was said, in the previous sections, about the search for asymmetry should more correctly be said about the search for causal explanation. Most importantly, the statement that there is a primary drive to seek asymmetry should now, more correctly, be restated thus:

CAUSAL SEARCH PRINCIPLE. *There is a primary drive to seek causal explanation; i.e., a drive independent of, and not subservient to, any other goal. In other words, the history-recovery problem is a primary drive.*

8.10 The Fundamental Aesthetic Principle

Before we explore the Causal Search Principle, let us go back to the assumption held by researchers that there is a primary drive to seek complexity, i.e., asymmetry. From now on, we shall use the term *asymmetry-seeking* rather than *complexity-seeking*, even when we are discussing the work of past researchers who explicitly used the term *complexity-seeking*. The reason for this is to maintain a language consistent with the rest of this book. In particular, we shall rename the Complexity-Level Hypothesis, the Asymmetry-Level Hypothesis, and restate it thus:

ASYMMETRY-LEVEL HYPOTHESIS. *(1) Each individual has a preferred amount of environmental asymmetry, an amount that the individual seeks, and finds extremely appetitive. (2) As the amount of environmental asymmetry is raised or lowered from this level, the environment becomes more and more aversive.*

Almost as soon as experimental psychologists discovered the phenomenon of asymmetry-seeking, it was linked to the phenomena of art and aesthetics. For example, the first major book that reviewed the extensive work on asymmetry-seeking was entitled *Psychobiology and Aesthetics* (Berlyne, 1971). The author of that book, D. E. Berlyne, who was himself probably the most prolific contributor to research on asymmetry-seeking went on to found a journal called *Scientific Aesthetics*. Again, in a more recent book reviewing research on asymmetry-seeking—a book entitled *Psychological Complexity and Preference*—the final chapters are devoted to aesthetics (Walker, 1980).

Why should researchers link asymmetry-seeking and aesthetics?

The answer is as follows: Researchers argued that asymmetry-seeking is a *primary*, rather than a secondary, drive. Under this drive, the animal attempts to increase the amount it perceives not in order to satisfy other goals but because the perceptual input is valuable in its own right; i.e., the input is *intrinsically* valuable. The reason why this behavior has been linked to aesthetics is that aesthetics is generally defined as the valuing of a percept for its *intrinsic* content. That is, when one removes from a percept all the value it can have with respect to obtaining other goals such as food, sex, etc., one is left only with the value it has in its own right. This is supposed to be its aesthetic value. For example, looking at a Van Gogh in a museum is appetitive not because the painting indicates where food or sex can be obtained, i.e., its value exists not because it points to goals *external* to itself. The painting is pleasurable because of its *intrinsic* content.

With these considerations in mind, we can therefore see that the link between aesthetics and asymmetry-seeking was made because of the distinction between *primary* and *secondary* drives, as follows: Since the aesthetic response is the response to a percept by virtue of its intrinsic content, aesthetics concerns perception as a *primary* drive. In other words, aesthetics concerns those aspects of perceiving that are not secondary to other goals. This is exactly what asymmetry-seeking research is concerned with; i.e., perceiving as a primary drive—not as secondary to other goals.

The link between aesthetic value and asymmetry is strengthened by examining data both in the history of art and in laboratory experiments; as follows:

First, observe that, in the historical acceptance of art-works, the successive emotional responses accord with the Asymmetry-Level Hypothesis. When a revolutionary art-work is introduced, its quantity of asymmetry is greater than that which can be comprehended by the audience. Consequently, in accord with the Asymmetry-Level Hypothesis it causes anxiety in the audience. For example, Beethoven's middle-period string-quartets aroused so much anger that even the musicians who played them trampled the manuscripts under foot. Observe that, at this stage, such an art-work is described as *ugly*. That is, at this stage, there is a consistent *aesthetic* judgment.

Now, over time, as the public becomes more familiar with the art-work, its asymmetry starts to become more comprehensible, and, therefore the work enters the maximal asymmetry level that is comprehensible by the audience. Correspondingly, in accord with the Asymmetry-Level Hypothesis, the work becomes highly appetitive. Observe now, that, at this stage the work is described as *beautiful*. That is, once again, a consistent *aesthetic* judgment is elicited.

As time progresses, works become too easily comprehensible; for example, to most people today, a musical piece by a 15th Century composer is only mildly interesting. Over still further time, works tend to arouse almost no pleasure; for example, music of ancient Egypt attracts only those people interested in ethnically remote music. At this stage, the works are regarded as *dull*—an evaluation that one can take as an *aesthetic* one.

These consistent aesthetic responses to the different levels of asymmetry have been replicated in laboratory experiments. For example, Eckblad

(1980) examined a range of laboratory studies in which subjects were presented with stimuli of varying levels of asymmetry, and found that stimuli that have too much asymmetry to be comprehensible are described as *ugly*; stimuli that have the maximal amount of comprehensible asymmetry are described as *beautiful*; and stimuli that are too simple are described as *dull*.

In conclusion, we see that both—the historical data on art and the data from laboratory experiments—show that there is a consistent correspondence between aesthetic response and quantity of asymmetry: Stimuli with the maximal quantity of comprehensible asymmetry are described as *beautiful*. Stimuli with more than this quantity are described as *ugly*; and stimuli with less than this quantity are described as *dull*.

The consequence of this is that we see that aesthetic pleasure is maximal at the maximal level of comprehensible asymmetry and aesthetic aversion increases with distance from that level. But this pattern of pleasure and aversion, with respect to asymmetry level, is the same as the pattern described in the Asymmetry-Level Hypothesis—which states that there is a level of asymmetry that an individual finds highly pleasurable and that aversiveness increases with distance from that level. The obvious conclusion that follows from this is that aesthetic pleasure and aversion are the pleasure and aversion described in the Asymmetry-Level Hypothesis.

Now, the Asymmetry-Level Hypothesis and its supporting data imply that pleasure increases with increasing asymmetry up to the maximal *comprehensible* level of asymmetry, and then decreases. Recall our conclusion that, since there is a limit to the quantity of asymmetry that is pleasurable, and since this limit equals the limit of the individual's ability to assign causal explanation, the pleasure evoked by asymmetry is in fact not due to the asymmetry itself but due to the assigned *causal explanation*. That is, the greater the causal explanation, the greater the pleasure; and the less the causal explanation, the less the pleasure.

However, we have just seen that the pleasure involved in asymmetry-seeking is *aesthetic pleasure*. This leads us to the following profound conclusion, which not only explains the purpose of aesthetic pleasure but shows that it has a central role in cognition:

FUNDAMENTAL AESTHETIC PRINCIPLE. *Aesthetics is the evaluation of amount of causal explanation. Aesthetic pleasure increases with increasing causal explanation and decreases with decreasing causal explanation.*

This conclusion implies that aesthetic pleasure is not an incidental product of evolution. It has a role that is basic to survival, as we shall now see.

8.11 Aesthetics and Cognition

The Fundamental Aesthetic Principle states that there is a profound relationship between aesthetics and causal explanation. One further principle forces this relationship into the very foundations of cognition itself: the Representation-is-Explanation Principle. This latter principle states that *any cognitive representation is a causal explanation.* Thus, since aesthetics is the intrinsic evaluation of causal explanation, aesthetics is the intrinsic evaluation of any cognitive representation.

This makes aesthetics intertwined with the very substance of cognition. As we said, in section 4.1, the Representation-is-Explanation Principle implies that a stimulus has no cognitive existence independent of the causal explanation assigned to it. Any role that the stimulus has, in the cognitive system, is by virtue of its assigned causal explanation. Thus, aesthetics evaluates a stimulus with respect to its role in the cognitive system, i.e., its assigned causal explanation. Since the Representation-is-Explanation Principle makes causality the very *content* of all cognition, aesthetics is the evaluation of cognitive content.

8.12 Summary

The main proposal to which the argument of this chapter led is that aesthetic response is the evaluation of causal explanation. This proposal was instantiated first with respect to art-works. That is, we showed that the aesthetic response to an art-work is a response to the causal explanation assigned to it. In fact, we proposed and corroborated the First Art-Work Principle, which states that an art-work is the process-history inferred from it. Furthermore, we proposed and corroborated a corollary to this principle, which states that the meaning of an art-work is the process-history inferred from it, and the experience of an art-work is the experience of inferring that history, i.e., the experience of solving its process-recovery problem.

We carried out our corroboration by analyzing, in considerable detail, four very different paintings. We found that, in each case, what one regards as the *composition* and *meaning* of each painting is the process-history inferred from it.

We then observed that, since this book claims that the perception of *anything* is the process-history inferred from that thing, the First Art-Work Principle claims simply that perceiving an art-work is not different from perceiving anything else. Thus, the First Art-Work Principle does not create a distinction between art-works and other things. We argued that two additional principles provide the required distinguishing criteria: (1) the Second Art-Work Principle, which states that an art-work allows the maximal amount of causal explanation that a person is capable of assigning to any percept; and (2) the Third Art-Work Principle, which states that the entire set of histories in an art-work are symmetrically organized.

The Third Art-Work Principle was corroborated in our analyses of the four paintings. The Second Art-Work Principle was corroborated as follows:

We first reviewed the extensive data, from experimental psychology, that support what we call the Asymmetry-Level Hypothesis: (1) Each individual has a preferred amount of environmental asymmetry, an amount that the individual seeks, and finds extremely appetitive; and (2) as the amount of environmental asymmetry is raised or lowered from this level, the environment becomes more and more aversive. The data show a powerful need for asymmetry as a primary goal in competition with the other primary goals of an animal, e.g., the goals of food and sex. Furthermore, the data are drawn from studies of many species, and from studies in many research areas—animal learning studies, child studies, sociological studies, etc.

We then considered the question of what it is that determines the limit to an animal's attempts to increase the amount of asymmetry, and saw that the maximal level sought by the animal is the maximal level to which the animal can assign *causal explanation*. We concluded therefore that the search for asymmetry should more correctly be understood as the search for causal explanation. Most importantly, the statement that there is a primary drive to seek asymmetry is, more correctly, restated in our Causal Search Principle, thus: There is a primary drive to seek causal explanation; i.e., a drive independent of, and not subservient to, any other goal. In other words, solving the history-recovery problem is a primary drive.

Before we explored this principle, we went back to the assumption held by past researchers that there is a primary drive to seek complexity—a drive we called the asymmetry-seeking drive. We noted that the link between aesthetics and asymmetry-seeking was made by these researchers because aesthetic response is the response to a percept by virtue of its *intrinsic* content—i.e., aesthetics concerns perceiving as a *primary* drive—and this is exactly what asymmetry-seeking research is concerned with; i.e., perceiving as a primary drive, not as secondary to other goals.

We also observed that the link between aesthetic value and asymmetry is strengthened by examining data both in the history of art and in laboratory experiments—data which show that there is a consistent correspondence between aesthetic response and quantity of asymmetry: Stimuli with the maximal quantity of comprehensible asymmetry are described as *beautiful*; stimuli with more than this quantity are described as *ugly*; and stimuli with less than this quantity are described as *dull*. This correspondence seems to show that aesthetic response is the evaluation of amount of asymmetry.

However, the argument leading to our Causal Search Principle indicated that what is regarded as the evaluation response to amount of asymmetry should more correctly be regarded as a response to amount of *causal explanation*. This led us to propose the Fundamental Aesthetic Principle, which states that aesthetics is the evaluation of amount of causal explanation; i.e., aesthetic pleasure increases with increasing causal explanation and decreases with decreasing causal explanation.

This principle led us to conclude that aesthetic pleasure is not an incidental product of evolution, but has a basic role in survival, as follows: The Representation-is-Explanation Principle states that *any cognitive representation is a causal explanation*. Thus, since aesthetics is the intrinsic evaluation of causal explanation, aesthetics is the intrinsic evaluation of cognitive representation.

As we observed, this makes aesthetics intertwined with the very substance of cognition. That is, since, by the Representation-is-Explanation Principle, causality is the *content* of all cognition, we conclude that aesthetics is the evaluation of cognitive content.

9 Political Prisoners

9.1 Introduction

We are all confined to the "prison" of the present. However, in political subjugation, the walls enclosing the present are often instantiated by physical walls. These walls have a particular role, with respect to the inference of history, that reveals, in an exceptionally clear way, the central role of the history-recovery problem to the phenomenon of mind.

Therefore, in this final chapter, we are going to use the basic principles of this book to analyze political subjugation. In order to do so, it is necessary to begin by examining a topic that might not at first seem a central one to politics: death.

9.2 Death

It is sometimes thought that, because the human body is reflectionally symmetric between left and right, the body represents a high degree of symmetry. The opposite is true. The body is hardly symmetric at all, and represents a massive degree of *asymmetry*. This can be emphasized by comparing the body to objects that are actually symmetric, e.g., a sphere or homogeneous flat space. Unlike the latter examples, the body is asymmetric between front and back, and between top and bottom. Furthermore, the outline of the body is completely suffused with curvature variation—a form of asymmetry we studied in Chapter 1. In fact, asymmetry continues on every level of scale: The individual limbs and organs are asymmetric; so are the individual cells, e.g., nerve cells are organized in one direction. Even the important components of a cell, e.g., the DNA molecules, are asymmetric.

The functions of living, e.g., functions such as eating and self-protection, are to maintain these asymmetries. At death, these functions can no longer be carried out and the body begins to decay. The important thing to understand is that the decay is a process of *symmetrization*. The highly asymmetric living body degrades into material that is successively more and more homogeneous. Quite simply, death is a example of symmetry.

Observe how the principles of this book apply to the symmetry created by death. For example, note that the increasing symmetry, in the decaying body, brings with it an increasing loss of memory. For instance, as decay proceeds, it becomes more and more difficult to infer who the living person was, what their features were, etc. This is a clear example of our

Symmetry Principle, which states that symmetry prevents the inference of history.

9.3 Politics and Death

Death has immense value for political systems. This becomes evident when one considers, for example, the number of deaths that have occurred for political reasons in this century. Not only are there the millions of deaths that have occurred at the battle fronts of two world wars and the many other wars of this century, but there are the millions of deaths that have occurred in concentration camps and labor camps. There are the millions of deaths that have occurred when governments have suppressed uprisings—through massacres, tortures, executions, etc. There are the millions of deaths that occurred in Stalin's programs for mass-starvation.... The list is almost endless.

Why should death be so highly valued by governments? As we shall see, the answer is that, more increasingly, governments require the creation of symmetry, both within individuals and across individuals. Death is an example of symmetry. Thus, there is a direct relation between the requirements of government and the structural properties of death.

Although death is a particularly reliable means of achieving the symmetry required by modern political systems, such systems choose still further means of enforcing symmetry. We now examine these other means in order to gain greater insight into the centrality of the history-recovery problem to the phenomenon of mind.

9.4 Prisoners of an Emptied Present

As was stated before, the fundamental problem examined in this book is that each of us is completely confined to the present—and thus the present is, in a very real sense, a prison that isolates us from the past. It is only from the contents of this prison that we might be able to infer prior events. In particular, since we are confined to the present, we are separated even from our own past. And thus, it is only by an examination of the objects within this prison that we can discover what we have ourselves undergone.

These considerations lead one to the conclusion that, if the prison is *emptied* of objects, the past cannot be reconstructed. That is, when all objects are removed, and none is therefore available for examination, the past cannot be inferred—that is, no memory is possible. Furthermore, since the past cannot be inferred, the future, which can be constructed only from the past, also cannot be produced.

Death ensures this condition by preventing the observer's own body from being memory; e.g., the brain stops functioning, and thus cannot extract memory from the body's own physical material.

However, even without death, the removal of physical objects from the observer's *external* environment has the following disastrous consequences:

We argued, in Chapter 4, that the use of the external environment as memory, i.e., by perception, allows the computational system, defined by the brain, to include the external environment. The metaphor of a computer is valuable to understand this. A computer consists of several interconnected memory stores that are accessed during a computation. We can consider the mind as a computer, many of whose memory stores lie *outside* the brain. That is, the brain is only a small part of the system of interconnected memory stores that define a single mind—i.e., the brain, as a *functioning* entity, is embedded in this larger interconnected system. In particular, the brain cannot operate if the external stores are removed—in the same way that a central-processing unit of a computer cannot function if the memory stores, to which it is connected and which it accesses, are removed. The brain is left with little out of which to form a computation, and it undergoes computational death.

We shall see that the main purpose and effect of political subjugation is to induce computational death in this way. That is, the political system removes from an individual the external environment that comprises much of the computational mechanism embodying that individual's mind. In other words, the political system removes part of the actual *hardware* necessary for computation to take place. In effect, the political system induces a type of brain damage.

Now, as we said, any part of a computational device interacts with any other part as *memory*, and the removal of the external environment of a brain removes a significant part of the latter's memory. The consequence is that, while *everyone* is a prisoner of the present, the individual without an external environment is a prisoner who cannot carry out significant computation, and, in particular, cannot infer significant parts of his or her own past because the hardware embodying the past is removed. Quite simply, the prisoner is in a perpetual state of not knowing who he or she is.

Political systems are able to carry out this removal by different methods and we shall examine three significant methods that are used. We claim that the precise definition of the three methods exactly characterizes three large categories of political victims: (1) refugees, (2) political prisoners, and (3) slaves. We now examine and characterize each of these categories in turn.

9.5 Refugees

A refugee is the prisoner of a present from which all external objects of memory have been removed by the following method: The refugee has been forced to leave the country in which he or she was raised. Because the refugee was raised in this country, this is the *only* country that contains the external objects from which he or she can reconstruct the past, including his or her own past. In arriving in another country, the refugee is presented with objects that contain memory only of other people; i.e., from which the refugee cannot infer his or her own past.

The consequence is a vast loss in the refugee's computational capacity. As we said before, the external environment from which a person can extract memory is part of the computational system that defines the person—literally part of the person's computational hardware. It is exactly this computational hardware that the refugee loses.

In the new country, the refugee is presented with external hardware that is incompatible with that which he or she holds *internally*. It is incompatible not only because it contains no memory of the refugee, but because it does not complement the computational hardware contained within the refugee, as follows: A computational system is a network of interacting memory stores. The use of one store necessarily involves the use of the others within the system. As any individual grows up, the external hardware and the internal hardware are built together so that they interlock and the entire system can function as a single computational system; i.e., a single mind. When a refugee leaves his or her own country, the external hardware is lost and much of the internal hardware cannot be used because it was built to interlock with that external hardware. In the new country, the refugee is presented with external hardware that was not developed to interlock with his or her internal hardware. It is as if an attempt is made to replace part of a machine by a part that belongs to an entirely different machine.

The physical objects that are part of an individual's computational hardware include home, money, family, social networks, language, information resources, institutions of protection, employment networks, financial services, documents of legal status, government, etc. The refugee has often lost all of these computational components. To replace any one of these requires an immense amount of work for years. Furthermore, the replacement of one of these is dependent on the replacement of all the others. For example, to obtain employment requires learning the new country's language, which, as an adult, requires attending (at least part-time) a college, which requires money, which requires financial services in the form of loans, which require employment—the very thing being sought. Thus, because these different sources of memory are mutually dependent, the refugee remains unable to acquire any one of them and is forced to live at the lowest levels of the economic structure, in occupations that require none of these external objects as memory. For example, immigrants often wash dishes because the occupation does not require learning a new language, employment networks, social and legal status, etc.

Not only are these various physical objects unavailable to the refugee as sources of memory because, as sources of memory, they are mutually dependent, but they are unavailable also because those who control them are often racially prejudiced and therefore actively prevent the refugee from having access to them. Indeed, the host population often subjects a refugee to physical attacks causing bodily injury, i.e., the destruction of the individual's internal memory, and attacks on the refugee's home or business, i.e., the destruction of any external memory which the refugee has been able to build.

Confronted with a vast physical world from which they can extract no memory, refugees attempt to seek physical objects such as fellow countrymen, ghettos, places of worship, etc., that increase a sense of ethnic identity. While making the refugees more conspicuous, and thus putting them at risk from attacks by the host population, this behavior has a strong survival value. The refugees are seeking objects that possess memory for them, and therefore give them a more complete computational system. These external objects of memory are used for the informational value they give, and they complement the internal objects of memory that cannot be used without the external objects. As we said earlier, a person consists of a computational system in which the external hardware interlocks with the internal one; and the internal one cannot function without the external one. The search for objects that increase the sense of ethnic identity is exactly the search for this crucial external hardware.

9.6 Political Prisoners

The political prisoner is a prisoner of a present that is emptied of the objects of external memory by enclosing the prisoner within physical walls. The walls serve the following two crucial functions with respect to memory:

(1) The walls physically separate the prisoner from his or her external memory and therefore cause a substantial reduction in the computational capacity of the prisoner. This is due to the fact that external memory, as we said earlier, interlocks with internal memory and therefore the latter can hardly function without the former. Observe that the physical walls serve the same role—a barrier between external and internal memory—that distance from the homeland serves with respect to the refugee.

(2) The walls are non-transformable; i.e., they cannot be deformed, segmented, rearranged, etc. This means they cannot be used to form objects of external memory because they cannot register the effects of actions. The consequence of this is two-fold: (a) The prisoner has no access to the history of the prisoners who previously occupied the cell, because the previous prisoners could leave no external memory. Thus the prisoner cannot take advantage of any history that occurred in the cell prior to his or her arrival, e.g., the prisoner cannot find out what the authorities learned from any previous prisoners who were his or her collaborators. (b) The prisoner cannot accumulate any memory of his or her own, for the following reason: Unless events intrude from outside the cell, one day is indistinguishable from the next because there are no effects of interactions with the environment. This means that neither *internal* memory of the world, nor *external* memory of the world, can be accumulated, because, quite literally, nothing happens. The consequence of this is that, even after years of imprisonment, the prisoner remains as computationally diminished as when he or she entered the cell, and thus as vulnerable to threats against survival.

When the prisoner enters the cell, he or she does contain internal memory. In fact, a prisoner is literally the internal memory in his or her physical tissue—because there is no external memory. The maintenance of this internal memory, i.e., the maintenance of the prisoner's survival, is dependent on the imprisoning authority; that is, food and physical protection come from the authority, and are given at the authority's discretion. They can be removed at any time, thus allowing the destruction of the prisoner's internal memory, i.e., the destruction of the prisoner's life.

Not only is the survival of the internal memory of the prisoner at the discretion of the authority, but the authority is often interested in controlling access to and even altering that internal memory.

This is achieved through torture. The primary role of torture is *computational.* One can see this by examining what are understood to be the two main purposes of torture: (1) The authority is attempting to extract information from the prisoner, e.g., the identity of the prisoner's collaborators. In this case, the authority is trying to access the internal memory within the prisoner, i.e., find out the past of the physical object which is the prisoner. (2) Alternatively, the authority is attempting to force the prisoner to confess to crimes he or she did not commit. In this case, the authority is trying to alter the history extractible from the prisoner; i.e., literally to write the internal memory of the prisoner, in the same way that a computer writes the content of a particular memory store.

In either case, the authority is interested in making the internal memory of the prisoner part of the external memory of the authority. Thus the authority, as a computational system, survives at the expense of the prisoner, as a computational system.

9.7 Slaves

A slave is the prisoner of a present emptied of the objects of external memory by the following method: The slave is forced to produce the history only of others. This is accomplished in the following way: (1) In contrast to the political prisoner—an individual who is not allowed to produce objects—the slave is forced to produce objects. However, the specifications of these objects are determined by the subjugator. Thus the slave cannot produce the full range of external memory that he or she requires. (2) The objects produced cannot become external memory for the slave because they are removed from the slave as soon as they are produced. That is, the slave lives in a present that is *continually* emptied by others. Furthermore, objects removed from the slave's present become the external memory of the subjugator. In short, the slave is forced to produce only the external memory of others.

Any memory that a slave possesses is purely internal. This makes it impossible for slaves to build up any shared external memory, and thus to gain any power. For example, a slave family cannot build shared external memory in the form of a home or savings. Indeed, whereas, within a free family, each individual uses the other family members as external memory, in a slave family, an individual cannot use the other family members as such because the other members—spouse, parents, children—are removed so that they can be external memory only for the subjugator. In particular, note that the slave's children are physical objects *produced* by the slave—and, as we observed above, all physical objects produced by a slave are removed to become the external memory of only the subjugator.

The fact that a slave's entire memory is internal, means that, when a slave dies, his or her entire memory disappears. This means that a population of slaves cannot accumulate any memory. The total memory in the population, no matter how many successive generations the population has existed for, is only the totality of the internal memory of the current individuals. In contrast, the subjugator has, through the successive generations, accumulated the external memory of those generations; i.e., expanding property, expanding economic and political networks, expanding culture, etc.

The fact that the entire memory of a slave population is the internal memory of only the current individuals means that there is almost a total and permanent wall that prevents the slave from obtaining knowledge of the past of the population. The slave is born into a present for which there exists no historical context, and therefore no computational framework.

The consequence is that, when a slave is freed, the slave is essentially a refugee. Recall that, by our analysis, a refugee is someone forced to live in an environment that is not their own external memory and that is therefore incompatible with their internal memory; i.e., that cannot form, together with the internal memory, a complete and viable computational system. In the same way, a slave is freed into a society that consists only of the external memory of others. Quite literally, a freed slave is in a foreign country.

The result of this is that the existence of a freed slave has all the features and problems we described of the existence of refugees. Thus the reader might find it worthwhile re-reading our discussion of refugees with an understanding that it applies also to freed slaves.

In particular, we should note that the freed slave, like any refugee, is usually obsessed with his or her own ethnic roots. Thus American blacks seek to trace, and identify with, their roots in Africa. As we observed, in our discussion of refugees, ethnic identification is a strongly rational attempt to solve the problem of dealing with incompatible external memory; i.e., the solution is that of seeking those external objects that contain the *actual* memory of the individual. Over the centuries in which blacks were slaves in America, they were not allowed to build external memory. Thus, in identifying with Africa, the freed slave is attempting to see over the wall erected by the period of slavery to the last objects, in his or her history, that held actual external memory—i.e., actual external hardware with which a complete computational relationship was formed.

9.8 Subjugators

People and cultures who have a long history of being subjugators possess vast amounts of external memory that has been obtained at the expense of those whom they have subjugated. Thus, for example, colonial powers have amassed their considerable external memory at the expense of colonies, men have amassed theirs at the expense of women, and whites at the expense of blacks.

Because this is external memory derived by subjugation, it is premised upon subjugation. That is, since this memory is literally the *memory* of a subjugating relationship, the *use* of this memory inevitably re-expresses the subjugation from which it was derived.

However, no one can extract themselves from using their external memory. Their internal memory is inextricably linked and dependent upon their external memory. Furthermore, the use of external memory by internal memory proceeds automatically, involuntarily, and unconsciously. Even self-conscious attempts to reject that memory are actually formed out of it. As Heidegger (1927) says: "In it, out of it, and against it, all genuine understanding, interpreting, and communicating, all re-discovering and appropriating anew, are performed."

Therefore, the subjugator has no choice but to be a *habitual* subjugator, since the subjugator cannot extract himself or herself from using the external memory that was derived at the expense of the subjugated, and is memory of the subjugating relationship.

9.9 Technology and History

We have examined two methods by which political systems achieve assent:

(1) the destruction of *internal* history (i.e., death),

(2) the destruction of *external* history.

What we should observe now is that these two forms of destruction are achieved mainly through *technology*. The destruction of internal memory tends to be achieved through the action of weapons, and the destruction of external memory through the threat of weapons. In either case, it is the power gained over the individual by the weapons that allows the destruction to take place against the individual's wishes. Weapons are produced by, and are examples of, technology. Therefore, people with greater technological power gain the greater ability to destroy the internal and external memory of others.

Do those with advanced technology tend to use it destructively? It is often thought, by members of the advanced technological societies, that the apparently greater civilization brought by the technology brings with it a greater level of morality. This is not the case. The two world wars in this century were events producing greater numbers of murders than any other punctate events in previous history—and yet they were brought about by the technologically advanced societies, and could be put into effect only with their technological advances. What is evident is that greater technology brings with it a greater opportunity for murder, and that this opportunity is taken. Not only has technology brought with it the destruction of internal memory—i.e., death—on a scale greater than that seen before, but it has brought with it the destruction of external memory—i.e., refugees, prisoners, and slaves—on a correspondingly more vast scale. For example, the mass enslavements of American blacks, the concentration camp system during the last war, and the labor camp system in Russia, were all made possible, and were realized, by technological power.

If there is any sense that advanced technology brings with it advanced morality, it is simply because history is written by the conqueror. More strongly, our discussion in the last few sections has shown that conquering is nothing else but the *manipulation* of history. To repeat: killing is the destruction of internal memory; and the other forms of subjugation—

driving people from their homeland, imprisoning, and enslaving—are the destruction of external memory.

9.10 The Symmetrizing Effects of Technology

We have seen that a major effect of technology has been the mass destruction of *history*, and this therefore means the mass destruction of people as *computational* systems—since all computational systems are based on inferring history; i.e., the use of memory. However, one might argue that technology has introduced, simultaneously, great benefits to society in terms of the production of goods and services. Nevertheless, we shall now argue that these apparently beneficial effects are formally indistinguishable from the destructive effects. That is, the so-called beneficial effects induce the large-scale destruction of history and therefore the mass destruction of people as effective computational systems. It would be unnecessary to add that the mass destruction of people as effective computational systems can lead only to the end of human life on this planet.

While most of the Western population believes strongly in the benefits of technology, there have been many critiques of technology—in fact, too many to review here. However, our purpose in the following sections is not to aim at any comprehensive critique of technology. Quite the contrary. We wish to make only a single observation: that the effects of technology are an example of our Symmetry Principle.

Let us therefore consider the technological society in its apparent beneficial capacity—as an economy of thriving and growing production.

The first thing to observe is that the technological society is a society of standardization. This is because, in a technological society, the manufacturing and marketing of a product is economically viable only if the product can be sold *en masse*; i.e., only if the product is duplicated in sufficient numbers to economically justify creating the technology to manufacture and market the product. The mass duplication of a product is, of course, an example of symmetry.

However, the symmetry does not stop here. The *use* of a product by the consumer requires a particular set of actions from the consumer. Thus, if products are duplicated, the actions must be duplicated across the consumers; i.e., symmetry must be created across consumers. Thus, a technological society requires its population to be symmetric.

If the actions of people become standardized, then their histories become standardized. This is because their histories are formed out of those actions. Thus a technological society requires the standardization of histories.

The necessary standardization of histories is performed by the media. Advertising attempts to persuade people to adopt histories in which the product fulfills a role. The more effective the advertising, the more universal this adoption becomes, i.e., the more symmetric the adoption becomes across people. Thus the goal of advertising is to create greater and greater symmetry in the behavior, and therefore, history of a population.

However, in a technological society, people must become symmetric, i.e., indistinguishable, not only as consumers, but also as producers, i.e., in their work-environment. This is because (1) the work-environment itself functions as a consumer of products—e.g., computers, office furniture, commuter transportation—all of which are available only in standardized forms; and (2) the work-environment is itself a producer of standardization and is therefore repetitive.

However, it is not only producers who are forced to behave repetitively. Because mass production forces symmetry across objects, entire environments become repetitive, and this forces the behavior of an individual in an environment to be repetitive; i.e., repetitiveness exists not only across individuals but within each individual. For example, a freeway has a repetitive structure because it is constructed from materials and objects that are mass produced—e.g., the surface of the freeway is constructed from mass produced, and therefore homogeneous, material; the successive components of the fence and the successive lamps, along a freeway, are the same because they are mass produced. Thus the actions of the driver are basically the same at any point along the freeway. Furthermore, because freeways are themselves mass produced, one freeway is indistinguishable from another, and the behavior on one is indistinguishable from the behavior on another. Mass production therefore creates an environment that is a hierarchy purely of symmetries, and this ensures that behavior is a hierarchy purely of symmetries.

These hierarchies of symmetries constitute all environments, from supermarkets, to office buildings, to mass transit systems, to housing developments. The purpose of architecture, in a technological society, is the imposition of repetitiveness on all levels of behavior. This repetitiveness of behavior, in turn, ensures the consumption of an environment that is symmetric on all levels.

9.11 The Symmetrizing Economy

We have seen that the structure of production, marketing, and consumption, in the technological society, requires the mass symmetrization of the human population, and the mass symmetrization of the physical environment.

It is at this point that our Symmetry Principle becomes crucial. The principle states that history cannot be extracted from a symmetry. The consequence, therefore, of the mass symmetrization created by the technological society, is the mass destruction of history. That is, because both the population and the physical environment are symmetrized, history can be extracted from neither of these.

Therefore, the supposedly beneficial consequences of a technological society—a thriving and expanding economy—are formally indistinguishable from the detrimental effects—genocide and mass political subjugation in the form of refugee populations, concentration camps, labor camps, slave populations, etc. Both the beneficial and the detrimental effects lead to one end: the mass destruction of history.

But one might therefore argue that, since both the beneficial and detrimental effects lead to the destruction of history, the destruction of history should not be regarded as necessarily damaging. However, this is not the case, as follows:

All computation, we have argued, proceeds by the extraction of history. Therefore, in a society in which history is destroyed, computation cannot be carried out. Quite simply, the population ceases to be able to think. The consequence of this is that problems grow to a massive scale without being examined or handled. For example, western society has, virtually un-impeded, created pollution that has extensively destroyed our water supplies, plant and fish life, the atmosphere, etc. It has created, virtually un-impeded, an energy crisis that forced it to go to war in the Middle East—a war that will inevitably be repeated, with greater destruction, as energy supplies diminish. It has ignored the spread of serious diseases such as AIDS, until they have reached epidemic proportions. All these problems—mass pollution, the energy crisis, disease epidemics—have occurred because of a failure simply to think. Thinking is the use of memory, and this failure to think has come about because of the mass destruction of memory that the technological society installs by the symmetrization of its population and environment. Quite simply, an economy that induces mass conformity, e.g., through advertising, loses the benefit of the individual memory, and therefore the individual intelligence, of each of its citizens—i.e., loses their responsible and careful participation in the daily realization of democracy.

9.12 The Expansion of the Symmetrizing Economy

As we observed earlier, the enormous economic profit of mass production is based on the fact that mass production means the duplication of a product, i.e., symmetrization of the process of production, and the duplication of consumption, i.e., symmetrization of the population. That is, costs are lower because the product is simply duplicated, and financial returns are higher because more consumers buy the product.

This means that those who participate in symmetrizing production and consumption become economically more successful than those who do not. Since symmetrization means, by our Symmetry Principle, the destruction of history, we conclude that profit accrues to those who participate more in the destruction of history.

There is no real way out of the mass destruction of history once it starts. Those who participate in it, and therefore reap greater rewards because they have created greater symmetry, re-invest their profits in technology capable of creating still greater symmetry. Powerful countries invest more and more in weapons of mass destruction; successful companies invest more and more in advertising. In either case, the intended consequence is the same: greater symmetrization of a larger population, and thus the greater destruction of history.

9.13 The Homeless

Within modern society, success comes most to those who are willing to give up their own individuality, their distinguishability, i.e., their asymmetry with respect to the rest of the population, and submit to symmetrization.

Those who cannot submit to symmetrization, either because of temperament or inability, become economically disadvantaged. The most plentiful cases of the inability to submit to symmetrization are the *homeless*.

The homeless are allowed to be plentiful because, unlike, for example, political dissenters, they are not a threat. Whereas political dissenters must be robbed of the objects they use for external memory, e.g., their printing presses, the homeless possess virtually no objects. A homeless person has no store in which objects can be retained, protected, and accumulated; e.g., no home, place of work, or bank account. Because a homeless person is restricted to possessing only what he or she can physically hold at any one time, the person is forced to immediately lose any objects that might enter their possession.

The consequence is that the homeless person can never form any external memory, i.e., any of the external hardware that makes it possible to build up an effective computational existence.

9.14 The Glorification of the Standardized Individual

In an economy dependent on symmetrizing the population, the values of conformity are expressed over the values of non-conformity. Advertisements offer conformity as the ideal state, declaring "You too can own a . . . " or "Don't be the only one without a . . . ."

To a similar end, societies idealize people in uniform. The term "uniform" literally means symmetry; and the wearing of a uniform means that the wearer has actually escaped the courage of individual expression and responsibility, for the camouflage of mass action and replicable legitimacy. Endless war movies celebrate the power of uniformed individuals to extend their uniformity across the environment—by reducing buildings to uniform rubble, and reducing people to the uniformity of death. It is no coincidence that, being a military man, Oliver North shredded documents required as evidence; i.e., destroyed the records of history, and did so by reducing the documents to the symmetry of indistinguishable shreds. It is no coincidence that, when called as a witness, North appeared in military uniform, hiding behind symmetry, and exploiting the destructive effect that symmetry has on history.

But it is not only military men who are glorified because of their uniformity and because of their capacity to reduce everything to uniformity. The ideal individual promoted by modern fashion and the media industries has a physique that first emerged in the sculptures of ancient Greece, and was re-affirmed in the Renaissance period, for example, in Michelangelo's sculpture of David: the impervious stare that advances unhindered to infinity, the body unscarred from the external blows of history, the absence also of personal deformities and birthmarks, an individual from whose appearance nothing of their past can be inferred, the ahistorical individual, the symmetrized polished image of resilient vacuity.

The celebration of the unmarked, ahistorical, individual did not always exist. In ancient Mexico, a figure of great value and respect was the *hunchback*. Indeed, sculptures were made of hunchbacks, indicating that such figures were taken to represent ideal states, and were celebrated.

Two crucial aspects of this respect for hunchbacks should be observed: (1) A hunchback is characterized by deformity, which makes the individual different with respect to the rest of the population. Therefore, the celebration of the hunchback is a celebration of deviance, i.e., asymmetry, as opposed to conformity, i.e., symmetry. (2) According to our Asymmetry Principle, an asymmetry is assigned a history that explains it. Thus, the asymmetry represented by the hunch is assigned a process-history; in fact, a history of growth in accord with our rules for extracting history from curvature extrema—a hunch being a curvature extremum. Thus the hunchback embodies the concept of history; and the celebration of the hunchback means the celebration of history.

To restate: The hunchback embodies (1) asymmetry, and (2) history; and it is for these factors that the hunchback was celebrated.

How different from the ideal of the modern Western society—the military man as the perfect citizen, the general-purpose conformist, the body untainted by the vicissitudes of time, memoryless, an instrument for enforcing conformity, and therefore enforcing memoryless-ness, in the surrounding population.

9.15 The History Ethic

Cognition proceeds by historical inquiry. This is because it uses objects as memory; i.e., it is the interaction between objects by virtue of assigned histories. In particular, we have argued that the human cognitive system is, from the ground up—even at its most involuntary perceptual level—a history-inferring system. Even the lowest involuntary level, even perception, is a means of historical inquiry.

However, we have seen that, in the modern world, considerable economic and political rewards are given to people if they extinguish history—their own history, the history of others, the history of both the animate and inanimate environment. Thus, we have the dichotomous situation where, although cognitive systems function by the creation of history, those parts of a cognitive system that are under voluntary control can be turned against history and therefore become non-cognitive, i.e., fail to interact with objects, even themselves, as memory. In such cases, the individual becomes literally *mindless*, because to have a mind is to respond to objects by virtue of their history.

The modern world offers profit to those who dissolve memory, i.e., those who destroy the ability to extract history from objects. A murdered person is not someone who has not had a history. The person is someone from whom history cannot be extracted. Thus, murdering a person, i.e., initiating their symmetrization, is the attempt to prevent the *inference* of history from them. The same is true of the symmetrization of the rest of the environment—from the Amazon rain forests to human cultures. What is being rewarded is the extinction of objects as memory, and thus the destruction of mind.

In a world in which the full might of technological and economic power is being used to destroy history, it is not surprising that a new ethical principle should be proposed. It is such a principle that we wish now to offer:

THE HISTORY ETHIC. *The history ethic is the commitment to revealing, to the fullest, the history of an entity—a person, a culture, the animate and inanimate environment, oneself—and to interact with the entity by virtue of its fullest history.*

By our Asymmetry Principle, the ethic involves (1) a commitment to the individuality of any entity, i.e., to its asymmetry with respect to others—for example, a refusal to force stereotypy on the individual either by classifying the individual within a stereotyping system, or by forcing the individual to act in a stereotyped way; (2) a commitment to the complexity of an entity, i.e., to the asymmetry it contains between its parts—for example, a refusal to simplify a person either by perceiving them to be simpler than they are, or by forcing simplicity upon them.

It is possible that all the great ethical principles of the past come from the History Ethic; e.g., the commandment "Thou shall not kill" is an injunction not to cause the physical symmetrization of a person and therefore an injunction to maintain our capacity to extract history from the person. But the History Ethic would place these principles within a single framework, a framework in which death, robbery, political subjugation, conformity, are seen as examples of the same phenomenon—the loss of history. Mind begins with the extraction of history, and the previous ethical principles would then be injunctions not to destroy mind. Thus a principle such as "Thou shall not kill" is an injunction to not destroy our own capacity to extract history from a person and therefore an injunction to enhance our own capacity to think by using that person as memory. The great ethical principles of the past would therefore be injunctions against different modes of destroying history or equivalently against different modes of destroying our capacity to think.

Thus one would argue that the History Ethic explicitly announces a single common ground for ethics, a previously unstated core to ethical action: Ethical action is the elaboration of historical inquiry and the adherence to the results of that inquiry.

9.16 The Purpose of this Book

There has been a political purpose to writing this book. In developing a system of rules for the inference of history, our purpose has been to develop tools by which a victim of political subjugation can counter the subjugator's primary goal: the destruction of the victim's capacity to recover history.

The intention to erase history is often stated by subjugators, *explicitly*. Nazi officers were reported to have declared to their prisoners:

> However this war may end, we have won the war against you; none of you will be left to bear witness, but even if someone were to survive, the world will not believe him. There will perhaps be suspicions, discussions, research by historians, but there will be no certainties, because we will destroy the evidence together with you. (Quoted in Levi, 1989, p11).

Anyone who has undergone political subjugation becomes fundamentally concerned with the recovery of history. And it is to assist in this fundamental concern that the tools of this book have been developed. A single sentence by Milan Kundera (1978, p3), the Czech writer who survived subjugation, states the problem in its most succinct form:

> the struggle of man against power is the struggle of memory against forgetting.

9.17 Where is History?

We have seen that modern Western society has induced a mass symmetrization of people and environments, and that the economic benefits of this symmetrization are so great that it can only increase. Symmetrization brings with it the destruction of history, i.e., memory, and therefore the destruction of thought—since all thought is the use of memory.

Given this mass erosion by symmetry, where, in the present world, is it possible for us to find the vestiges of history, and thus maintain the possibility of thought?

We began this book with a description of a subway station, and it is to this subway station that we now return. Here—in the dented garbage

Figure 9.1
A photograph by Bruce Davidson from his collection *Subway*, published by Aperture, New York (1986). © Bruce Davidson/Magnum.

cans, the graffiti, the scratched surfaces, the crumpled newspapers, the festering splashes, the broken walls and bottles, the squashed containers, and the torn clothes—here, in these objects, one can find history. For here one can engage in that genuine interchange with the environment that we call cognition.

And in the prison of the subway, there are citizens, some remaining un-named guardians of history. In them, history has escaped destruction by the modern economy simply because it was too disadvantaged to pose any economic threat. One of these guardians is the woman in the photograph by Bruce Davidson, shown in Fig 9.1.

On the walls of her prison, we see the external markings of her history—markings that escaped the erasure of the symmetrizing economy. But within her head, there is an internal history that also escaped erasure by that economy.

Any act of cognition begins with the inquiry "What happened?" Any act of thought is an attempt to recover history, because it is the use of memory. If we are to be able to think, we must find history. This woman contains history, and thus she contains the single key that can unlock, in us, our own capacity to think. We must therefore lean over to her, lean over in the act of inquiry, and begin.

Notes

Chapter 1

1. In mathematics, there is a small, third branch, the ordinal structures.

2. The original statement of the Symmetry Curvature Duality Theorem (Leyton, 1987b) is this: Any segment of a smooth planar curve, bounded by two consecutive curvature extrema of the same type (either both maxima or both minima) has a unique differential symmetry axis, and this axis terminates at the curvature extremum of the opposite type (minimum or maximum, respectively).

3. Our conventions for the specification of curvature on a planar curve are as follows: (1) In going around a curve, positive curvature is the rate of anticlockwise rotation, and negative curvature is the rate of clockwise rotation, that a traveler along the curve would have to make. (2) The direction chosen for going around a curve is that which keeps the inside of the figure to the left of the curve and the outside to the right.

4. Hayes & Leyton (1989) have done work on the formation of the corners themselves, as will be seen in our chapter on art.

Chapter 2

1. Freyd (1987) has shown that, when looking at handwriting, human beings deduce subtle differences in the rule-systems that were used in the writing process.

2. Recall that we assumed that the outlines were smooth; i.e., without corners. If corners existed in the outlines, then the starting shapes would be regular polygons.

3. The Externalization Principle demands that the internal structure is also nested. The circle satisfies this requirement trivially: the nesting is the simplest possible, consisting of only one level. Note also that, if the shapes contained corners, and thus had starting shapes that were regular polygons, the starting shapes would also be nested, repetitive, Euclidean hierarchies.

4. The symmetry analyses of Blum (1973) and Brady (1983) would put the three arrows all on the same side of the curve and would thus not be able to explain the flattening.

5. It is sufficient to consider cases only where e has the same sign as E. This is because the other cases can be generated from these via the continuation-rules given earlier. Thus, for each E, there is a unique e.

6. At the risk of sounding facile, we should note that there is a possibility that a third term *information* is equivalent to these other two. Certainly, in statistical information theory, the term information is defined as *variety*, and this makes the term similar to the term *asymmetry* which we are defining as *distinguishability*. Algorithmic information theory can also be regarded as a measure of the *variety* in a set. Thus, the ingredient of the present, from which one is extracting the past, might therefore be considered to be *information*, in the abstract sense of some information theory; e.g., statistical or algorithmic information theory. Therefore, we might be able to regard the terms *shape* and *information* as identical terms. That is, we might be able to regard the mathematical study of shape as a general, and more thorough, information theory than has been attempted in the current approaches to the study of information measurement.

7. To go any higher, the ball would have to gain a potential energy that is greater than its starting potential energy, which was its total energy.

Chapter 3

1. For research on the characterization of texture, see Rosenfeld (1971, 1975), Rosenfeld & Lipkin (1970), Rosenfeld & Strong (1971), Rosenfeld & Milgram (1972); and also Haralich (1978).

2. See Julesz (1971, 1975) for experimental research on statistical factors.

3. Related proposals have been made by Lowe & Binford (1981); Lowe (1985); Witkin & Tenenbaum (1983); and Rock (1983).

4. Brady and Yuille use parallel projection, and thus can ignore scaling.

5. Thus, whereas the vertical axis in Figs 37a–c should be considered as representing depth from the viewer, the vertical axes in Fig 3.37d should be considered as representing some other property in the image. The process of marking has created a change in this property rather than depth.

6. The equation of Blake & Zisserman (1987, pp98–99), for the one-dimensional surface, is the following:

$$\text{Energy} = \gamma \int (u'')^2 + \alpha Z_{step} + \beta Z_{crease} + \int (u - d)^2$$

$$= \text{gradient variation} + \text{steps} + \text{creases} + \text{noise}$$

The symbols are as follows: u is the reconstructed surface, d are the data values, Z_{step} and Z_{crease} are the number of steps and the number of creases respectively, and γ, α, and β are weighting constants.

7. See also related work by Kass & Witkin (1987).

Chapter 4

1. In a sense, we use Turing's Thesis but only in that it guarantees the realization of a well-defined history by *some* member of the *entire* class of formally defined machines (i.e., where a machine is defined as a state-transition function together with an input-output function). We will not need the real strength of Turing's Thesis which concerns the realization of a well-defined history in terms of a highly particular type of machine, a Turing machine. For example, the particularities we shall not need are the symbol-based nature of the device (i.e., that it reads, writes, corrects, etc.), the finiteness of the device state-space, etc. Thus, we use Turing's Thesis but in an extremely weak form.

2. We introduced the Representational Machine Analogy in Leyton (1986a, 1986b, 1987a), where it was called the Content Machine Analogy because it refers to the definition of cognitive *content*.

3. The Processing Machine Analogy follows from the Representational Machine Analogy in two ways. The first is rather obvious and the second is less so:

The first is simply this: The Representational Machine Analogy states that cognition represents anything as a machine. This means that, in particular, cognition has no alternative but to represent a cognitive system (e.g., the human mind) as a machine. In other words, the Processing Machine Analogy follows from the Representational Machine Analogy simply as an instantiation of the latter.

The second, less obvious, reason why the Processing Machine Analogy follows from the Representational Machine Analogy is this: The Representational Machine Analogy, which states that cognition represents any object as a machine, is a precise statement of the Representation-is-Explanation Principle, which states that objects are represented in terms of histories that causally explain them. Now, if an object is represented in terms of its causal explanation, then the object is interpreted as memory. However, when something is interpreted as memory, it becomes part of the mind, because it has a computational role within it. In short, we conclude that *any object that is described as a machine becomes part of the mind*; i.e., by virtue of the object being causal explanation and therefore memory.

This second argument shows that the Processing Machine Analogy follows from the Representational Machine Analogy not only because minds are *examples* of objects described as machines but because minds, and their parts, are the *only* objects that are described as machines.

4. Schmidt, Noice, Marsella & Bresina (1987) have argued that human beings use symmetry in planning and learning certain plans.

Chapter 6

1. This view has similarities to that which underlies a research program developed by Terzopoulos, Witkin & Kass (1988) who try to simulate processes in video graphics by developing first a model of the specific materials involved in a situation. (See also Witkin, Fleisher & Barr, 1987; and related work by Kass, Witkin, & Terzopoulos, 1987). Their goal of simulation is of course the opposite of the goal of the process-recovery problem; i.e., they are trying to derive an image from a causal structure whereas we are trying to derive a causal-structure from an image. Nevertheless, it would appear that we are both bound to establishing specific material domains in which to work out the dynamics.

2. In the case of the plate the initial surface has zero fourth order derivatives, and is therefore a flat plate or a low-order spline.

3. In brief, the argument we shall give is this: The Asymmetry and Symmetry Principles ensure the symmetrical nature of the past state. The Symmetry-to-Trace Conversion Principle converts the symmetries of the past state into a space of traces. The History Minimization Principle ensures (via the psychological definition of process as repetition) that these traces, as well as the external history leading back to the past state, have maximal translational symmetry across time, and indeed are decomposed into causal factors (processes) that have complete translational symmetry across time.

4. When we introduced $\mathbf{D}_4$ in section 5.4, we considered the operations to be acting on the entire square. Thus, t was regarded as the reflection about the vertical bisector of the square. However, we are now considering the operations to be acting on individual sides. Thus we re-define t as acting about the perpendicular bisector of a side. For example, consider the right side in Fig 6.5. The operation t reflects this side about its own perpendicular bisector, which is horizontal. Similarly, each of the other reflections, $r_{90}t$, $r_{180}t$, $r_{270}t$, is defined relative to each side. For example, $r_{270}t$ is defined relative to the diagonal axis through the right-hand end of the side, as judged by facing the side from the center of the square.

Notice also that the arrows are drawn pairs in Fig 6.5. For example, every r_{90} arrow is drawn together with a $r_{270}t$ arrow. Generally, each pair consists of a rotation and a reflection. The rotation arrow should be interpreted as dragging the side along that arrow. The reflection arrow should be interpreted as reflecting the side through the bisector of the arrow itself.

5. Note that r_{90} and r_{270} occur on the right and left (respectively) of the inner ring in Fig 6.7, and appear in those same positions in Fig 6.6. However, $r_{90}t$ and $r_{270}t$ appear on the right and left (respectively) of the outer ring in Fig 6.7, but on the left and right (respectively) in Fig 6.6; i.e., they have changed sides in going from Fig 6.7 to Fig 6.6. Thus the inner rings match in the two figures, and the outer rings are reflections of each other about the vertical axis.

This reflection occurs because the dotted links in Fig 6.7 represent reflection about the vertical axis, and therefore, whenever one traverses a dotted link in Fig 6.7, one must change sides in Fig 6.6.

This switching of sides is actually not a problem. The color graph represents only the *connectivity* of a group, and thus the graph can be pulled about and twisted in any manner provided that this connectivity is preserved. Thus, in laying the color graph over the square, we can simply twist the outer ring of the color graph around the inner ring as shown in Fig 6.9. The elements on the vertices now match the labels on the sides in Fig 6.6.

6. The reader might object that the *outer* ring of one color, in Fig 6.7, consists of the elements, t, $r_{90}t$, $r_{180}t$, $r_{270}t$. It does not. A true color graph does not have a labelling of its

vertices, as is explained in the next footnote. It has a labelling only of its links (i.e., via the coloring). The outer ring is therefore a copy of **Cycle(4)**.

7. In fact, a color graph is so homogeneous that any group element can label any of the vertices. That is, given any group element one can assign it arbitrarily to label any of the vertices, and generate the rest of the labels from this label. This arbitrariness, or homogeneity, means that one does not actually have to have any labels on the color graph; i.e., one does not have to pick a distinguished labeling on the graph. The labeling given in Fig 6.7 was introduced in the initialization phase (Stage 2) in realizing the Symmetry-to-Trace Conversion Principle.

8. Without loss of generality, we shall assume that the perceived transformations preserve the origin. The experiments in which the shape sequence was discovered did not test for this aspect (Leyton, 1986a).

9. Strictly speaking, Fig 6.22 shows the universal covering group of **Special–Linear**. The reader may wish to know that there is an entire book by Lang (1975) devoted to **Special–Linear** and the above decomposition. The book is entitled simply $SL_2(R)$. The letters *SL*, in the title, of course mean *Special Linear*.

10. The rigorous description of such a generator is as follows: A continuous process is a one-parameter subgroup of a Lie group. The generator for any such subgroup is a vector tangential to the subgroup. This vector is initially a tangent to the identity element of the group. The collection of vectors tangential to the identity element form the Lie algebra of the group. Any one-parameter subgroup, i.e., any continuous process, is generated by selecting a member of the Lie algebra.

11. Effectively, we need only choose a generator that is below the resolving power of the eye.

12. We normally involve all of $\mathbf{D}_4$ to emphasize that the stimulus allows both a history of rotations and a history involving reflections, as discussed in section 2.5.

13. The general definition of coset we use is this: Let **G** be a structure of some type. Let **H** be a substructure of **G** of the same type as **G**. Then the set **H**.*g*, within **G**, is called a coset of **H**, where "." is some operator within **G**.

In the example, with which we introduced the concept of coset in the text, the relevant structures were *groups*. That is, both **G** and **H** are groups, **H** is a subgroup of **G**, and "." is the binary operation of the group. In the case we consider at length in the text, i.e., grouping, the relevant structures are *nested group sequences*. Precisely, **G** and **H** are nested group sequences, **H** is a left subsequence of **G**, and "." is the nesting operator. Note that coset leaders *g* have two associated operators, "*", the operation within the group component from which *g* is selected, and ".", the nesting operator between components.

14. Since a history is the instantiated component of the explanation map (section 6.13), the Grouping Principle should, more strictly, state: A grouping is a coset of a left-subsequence of the instantiated explanation map.

15. The group of linear transformations, without reflection, that preserve area.

16. Any group sequence in the text should be interpreted as a nested hierarchy in which the symbol "." separates levels of the hierarchy and denotes the nesting relationship; i.e., the level to the left of "." is nested within the level to the right of it. Any level is a group. In some cases, a level is shown as decomposed into a direct product of subgroups. This product is indicated by the " × " sign, within the level. The direct product is a symmetric relationship, and in the cases in which we have indicated it, we have not investigated whether it is coincident, in the sequence, with a nesting relationship (which is asymmetric), thus causing a decomposition of the level into further levels. However, since a semi-direct product is an asymmetric relationship with a clear nesting structure, one might more easily assume that it coincides with a nesting relationship. However, this latter issue, which is presumably determined on a case by case basis, is not examined in the book. Finally, we observe that in some cases—where two

successive levels are subgroups of the same group—the nesting operator between the two subgroups might be coincident, in the sequence, with the use of the binary operator from the group.

17. One possible position for the second Cartesian axis is along the bottom side of the triangle. The evidence for this is that the pointing triangle is seen as allowing stretch in two directions: along the bottom side, i.e., by widening the latter, and along the direction of its pointing. Thus these directions are eigenspaces. This means that, via the Interaction Principles, the two Cartesian symmetry axes have been imposed along those two directions.

18. A vector space is an Abelian group.

19. By Corollary 1, the circles travel towards and reach the extremum. In the SLS and PISA, this means that the axis itself reaches the extremum. In the SAT, the axis stops before reaching the extremum because the final point of the axis is the *center* of the final circle. However, recall from Corollary 1 that the final circle is tangential only to the extremum. This means that the final circle can be arbitrarily small. Thus, letting the final circle shrink to zero radius, the symmetry point reaches the extremum even in the case of the SAT.

20. We should also observe that, with respect to *manipulation* issues, these points are the most likely *grasp-points* on the ellipse. This is no random coincidence: Grasping requires the same constraining action as squashing. Thus the inference of squash points is important not just for the analysis of deformation, but also for manipulation.

Chapter 7

1. In fact, currently, the tentative assumption is that subcategorization information is derivable from case-role information, and is therefore unnecessary (Chomsky, 1986a). We shall not consider these issues here.

2. The full statement is that movement cannot occur across more than one bounding node where the bounding nodes are S, S′, and NP.

3. We should note that Wickelgren (1979, p340–4) puts forward a very subtle view of the above constructs, using an information-processing approach.

4. Mathesius used the terms *basis* and *nucleus* instead of *topic* and *comment*, respectively.

5. Givon (1979) gives a profound analysis of the relationship between syntax and presuppositionality.

Chapter 8

1. Technically, this operation inserts an infinite spike into the curvature function of the shape. The reason for this is that curvature is the rate of change of rotation (of the tangent) per movement along the curve. Since, at a kink, one can rotate a tangent without moving any distance along the curve, one has finite rotation per zero distance. In this sense, one can regard the kink as having infinite curvature.

2. We should observe that asymmetry is a goal not only on the large-scale behavioral and motivational level of the animal, but on the small-scale involuntary level. That is, even on the local level, the nervous system is constantly seeking asymmetry. For example, the eye-ball is in perpetual involuntary movement, thus ensuring that the neurons in the retina are receiving constantly changing stimuli. When the image on the retina is forced to be still, as can only happen by some imposed artificial means, the person goes "blind", i.e., the person sees only a completely grey empty field—despite the fact that a perfect image of the world is arriving at the retina (Riggs, Ratliff, Cornsweet, & Cornsweet, 1953; Kowler, 1990). Thus, one can conclude that the visual system does not seem to be interested in an image of the static world: it is interested only in change, i.e., asymmetry.

References

Aloimonos, J. (1988). Shape from texture. *Biological Cybernetics, 58,* 345–360.

Attneave, F. (1954). Some informational aspects of visual perception. *Psychological Review, 61,* 183–193.

Attneave, F. (1968). Triangles as ambiguous figures. *American Journal of Psychology, 18,* 447–453.

Attneave, F. (1971). Multistability in perception. *Scientific American, 225,* 62–71,

Attneave, F. (1982). Pragnanz and soap-bubble systems: A theoretic exploration. In J. Beck (Ed.) *Organization and Representation.* Hillsdale, NJ: Erlbaum.

Bajcsy, R., & Lieberman, L. (1976). Texture gradient as a depth cue. *Computer Graphics and Image Processing, 5,* 52–67.

Barr, A. (1984). Global and local deformations of solid primitives. *Computer Graphics, 18,* 21–30.

Bennet, C. H. (1988). Dissipation, information, computational complexity and the definition of organization. In D. Pines (Ed.), *Emerging synthesis in science.* New York: Addison-Wesley.

Berlyne, D.E. (1971). *Aesthetics and psychobiology.* New York: Appleton-Century-Crofts.

Bexton, W. A., Heron, W., & Scott, T. H. (1954). Effects of decreased variation in the sensory environment. *Canadian Journal of Psychology, 8,* 70–76.

Binford, T. O. (1971). *Visual perception by computer.* Presented at IEEE Systems, Science, and Cybernetics Conf., Miami, FL.

Blake, A., & Zisserman, A. (1987). *Visual reconstruction.* Cambridge, Mass: MIT Press.

Blum, H. (1973). Biological shape and visual science (part 1). *Journal of Theoretical Biology, 38,* 205–287.

Bohm, C., & Jacopini, G. (1966). Flow diagrams, Turing machines, and languages with only two formation rules. *CACM, 9,* 366–371.

Bolinger, D. (1952). Linear modification. In D. Bolinger (Ed.), *Forms of English.* Cambridge, Mass: Harvard University Press, 1965.

Brady, M. (1983). Criteria for representations of shape. In A. Rosenfeld & J. Beck (Eds.), *Human and Machine Vision: Vol 1.* Hillsdale, NJ: Erlbaum.

Brady, M., & Yuille, A. (1983). *An extremum principle for shape from contour* (AI Memo No. 711). MIT.

Bryant, S. V., French, V., & Bryant, P. J. (1981). Distal regeneration and symmetry. *Science, 212,* 993–1002.

Buffart, H., Leeuwenberg, E., & Restle, F. (1981). Coding theory of visual pattern completion. *Journal of Experimental Psychology: Human Perception and Performance, 7,* 241–274.

Butler, R. A. (1953). Discrimination learning by rhesus monkeys to visual exploration motivation. *Journal of Comparative and Physiological Psychology, 46,* 95–98.

Butler, R. A., & Harlow, H. F. (1954). Persistence of visual exploration in monkeys. *Journal of Comparative and Physiological Psychology, 47,* 258–263.

Canny, J. (1984). Collision detection for moving polyhedra (AI Memo No. 806). MIT.

Carlton, E. H., & Shepard, R. N. (1990a). Psychologically simple motions as geodesic paths: I. Asymmetric objects. *Journal of Mathematical Psychology, 34,* 127–188.

Carlton, E. H., & Shepard, R. N. (1990b). Psychologically simple motions as geodesic paths: II. Symmetric objects. *Journal of Mathematical Psychology, 34,* 189–228.

Chaitin, G. (1969). On the length of programs for computing finite binary sequences. *Journal of the Association of Computing Machinery, 16*, 145–159.

Chomsky, N. (1957). *Syntactic structures*. The Hague: Mouton.

Chomsky, N. (1965). *Aspects of the theory of syntax*. Cambridge Mass: MIT Press.

Chomsky, N. (1974). *The Amherst lectures*. Unpublished lecture notes distributed by Documents Linguistiques, University of Paris VII.

Chomsky, N. (1981). *Lectures on government and binding*. Dordrecht, Holland: Foris.

Chomsky, N. (1986a). *Knowledge of language*. New York: Praeger.

Chomsky, N. (1986b). *Barriers*. Cambridge, Mass: MIT Press.

Creider, C. A. (1979). In T. Givon (Ed.), *Syntax and Semantics*: *Vol. 12. Discourse and syntax*. New York: Academic Press.

DeJong, G. F., & Mooney, R. J. (1986). Explanation-based learning: An alternative view. *Machine Learning, 1*, 145–176.

Dember, W. N., Earl, R. W., & Paradise, N. (1957). Response by rats to differential stimulus complexity. *Journal of Comparative and Physiological Psychology, 50*, 514–518.

Dinnerstein, D., & Wertheimer, M. (1957). Some determinants of phenomenal overlapping. *American Journal of Psychology, 70*, 21–37.

Do Carmo, M. (1976). *Differential geometry of curves and surfaces*. Englewood Cliffs, NJ: Prentice-Hall.

Duncker, K. (1929). Uber induzierte Bewegung. *Psychologische Forschung, 12*, 180–259. Translated and condensed in W. D. Ellis (Ed.), *A source book of gestalt psychology*. New York: Harcourt Brace.

Eckblad, G. (1980). The curvex: Simple order structure revealed in ratings of complexity, interestingness, and pleasantness. *Scandinavian Journal of Psychology, 21*, 1–16.

Fillmore, C. J. (1968). The case for case. In E. Bach and R. T. Harms (Eds.), *Universals in Linguistic Theory*. New York: Holt, Rinehart & Winston.

Fodor, J. A., & Pylyshyn, Z. W. (1988). Connectionism and cognitive architecture: A critical analysis. In S. Pinker & J. Mehler (Eds.), *Connections and Symbols*. Cambridge, Mass: MIT Press.

Foley, W. (1976). *Comparative syntax in Austronesian*. Ph.D. dissertation. University of California, Berkeley.

French, V., Bryant, P. J., & Bryant, S. V. (1976). Pattern regulation in epimorphic fields. *Science, 193*, 969–994.

Freyd, J. J. (1987). Dynamical mental representations. *Psychological Review, 94*, 427–438.

Gibson. J. J. (1950). *The perception of the visual world*. Boston, Mass: Houghton Mifflin.

Givon, T. (1979). *On understanding grammar*. New York: Academic Press.

Goldmeier, E. (1936). Uber Anlichkeit bei gehenen Figuren. *Psychologische Forschung, 21*, 146–209. English translation by E. Goldmeier, Similarity in visually perceived forms. *Psychological Issues*, 1972, *8*, no. 1, Monograph 29. New York: International Universities.

Grimson, W. E. L. (1981). *From images to surfaces*. Cambridge, Mass: MIT Press.

Halliday, M. A. K. (1967/1968). Notes on transitivity and theme in English: Parts 1, 2, and 3. *Journal of Linguistics, 3.1, 3.2, & 4.2*, 370–381, 199–244, & 179–215.

Halliday, M. A. K. (1985). *An introduction to functional grammar*. Baltimore, Maryland: Edward Arnold.

Haralick, R. M. (1978). Statistical and structural approaches to texture. *Proceedings of the International Joint Conference on Pattern Recognition*, 45–60.

Hassenstein, B., & Reichardt, W. (1956). Systemtheoretische Analyse der Zeitreihenfolgen- und Vorzeichenauswertung bei der Bewgungsperzeption der Russelkafers. *Chlorophanus. Z. Naturf., IIb*, 513–524.

Hayes, P. J. (1985). Naive physics I: Ontology for liquids. In J. R. Hobbs & R. C. Moore (Eds), *Formal theories of the commonsense world.* Norwood, NJ.: Ablex.

Hayes, P. J., & Leyton, M. (1989). Processes at discontinuities. *Proceedings of the International Conference of Artificial Intelligence*, Detroit, 1267–1272.

Heidegger, M. (1927). *Being and time*. English translation by J. Macquarrie and E. Robinson. New York: Harper and Row, 1962.

Hoffman, D. D., & Richards, W. A. (1985). Parts of recognition. *Cognition, 18*, 65–96.

Hollerbach, J. M. (1975). *Hierarchical shape description of objects by selection and modification of prototypes* (AI Memo No. 346). MIT.

Hollerbach, J. M. (1981). An oscillation theory of handwriting. *Biological Cybernetics, 39*, 139–156.

Horn, B. K. P. (1977). Understanding image intensities. *Artificial Intelligence, 8*, 201–231.

Hunt, H. F., & Brady, J. V. (1955). Some effects of punishment and intercurrent "anxiety" on a simple operant. *Journal of Comparative and Physiological Psychology, 48*, 305–310.

Ikeuchi, K. (1984). Shape from regular patterns. *Artificial Intelligence, 22*, 49–75.

Ikeuchi, K., & Horn, B. K. P. (1981). Numerical shape from shading and occluding boundaries. In J. M. Brady (Ed.), *Computer Vision*, New York: North-Holland.

Jacobs, D. W. (1988). *The use of grouping in visual object recognition.* Masters thesis, MIT.

Johansson, G. (1950). *Configurations in event perception.* Stockholm, Sweden: Almqvist and Wiksell.

Julesz, B. (1959). Method of coding television signals based on edge detection. *Bell Systems Technical Journal, 38*, 1001–1020.

Julesz, B. (1971). *Foundations of cyclopean perception.* Chicago: University of Chicago Press.

Julesz, B. (1975). Experiments in the visual perception of texture. *Scientific American, 232*, 34–43.

Kanade, T. (1981). Recovery of the three-dimensional shape of an object from a single view. In J. M. Brady (Ed.), *Computer vision.* New York: North-Holland.

Kass, M., & Witkin, A. (1987). Analyzing oriented patterns. *Computer Vision, Graphics, and Image Processing, 37*, 362–385.

Keenan, E., & Comrie, B. (1977). Noun phrase accessibility and universal grammar. *Linguistic Inquiry, 7*, 63–99.

Kender, J. R. (1979). *Shape from texture.* Ph.D. thesis, Carnegie-Mellon University.

Kirsner, R. S. (1973). Natural focus and agentive interpretation: On the semantics of the Dutch expletive *er*. *Stanford Occasional Papers in Linguistics, 3*, 101–114.

Kirsner, R. S. (1976). On the subjectless pseudo-passive in standard Dutch and the problem of background agent. In C. Li (Ed.), *Subject and topic.* New York: Academic Press.

Kish, G. B. (1955). Learning when the onset of illumination is used as reinforcing stimulus. *Journal of Comparative and Physiological Psychology, 48*, 261–264.

Koenderink, J. J. (1984). The structure of images. *Biological Cybernetics, 50*, 363–370.

Koffka, K. (1935). *Principles of gestalt psychology*. New York: Harcourt, Brace & World.

Kohler, W. (1947). *Gestalt psychology: An introduction to new concepts in modern psychology*. New York: Liveright.

Kolmogorov, A. (1968). Logical basis for information theory and probability theory. *IEEE Transactions on Information Theory, IT-14*, 662–664.

Kowler, E. (Ed.). (1990). *Eye movements and their role in visual and cognitive processes*. Amsterdam: Elsevier.

Krechevsky, I. (1937). Brain mechanisms and variability: II. Variability where no learning is involved. *Journal of Comparative and Physiological Psychology, 23*, 139–163.

Kundera, Milan. (1978). *The book of laughter and forgetting*. English translation by M. H. Heim. New York: Penguin Books, 1981.

Lakoff, G. (1987). *Women, Fire, and Dangerous Things*. Chicago: University of Chicago Press.

Lang, S. (1975). SL_2R. London: Addison-Wesley.

Lazarus, R. S., & Folkman, S. (1984). *Stress, appraisal, and coping*. New York: Springer.

Levi, P. (1986). *The drowned and the saved*. English translation by R. Rosenthal. New York: Random House, 1989.

Leyton, M. (1981). *Artistic activity and human survival: Vol. 1*. Research report, University of Warwick, England.

Leyton, M. (1984). Perceptual organization as nested control. *Biological Cybernetics, 51*, 141–153.

Leyton, M. (1985). Generative systems of analyzers. *Computer Vision, Graphics, and Image Processing, 31*, 201–241.

Leyton, M. (1986a). Principles of information structure common to six levels of the human cognitive system. *Information Sciences, 38*, 1–120, entire journal issue.

Leyton, M. (1986b). A theory of information structure: I. General principles. *Journal of Mathematical Psychology, 30*, 103–160.

Leyton, M. (1986c). A theory of information structure: II. A theory of perceptual organization. *Journal of Mathematical Psychology, 30*, 257–305.

Leyton, M. (1987a). Nested structures of control: An intuitive view. *Computer Vision, Graphics, and Image Processing, 37*, 20–53.

Leyton, M. (1987b). Symmetry-curvature duality. *Computer Vision, Graphics, and Image Processing, 38*, 327–341.

Leyton, M. (1987c). A limitation theorem for the differential prototypification of shape. *Journal of Mathematical Psychology, 31*, 307–320.

Leyton, M. (1988). A process-grammar for shape. *Artificial Intelligence, 34*, 213–247.

Leyton, M. (1989). Inferring causal-history from shape. *Cognitive Science, 13*, 357–387.

Li, C. (1976). *Subject and topic*. New York: Academic Press.

Lloyd, S., & Pagels, H. (1988). Complexity as thermodynamic depth. In press.

Lowe, D. (1985). *Perceptual organization and visual recognition*. Holland: Kluwer.

Lowe, D., & Binford, T. O. (1981). The interpretation of three-dimensional structure from image curves. *Proceedings of the International Joint Conference of Artificial Intelligence*, Vancouver, 613–618.

Mach, E. (1897). *The analysis of sensations*. English translation. New York: Dover, 1959.

Marr, D. (1982). *Vision: A computational investigation into the human representation and processing of visual information*. San Fransisco: Freeman.

Marr, D., & Nishihara, H. K. (1978). Representation and recognition of the spatial organization of three-dimensional shapes. *Proceedings of the Royal Society of London, B200*, 169–294.

Martin, J., & McClure, C. (1985). *Structured techniques for computing*. Englewood Cliffs, NJ: Prentice-Hall.

Mathesius, V. (1915). O passivu v modern anglictine. *Sbornik filologicky*, *5*, 198–220.

Mitchell, T., Keller, R., & Kedar-Cabelli, S. (1986). Explanation-based generalization: A unifying view. *Machine Learning*, *1*, 47–80.

Mokhtarian, F., & Mackworth, A. (1986). Scale-based description and recognition of planar curves and two-dimensional shapes. *IEEE Transactions on Pattern Analysis and Machine Intelligence*, *8*, 34–43.

Montgomery, K. C., & Segall, M. (1955). Discrimination learning based upon the exploratory drive. *Journal of Comparative and Physiological Psychology*, *48*, 225–228.

Moon, L. E., & Lodahl, T. M. (1956). The reinforcing effect of changes in illumination on lever-pressing in the monkey. *American Journal of Psychology*, *69*, 288–290.

Morris, C. (1938). *Foundations of the theory of signs*. Chicago: University of Chicago Press.

Morris, D. (1962). *The biology of art*. London: Methuen.

Newell, A. (1980). Physical symbol systems. *Cognitive Science*, *4*, 135–183.

Newell, A., & Simon, H. A. (1976). Computer science as empirical inquiry: Symbols and search. *Communications of the ACM*, *19*, 113–126.

Oliver, W. R. (1989). Locomotion of white blood cells: A biophysical analysis. *Blood Cells*, *15*, 334–342.

Oppenheimer, F. (1934). Optische Versuche uber Ruhe und Bewegung. *Psychologische Forschung*, *20*, 1–46.

Pentland, A. P. (1983). Fractal-based description. *Proceedings of the International Joint Conference on Artificial Intelligence*, Karlsruhe, Germany, 973–981.

Pizer, S. M., Oliver, W., & Bloomberg, S. H. (1986). Hierarchical shape description via the multiresolution of the symmetric axis transform. *IEEE Transactions PAMI*, *9*, 505–511.

Pylyshyn, Z. W. (1984). *Computation and cognition: Toward a foundation for cognitive science*. Cambridge, Mass: MIT Press.

Reichardt, W. (1961). Autocorrelation, a principle for the evaluation of sensory information by the central nervous system. In W. A. Rosenblith (Ed.), *Sensory communication*. Cambridge, Mass: MIT Press.

Richards, W., & Hoffman, D. D. (1985). Codon constraints on closed 2D shapes. *Computer Vision, Graphics, and Image Processing*, *31*, 265–281.

Richards, W., Koenderink, J. J., & Hoffman, D. D. (1987). Inferring three-dimensional shapes from two-dimensional silhouettes. *Journal of the Optical Society of America A*, *4*, 1168–1175.

Riggs, L. A., Ratliff, F., Cornsweet, J. C., & Cornsweet, T. N. (1953). The disappearance of steadily fixated visual test objects. *Journal of the Optical Society of America*, *43*, 495–501.

Rock, I. (1973). *Orientation and form*. New York: Academic Press.

Rock, I. (1975). *An introduction to perception*. New York: Macmillan.

Rock, I. (1983). *The logic of perception*. Cambridge, Mass: MIT Press.

Rosch, E. (1975). Cognitive reference points. *Cognitive Psychology, 7*, 532–547.

Rosch, E. (1977). Human categorization. In N. Warren (Ed.), *Studies in Cross-Cultural Psychology: Vol. 1*. London: Academic Press.

Rosch, E. (1978). Principles of categorization. In E. Rosch & B. B. Lloyd (Eds.), *Cognition and Categorization*. Hillsdale NJ: Lawrence Erlbaum.

Rosenfeld, A. (1971). Isotonic grammars, parallel grammars, and picture grammars. In B. Meltzer & D. Michie (Eds.), *Machine intelligence: Vol. 6* (pp. 281–294). Edinburgh: Edinburgh University Press.

Rosenfeld, A. (1975). A note on automatic detection of texture gradients. *IEEE Transactions on Computers, C23*, 988–991.

Rosenfeld, A. (Ed.), (1984). *Multiresolution Image Processing and Analysis*. Berlin: Springer-Verlag.

Rosenfeld, A., & Lipkin, B. S. (1970). Texture synthesis. In B. S. Lipkin & A. Rosenfeld (Eds.), *Picture Processing and psychopictorics*. New York: Academic Press.

Rosenfeld, A., & Milgram, D. L. (1972). Web automata and web grammars. In B. Meltzer & D. Michie (Eds.), *Machine Intelligence: Vol. 7* (pp. 307–324). Edinburgh: Edinburgh University Press.

Rosenfeld, A., & Strong, J. P. (1971). A grammar for maps. In J. T. Tou (Ed.), *Software Engineering*: Vol. 2. New York: Academic Press.

Ross, J. R. (1967). *Constraints on variables in syntax*. Unpublished doctoral dissertation, MIT.

Rubin, E. (1921). *Visuell wahrgenommene Figuren*. Glydendalske.

Sanders, G. (1969). *On the natural domain of grammar*. Mimeographed paper circulated by the Indiana University Linguistics Club.

Sanders, G. (1970a). *Invariant ordering*. Mimeographed paper circulated by the Indiana University Linguistics Club.

Sanders, G. (1970b). Constraints on ordering. *Papers in Linguistics, 2*, 460–502.

Sanders, G. (1971). On the symmetry of grammatical constraints. *Chicago Linguistic Society* (University of Chicago), *7*, 232–241.

Sanders, G. (1972a). *Equational grammar*. The Hague: Mouton.

Sanders, G. (1972b). *Precedence relations in language*. Mimeographed paper circulated by the Indiana University Linguistics Club.

Schank, R. C., & Abelson, R. P. (1977). *Scripts, plans, goals, and understanding*. Hillsdale, NJ: Lawrence Erlbaum.

Schmidt, C. F., Noice, H., Marsella, S., & Bresina, J. (1987). Preference for and transfer of problem-solving methods. *Psychonomic Society Conference*, New Orleans.

Schumann, F. (1900, 1902, 1904). Beitrage zur Analyse der Gesichtswahrnehmungen: I–IV. *Zeitschrift für Psychologie, 23, 24, 30, 36*.

Shepard R. N., & Cooper, L. (1982). *Mental images and their transformations*. Cambridge, Mass: MIT Press.

Solomonoff, R. (1964). A formal theory of inductive inference: Part I. *Information and Control, 7*, 1–22.

Spitz, R. A. (1945). Hospitalism: An inquiry into the genesis of psychiatric conditions in the early childhood. *Psychoanalytic Study of the Child, 1*, 53–74.

Spitz, R. A. (1946a). Hospitalism: A follow-up report. *Psychoanalytic Study of the Child, 2,* 113–117.

Spitz, R. A. (1946b). Anaclitic depression. *Psychoanalytic Study of the Child, 2,* 313–342.

Staal, J. F. (1967). Word order in Sanskrit and universal grammar. *Foundations of Language* (Dordrecht), suppl. series no. 5.

Stevens, K. A. (1981). The information content of texture gradients. *Biological Cybernetics, 42,* 95–105.

Stowell, T. (1981). *Origins of phrase structure.* Ph.D. thesis, MIT.

Terzopoulos, D. (1984). *Multiresolution computation of visible-surface representations.* Ph.D. dissertation, MIT.

Terzopoulos, D., Witkin, A., & Kass, M. (1988). Constraints on deformable models: Recovering 3D shape and nonrigid motion. *Artificial Intelligence, 36,* 91–123.

Traub, J. F., Wasilkowski, G. W., & Wozniakowski, H. (1983). *Information, uncertainty, complexity.* Reading.

Turing, A. M. (1936). On computable numbers, with an application to the Entscheidungsproblem. *Proceedings of the London Mathematical Society, Ser. 2-42,* 230–265.

Ullman, S. (1979). *The interpretation of visual motion.* Cambridge, Mass: MIT Press.

Ullman, S. (1984). Maximizing rigidity: The incremental recovery of 3-D structure from rigid and nonrigid motion. *Perception, 13,* 255–274.

Varela, F., Thompson, E., & Rosch, E. (1991). *The embodied mind.* Cambridge, Mass: MIT Press.

Walker, E. L. (1980). *Psychological complexity and preference.* Monterey, California: Wadsworth (Brooks/Cole).

Wallach, H., & O'Connell, D. N. (1953). The kinetic depth effect. *Journal of Experimental Psychology, 45,* 205–217.

Wertheimer, M. (1923). Laws of organization in perceptual forms. English translation in W. D. Ellis (Ed.), *A source book of gestalt psychology.* New York: Harcourt Brace, 1938.

Wickelgren, W. A. (1979). *Cognitive psychology.* Englewood Cliffs, NJ: Prentice-Hall.

Wildes, R. P., & Richards, W. A. (1988). Recovering material properties from sound. In W. A. Richards (Ed.), *Natural computation.* Cambridge, Mass: MIT Press.

Williams, E. S. (1977). Discourse and logical form. *Linguistic Inquiry, 8,* 101–139.

Winston, P. H., Binford, T. O., Katz, B., & Lowry, M. (1984). Learning physical descriptions from functional definitions, examples, and precedents. In M. Brady, & R. Paul, *Robotics Research.* Cambridge, Mass: MIT Press.

Witkin, A. P. (1981). Recovering surface shape and orientation from texture. In J. M. Brady (Ed.), *Computer Vision.* New York: North-Holland.

Witkin, A. P. (1983). Scale-space filtering. *Proceedings of the International Joint Conference on Artificial Intelligence* (Karlsruhe, Germany), 1019–1022.

Witkin, A. P., & Tenenbaum, J. M. (1983). On the role of structure in vision. In A. Rosenfeld & J. Beck (Eds.), *Human and machine vision: Vol 1.* Hillsdale, NJ: Erlbaum.

Witkin, A. P., Terzopoulos, D., & Kass, M. (1987). Signal matching through scale space. *International Journal of Computer Vision, 2,* 133–144.

Yuille, A., & Poggio, T. A. (1986). Scaling theorems for zero crossings. *IEEE Transactions on Pattern Analysis and Machine Intelligence, 8,* 15–25.

Zubek, J. P. (Ed.). (1969). *Sensory deprivation.* New York: Appleton-Century-Crofts.

Zucker, S. W. (1976). Toward a model of texture. *Computer Graphics and Image Processing, 5,* 190–202.

Zucker, S. W. (1983). Computational and psychophysical experiments in grouping: Early orientation selection. In A. Rosenfeld & J. Beck (Eds.), *Human and machine vision: Vol 1.* Hillsdale, NJ: Erlbaum.

Zucker, S. V. (1985). From orientation selection to optical flow. *Computer Vision, Graphics, and Image Processing, 37,* 196–220.

Zucker, S. W., & Hummel, R. A. (1986). Receptive fields and the representation of visual information. *Human Neurobiology, 5,* 121–128.

Index

NORTHERN MICHIGAN UNIVERSITY LIBRARY
3 1854 004 306 018